OIL ON PUGET SOUND

An Interdisciplinary Study in Systems Engineering

OIL ON PUGET SOUND

An Interdisciplinary Study in Systems Engineering

Supervised by JURIS VAGNERS

Coordinated by PAUL MAR

A Washington Sea Grant Publication

Distributed by
UNIVERSITY OF WASHINGTON PRESS
SEATTLE AND LONDON

Library of Congress Catalog Card Number 72-738
ISBN 0-295-95239-3
Printed in the United States of America

PREFACE

As part of the University of Washington's goal to adapt academic talent to solve real world problems, a six-month interdisciplinary course was formulated to study the prevention and control (containment and cleanup) of oil spills in Puget Sound. Administered by the College of Engineering Committee on Ocean Engineering as OE 551 and 552, students and faculty from various disciplines within the campus were integrated to study this timely and controversial topic. The graduate schools of business, fisheries, geography, law, and the various schools within the College of Engineering were represented.

The objectives set forth at the beginning of the study were to define the oil spill problem in Puget Sound and to formulate a model for the solution of the problem. The study group discovered, as the study progressed, that identifying the sources and consequences of oil spills was a most time-consuming task. If solving a problem warrants a ten dollar reward, then definition of that problem should be worth at least one hundred dollars. At least this was the sentiment of the group upon completion of their study. Thus, the primary efforts of the students were in collecting, analyzing, and evaluating data on Puget Sound and its related oil industry. It was only after completion of this tedious task that a meaningful solution could emerge.

The tempo of this report is to critically evaluate the current status of oil spill prevention and control in Puget Sound, and to suggest recommendations to alleviate any noticable deficiencies. In this day of public outcry for environmental conservation, many have been accused of casting stones upon previously accepted and long established practices. This is evident in this report in terms of the evaluations. But it also has been a demonstrated fact that recommendations for altering existing institutions go unheard until these stones have been thrown.

The study has been performed with a minimum of bias, and as much as possible, without undue influence or pressure from self-interest groups. As the reader progresses through the text, he will quickly perceive that there are no pure angels, no pure devils. All the groups evaluated in the report are aware of the oil pollution problem, though their actions frequently reflect only their narrow view of the problem.

For the results of this study to evolve into an effective solution to the oil spill problem, the organizations involved, and the people within them, must consent willingly to change. Nothing could be worse than blindly instituting some legislation, technique, or procedure without fully understanding or accepting the overall impact of its activation. Too often, such fragmented actions are taken to obtain short run, narrow solutions to problems without evaluating the bigger picture.

Historically, plans to cope with oil spills have emerged as the aftermath of disasters, or have been formulated in a vacuum, without considering the implication or consequence of such plans. Therefore, the most important message of this report is: The environmental preservation of the Puget Sound region requires the coordinated effort of all responsible and capable parties, whether they represent society or industry.

Endnotes have been used throughout the reports and have been tabulated at the end of the various subsections, rather than inconvenience the reader with one gigantic list. A bibliography of the data sources is listed in Section V.

The status of the technical information in this document is as reported up to August 30, 1971.

ACKNOWLEDGMENTS

This report is the result of a two-quarter graduate systems engineering course, Ocean Engineering 551 and 552, conducted by Juris Vagners, Assistant Professor of Aeronautics and Astronautics at the University of Washington, during the Winter and Spring quarters of 1971. Special thanks are accorded to Dr. Vagners for his role as coordinator and advisor, and to his faculty team. Also, a word of praise is due the two very capable student leaders from OE 551 and 552, Robert Stockman and Roger Harman, respectively. Lastly, a very deserving word of thanks is extended to the many members of industry, government, and academic institutions who provided the valuable inputs for this study.

As is typical with most studies, documentation of the study efforts for this project became a monumental task. Special thanks are given to the secretaries of the Department of Aeronautics and Astronautics and the Division of Marine Resources for their long hours. As with most scientific reports, the final writing efforts are usually borne by a very few. The sweat and toil of composing the contents between these covers belongs to the following group: Greg Bargmann, Robert Behm, John Cathey, Jacques Duringer, Charles Gundersen, Roger Harman, Charles Klabunde, Tim Larson, Mark Longmeier, Paul Mar, Terje Melkstavik, Michael Redfield, Chris Rusch, Robert Stockman, John Von Ries, and James Walsh.

Financial support for this course came from the University of Washington College of Engineering and from the National Oceanic and Atmospheric Administration through the Washington Sea Grant Program. Funds to publish this report were made available by the University of Washington College of Engineering, the Oceanographic Commission of Washington, the University of Washington Press, and the National Oceanic and Atmospheric Administration through the Washington Sea Grant Program.

FACULTY PARTICIPANTS

Bruce H. Adee	Assistant Professor	Mechanical Engineering
Clifford A. Barnes	Professor	Oceanography
William J. Love	Professor	Mechanical Engineering
Juris Vagners	Assistant Professor	Aeronautics & Astronautics
Karl H. Vesper	Associate Professor	Management & Organization Mechanical Engineering

STUDENT PARTICIPANTS

Name	Discipline	Program
Greg Bargmann	Fisheries	Master
Robert J. Behm	Business Administration	Master
John B. Cathey	Law	Marine Affairs, LLM
Raymond C. Clark	Mechanical Engineering	Master
Bart Dawson	Business Administration	Master
Jacques Duringer	Aeronautical Engineering	Master
Tom Ferenchak	Mechanical Engineering	Master
Charles Gundersen	Ocean Engineering Program	Master
David Gunderson	Electrical Engineering	Bachelor
Roger A. Harman	Business Administration	Master
Charles Klabunde	Computer Science	--
Timothy Larson	Chemical Engineering	Master
Jeff Layton	Civil Engineering	Master
Mark Longmeier	Business Administration	Master
Mark Maher	Civil Engineering	Master
Paul Mar	Business Administration	Master
Terje Melkstavik	Civil Engineering	Master
Dennis Morden	Aeronautics & Astronautics	Master
Michael Redfield	Law	Marine Affairs, LLM
Bill Ross	Geography	Master
Chris Rusch	Business Administration	Bachelor
R. M. Schusler	Ocean Engineering Program	Master
Robert Stockman	Ocean Engineering Program	Master
John Von Ries	Law	Marine Affairs, JD
James P. Walsh	Law	Marine Affairs, JD
Harold N. Wiren	Librarian, Engineering Library	-- --
Kirk Youngman	Law	Marine Affairs, LLM

CONTENTS

FIGURES

FIGURES IN APPENDICES

TABLES

SECTION II-A:

SECTION II-B:

SECTION II-C:

SECTION II-D:

TABLES IN APPENDICES

APPENDIX 8:

APPENDIX 9:

APPENDIX 13:

APPENDIX 14:

SECTION I

INTRODUCTION

SECTION I. INTRODUCTION

"Oil and Puget Sound" - to the consumer, it means the comfort of motorized transportation, heating, and mechanized leisure; to the oil industry, it signifies a profitable business; to the conservationist, it means a threat to the environment quality; to the commercial fishing industry, it represents a threat to the long-term benefits of its business; and to the tourism industry, it represents possible degradation to those attractions upon which its revenue and livelihood depend.

Five years ago, the topic of oil and Puget Sound drew nothing more than a passing comment or a two line article on page 46 of the local newspaper. In fact, not a single bill concerning oil-on-water issues in Puget Sound was introduced in the 1965 state legislative session.[1] Now, besides TV specials and front page headlines, literature ranging from sloganed leaflets to lengthy legislative acts are available on the subject.[2] Why this recent on-rush of public interest?

Environmental and petroleum demands have contributed greatly to the oil-on-water controversy. The concern in the late 1960's for the preservation of the environment led to publication of study upon study analyzing the air and water quality degradation.[3] On the other hand, there is the realization that man's ever-increasing need for petroleum products has all but depleted the existing oil reserves, resulting in the relentless search for potential oil fields. By 1985, 45% of domestic oil production will have to come from reserves yet to be found.[4]

OIL SPILL STATISTICS

Man's increased consumption of oil has contributed to his productivity and wealth. This appetite has led to heightened oil production levels and heavy oil traffic on land and sea, which has helped cause near-saturation of the distribution channels. This enormous growth in the amount of oil transported and transferred has not as yet affected Puget Sound. Based on Corps of Engineer data, the quantity of oil transported annually on the Sound has ranged between 12.3 and 15.8 million short tons since 1965 with the 1970 volume at 12.9 million short tons (Section III-A). Projections for 1980, 2000, and 2020, accounting for the Alaskan oil effects, show 20.2, 32.5, and 55.3 million short tons transported, respectively. (Section II)

Since the predominant process of polluting water with oil is due to spillage, minimizing the number of shippings and transfers will have an impact on the frequency of potential spill incidents. Study team members attempted to accumulate data on frequency of shipments or transfers, number of spill incidents, and quantities of oil shipped, transferred, and spilled. This was pursued at the local, state, federal, and worldwide echelons. (Section II-A.) Surprisingly, very little of this data is available. For example, it is a known fact that collisions involving barges or tankers totaled 17 since 1965, in Puget Sound.[5] But the quantity of oil spilled per incident is unavailable. A request to the Environmental Protection Agency (EPA) for both U.S. and worldwide oil spill statistics elicited a lukewarm response. Apparently, the data is for in-house use and stored in computers.[6]

Further research showed that well documented statistics were available for a British oil port, Milford Haven, Southwest Wales for the 1965-1970 period. Three oil refineries and one tank farm are located on the shores of this port, whose annual oil traffic is between 2.5-3 times that of Puget Sound, on a quantity basis. Though the incidents of oil spills have increased, the probability of a spill incident has not changed; in fact it is diminished slightly. The eight year average for amount of oil spilled is 0.00018% of the total oil transferred. (Section III-A.)

CONCERN FOR OIL SPILLS

Comprehensive mechanisms for regulating oil discharges into water have only surfaced in the past five years. International and national contingency plans were formulated to combat oil spills subsequent to the _Torrey Canyon_ incident in 1967.[7] These plans to cope with oil spills center on one theme: Man's efforts to clean up spills are controlled through legally responsible organizations. The procedures set forth in these contingency plans have been very general, and the inherent environmental quality criteria are very vague. Regulations governing oil discharge into navigable waters consider it illegal to "...cause a film or sheen upon...the surface of the water..."[8] Attempts to obtain quantitative interpretation of such standards to those related to oil spill cleanup criteria did not clear up the situation.[9]

At the regional level, the Coast Guard has formulated a contingency plan which delineates the legal responsibilities and chain of command in event of a spill.[10] What is lacking are guidelines for on-scene decision making and operational controls. (Section III-B, Part 1, and Appendix 7.) The State Department of Ecology has been quoted having a response plan, but as of publication date, this document was in the developmental stages. They have published a document summarizing the Washington Statutes for oil spills.[11] (Appendix 2 critiques these statutes.)

Regulatory plans and standards applicable to one region are not universally applicable due to regional uniqueness. Thus, there is a need for local awareness of oil pollution. State, federal, and local agencies should seek a common definition for the environmental quality standards for Puget Sound, and base these on scientific analysis.

In the private sector, industry's stand on environmental quality is not clearly defined, and where its position is evident, it does not necessarily coincide with that of today's diverse society. Thus, a demand arises here for concern of oil pollution and its consequences, and this local awareness applies to all segments of society. All groups should adapt the "if you help me, I'll assist you" attitude, instead of the currently popular "don't tread on my territory, and I'll stay clear of yours" outlook.

OIL SPILL CONCERN IN PUGET SOUND

The discovery of oil in Alaska's North Slope has been a significant, if not the most important, catalyst of local awareness in Puget Sound. The breakthrough in the oil reserve crisis awakened the possibility of increasing the oil refinery industry in the Pacific Northwest. Using the oil transport values as substitutes for oil refinery output, volume has decreased from 15.8 to 12.9 million short tons between 1965 and 1970. Because of Puget Sound's close proximity to the Alaskan oil fields, plus its favorable conditions (deep water facilities and dependable inexpensive electrical power) for oil refining, it made sound business sense to propose transporting at least a portion of Alaskan crude oil to Puget Sound. With the addition of the ARCO refinery, oil transport figures for 1980 are projected at 20.2 million short tons. At the time of the Prudhoe Bay discoveries in 1968, this area had not experienced any major oil spill.[12] There was, however, a lot of discussion on the subject.

The Santa Barbara channel incident in January 1969,[13] aroused local interest with respect to the beach and wildlife damage. The San Francisco Bay tanker collision in January 1971,[14] cast the tanker traffic aspect into the forefront of the oil-on-water issue. Even in light of these disasters, there remained an air of optimism as to a possible oil spill occurrence in Puget Sound.

Fate finally turned its hand onto Puget Sound. Less than four months after the San Francisco spill, this area witnessed its first major oil spill, 242,000 gallons of diesel fuel at Anacortes.[15] Less than one month before this accident, oil industry firms located in Puget Sound had formed a co-operative to clean up spills,[16] and several third party contractors had pooled their resources to stockpile equipment.[17] A regional contingency plan had also been drawn up for the Puget Sound area. How effective any or all of these efforts have been can be evaluated in the light of the following facts: of the 242,000 gallons spilled at Anacortes, less than 450 gallons have been recovered to date.[18] Last but not least, no visible effort of any sort was made to divert the oil from the valuable sea life in the vicinity of the spill. Only after oil was sighted near valuable areas were attempts made to set up booms. (Appendix 12.)

In order for any model or plan to prevent and control[19] oil spills effectively, a baseline[20] definition on what is being protected - the environment - must be fully understood by all parties (society as well as industry). Thus, there is an urgent need for this assessment in the Puget Sound region. Section II-A documents such an analysis. It is highly recommended that a continuous update and refinement of this or similar inventory be documented. Areas of critical economic and social value should be identified. Next, potential sources of oil spills on Puget Sound should be singled out. Included in these evaluations are the industrial, societal, and political consequences. Then, and only then, can optimal response plans for cleanup and prevention be realized.

The responsibility for maintaining the quality of marine environment in and around Puget Sound is currently vested among three organizations: United States Coast Guard (USCG), Environmental Protection Agency (EPA), and the Washington State Department of Ecology (DOE). This responsibility includes the prevention and control of oil spills. Though broad guidelines and divisions of responsibility are established, the operational aspects of these directions are far from effective. (Section III-B, Part 1; Appendices 2, 3, and 7.)

It remains the task of both federal and state governments to iron out jurisdictional conflicts between each government's control over water pollution. For the Puget Sound area, this was evident in the oil spill at Anacortes. For this incident, DOE found United Transportation guilty of negligence in allowing oil to be spilled into water, and fined the firm \$15,000.[21] It also ruled that Texaco was absolved of any blame. On June 26, 1971, the U.S. attorney's office filed suit against both firms for damages to the environment.[22]

Thus, the federal government's efforts should strive to insure that in possible future oil pollution rulings, both its aim and that of the state are in concert.

Identification of potential spill sources is a necessary starting point for oil spill prevention. For each source, an evaluation of the economic and social consequences of a spill should be closely examined. From this information, water quality standards should be established by the responsible agency, and enforced utilizing state-of-the-art techniques. In certain situations, more than one agency may have jurisdiction. In these cases, the responsible parties should endeavor to search for an equitable solution for society, instead of striving for individual self-interests.

Section III-A of this report delineates the sources of potential oil spills for the Puget Sound region. In addition to the oil industry, vessels, municipalities, factories and plants, and private citizens discharge oil into the water. One other potential future source is submarine drilling, which is not presently active in the Sound. By identifying the source, necessary and proper action can be initiated to:

1. prevent spills from occurring, and
2. control any spill incidents that might happen.

Figure I-1 presents a simplified model of oil on Puget Sound. On the top of the diagram are listed the potential sources. Prevention efforts will minimize, if not completely abate, the discharge of oil into the Sound. If oil does reach the water, then capability must be available to remove this discharge. Otherwise the varied resources of the Sound will suffer the consequences.

OIL SPILL PREVENTION

On the prevention side, there are presently several refineries and processing plants in the Puget Sound area that handle industrial and private waste oil. (See Section III-A.) DOE and organizations such as the Municipality of Metropolitan Seattle (Metro) have expended much effort on municipal oil wastes. (See Section III-A.) In addition, a comprehensive analysis on submarine drilling has been published.[23] Therefore, this report is primarily concerned with prevention of oil spills from waterborne traffic and oil transfer operations.

An analysis of past, present, and future commercial waterborne traffic in Puget Sound is presented in Section II-B. Data for ship movements, quantities of petroleum, and other relevant factors are documented. Results of statistical analysis for ship collisions and resultant oil spills also appear, with the model described in detail in Appendix 13. Trends in vessel size and capacity are also reported in Section II-B; they reveal the possibility of one million deadweight ton (dwt) tankers by 1980.

Prevention techniques for oil tankers and other vessels are relevant to Puget Sound from two viewpoints.

1. Tanker design, and
2. projected tanker operation on the Sound. (Section III-C, Part 1.)

While a complete re-design of tankers is a long run solution, certain short-term measures can be undertaken to aid the cause. Standard Load on Top (LOT) procedures should be instituted in all oil carrying vessels. (Section III-C, Part 2.) Since LOT requires settling time to be effective, vessels primarily used for intra-sound cargo should be provided with waste oil reception facilities. One such plant has recently begun operation at Pier 91, Seattle.[24]

The increasing concern, in recent years, for tanker and barge mishaps has signaled a need for a waterborne traffic control system, integrating

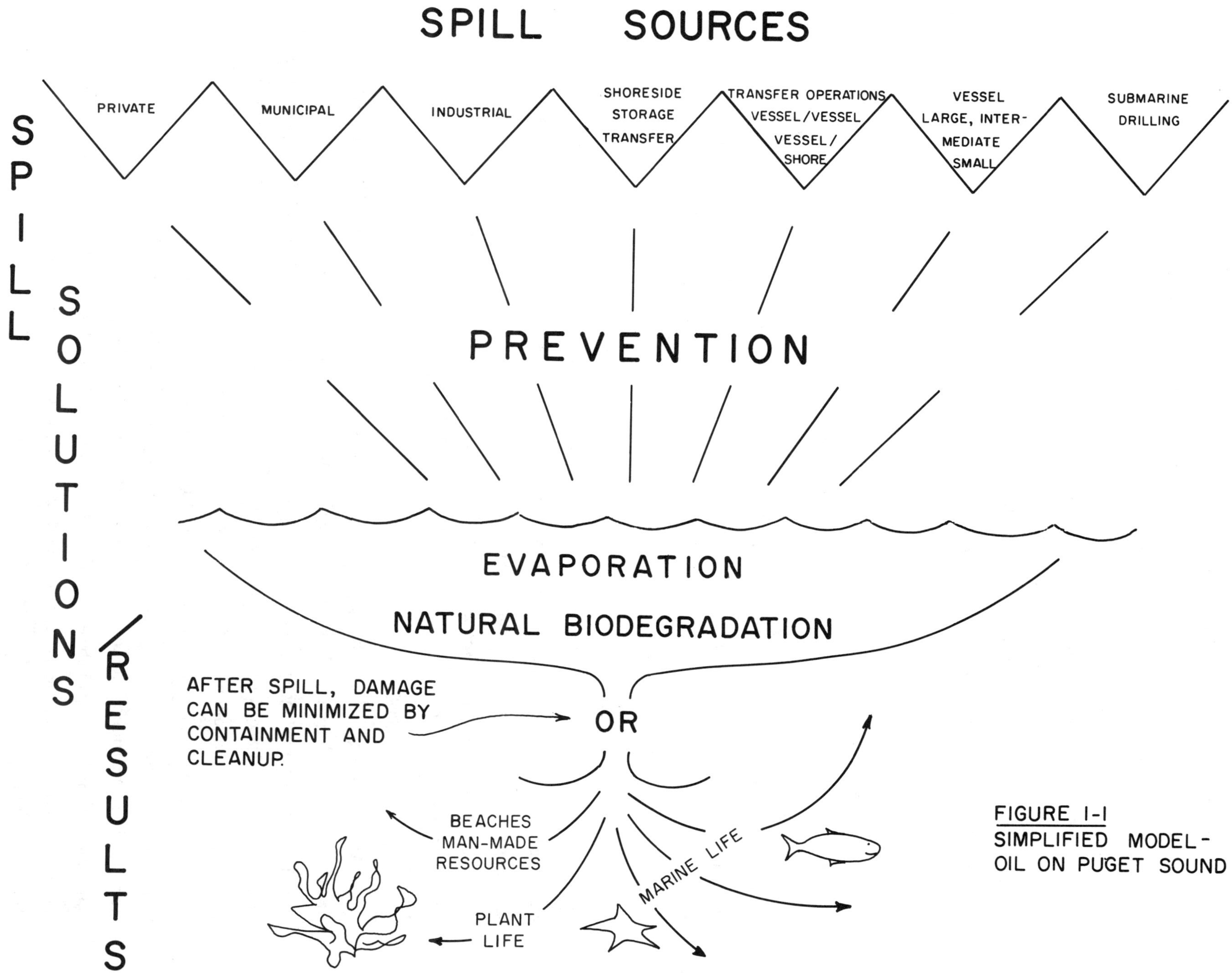

FIGURE I-1
SIMPLIFIED MODEL- OIL ON PUGET SOUND

commercial schedules and information, and existing marine traffic, to that of the oil industry. The ideal long term solution, as presented in this report, is an automated traffic advisory system tailored for Puget Sound. (Section III-C, Part 2; and Appendix 8.) Its value is in the future protection, after determination of cost effectiveness. Presently available measures include adapting international regulations to the Sound, improving dissemination of information, and updating seamanship techniques and equipment. (Section III-C, Part 2.)

Surveys of current operations at eleven Puget Sound oil handling facilities by the study group indicated compliance with Coast Guard requirements, but these measures were proven inadequate for preventing a dockside spill. (Section II-C, and Appendix 1.) Recommendations presented include revising regulations, using state-of-the-art devices and techniques, and training of personnel (Section III-C, Part 4). Most of the facilities surveyed are operated by oil firms and are responsible to the USCG, plus EPA; hence, a coordinated level of effort, utilizing manpower and funding, is imperative to the prevention of spills.

One other conclusion reached in the study is related to the oil industry's efforts in conserving the resources of Puget Sound. Available data on expenditures revealed that the industry's annual dollar outlay for this area is about $1.4 million. When compared with the benefits gained from these efforts, the industry appears guilty of shortchanging the Pacific Northwest in favor of its other interests (Section II-E).

OIL SPILL CONTROL

Assuming that all the prevention measures have been implemented, one still must anticipate that an oil spill will occur. In the event of a spill, the authority and capability of each resonsible party must be clearly and conclusively defined. This includes the operational personnel, as well as those in management roles. Once oil has been spilled into water, time is of the essence and jurisdiction questions cannot be argued. Furthermore, authority to act is not sufficient; one must also have the capability to act. This includes having the necessary skills, material, equipment, and manpower readily available. One also needs the authority to exercise judgement without fear of reprisal except under due process in a public tribunal. (Section III-B.)

At the present time, the capabilities of any one party to respond to an oil spill emergency of any magnitude in Puget Sound are essentially non-existent. (Section III-B.) The USCG has limited manpower and no specialized equipment. Similar conditions exist with EPA and the DOE.

A 1969 oil industry survey[25] revealed that, of the 54 reporting facilities, only four possessed on-site oil spill cleanup capability. In addition, 17 claimed to use third party cleanup services. There were five third party contractors listed. The Washington Oil Spill Cooperative - a collection effort of ten oil firms - has claimed capability and preparedness to cope with spills. As yet, no supporting evidence of these claims is evident.* The key to the solution once again lies with an integrated inter-agency effort, receiving the necessary financial and manpower backing. Several specific recommendation in this area are documented in Section III-B, Part 1.

In order to enhance a given emergency response, a suggested decision matrix for economic priorities in the Puget Sound area has been developed by the study group (Section III-B, Part 2). This will enable the responsible party to base his actions on technical information rather than hearsay or conjecture. The matrix can also be useful as an indicator of the social values placed on this region's resources.

An overriding factor to response effectiveness is the ability to track an oil spill. The state-of-the-art of oil detection is discussed in Section III-C, Part 3. Currently available techniques are not cost-effective. Appendix 8 presents a computer model with potential for predicting the behavior of an oil slick.

SUPPORTIVE ANALYSIS

To support the recommendations for oil spill response capability, evaluations of containment devices (Appendix 5) and cleanup devices (Appendix 6) were performed. A critique of the equipment and technology displayed at the 1971 American Petroleum Institute Oil Spill conference appears in Appendix 9. A theoretical discussion on booms (Appendix 10) supplements the discussion of containment devices.

The impact of international laws on oil pollution in Puget Sound are documented in Appendix 4. This discussion augments the state and federal

*See Endnote 26, this section.

statute analysis of Appendices 2 and 3. Together with the critique of the regional contingency plan (Appendix 7), the legal aspects of oil pollution in the Sound are well covered. In addition, two pollution abatement proposals related to vessels are evaluated in Appendix 10.

The effects of oil on water and the environment are detailed in Appendix 14. Included is a glossary of terms, and a conversion table relating the various units for measuring quantity of oil. Short- and long-term biological effects are discussed, primarily from a survey of the available literature. Appendix 8 presents a computer model predicting the behavior of an oil slick on Puget Sound. It is hoped that such a model could be integrated into an overall cleanup response concept.

The study group had the benefit of first hand analysis of two recent oil spill incidents, at San Francisco and Anacortes. Though the former was of interest, it did not directly apply to the Puget Sound study. Data on the response capability to cleanup the Anacortes spill and the subsequent damage assessment was accumulated by the study group. Appendix 12 presents a documentation of these events. This case analysis brings into focus those concepts covered in the bulk of this report. Needless to say, the Anacortes results are not encouraging. No attempt was made to analyze the causes of the incident.

The myriad of conclusions and recommendations are summarized in Section IV.

OBJECTIVE OF STUDY

It is the goal of the study team that each and every topic covered in this report be critically reviewed by all parties. It is further desired that those actions necessary to upgrade the water quality be immediately initiated. Only when these tasks are accomplished can the Puget Sound region enjoy an oil pollution free environment in the future. <u>These efforts require the co-ordinated efforts of all responsible and capable parties</u>. It is also hoped, as a residual benefit, that the gap between the various industrial and social standards of water quality be narrowed, and ultimately eliminated.

ENDNOTES SECTION 1

[1]Legislative Digest and History of Bills of Senate and House of Representatives for Washington State, 39th Session, 1965.

[2]For the first half of the 1971 legislative session, fifteen bills had been introduced on pollution of water, of which six were addressed to the oil pollution problem. Source: Legislative Digest and History of Bills of Senate and House of Representatives for Washington State, 42nd session, 1971.

[3]An example is the extensive multi-volume study on Puget Sound and adjacent waters, published by the Puget Sound Task Force.

[4]"U.S. Energy Demands & Supply - 1960 to 1985", Ocean Industry, June, 1971, p. 15.

[5]Data provided by Commander Greiner, Jr., 13th Coast Guard District.

[6]A letter dated July 20, 1971, from Mr. G. Moeins of the Division of Oil and Hazardous Materials, EPA to the editor of this report, indicated that records of U.S. oil spills are stored on computers for that agency's internal use.

[7]The Torrey Canyon spill occurred March 18, 1967 off Lands End, England. The Intergovernmental Marine Consultative Organization (IMCO) published its resolutions for oil spill prevention in 1968 and 1969; the U.S. formulated its national Multi-agency Contingency Plan in November, 1968.

[8]18 C.F.R., Chapter V, Section 610.3.

[9]Numerous questions and answer sessions at the Oil Spill Conference sponsored by API, Washington D.C., June 15-17, 1971.

[10]Seattle Coastal Region, Oil and Hazardous Materials Pollution Contingency Plan, U.S. Coast Guard, dated December 1, 1970.

[11]Laws and Oil Spill Emergency Procedure, Department of Ecology.

[12]A major spill for Puget Sound, as defined by U.S. Coast Guard, is one where more than 10,000 gallons are discharged.

[13]Union Oil platform blowout, Santa Barbara Channel, January 28, 1969.

[14]Collision of Arizona Standard and Oregon Standard tankers in San Francisco Bay, January 18, 1971.

[15]United Transportation barge leak at Texaco dock, March Point, Anacortes, April 26, 1971, during transfer of diesel fuel.

[16]Washington OIl Spill Cooperative.

[17]Marine Oil Pickup Service, a joint venture between Puget Sound Tug and Barge and Pac-Mar Services, Inc.

[18]"Stations Seattle Pollution Report #043-71", by C. M. Stone, LTJG, U.S. Coast Guard, Captain of the Port of Seattle, Washington, Enclosure (1), p. 60.

[19]Control, as used in this report, is the containment and cleanup of spilled oil.

[20]Baseline is defined as a starting point. With respect to resources, this means detailed knowledge of what exists (quantity and state) under nominal conditions.

[21]Docket No. DE 71-141, In the Matter of Assessment of Penalty Against United Transportation Company, June 17, 1971.

[22]U.S. District Court, Western District of Washington, No. 9779, Complaint for Injunction, Environmental Clean-up Expenses and Penalty for Unlawful Discharge, June 24, 1970.

[23]Dolan, Savage, and Clark, Environmental Statement on the Proposed Leasing of Puget Sound Shorelines and Beds of Navigable Waters for Oil and Gas Exploration, 1970.

[24]Chemical Processors, 5501 Airport Way South, Seattle.

[25]Oil Spill Control Survey for Onshore and Offshore Facilities, American Petroleum Institute, March, 1970. Report No. 4023, Section V, Tables 5, 6, 7.

[26]In a conversation on November 12, 1971 with Mr. John C. Doolittle of the Washington State Oil Spill Cooperative, the following information was obtained:

> "In the last quarter of 1971 the Washington State Oil Spill Cooperative (WSOSC) has progressed toward their intended goal. They have committed $100,000 toward the purchase of 1000 feet of off-shore Bennett boom and a 40-foot long Huskey SBG floating skimmer. This equipment is to be stored in the Port of Bellingham with the skimmer arriving in early November and the boom in late December. The Coop has also retained for three months a technical and managerial advisor/coordinator to provide the WSOSC membership with a three year illustrative budget and to draft an oil spill response plan. The contingency plan draft will be completed by late December. The Coop is currently considering hiring a full time manager to formalize the organizational plans currently being developed and to coordinate the activities in the oil spill response plan."

SECTION II

DESCRIPTION OF PUGET SOUND AND ITS INDUSTRIES

The necessary prerequisite to maximum effectiveness for any water conservation program related to the oil pollution problem is the evaluation of the consequences for specific techniques on the ecosystem. To the extent that this study is concerned with the Puget Sound region, it is imperative that this area's characteristics be accurately defined, so that the effects of oil spills, their prevention, containment, and cleanup can be analyzed accurately. Thus, this section is devoted to establishing such a definition.

The physical environment and resources of the region are described in Section II-A. Included are both general and detailed analyses of the location of resources. Sections II-B and II-C discuss waterborne commerce and oil handling on the Sound, respectively. This portion of the report concludes with an economic evaluation of the various resources, and the oil industry's efforts in conserving these resources in the Puget Sound region, which are presented in Sections II-D and II-E.

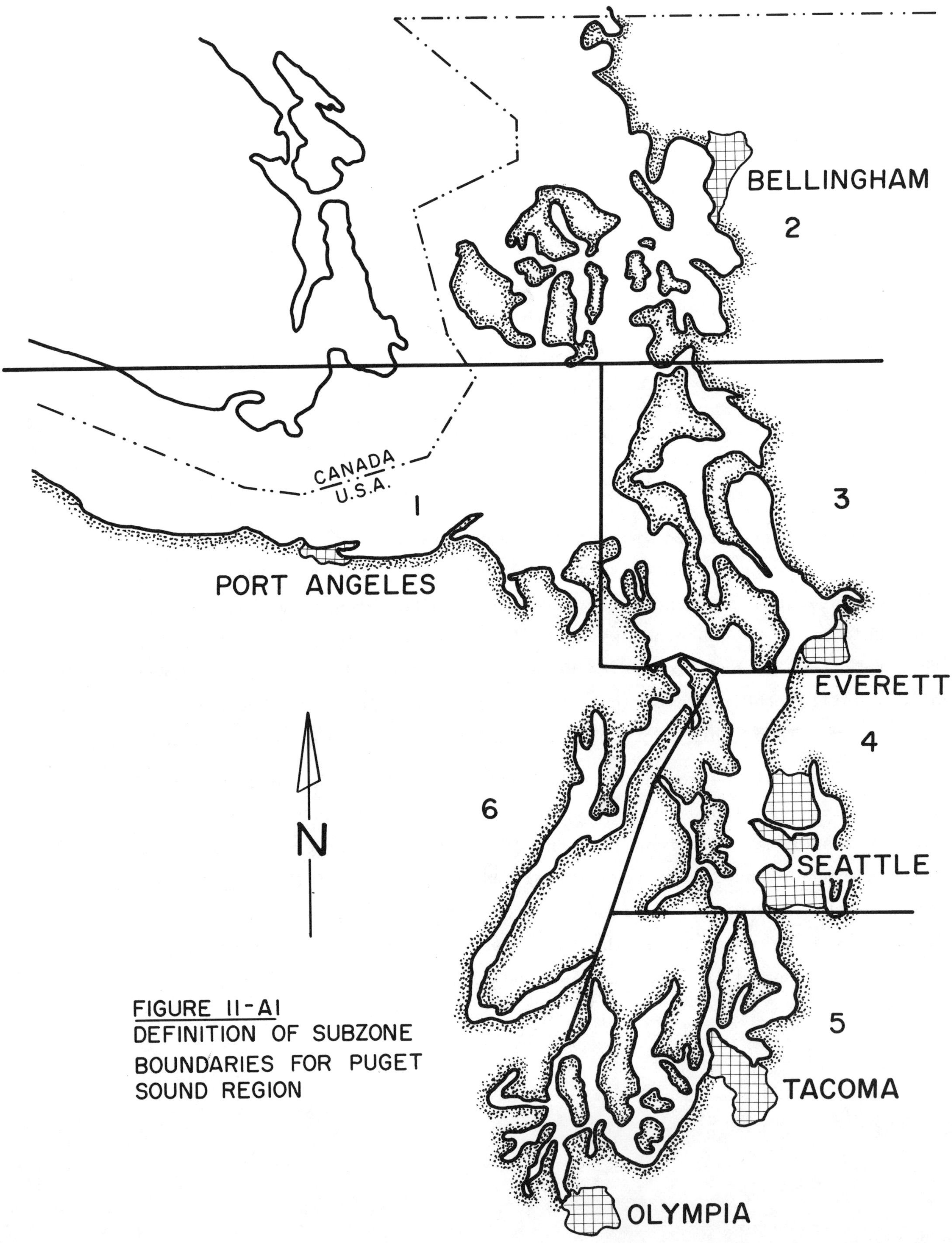

FIGURE II-A1
DEFINITION OF SUBZONE BOUNDARIES FOR PUGET SOUND REGION

II-A. PHYSICAL DESCRIPTION

1. INTRODUCTION

In order to place the possible oil pollution problem on Puget Sound in proper perspective, an inventory of the prevailing environmental conditions and significant resources must be made. A more complete assessment of the value of resources that might be threatened by an oil pollution incident, and of the potential operational difficulties resulting from adverse environmental conditions, may provide justification for more effective preventive measures. In the event of an oil spill, environmental and resource information can provide valuable assistance to the on-scene decision maker to direct protection, containment, and cleanup measures in an optimal manner.

The purpose of this section is to delineate the general environmental conditions and resources present on Puget Sound. For the purpose of this analysis, Puget Sound will be considered to include Puget Sound proper and all contiguous saline state waters east of Cape Flattery including the Strait of Juan de Fuca, the San Juan Islands, and Hood Canal.

After a few general remarks about Puget Sound as a unit, a more detailed analysis will follow with the study area arbitrarily divided into subzones to facilitate data presentation. The subzones will be represented by the following six units (see Figure II-A1):

1. Strait of Juan de Fuca
2. San Juan Islands including Bellingham and Anacortes
3. Whidbey Island to Everett
4. Northern Puget Sound
5. Southern Puget Sound
6. Hood Canal

It must be emphasized that although the information used for the subzone analysis was drawn from all available sources, it is intended to serve as an interim compilation until a comprehensive investigation is accomplished. The understanding of the environmental processes and trends is particularly in need of further study, and more sophisticated environmental observations will be necessary to enhance the prediction of the behavior of oil spilled on Puget Sound.

2. PHYSICAL ENVIRONMENT - GENERAL

a. Description of Area

Puget Sound is characterized by numerous channels, sounds, and inlets. It occupies an area of 767 square nautical miles, compared to the whole basin area of 11,000 square nautical miles. The entire system lies within an area of about 40 by 90 nautical miles. Through the Middle Point - Point Partridge entrance passes approximately 95% of the volume of water entering or leaving Puget Sound; 5% flows through Deception Pass, while the contribution of water from Swinomish Slough may be considered negligible. The Strait of Juan de Fuca interposes between Washington Sound to the north and Puget Sound to the south, as well as between Vancouver Island and the Olympic Peninsula. That section of Puget Sound connecting with the Strait, and lying between Point Wilson and Point No Point on the west and Whidbey Island on the east, is known as Admiralty Inlet. Hood Canal, a long narrow arm of Puget Sound, extends southward about 50 nautical miles. As the result of its great length, narrowness, depth, and weak tidal currents, Hood Canal has many characteristics which distinguish it from the remaining waters of the region.

Puget Sound proper is connected to the Strait of Juan de Fuca by a series of waterways to the east of Whidbey Island: Possession Sound, Saratoga Passage, Skagit Bay, and Deception Pass. The width of these deep waterways decreases with progress northward, but shoals and tidewater flats frequently give a total width of as much as five nautical miles. Deception Pass separates Whidbey Island from Fidalgo Island and connects the Strait of Juan de Fuca with Skagit Bay. Although two miles in length, it is extremely narrow and shallow - in some places only 200 yards across with a depth of 96 feet. The Swinomish Slough, near Deception Pass, connects Skagit Bay with Padilla Bay to the north.[1]

b. Shoreline Features

The shoreline of Puget Sound and the Pacific Coast is very dissimilar in character compared with that of the Atlantic Coast. The Atlantic Coast is largely composed of mud and sand formations with comparatively little elevation, whereas the Pacific Coast of Washington is predominantly rocky and precipitous.

The total shoreline length of Puget Sound enclosed by the northern boundary to Admiralty Inlet (Middle Point - Point Partridge), Deception Pass, and Swinomish Slough is 2167 miles, including all islands. Of this distance, the greater portion is faced with bluffs ranging from 50 to 500 feet in height that are composed of glacial till and material deposited during the past ice age. A generally narrow beach line is experienced at the base of these bluffs with depths of water exceeding 50 to 100 feet close to the shore along the portions of the Sound. Near river deltas, tidal flats extend for several miles in some cases; as for example, near the mouth of the Skagit River. At the extreme southern end of Puget Sound, prairie lowlands face the water as a continuation of the Puget lowland. Throughout the region dense coniferous forests are encountered. Except where man has cut away the forests to build farms, resorts, and towns, the Douglar Fir, cedar, and hemlock trees grow out to the beach line or overhand these bluffs so that it is difficult to identify the several points due to their similarity in appearance.[2]

Puget Sound, while faced by high bluffs, has a considerable number of beaches. In general,Puget Sound beaches consist either of gray, coarse, sand or a mixture of such sand with small gravel, which form narrow strips along the shore.[3]

c. Climate

The climate of the area is typically maritime, the summer is cool and the winter mild. At Seattle, the monthly average temperature is 39.8^{o}F in January and 63.8^{o}F in July. The mean annual temperature is about 50°F.

The prevailing air movement is from the open ocean except during the winter, when storms may draw cold continental air into the area. Winds may exceed 60 knots during winter storms but because of the limited fetches, the waves rarely exceed 6 feet in height, even in the larger open areas. During most of the year prevailing winds are from a southerly direction, but during July and August the prevailing winds at Seattle are from the north. Wind directions are closely related to the local land topography. Average wind speeds are between 7 and 10 knots. Winds exceeding 30 knots have been recorded in Seattle for all months of the year.

As the warm maritime air masses rise over the Olympic and Cascade mountains, heavy orographic precipitation takes place. But as these air masses descent and blow over Puget Sound and its adjacent lowland, the air masses become warm and only a moderate precipitation occurs, averaging between 35 and 50 inches. Most of the rain falls during autumn and winter with minimum precipitation during July and August. The days with appreciable precipitation average 40% for the year indicating that the rains are frequent but rarely heavy. Overcast skies occur some 80% of the time during the winter.

Snowfall on the lowlands is relatively rare and although it may snow several days during the winter, snow rarely remains on the ground for more than a few days.[4]

d. Tides and Tidal Currents

The tides in Puget Sound are of the mixed type showing a large diurnal inequality between the heights of succeeding low tides. The average diurnal range at Seattle is 11.3 feet. The maximum range of spring tides rarely exceeds 16 feet. There is a general increase in range as the tide progresses from the Strait of Juan de Fuca to the heads of the inlets with a delay in time of high water about one hour. Due to the large area of Puget Sound and its narrow entrances, tidal currents have exceptionally strong tropic velocities, attaining 4.7 knots in Admiralty Inlet, 5.1 knots in Tacoma Narrows, and 7.2 knots in Deception Pass. Elsewhere in Puget Sound the tidal currents are generally less than one knot. In addition to their importance in navigation, the strong tidal currents greatly affect the temperature and salinity distribution of the region through their mixing action and their influence on the water exchange.[5]

e. Water Temperature and Salinity

Water temperatures in Puget Sound are relatively uniform throughout the year. Because of the mixing produced by the tidal currents, surface temperature rarely falls below 40°F even in winter except in the head waters of Puget Sound, where the accumulation of cold river waters and relatively quiet conditions will permit the formation of ice during extreme cold weather. The same mixing processes tend to maintain relatively low temperatures during the summer and, except in shallow and isolated areas, surface temperatures above 60°F are rare. Because of sills at moderate depths, and the mixing that occurs over such sills, the water in the deeper basins is always relatively warm with temperature between 44°F and 55°F.

Fresh water in large amounts is emptied into Puget Sound from a number of rivers and streams. The greatest amounts are discharged by the Skagit, Stillaguamish and Snohomish Rivers, all located on the northeastern shore. Despite this, the salinity of the waters is remarkably uniform averaging between 29 and 31 parts per thousand, except very near the surface where following heavy runoff and light winds, a dilute surface layer may occur. Seasonal variations are small. The relatively high salinities are maintained by active exchange of water between Puget Sound and the Strait of Juan de Fuca. Puget Sound is characterized by a net average outflow of the diluted surface water and a net average inflow of a more saline water at depth. The salinity of the water in the deeper basin reaches a maximum in late summer or autumn, after the subsurface waters in the Strait of Juan de Fuca have reached their greatest salinity and lowest temperatures of the year which normally occurs in August.[6]

f. Mixing and Flushing

The mixing and flow across the sills which tends to equalize the temperature and salinity in the region are effective in distributing dissolved oxygen and nutrient salts. The depths of the major basin with the possible exception of Hood Canal in summer are well ventilated during all seasons of the year. Subsidiary basins behind deep sills at the extremities of the various arms may not flush from winter until the following autumn, and more infrequently not until the autumn of a succeeding year. These stagnant waters may become almost entirely devoid of oxygen and considerable concentrations of dissolved nutrients such as soluble phosphates and silicates accumulate.

The bottom water in the shallow extremities of the inlets where both the tidal currents and runoff are small also become impoverished in oxygen and enriched in nutrients from spring to fall as a result of biological activity.[7]

The flushing time for any particular harbor or enclosed body of water is a theoretical indication of the time required to remove a parcel of water containing a concentration of any dissolved or suspended contaminant.[8] It is emphasized that the contaminant is dispersed uniformly throughout the water column and that this measured, flushing time does not necessarily bear a direct relation to the behavior of a surface film such as an oil slick. However, certain refined oils do dissipate in the water column to variable degrees due to turbulence and an examination of the flushing of Puget Sound may suggest the order of magnitude of resident times for spilled oil neglecting any natural degradation.

A generalized, analytical procedure for determining tidal flushing, the tidal prism method, is defined as:[9]

> A theoretical procedure for determing the flushing time of a harbor or estuary. The method assumes that the contaminant is initially distributed uniformly throughout the harbor or estuary, and that during each tide cycle a volume of water and contained contaminant equal to the tidal prism is removed from the harbor, and replaced by a new volume of sea water, which mixes completely and uniformly with water present in the estuary at low water level. Therefore, the amount of contaminating material removed on each tidal cycle may be expressed as a percentage of the contaminant in the harbor during the previous tidal cycle:
>
> $$\frac{\text{Tidal Prism Volume}}{\text{High Water Volume of}} \times 100$$
>
> = Percent of Contaminant Removed from Harbor.
>
> This method is quite crude because of the fact that very unrealistic assumptions are required concerning total mixing and total replacement in succeeding tidal cycles.

A more sophisticated study[10] utilizing a water budget analysis provides flushing rates for Puget Sound and some subregions. This method employs data recording the volumes and salinities of fresh and salt water and their resulting mixture. Replacement time is defined as:

$$\frac{\text{the basin volume}}{\text{average rate of outflow from the basin to sea}}$$

and is depicted for various Puget Sound basins by month in units of days.

Although some of the parameters used to determine the listed ratios are accurate to ± 30%, and the data does not include the San Juan

Archipelago and the Strait of Juan de Fuca, this flushing data is the best available, and indicates a reasonable estimate of the persistence of dissolved contaminants. It must be re-emphasized that the boundary effects of pollutants as a surface film or absorbed on shoreline sediments will not behave the same as the contaminants uniformly dissolved in the water column. Table II-A1 shows monthly replacements times for the waters of three regions in the Sound, and times for the total Sound.

TABLE II-A1: REPLACEMENT TIME, BY MONTH (DAYS)[11]

	Whidbey Basin	Southern Puget Sound	Hood Canal	Entire Puget Sound Area
Jan	46	33	272	113
Feb	32	45	85	62
Mar	54	41	202	582
Apr	30	28	97	166
May	44	114	672	184
Jun	25	54	328	99
Jul	30	80	152	124
Aug	47	70	191	132
Sep	93	80	149	407
Oct	57	51	215	120
Nov	76	174	-183	480
Dec	18	35	101	146
MEAN	40	56	177	152

$$\text{Replacement time} = \frac{\text{basin volume}}{T_o}$$

3. PHYSICAL RESOURCES - GENERAL

a. Natural Physical Resources

1. Introduction

The biological resources within Puget Sound represent one of of the most, if not the most, important resources of the State of Washington. Abundant and varied forms of life are present. Crabs, clams, oysters, five species of salmon, and many species of bottom fish are important economically. A great deal of recreation is derived from these and other organisms. Destruction or elimination of any of these resources would represent a severe blow to the quality of life of the citizens of Washington. An economic analysis of the region's resources is presented in Section II-D.

The ability of Puget Sound to support plentiful flora and fauna is dependent upon many factors. Puget Sound is largely protected from ocean waves and its waters are relatively calm, yet it receives a constant flow of clean ocean water from the Strait of Georgia and the Strait of Juan de Fuca. This protection plus the flow of new water into the area enables many organisms to exist and flourish in Puget Sound which could not endure the pounding of the open ocean.

Perhaps the most important feature that enables the great productivity of the region is the great length of shoreline. The waters are bordered by some 2167 miles of shoreline which is roughly equal to the rest of the western coastline of the United States.[12] This elongated shoreline, with many small bays and channels, is largely responsible for the rich fauna that exists. It provides ideal habitat for organisms and enables the large population of aquatic organisms to grow. Life in the Sound can be divided into:

a. shallow bottom
b. surface water
c. deep bottom

The shallow water near the shore is the most productive area. Here, the bottom is well within the reach of sunlight which permits plants to flourish. The plants, either attached or floating, enable the large population of fish and invertebrates to exist. Especially important shoreline areas are estuaries.

These nutrient-rich waters provide food and shelter for the young of many organisms. Pollution or destruction of an estuary could have serious long range implications.

The surface waters of Puget Sound also contribute to the productivity of the area. Sunlight enables plankton to grow, which forms a partial basis for the food chain. In addition, juvenile or larval stages of many organisms mature in the plankton at the surface of the water. Dungeness crab, and many species of flounder are examples of marine life that spend at least part of their life cycle floating more or less passively at the surface of the Sound. Generally speaking, larval stages of most organisms are more sensitive to damage from oil than adults.

The waters of Puget Sound are very deep, exceeding 100 fathoms in many locations. The bottom in these areas is far beyond the reach of sunlight and, as far as known, do not support much life. The organisms which occur in this zone are found more or less uniformly throughout Puget Sound. The most common species include dogfish, ratfish, skates, and turbot.

In delineating areas of biological importance, it is difficult to draw precise boundaries. Many of the organisms are mobile, and many of the immobile species depend on mobile ones for their survival. Oil pollution in an area represents a threat not only to the present occupants, but also to nearby organisms, and to organisms that may occupy the area at some future date. The inter-relationships of the organisms present in different areas are complex and not presently understood. Thus, no area of Puget Sound should be considered a "waste area" and oil be directed there as a cleanup measure unless all other possibilities for control have been exhausted. However, it is apparent that some areas are more valuable than others. In view of the limited cleanup capability, it is advisable to protect these areas at the expense of other areas, although all areas should be protected if possible.

Most of these critical areas are associated with shorelines, and this report will concentrate on shoreline areas. It should be remembered that other areas not considered here are also important and should not be neglected.

The vulnerability of a certain area to damage may vary considerably with season, and this report will attempt to account for this change, although knowledge of such changes is incomplete.

The great size and many different physical environments that exist within Puget Sound cause considerable variation in the types of organisms that live in different parts of the Sound. Some organisms may be extremely common in one portion of the Sound where environmental conditions favor its presence, and be scarce or non-existent in other areas. Organisms present in the Sound dwell in distinct communities. These groupings in the different parts of the Sound occur mainly at the surface, the inter-tidal, and subtidal zones. Deeper water groups of organisms which encounter a more uniform environment tend to be located throughout the Sound.

2. Shellfish

The Hood Canal region provides an ideal environment for many types of shellfish. Pacific oysters are found in abundance throughout the region and are used for both commercial and sport fisheries. Clams, especially littleneck, butter, softshell and geoducks, are abundant, with other varieties found in somewhat lesser number. Rock scallops are in moderate abundance. The Hood Canal area is used extensively for shellfish propagation. In 1965, over 610,000 lbs. of shellfish were commercially harvested from Hood Canal. Over one half of this total consisted of oysters with the bulk of the remainder clams. Over 50,000 lbs. of shrimp are harvested annually. The harvest of crabs is insignificant in this area.[13]

The Strait of Juan de Fuca generally has exposed shorelines that offer little shellfish habitat. The numbers of shellfish found in this area are not as great as that found in other areas; nevertheless, a considerable population of shellfish, especially crabs, does exist. Oysters are found in moderate quantities in this area and clams of many types are relatively common. Octopi are very common also. Most of the shellfish locate in the more protected areas, such as Port Discovery or Neah Bay. The Strait provides the entrance to Puget Sound for nearly all of the salmon that return to the Western Washington spawning grounds.[14]

The San Juan area is another rich zone, both in shellfish and finfish. The bulk of the commercial bottomfish fishing occurs in this zone, as well as a great deal of the salmon fishing, both sport and commercial.

Shellfish of many varieties are plentiful throughout the region, although not as plentiful as in Hood Canal. Dungeness crab are in especially heavy abundance throughout the area. Clams, especially littleneck, butter and horse are available throughout the tidal zones. This area is one of the few places in the world where the pinto abalone is found. Scallops, shrimp, and oysters occur in moderate quantities, while oysters are very abundant near the mouth of the Nooksack River.

The central Sound area has the highest human population in the area and also the most prolific source of municipal and industrial wastes.[15] Industrialization and residences have reduced the amount of shoreline habitat available for marine life. The amount of shellfish produced is not as great as other areas of the Sound. Crabs are found throughout the zone, although they are not especially plentiful. Oysters and many types of clams also are scattered throughout the region, clams being most plentiful near Whidbey Island. Shrimp, squid, and octopus are also found in this region.

The southern Sound zone is a very productive area for shellfish. In fact, it produces the greatest amount of shellfish of any region of Puget Sound. In 1965, more than 3 million pounds of shellfish were harvested, all but 500 lbs. of the total being clams and oysters.[16] This area contains important commercial and sport fisheries for shellfish. Most of the shellfish production occurs in the protected bays and inlets, such as Budd Inlet.

Oysters, especially Pacific and native, are very abundant, as are all varieties of clam, squid, and octopus are also plentiful. Shrimp are found in moderate quantities. Scallops are not plentiful in this area.

3. Fish

In addition to shellfish, fish also form a valuable resource in the Puget Sound. Salmon are far and away the most important fish, economically, in the area. The value of salmon to the economy is reckoned in tens of millions of dollars annually.[17] Salmon are found in all parts of the Sound the year round, though their numbers vary dramatically during certain seasons. In general, they are most plentiful in the summer and fall. Most of the salmon pass through the Strait of Juan de Fuca on their journey to the Sound but a sizeable number also passes through the Strait of Georgia. The San Juan area experiences the highest number in commercial catch, averaging well over 4 million salmon of all types annually.[18] The Strait of Juan de Fuca produces about one-half million salmon annually. This figure could be much higher if the majority of the area were not closed to commercial salmon fishing. Hood Canal is entirely closed to commercial fishing, thus the numbers of catches in Hood Canal also tend to be relatively low. The central and south zones of Puget Sound do not have as high a population of salmon as other areas. South Sound has an annual catch of about 50,000 salmon and central Sound has had about 400,000 salmon annually caught. The addition of a large sockeye salmon run to Lake Washington should increase the annual number of salmon catches in the central zone by hundreds of thousands.

In addition to salmon, groundfish provide an important resource. In 1969, over 16.5 million pounds of these bottomfish were caught in Puget Sound.[19] The bulk of the harvest went for reduction and animal food, while about 4 million pounds were sold for human consumption. The two most important species of food fish are English sole and true cod, which together comprise about 75% of the weight of bottom fish sold for human consumption.

The bulk of the commercial catch, at least for food fish, occurs in the San Juan area. In 1969, over 2 million pounds of food fish were harvested here. True cod comprised about one-half of this total. Almost 1/2 million pounds of English sole was captured in this zone. Starry flounder were caught in approximately the same proportion as English sole. Ling cod is another important species in this area.

The central Sound area produces about 1.3 million pounds of food fish annually. The most commonly caught species is English sole, of which over 1 million pounds was caught in 1969. Starry flounder made up most of the remaining weight. In addition, this area yields large amounts of non-food fish which are reduced to fish protein concentrate. About 7.7 million pounds of fish were reduced to protein concentrate from this area in 1969. This consisted almost entirely of hake caught in the Port Susan area during the late winter and spring.

The southern zone of Puget Sound does not yield an abundance of commercial fish. Less than one-half million pounds was harvested in 1969, and of this only 41,000 consisted of food fish. The two most commonly caught food fish were sand sole and starry flounder.

The Strait of Juan de Fuca yields about 1 million pounds of fish annually of which 90% is food fish, primarily due to the abundance of true cod. 700,000 pounds of live cod were caught in 1969. Next in importance is English sole.[20] It is important to note that all commercial fishing occurs in the eastern half of the Strait and mainly during the winter.

The Hood Canal area is not heavily fished commercially and does not yield many fish - less than 10,000 pounds total in 1969. This lack of effort is due:

a. to the great distance from the major processing plants;
b. many of the fish in this region are heavily infested with a nematode parasite which renders the fish useless for human consumption

4. Waterfowl

Waterfowl are also an important resource in Puget Sound. Puget Sound is an important wintering range for many species of duck and geese. There is a resident population of waterfowl present the year round. The most important wintering areas for waterfowl appear to be around Dungeness Spit, in the Skagit Game Preserve and in the estuary of the Nisqually River. The largest flock of snow geese in the United States winters in the Skagit Flats.

b. Man-Made Resources

The Puget Sound region, bounded by the Olympics in the west, Cascades in the east, the Canadian border to the north, and Olympia in the south, constitutes the state's most impressive recreational area. The vast majority of the activities are either water of shoreline related. This area has been called the pleasure boating capital of the world.

To enhance the enjoyment of the Puget Sound waters, man has developed all types of facilities along the shoreline. These include state parks, launching ramps, moorages, and marinas. Beaches and tidelands have also been set aside. In the more populous areas, city and county parks are plentiful. Listed below is a summary of the primary man-made resources in Puget Sound. Details on each subzone are presented in the next part of this section.

TABLE II-A2: MAN-MADE PHYSICAL RESOURCES[21]

	State Parks	Launching Ramps	Rental Moorages
Subzone 1	3	16	11
Subzone 2	12	29	21
Subzone 3	11	31	15
Subzone 4	4	51	68
Subzone 5	6	30	39
Subzone 6	6	13	7
TOTAL	42	170	161

4. SUBZONE DESCRIPTION

The following pages include a detailed analysis of the parameters in each subzone which can affect or be affected by an oil spill incident. Categories included are meteorology, hydrology, biology, and technology. Listed below are the topics covered.

1. Geography.
2. Currents.
3. Winds
4. Waves
5. Natural physical resources
6. Man-made physical resources

No attempt has been made, in this section, to evaluate the consequences relating these parameters with an oil spill.

The text on each subzone is supplemented by tables and charts. For a given subzone, the geographical outline (orientation and scale) in the various charts is identical. Thus, the reader, by comparing charts, can attain an overall picture of how the various parameters are distributed in any locale. When compared to the set of charts delineating oil handling facilities in Section II-C, it is interesting to note that those areas which are abundant in resources are also heavily populated by oil operations.

a. Subzone 1: Strait of Juan de Fuca

1. Geography

The Strait of Juan de Fuca separates the southern shore of Vancouver Island, Canada, from the northern coast of the state of Washington. This important body of water is the connecting channel between the Pacific Ocean and the inter-island passages extending southward to Puget Sound and northward to the inland waters of British Columbia and southeastern Alaska.

At its entrance, and for 50 miles eastward to Race Rocks, the Strait has a width of about 12 nautical miles, and then it has a width of about 16 nautical miles for 30 miles to Whidbey Island. The waters are generally deep until near the shore with few outlying dangers, most of which are in the eastern part. The shores on both sides are heavily wooded, rising rapidly to elevations of considerable height, and except in a few places, are bold and rugged.

The shores of the Olympic Peninsula bordering the Strait and characterized by relatively virgin areas, especially west of Port Angeles. Most near-shore lands are privately owned, with the Makah Indians occupying the northwest tip of the area, and the Department of Natural Resources holding several tracts spaced intermittently. West of Port Townsend the density of privately owned individual tracts increases more, particularly around Sequim Bay and Discovery Bay and in the vicinity of Port Townsend.

In this general area, the prevailing uses can be categorized as either recreational or small residential in nature. The region supports relatively few residents, but is very popular with the weekend visitor from the more populous parts of Washington. Built-up areas are few, occurring primarily in the vicinity of the larger towns such as Port Angeles, Port Townsend, and Neah Bay. Minor ports are located at Sekiu, Pysht, and New Dungeness Bay.

The coastline is generally steep with narrow beaches of coarse sand and large boulders, except in the more protected areas and bays, where the gradient is gradual and the intertidal zone composition is mixed sand and gravel. A broad shelf of gravelly shorelines extends from Port Angeles to Puget Sound with the shallow sandy-mud flats of Dungeness in between. Moving westward from Port Angeles, there are larger sweeping sandy beaches lined by pine forests and gentle slopes, and small bays with small rock formations within them.

Highway 101 services the general area but does not provide access to beach and shore areas. Access is quite restricted because of the proliferation of private ownership. The water can be reached at the larger towns or parks.

2. Currents

The currents in the Strait of Juan de Fuca may attain velocities of 2 to 4 knots varying with the range of the tide and are influenced by strong winds. Eastward of Race Rocks in the wider portion of the Strait, the velocity is considerably less. At Race Rocks and Discovery Island the velocity may be 6 knots or more.

The flood current entering the Strait sets with considerable velocity over Duncan and Duntze Rocks; but instead of running in the direction of the channels, there is a continued set toward Vancouver Island shore which is experienced as far as Race Rocks. The flood current velocity is greater on the northern shore of the Strait than on the southern. (See Figure II-A2.)

The ebb current is felt most along the southern shore of the Strait and between New Dungeness Light and Crescent Bay, as shown in Figure II-A3. There is a prevailing set southward and westward especially during large tides. With wind swell against the current, a short choppy sea is raised near the entrance to the Strait. Tide rips occur off the prominent points and in the vicinity of the banks. These are particularly heavy off Cape Flattery, Race Rocks, Dungeness Spit, and Point Wilson; at times becoming dangerous to small vessels.

3. Winds

The wind conditions in the Strait of Juan de Fuca, west of Port Angeles to Cape Flattery, are predominantly easterly with speeds exceeding 20 miles per hour (MPH) during the period from October to March. East of Port Angeles, winds are westerly with speeds to 20 mph during this period. From April until September, winds are generally westerly to 20 mph. Figure II-A4 summarizes the monthly wind patterns.

4. Waves

Waves generated at sea may enter the Strait of Juan de Fuca in the form of swells. In addition, shorter period waves may be initiated in the Strait due to the relatively long fetch and predominantly moderate wind stress on the water surface. During the period from October to March, the wind direction in the Strait normally tends to oppose the progress of swells from the ocean and waves up to 13 feet may occur during storms. From April until September, prevailing 20 mph winds acting along the 60 mile fetch over a long duration may produce wind waves of up to 6 feet in height which interact with the swell conditions to create even greater height.

5. Natural physical resources

The following outline describes the natural physical resources present in and near the waters of subzone 1. Figure II-A5 through II-A7 should be referred to for specific locations.

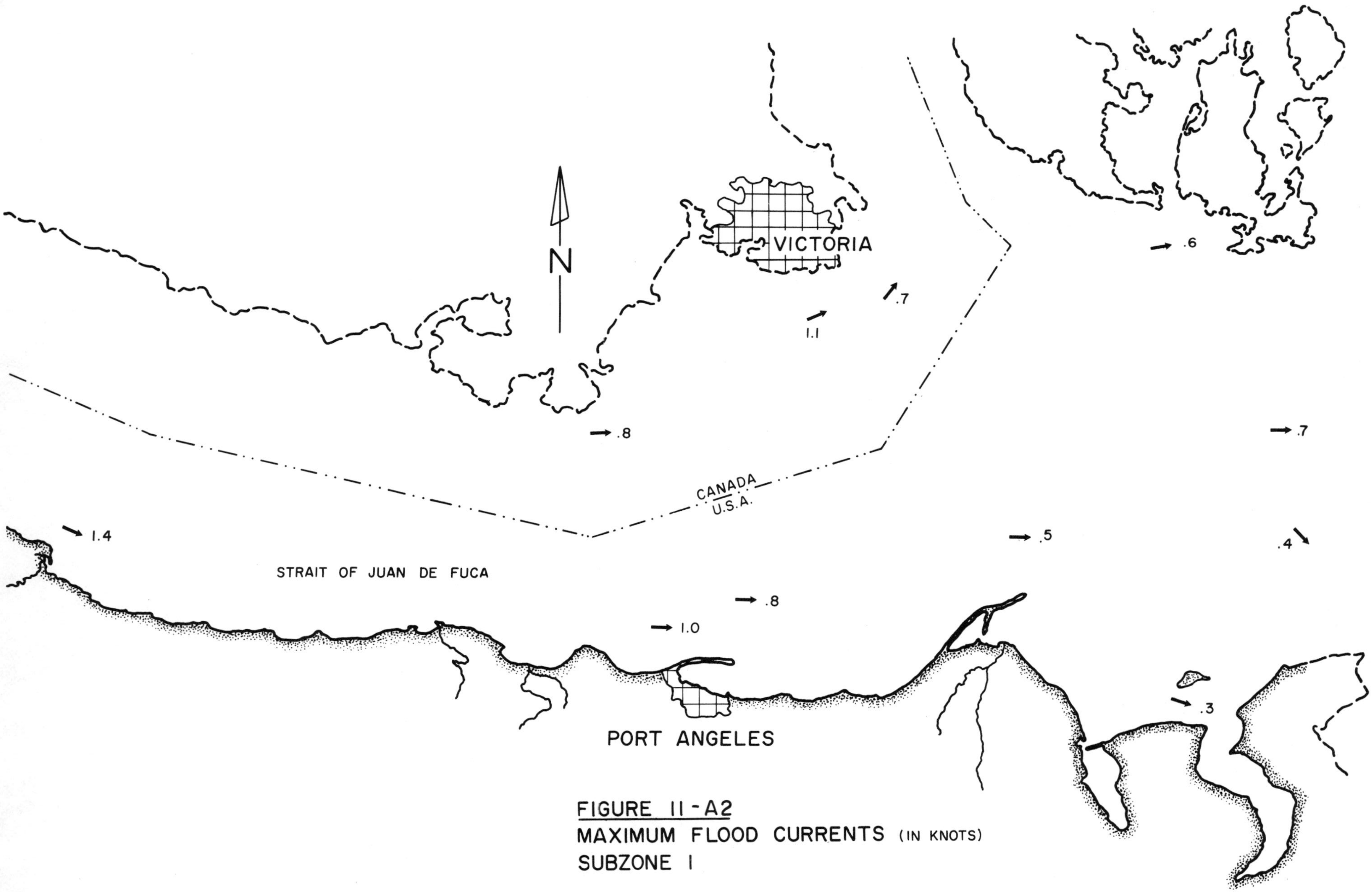

FIGURE II-A2
MAXIMUM FLOOD CURRENTS (IN KNOTS)
SUBZONE I

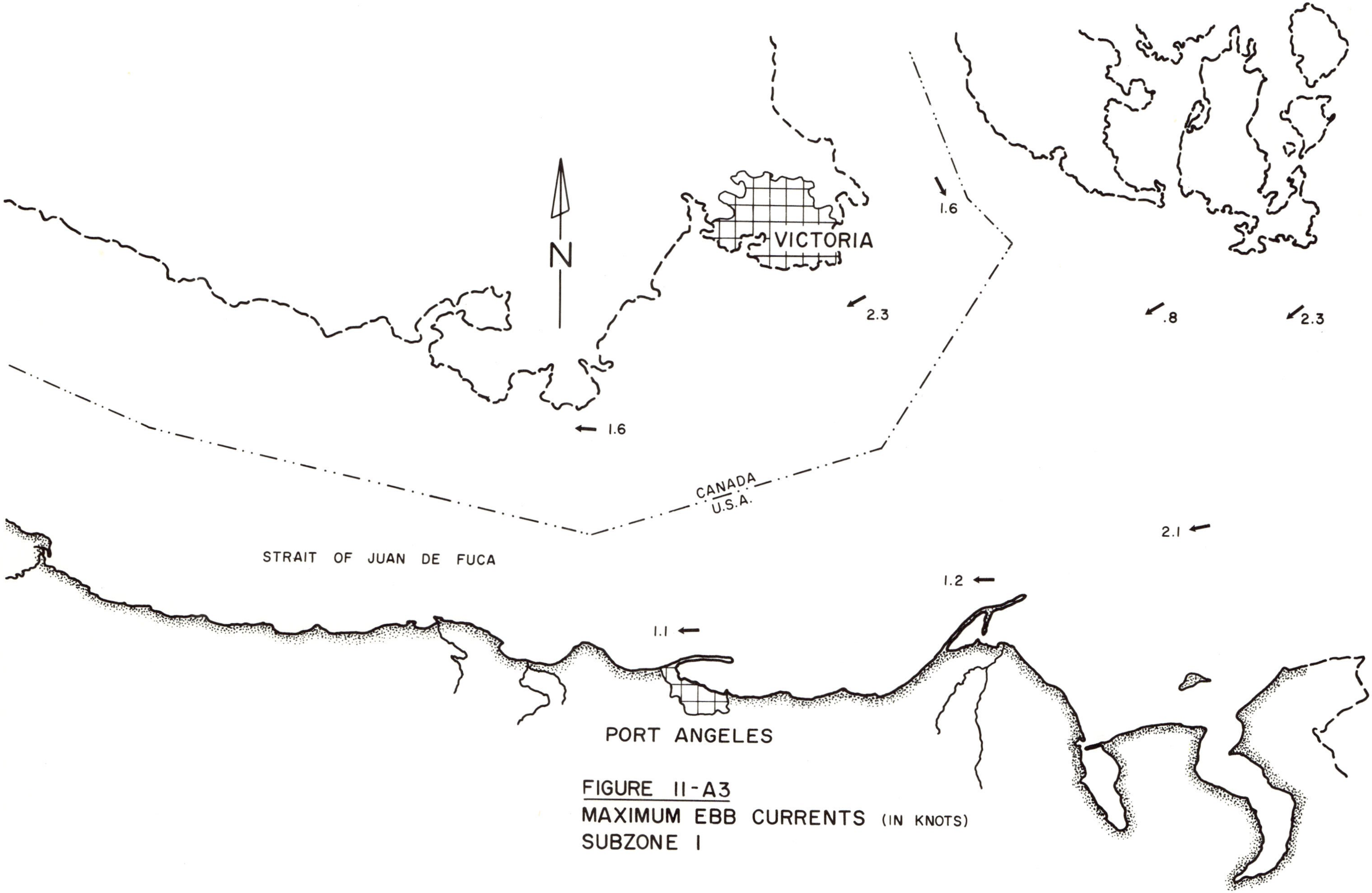

FIGURE II-A3
MAXIMUM EBB CURRENTS (IN KNOTS)
SUBZONE I

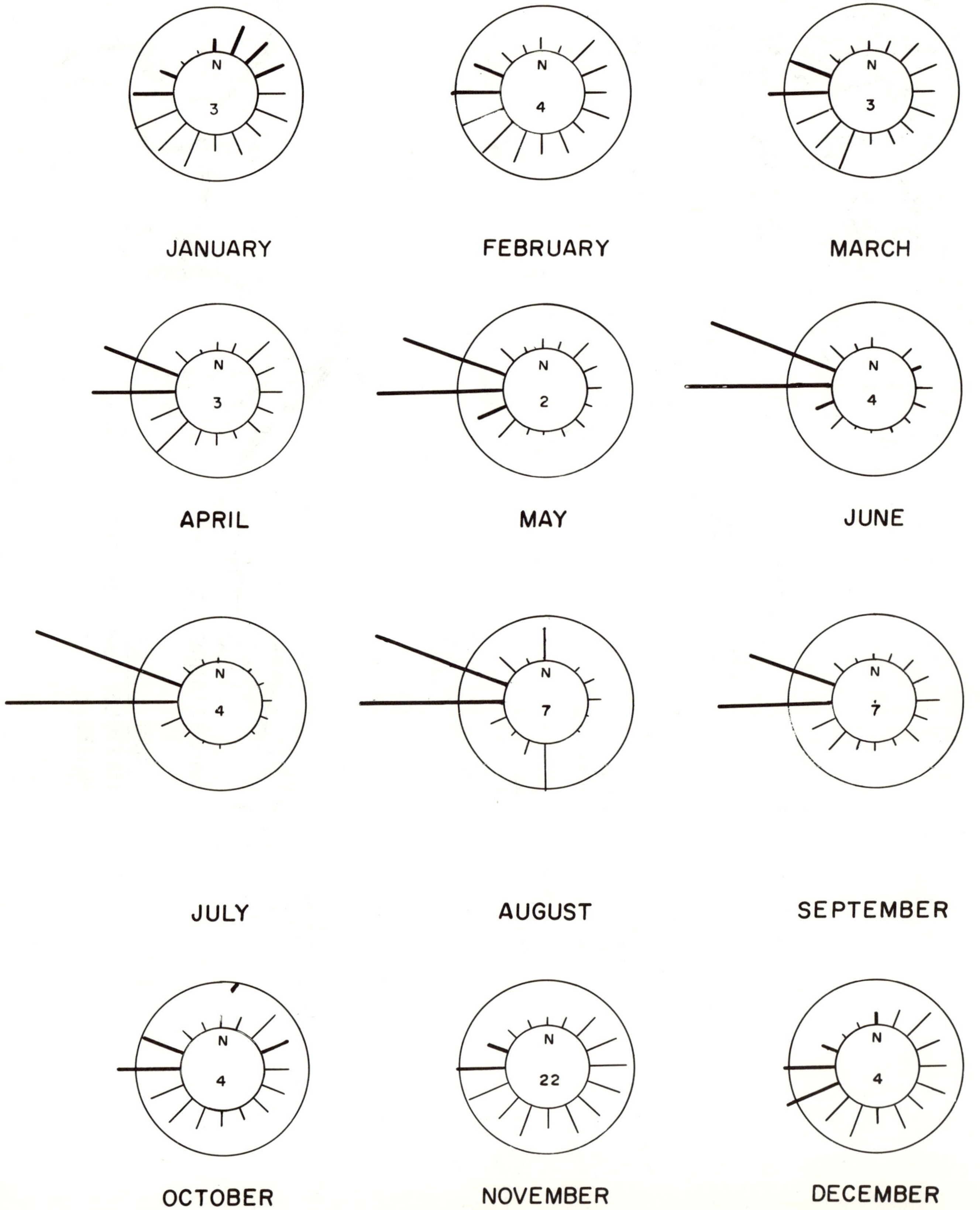

FIGURE II-A4, WIND PATTERNS FOR PORT ANGELES AFTER HARRIS, 1954. WIDTH OF BAR INDICATES MONTHLY AVERAGE WIND VELOCITY, EACH RADIAL UNIT INDICATES 10 % OF TOTAL OBSERVATIONS, CENTRAL NUMERAL INDICATES PERCENTAGE OF CALM WIND. 0 - 10 MPH ——, 11 - 19 MPH ——

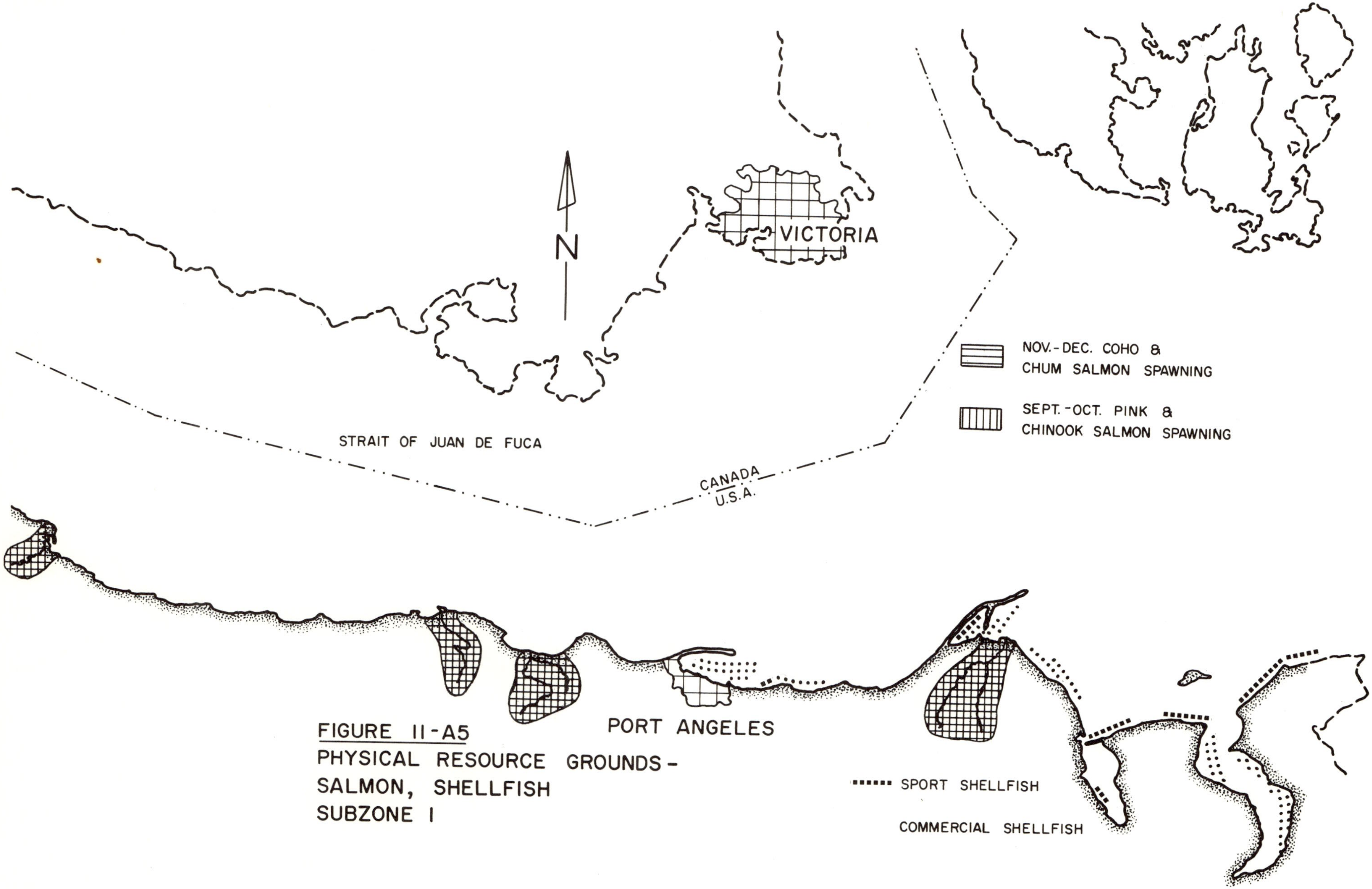

FIGURE II-A5
PHYSICAL RESOURCE GROUNDS-
SALMON, SHELLFISH
SUBZONE I

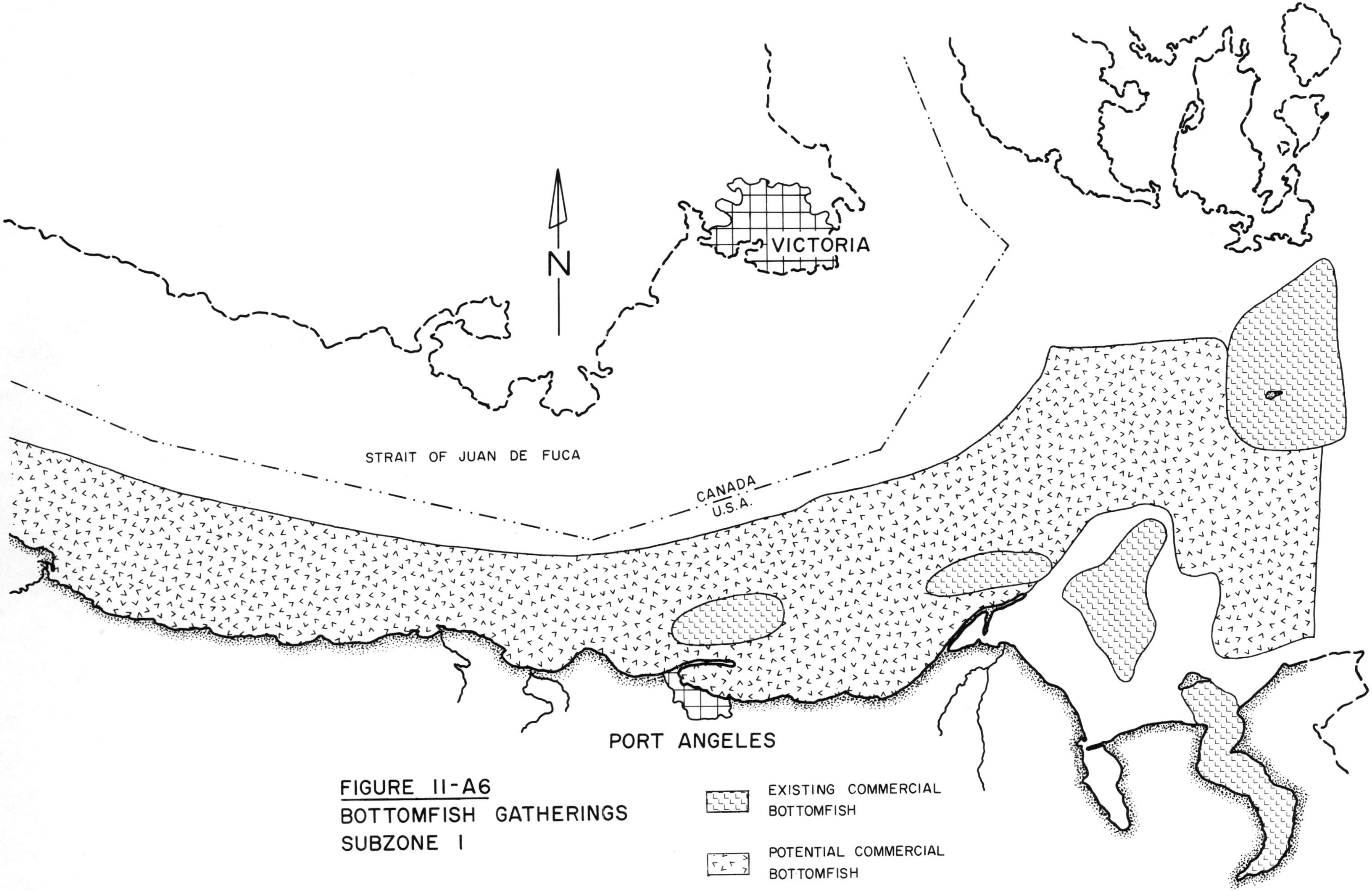

FIGURE II-A6
BOTTOMFISH GATHERINGS
SUBZONE I

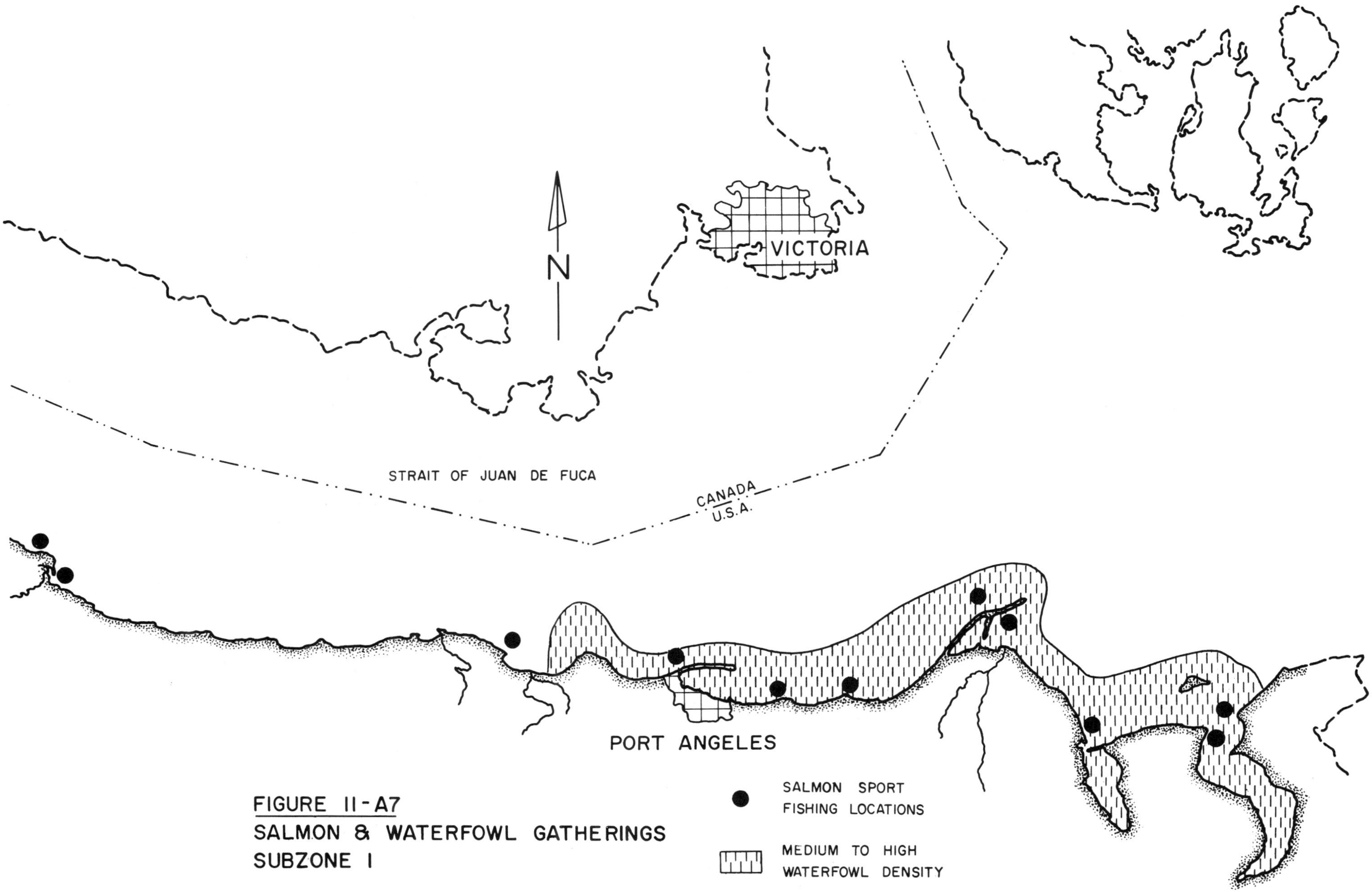

FIGURE II-A7
SALMON & WATERFOWL GATHERINGS
SUBZONE I

i. shellfish

Crabs: Dungeness crabs amply inhabit this zone, especially in sandy estuaries from the Elwha River to Port Townsend. Red and box crabs appear throughout the region.

Oysters: Pacific oysters are scattered lavishly near Dungeness, Port Townsend and Kilisut Harbor. Other species of oysters not common in this area.

Clams: Butter, littleneck, and Manila clams are common in sheltered areas such as Washington Harbor, Kilisut Harbor, Port Discovery, Oak Bay, Port Townsend, Neah Bay, and Sequim Bay. Other areas along the coast are generally too exposed for clam habitat.

Shrimp: Scattered beds of coon stripe shrimp are in the deep waters from Port Townsend to Dungeness Spit. Spot and pink shrimp are found near Protection Island.

Other shellfish: Moderate amounts of California mussel are located from Elwha River to Port Townsend. The blue mussel is common throughout the zone. Some pink and rock scallops occur from Port Angeles to Port Townsend. Rock scallops are found near the mean low water line. Seasonal migrations of squid occur in Port Discovery, Washington Harbor, Oak Bay, Port Townsend, and Sequim Bay. Large octopus populations are found near Port Angeles, Dungeness, Port Discovery, and Port Townsend.

ii. fish

Salmon: Salmon are common all year round. The largest number occur in spring and summer. Highest concentrations are between Waddah Island, Port Townsend, and at Neah Bay, Port Angeles harbor, and Sequim and Discovery Bays. (See Figure II-A7.)

Other fish: All commercial bottom fishing occurs in the eastern part of the Strait, with Port Discovery being the primary area. Figure II-A6 depicts the geographic distribution.

iii. waterfowl

This is an area of high waterfowl density throughout the area especially near Dungeness Spit as noted in Figure II-A7.

6. Man-made physical resources

The man-made resources of interest in this study are those which are adjacent to the waters of Puget Sound. These include state parks, launching ramps, and rental moorages for vessels. The ramps and moorages include both public and private facilities. The following table summarizes this information. Figure II-A8 depicts the locations in subzone 1.

TABLE II-A3: MAN-MADE RESOURCES, SUBZONE 1[23]

Location	State Parks	Launching Ramps	Rental Moorages
Neah Bay/Seiku	0	9	9
Port Angeles	0	1	2
Dungeness Bay	1	2	0
Sequim Bay	2	2	0
Discovery Bay	0	2	0
SUBZONE 1 - TOTAL	3	16	11

Three launching ramps and one rental moorage in this subzone are open to the general public.

b. Subzone 2: San Juan Islands-Bellingham

1. Geography

The San Juan Islands are bounded on the south by the Strait of Juan de Fuca, to the west by Haro Strait, to the north by the Strait of Georgia, and to the east by Rosario Strait. The straits and many channels between the islands have natural water depths of several hundred feet. The San Juan Archipelago consists of about 172 islands, although a total of 473 islands (rocks) are visible at low tide.[24] The islands are generally wooded and rocky. Numerous beaches of rock and gravel with some sand are visible only at low tide. The islands have about 376 miles of salt water shoreline.

The principal harbor facilities in this subzone are located at Bellingham, Anacortes and Friday Harbor. The islands have shorelines which are primarily steep and rocky, of limited accessibility, and which are bordered by rugged terrain of considerable relief. The major estuarine

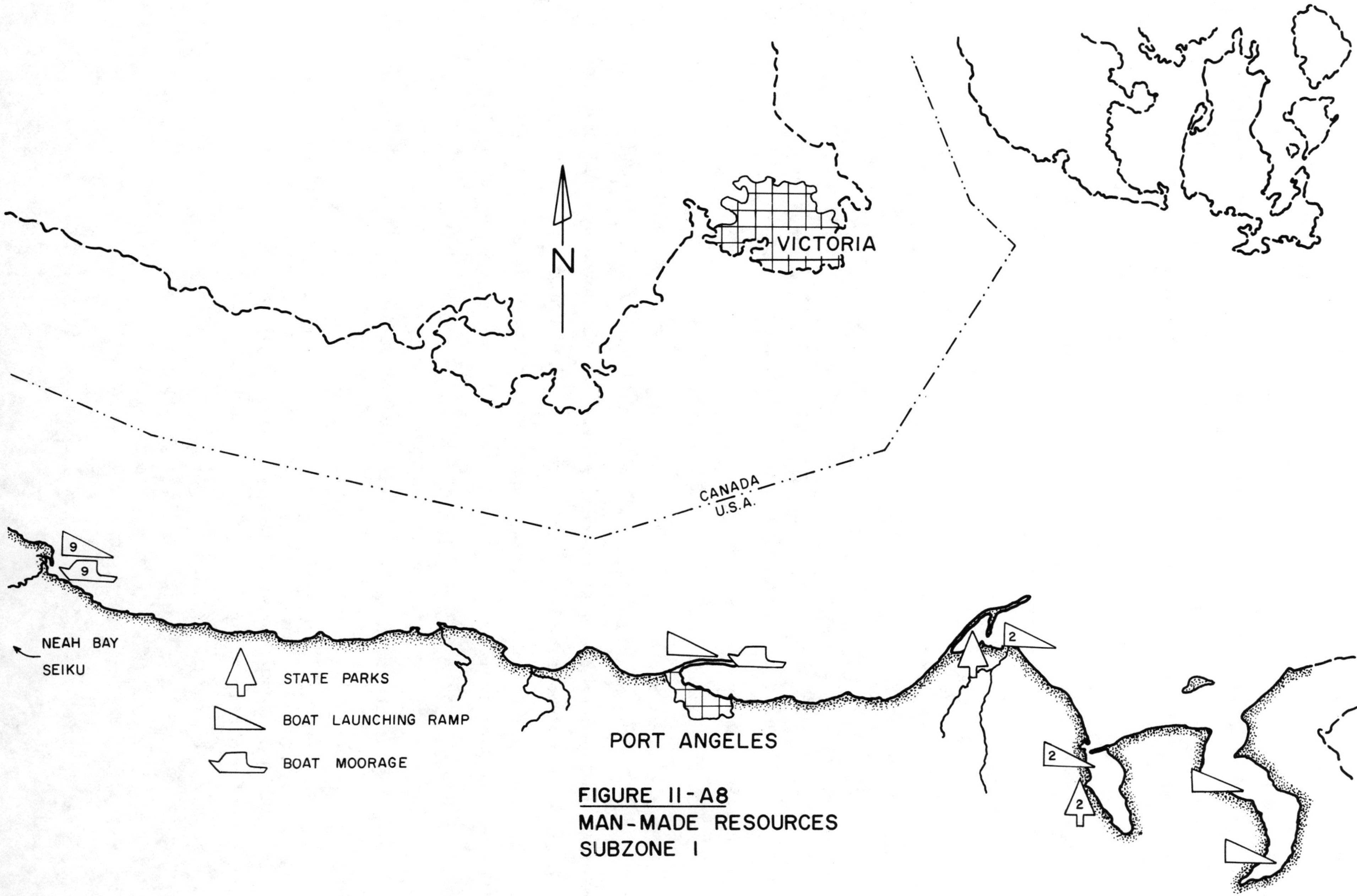

FIGURE II-A8
MAN-MADE RESOURCES
SUBZONE I

areas exist west of Bellingham, near Lummi Island and east of Anacortes in Padilla Bay. With the exception of the industrial, maritime, and commercial fishing operations at Bellingham and Anacortes, the remainder of the subzone is predominantly recreational and residential in use. Although there are 12 state parks and several public beaches in this region, much of the near-shore land is privately owned.

2. Currents

Due to the large number of constrictions to tidal flow, tidal currents in the San Juan Archipelago can attain very high local velocities.

In Haro Strait the flood current sets northward through the Strait, and the ebb in the opposite direction as shown on Figure II-A9 and II-A10. The ebb usually runs longer and has a greater velocity than the flood. At its northern entrance the flood sets eastward on both sides of Sucia Islands and eastward across Alden Bank. The velocity in Haro Strait is 1 to 5 knots. Off turn point the ebb may attain a velocity of 6 knots during large tides. The current develops moderate velocity between Sucia Islands and Orcas Island. Heavy tide rips occur on middle bank, northward of it, and around Discovery Island. Tide rips also occur between Henry Island and Turn point. Heavy, dangerous, tide rips are prevalent between East point and Patos Island and for 2 miles northward in the Strait of Georgia.

The currents in Lopez, Thatcher, and Obstruction Passes develop peak currents of 3 to 7 knots as they enter into Rosario Strait, creating hazardous conditions for smaller craft.

In the Strait of Georgia, the tidal currents are not nearly as strong as those in the channels leading to it from the Strait of Juan de Fuca. The currents in the Strait of Georgia attain velocities of about 3 knots at times, particularly during the freshets of the summer.

3. Winds

During the period from October to March, the winds are generally from the south-southwest to 20 mph. From April to September, winds tend to be southerly with velocities of 10 to 20 mph. Figure II-A11 shows monthly wind patterns for Bellingham.

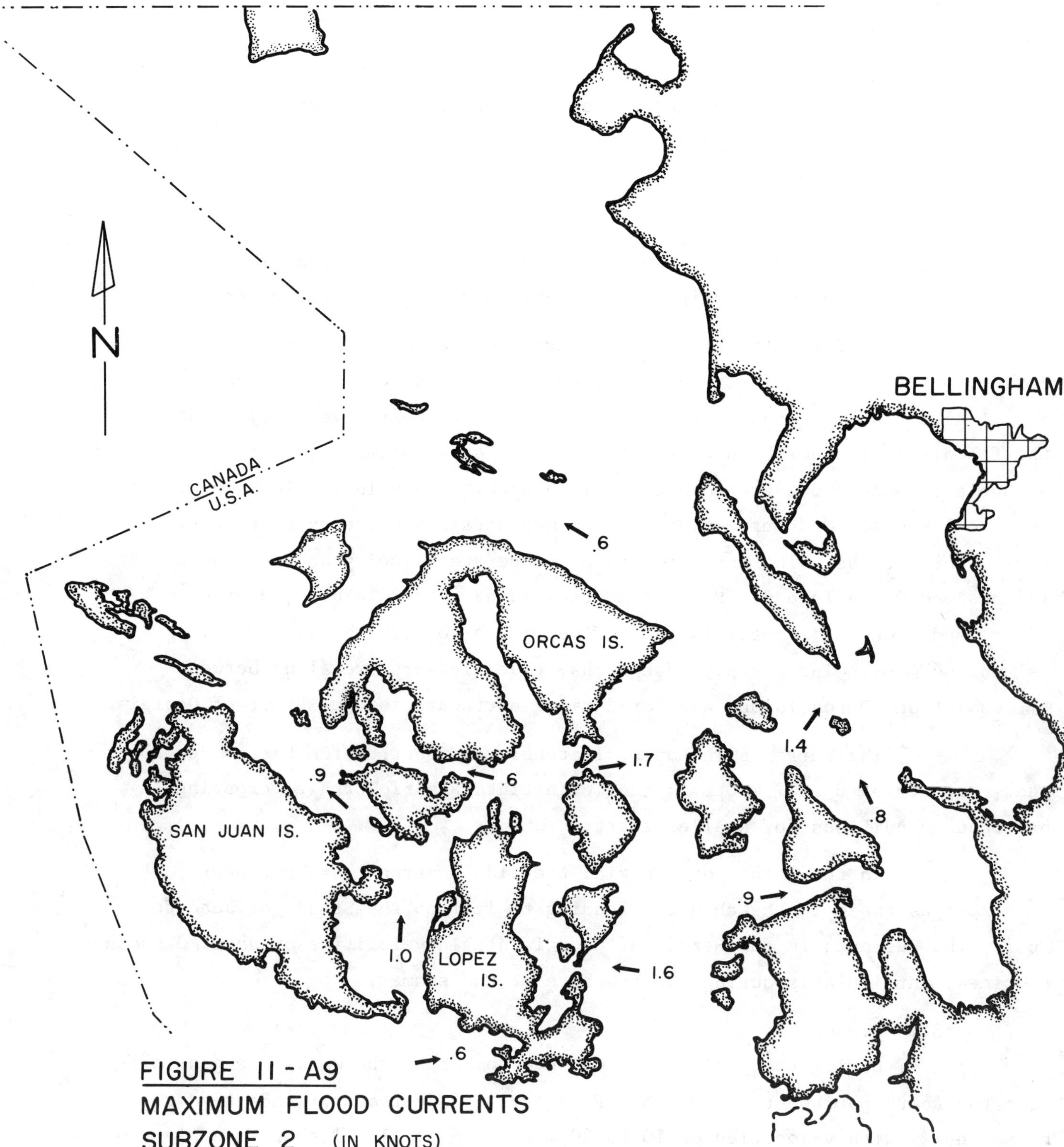

FIGURE II - A9
MAXIMUM FLOOD CURRENTS
SUBZONE 2 (IN KNOTS)

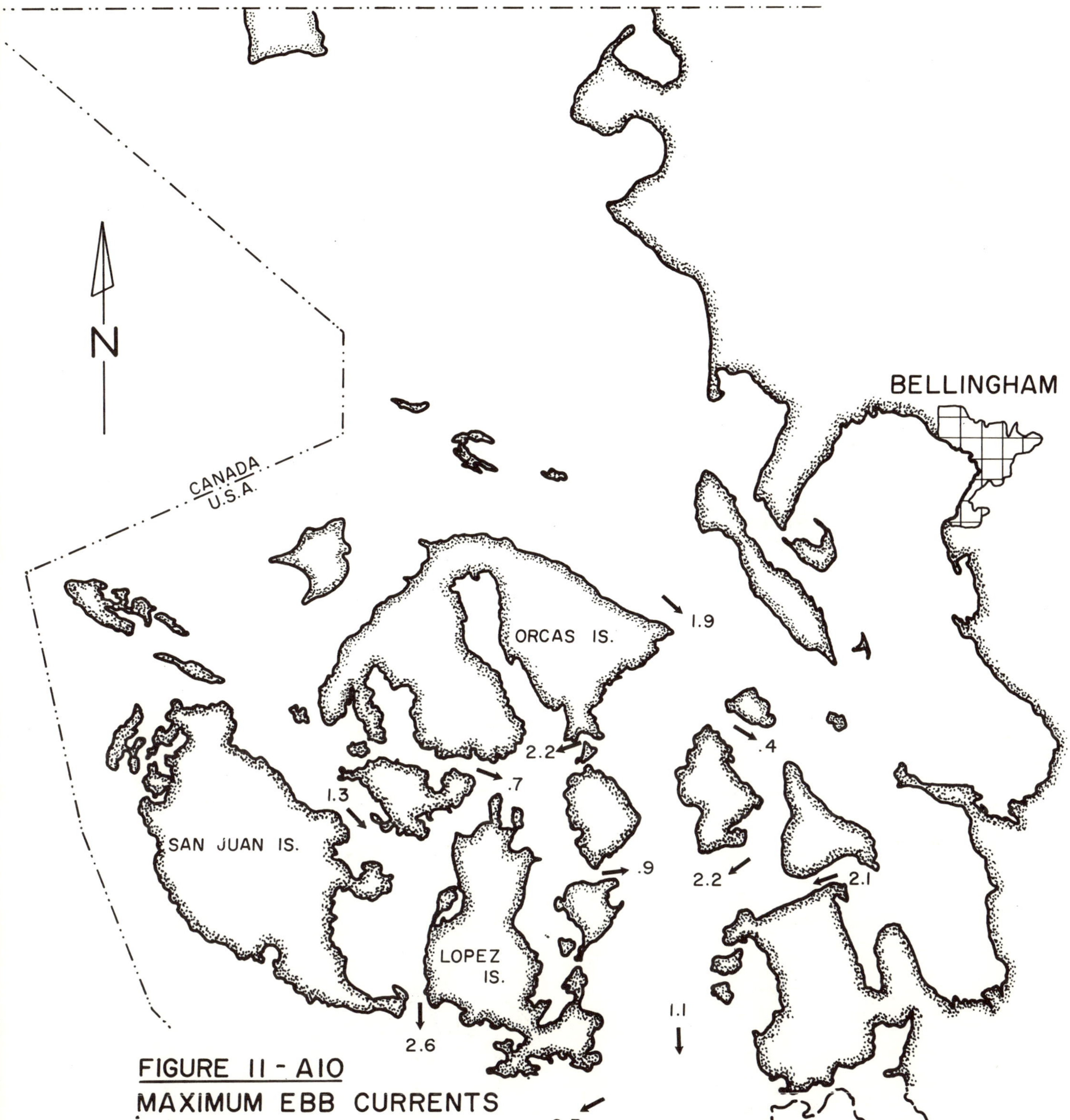

FIGURE II - A10
MAXIMUM EBB CURRENTS
SUBZONE 2 (IN KNOTS)

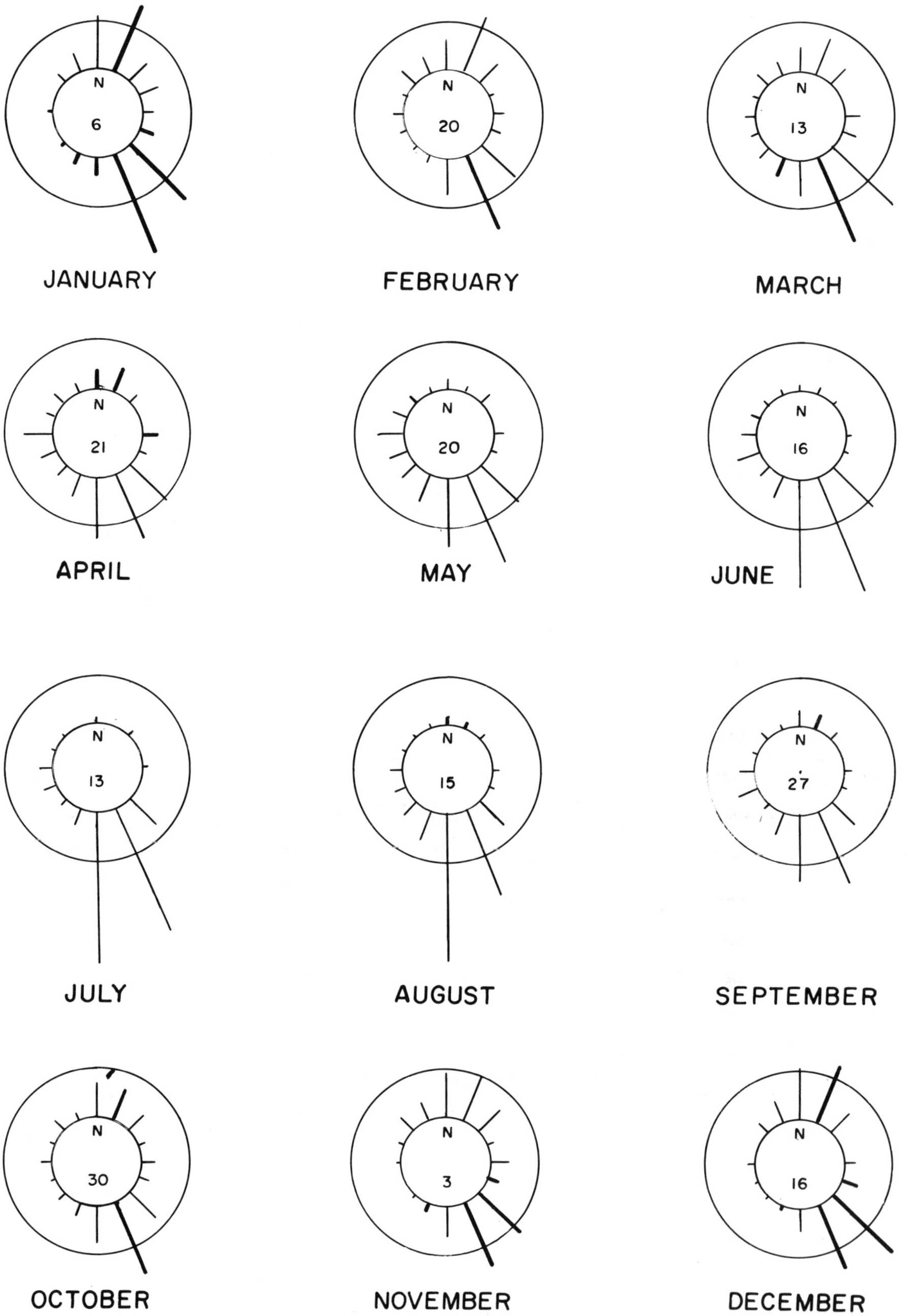

FIGURE 11-A11, WIND PATTERNS FOR BELLINGHAM AFTER HARRIS, 1954. WIDTH OF BAR INDICATES MONTHLY AVERAGE WIND VELOCITY, EACH RADIAL UNIT INDICATES 10 % OF TOTAL OBSERVATIONS, CENTRAL NUMERAL INDICATES PERCENTAGE OF CALM WIND. 0 - 10 MPH ——, 11 - 19 MPH ——[25]

4. Waves

During the period from October to May, the maximum wave height in the Bellingham area may reach 5-1/2 feet. In all parts of the San Juan Islands not exposed to waves from the Strait of Juan de Fuca, the maximum wave height is less than 4 feet. Areas exposed to the Strait of Juan de Fuca may experience waves and swell of the size prevalent in the Strait. From June-September, maximum wave height will be less than 2 feet in the Bellingham-San Juan Islands region.

5. Natural physical resources

The proliferation of natural physical resources related to this subzone of Puget Sound is easily seen from a quick glance at Figures II-A12 through II-A14. The following describes the areas presented on the maps:

i. <u>shellfish</u>

<u>Crabs</u>: Dungeness crab are abundant throughout the zone. Areas of highest quantity occur around Sucia and Waldron Islands, Lopez Sound, Hunter Bay, Mud Bay, San Juan Channel, Upright Channel and Griffin Bay. Other areas rich in not included are Padilla, Boundary, Semiahmoo, Birch, Lummi, Bellingham, Samish, and Mitchell Bays, as well as the channels between the islands and the mainland and Spencer Spit. They are also located near Similk Bay and near Hope Island-Seal Rocks in Skagit Bay. Red crabs are moderately abundant in shallow waters in Skagit Flats and near oyster beds.

<u>Oysters</u>: Pacific oysters grow in moderate quantities along Dewey Beach in Similk Bay, and on gravelly and mud beaches from Skagit Island to Swinomish Flats. Pacific oysters also are found from Drayton Harbor (Clayton Beach) south, to the southern part of Samish Bay; William Point to Bay View (Padilla Bay). Lesser amounts inhabit Birch, Lummi, and Bellingham Bays, as well as the west side of Samish Bay, and areas along rocky beaches. Olympia and Kumamoto oysters occur only in Samish Bay.

<u>Clams</u>: The major clam producing areas include Portage, Bellingham, Chuckanut, Samish, Padilla, Fidalgo, and Burrows Bays. Butter and littleneck clams are bountiful in the Blaine-Pt. Roberts area, Birch, Portage, Samish, and Padilla Bays, and along the shores of most islands.

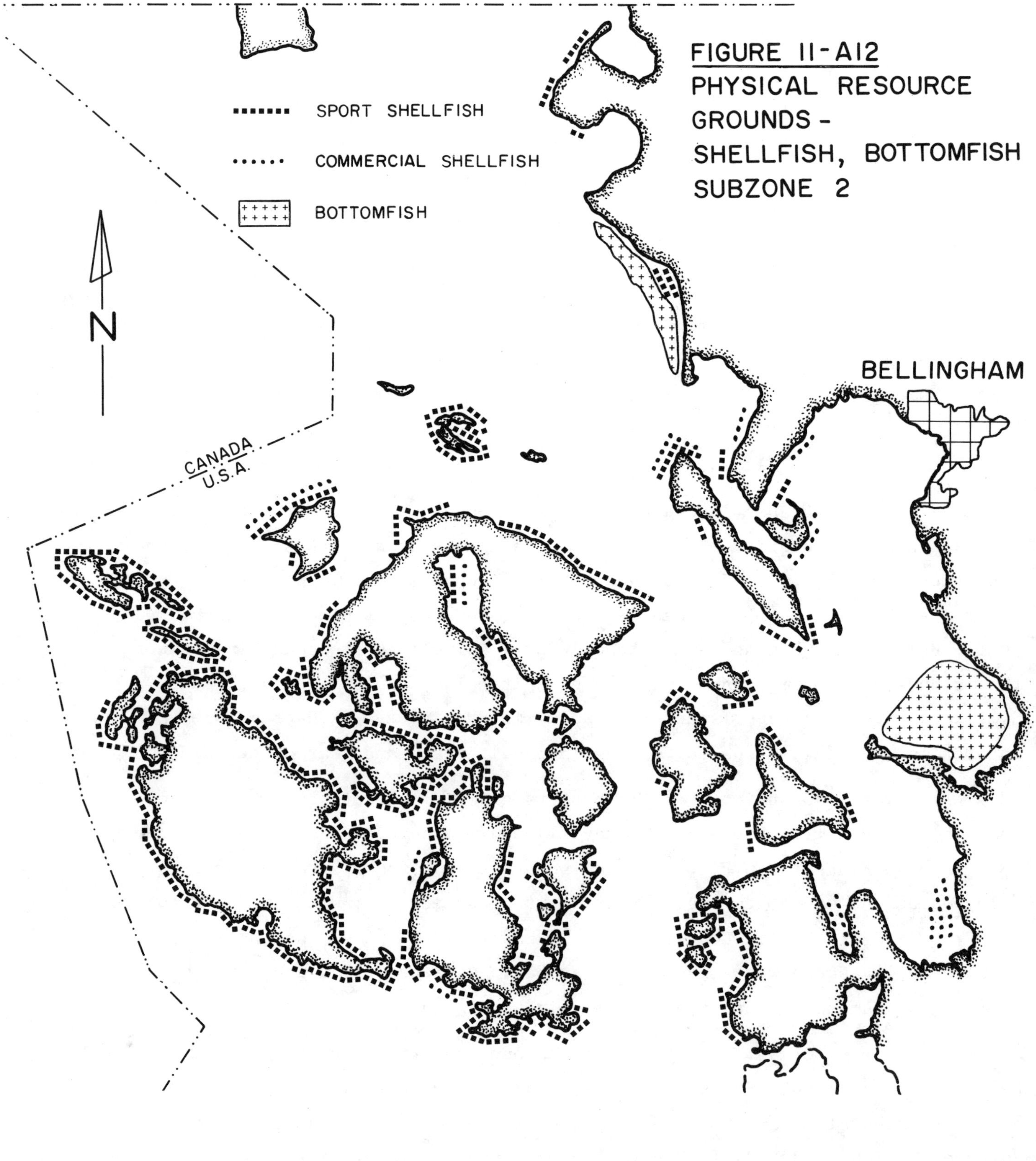
FIGURE II-A12
PHYSICAL RESOURCE GROUNDS - SHELLFISH, BOTTOMFISH
SUBZONE 2
SPORT SHELLFISH
COMMERCIAL SHELLFISH
BOTTOMFISH
N
CANADA
U.S.A.
BELLINGHAM

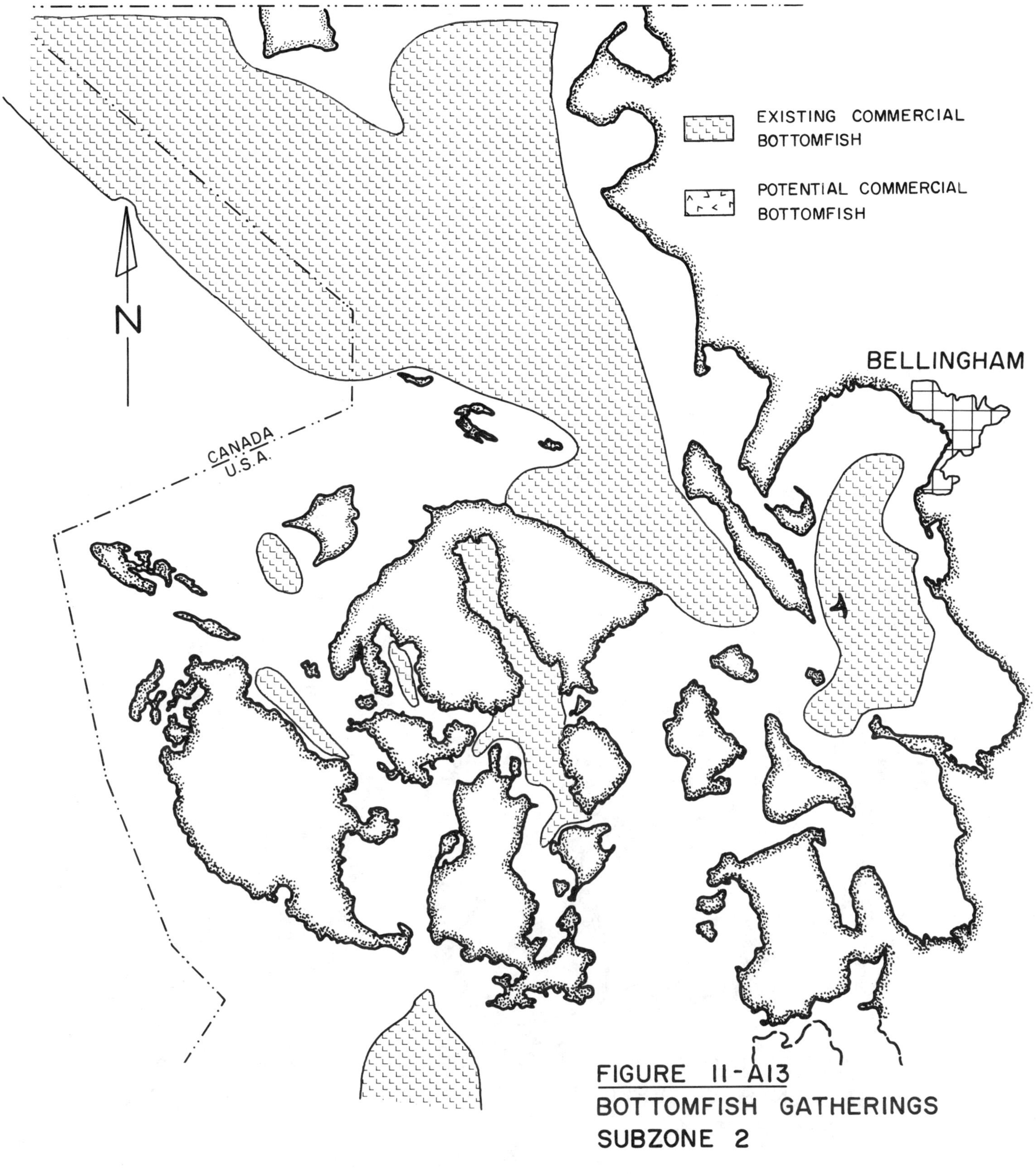

FIGURE II-A13
BOTTOMFISH GATHERINGS
SUBZONE 2

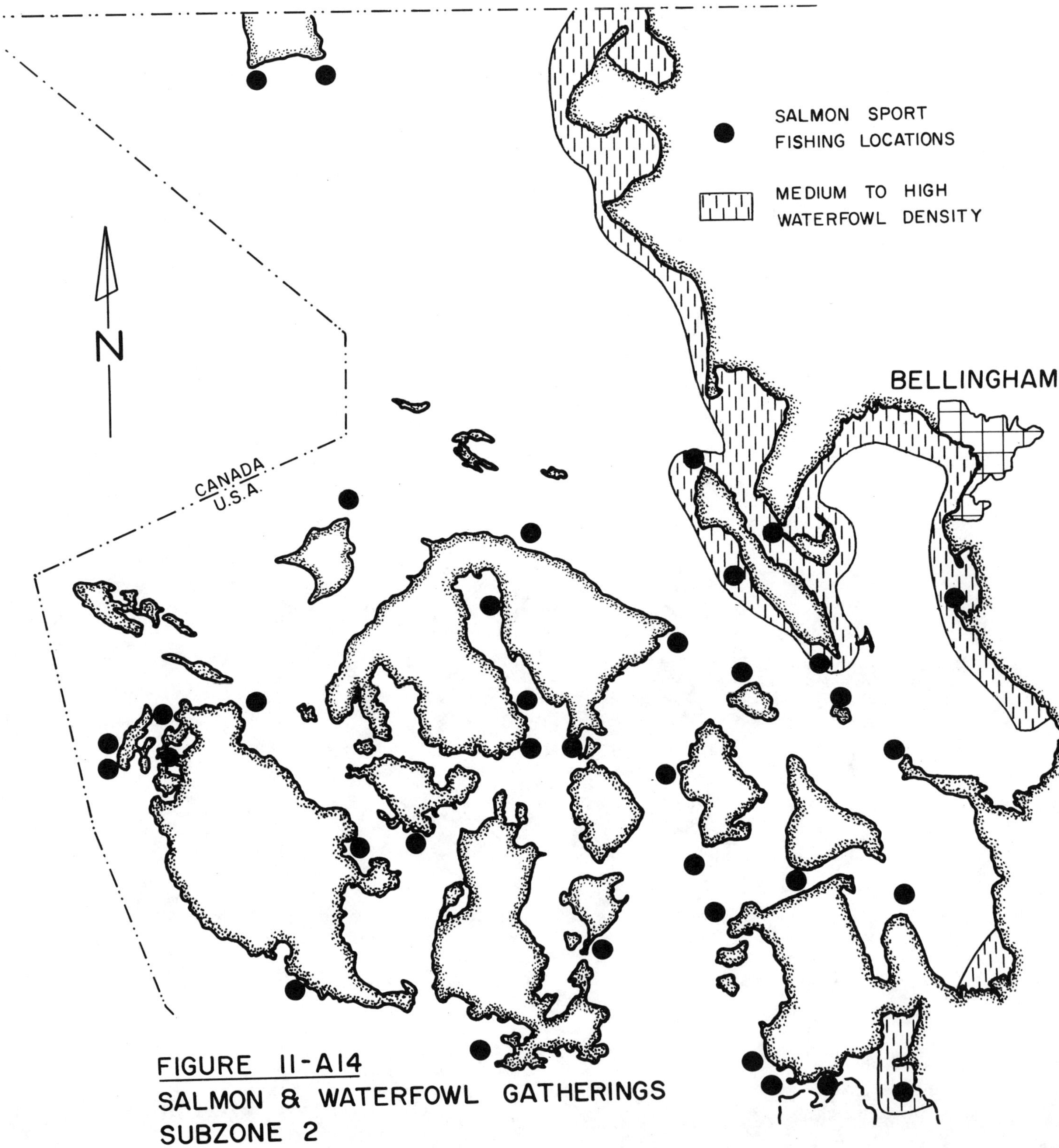

FIGURE II-A14
SALMON & WATERFOWL GATHERINGS
SUBZONE 2

They are also common, along with manila clams, in the northern part of Skagit Bay and from Dewey Beach to Swinomish Flats. Horse and softshell clams are plentiful on Skagit Bay. Geoducks are relatively scarce in this area.

Shrimp: Spot, pink and sidestripe shrimp are common in Bellingham Bay. These and coon stripe shrimp occur off Similk Bay in main channel, and in San Juan channel.

Other shellfish: Sporadic concentrations of squid occur around the San Juan Islands and in the general area of Skagit Flats. Pink scallop are found scattered around the larger bays. Rock scallops are common around the San Juan Islands. Pink scallops are found between Sucia and Waldron Island, and between Waldron and Stuart Islands, in parts of President Channel and in Upright Channel. Abalone, both pinto and red, occur around many of the San Juan Islands. Pinto abalone are most common in the northern San Juan Islands. Blue mussels are abundant on the rocky shores of San Juans and on most northern shorelines. California mussels are found in small numbers in much the same areas as the blue mussel. Octopi are very common throughout this area, especially around the San Juan Islands.

ii. fish

Salmon: Salmon are plentiful throughout the subzone the year round as shown on Figure II-A14. The San Juan Islands and Point Roberts are two of the most popular commercial fishing areas on Puget Sound. Salmon seem to congregate in many locations in this subzone, especially Harney Channel, Mosquito Pass, East Sound, Lopez Pass, and Guemes Channel near Anacortes.

Other fish: Principal commercial fishing locations include eastern Rosario Strait, Skagit Bay, south Samish Bay, outer Bellingham Bay to San Juan and north to Blaine, west to Rosario Strait, San Juan Channel, Saratoga Passage, and off Sucia Island and West Beach.

iii. waterfowl

Waterfowl density is generally low around the San Juan Islands. The common range of waterfowl usually extends from just west of Cypress Island north to the Canadian border, and south to Deception Pass. Medium to high waterfowl density in Padilla and Samish Bay, around Lummi Island and Bellingham Bay. Medium to high waterfowl density occur near the shore from Padilla Bay to the Canadian border. A lower waterfowl density occurs around Point Roberts. Figure II-A14 illustrates the location of waterfowl in subzone 2.

6. Man-made physical resources

In addition to the abundance of natural resources, man-made resources are also plentiful around the waters of Subzone 2. The following table lists these resources by location, and these are shown in Figure II-A15.

TABLE II-A4: MAN-MADE RESOURCES, SUBZONE 2[26]

Location	State Parks	Launching Ramps	Rental Moorages
Boundary Pass	3	0	2
Haro Strait	0	4	0
San Juan Channel	2	6	2
East Sound-Harney Channels	0	5	5
Strait of Georgia	3	1	1
Semiahmoo-Birch Bays	1	4	2
Lummi Is. (west side)	0	1	2
Hale Passage	0	1	1
Bellingham Bay	1	2	2
Burrows Bay	1	2	1
Guemes Channel-Padilla Bay	1	3	3
SUBZONE 2 - TOTAL	12	29	21

Twelve of the launching ramps and three rental moorages in this subzone are available to the general public.

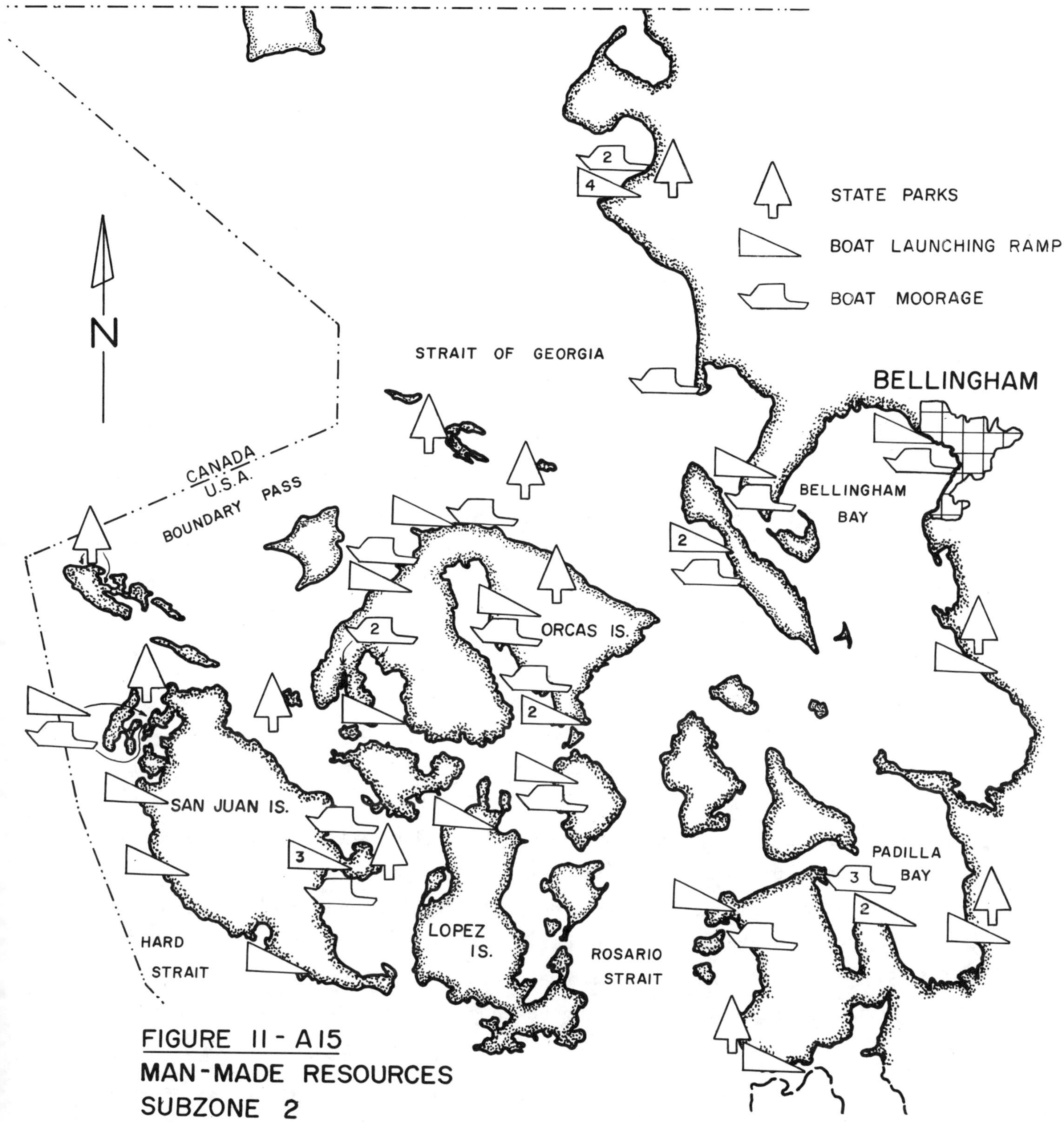

FIGURE II - A15
MAN-MADE RESOURCES
SUBZONE 2

c. Subzone 3: Whidbey Island-Everett

1. Geography

Located to the north of Puget Sound proper, this subzone includes several prominent geographical features. Whidbey Island, to the east of the Strait of Juan de Fuca, comprises 165 square miles of area and is the second largest island in the continental United States. Camano Island lies to the east of Whidbey Island and its 45 square mile area is enclosed by the bays and estuaries of Skagit Bay, Saratoga Passage, and Port Susan. Whidbey and Camano Islands are both agricultural and recreational in land use as are the mainland areas to the east of the subzone. Both islands are long and narrow, and no point on either is more than 2.5 miles from the sea. Most of the land surface is rolling uplands that range in altitude from 100 to 300 feet, but may exceed 500 feet in some places.

Whidbey Island has a number of minor harbors with ferry landings or piers for other local traffic. Minor harbors are located at Oak Harbor, Coupeville, Columbia Beach, Juan de Fuca, and Coronet Bays. Numerous single family recreational residences exist on the shorelands, usually on banks of considerable relief. The port of Everett is the most populated area of the subzone with industrial as well as recreational and residential land uses.

Most of the beaches in Similk and Skagit Bays are silty clay with some sand. Broad tide flats characterize Skagit Bay from Swinomish Slough to Stanwood. The Skagit Flats are exposed at mean lower low water (MLLW) for 1.5 to 2 miles bayward. Shallow mud flats extend into Port Susan and Skagit Bay 2 to 3 miles from the mouth of the Stillaguamish River. Broad tide flats also predominate in northern Port Susan. The shore between Kayak Point and Tulalip is shorter and steeper, with rock and gravel predominating as on the beaches exposed to considerable wave and current action on Whidbey and Camano Islands. Beaches from Everett south to Edmonds are predominantly faced by banks with narrow, sloping beaches consisting mostly of rock and gravel.

Accessibility from land to the beaches on Whidbey and Camano Islands as well as most of the tidelands of Skagit Bay to the north and from Everett to Edmonds to the south is hampered by the prevalence of steep banks or broad tidelands with only very intermittent connections by public roads.

The shoaling of the shorelines in many places restricts the accessibility of the beaches from the water by boat.

2. Currents

In Admiralty Inlet, the tidal currents are subjected to daily inequalities similar to those of the tides. Velocities of 2 to 5 knots occur from Point Wilson to Point No Point. In the more open waters of the Sound, southward of Point No Point, the velocities are much less. At Point Wilson and at Marrowstone Point, slack water occurs from one-half to one hour earlier near shore than in midchannel. In the winter when southerly winds prevail, there is generally a northerly surface drift which increases the ebb current and decreases the flood current. This effect is about .5 knot between Nodule and Bush Points. The tidal currents in the southern entrance to Possession are weak and variable.

With the exception of Deception Pass, tidal currents in the waters of Skagit Bay, Saratoga Passage and Port Susan are generally moderate to weak. Currents in the narrows of Deception Pass attain velocities of 5 to 8 knots and more, especially at times when strong eddies are formed along the shores. With westerly winds, heavy swells and tide rips form and make passage dangerous to small craft. Figures II-A16 and II-A17 summarize the flood and ebb current patterns, respectively.

3. Winds

During the period from October to March the winds are predominantly southerly to 20 mph. From April to September, the prevailing winds are northwesterly to 10 mps. Figure II-A18 shows monthly surface wind patterns for Everett.

4. Waves

From October to March, an effective fetch of over 30 miles to the east of Whidbey Island would cause wave heights of 6 feet at maximum in the area east of the southern part of Whidbey Island. In the region of Saratoga Passage and Port Susan, waves of a maximum height of less than 5 feet may be expected in the northern reaches. Waves in Possession Sound should be less than 2 feet.

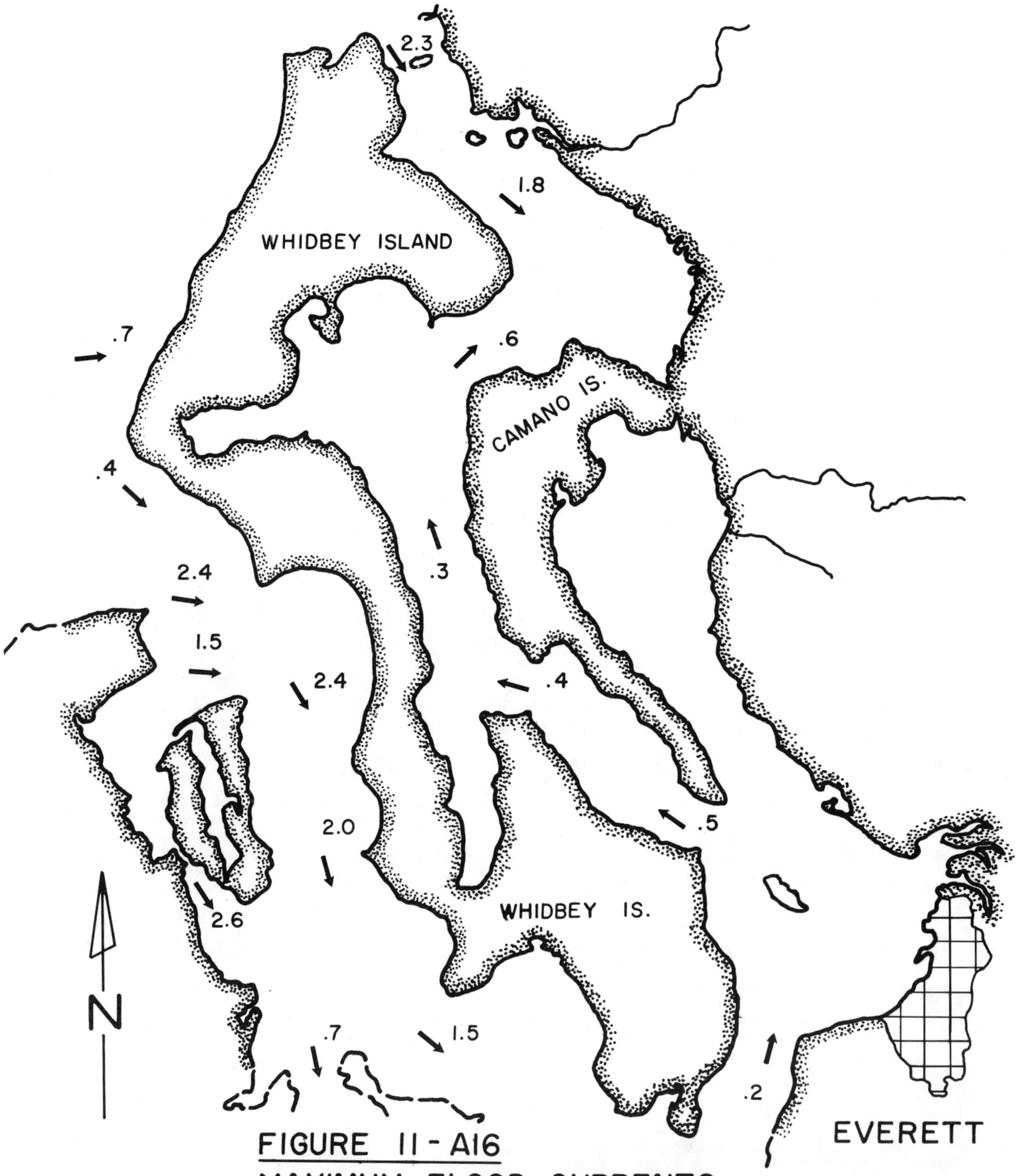

FIGURE II - A16
MAXIMUM FLOOD CURRENTS
SUBZONE 3 (IN KNOTS)

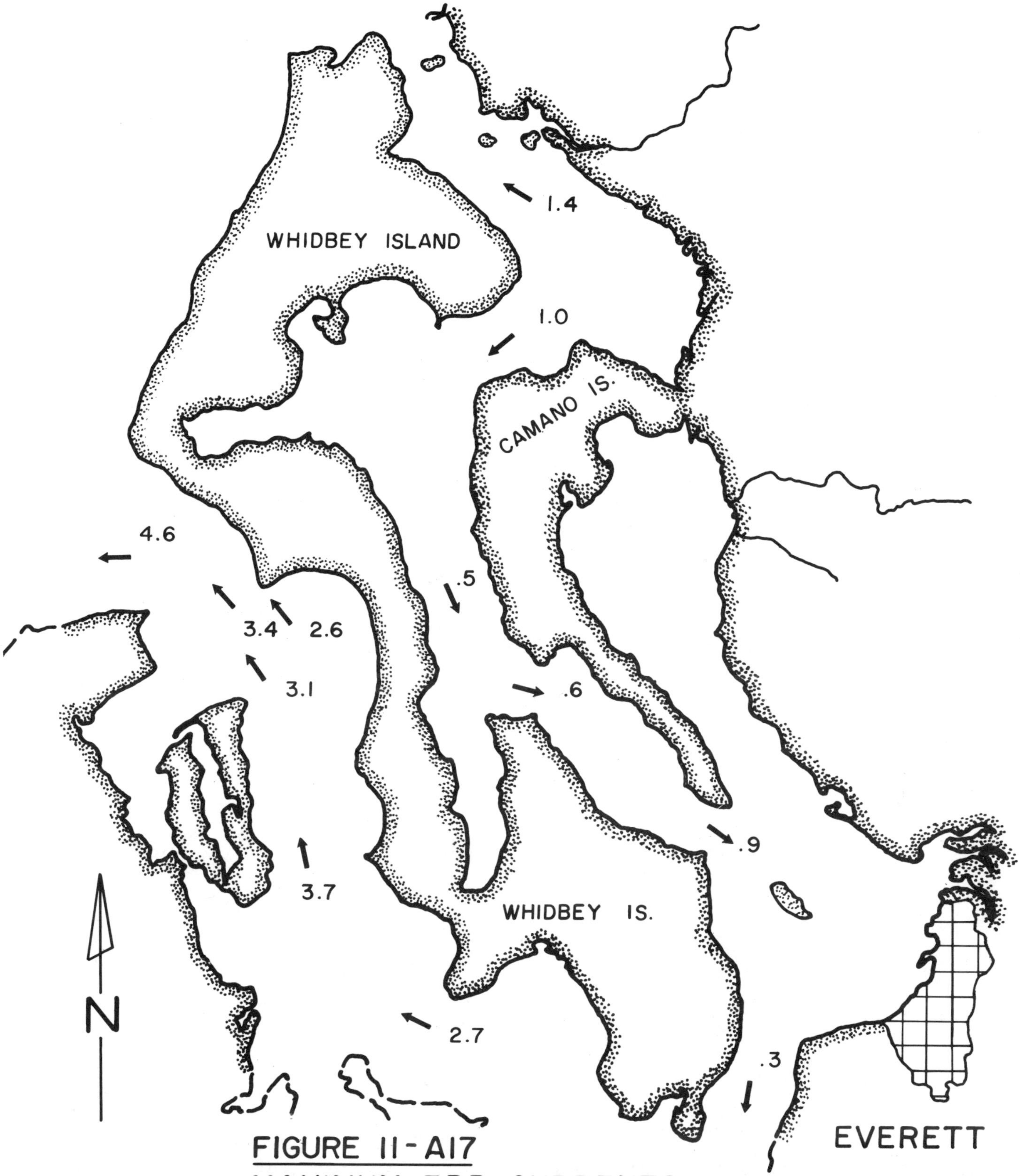

FIGURE II-A17
MAXIMUM EBB CURRENTS
SUBZONE 3 (IN KNOTS)

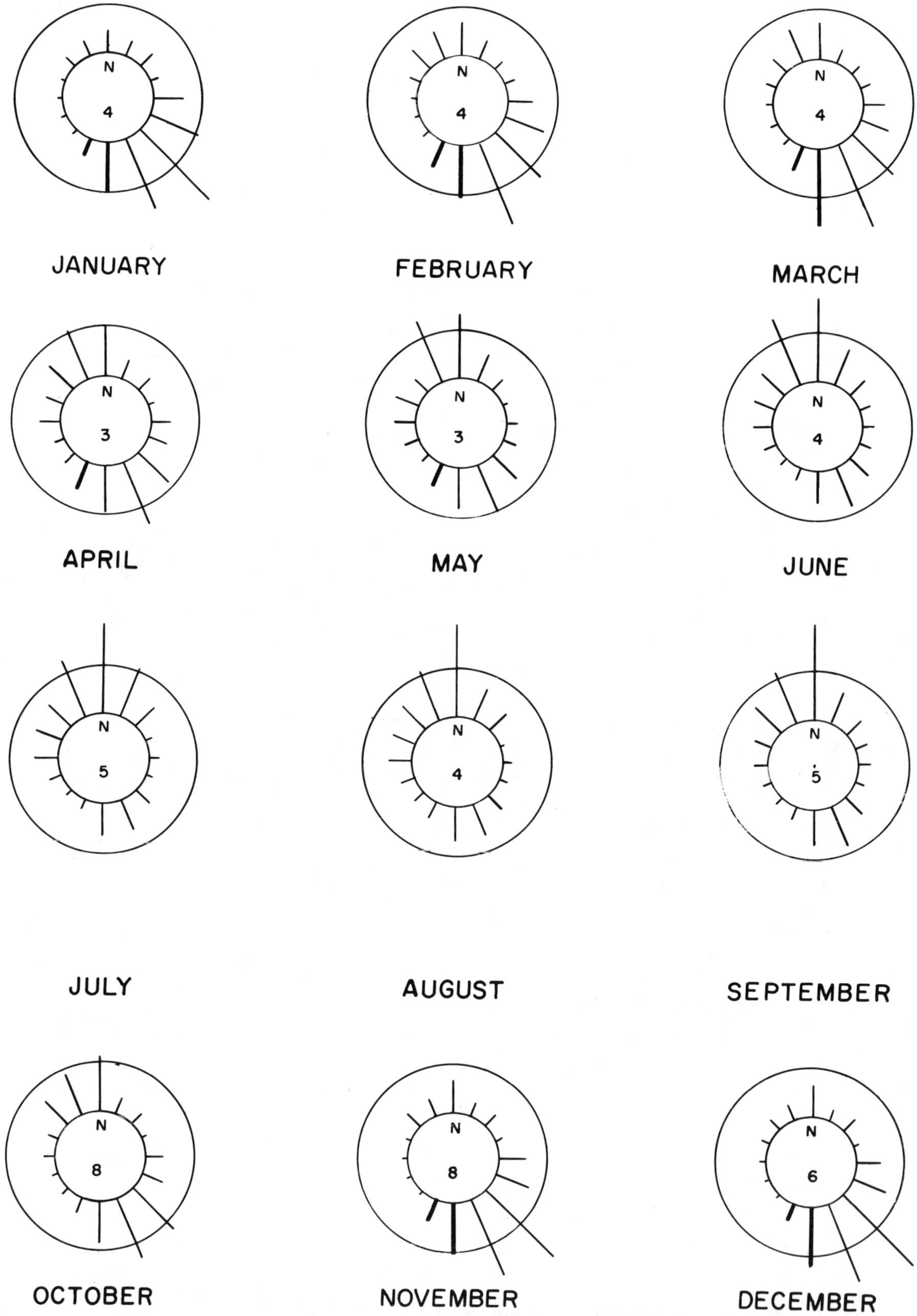

FIGURE II-A18, WIND PATTERNS FOR EVERETT AFTER HARRIS, 1954. WIDTH OF BAR INDICATES MONTHLY AVERAGE WIND VELOCITY, EACH RADIAL UNIT INDICATES 10 % OF TOTAL OBSERVATIONS, CENTRAL NUMERAL INDICATES PERCENTAGE OF CALM WIND. 0 - 10 MPH ——, 11 - 19 MPH —— [27]

From April until September, the waters of the subzone with the exception of those lying to the west and southwest of Whidbey Island should be predominantly less than 2 feet.

5. Natural physical resources

The Whidbey Island-to-Everett subzone is not as rich as other Puget Sound regions with respect to natural physical resources. The most plentiful species in this area are clams and marine fish. Figures II-A19 through II-A21 depict the distribution of the resources described below.

i. <u>shellfish</u>

<u>Crabs</u>: Dungeness crabs are found in only moderate quantities in this area, with the largest populations around Whidbey Island, especially the southern part. Large populations also occur in Saratoga Passage, Port Susan, Possession Sound, and in Cutlus, Useless, and Mutiny Bays. Dungeness crabs are also found in the Stillaguamish Flats, from Spee-bi-dah to Hermosa Point, from Mission Beach to Port Gardner, and near the shore of the Everett boat basin to Mukilteo past to Edmonds. Dungeness are common in the shallow waters around Gedney Island. Red crabs occur in the same general area.

<u>Oysters</u>: Generally speaking, oysters are not common in this subzone. Pacific oysters are in scattered locations in sheltered bays around Whidbey Island. Olympia and Kumamoto oysters are scarce.

<u>Clams</u>: The entire region is an excellent habitat for clams which are abundant on nearly all of the tidal zones. Littleneck and butter clams are especially plentiful from Mission Beach to Priest Point, around Edgewater Beach, and from Elliott Point to Edwards Point. Softshell clams and horse clams occur along the shore opposite Skagit Flats.

<u>Other shellfish</u>: Geoducks and scallops are found scattered in the deeper waters. Pink and spot shrimp are abundant in Saratoga Passage as far north as Oak Harbor; they also populate Possession Sound, Admiralty Inlet, and Port Gardner. Large number of squid are found in Saratoga Passage and Holmes Harbor. The greatest populations occur during the winter. Octopi are abundant throughout. Blue mussels are common throughout the region.

FIGURE II-A19
PHYSICAL RESOURCE GROUNDS -
SALMON & SHELLFISH
SUBZONE 3

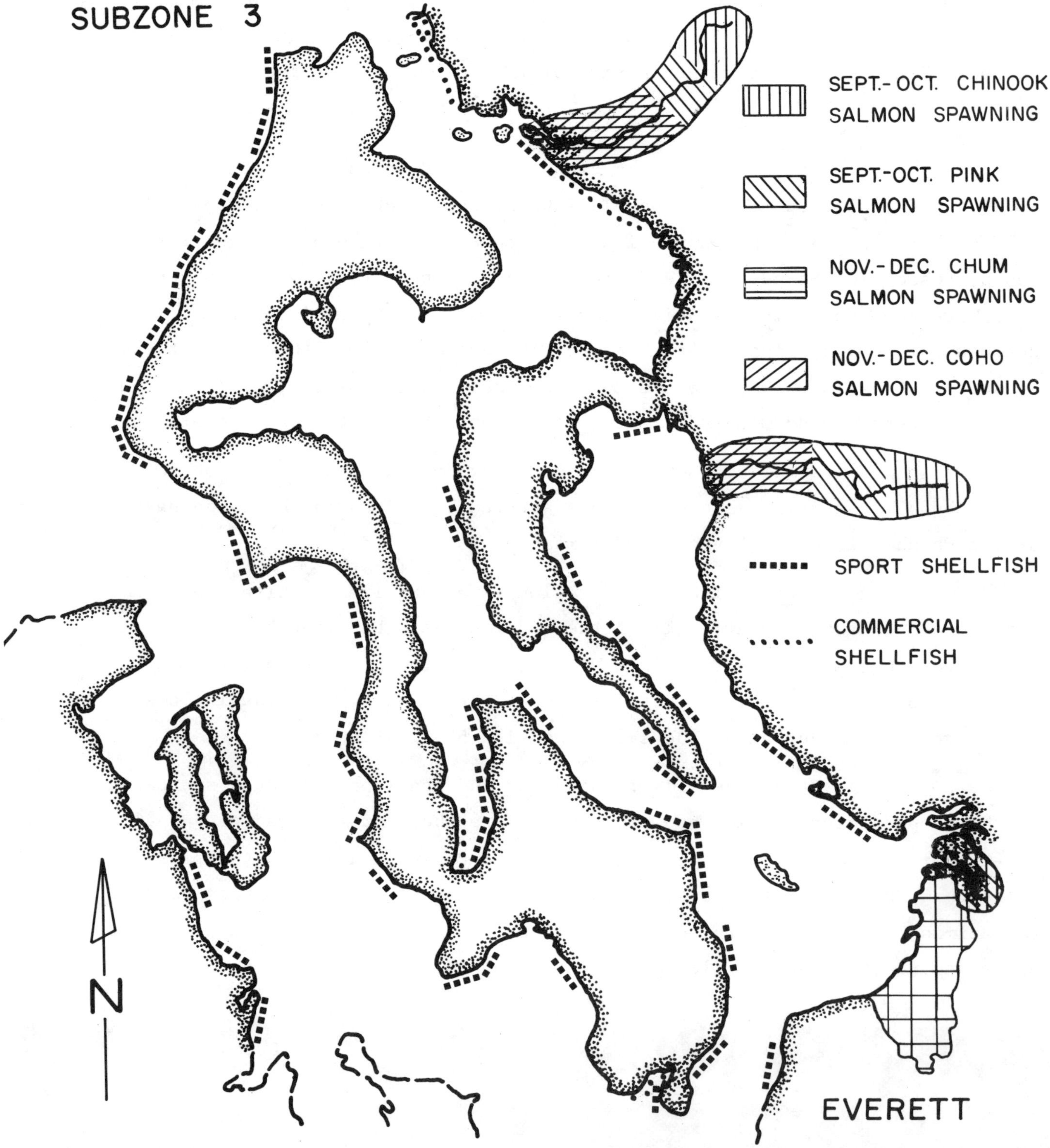

FIGURE II-A20
BOTTOMFISH GATHERINGS
SUBZONE 3

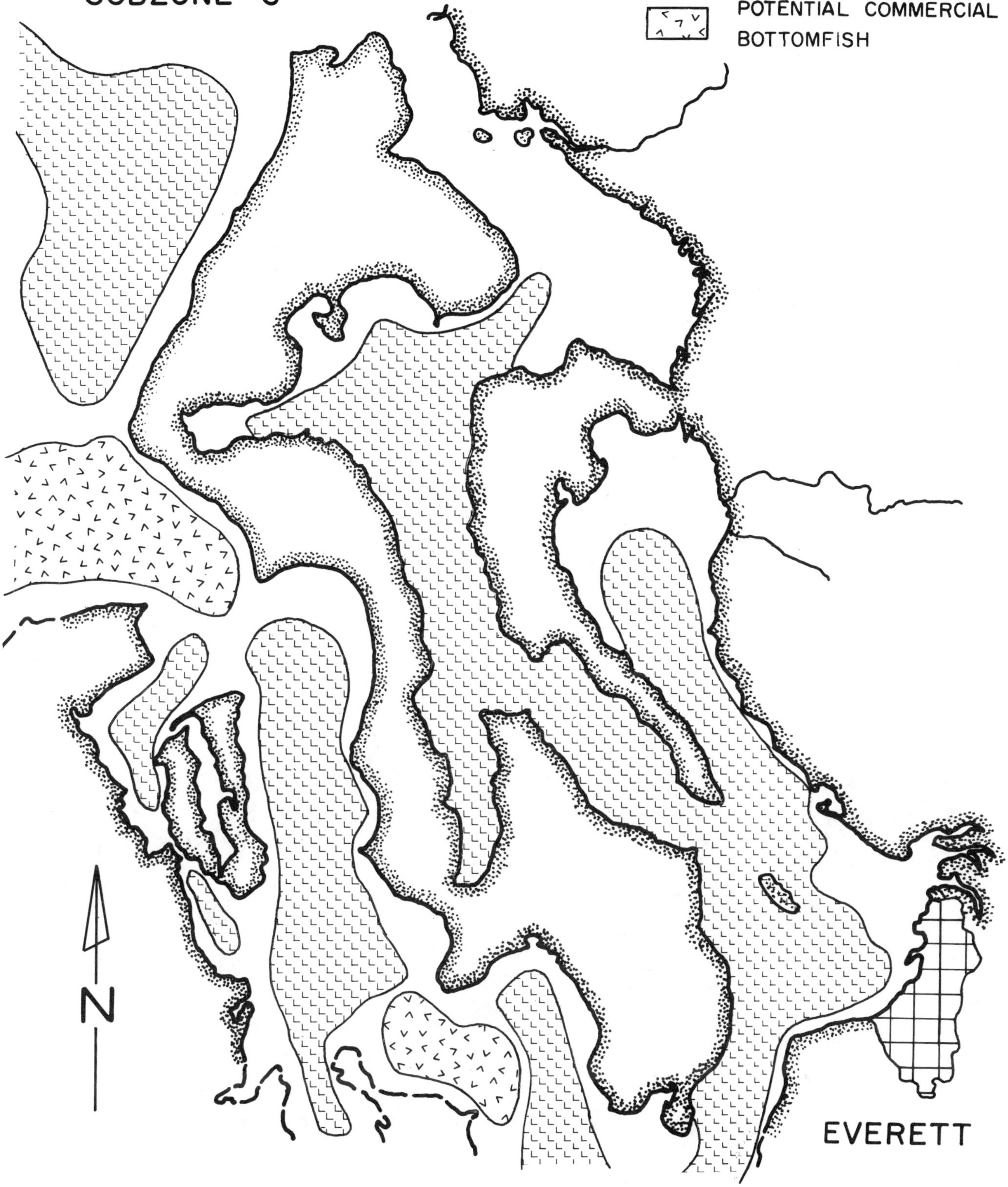

FIGURE II-A21

SALMON & WATERFOWL GATHERINGS

SUBZONE 3

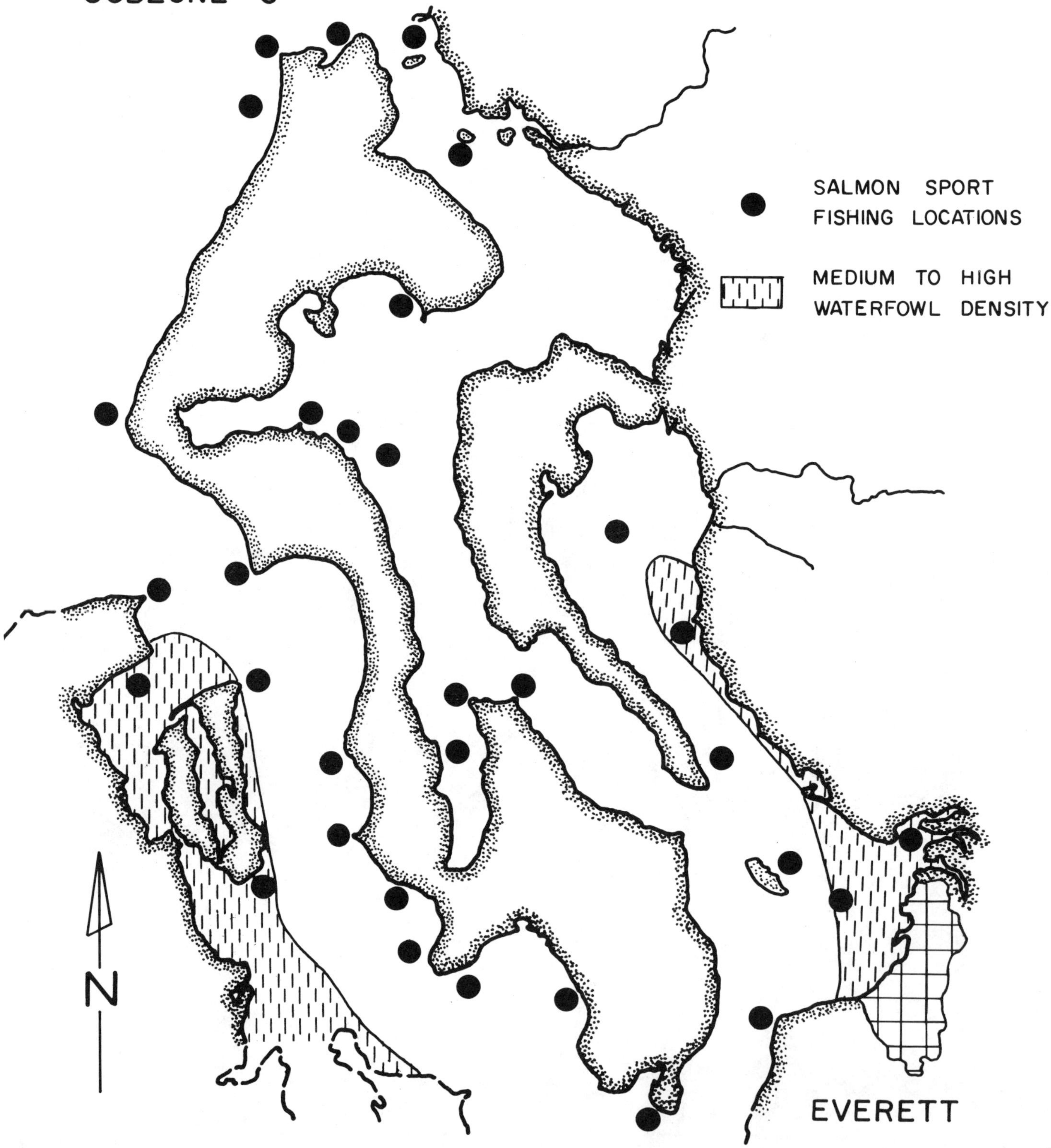

ii. fish

The waters around West Beach, Bush Point, Point No Point, Possession Bay, near Edmonds, Richmond Beach, Camano Head, Baby Inland, Holmes Harbor are amply stocked with salmon. (See Figure II-A21.) Groundfish, such as herrings, are found in Saratoga Passage and Holmes Harbor.

iii. waterfowl

Waterfowl are not common around Whidbey or Camano Islands. Medium to high waterfowl density exists in the entrance to Hood Canal and Port Townsend. High waterfowl density occurs in Port Susan and near the mouth of the Skagit River.

6. Man-made physical resources

Subzone 3, with its extensive shoreline, has a multitude of parks, launching ramps, and boat moorages. These are listed below, and their distribution is shown in Figure II-A22.

TABLE II-A5: MAN-MADE RESOURCES, SUBZONE 3[28]

Location	State Parks	Launching Ramps	Rental Moorages
Skagit River mouth	0	3	2
E. side, Admiralty Inlet	3	3	1
Deception Pass	2	4	1
Saratoga Passage-Holmes Harbor	1	10	3
Skagit Bay	1	0	1
N. Possession Sound	0	1	1
Port Gardner	0	2	4
S. Possession Sound	1	2	0
Port Townsend	3	6	2
SUBZONE 3 - TOTAL	11	31	15

Not included were city and county parks, and private beaches, since these data were not available.

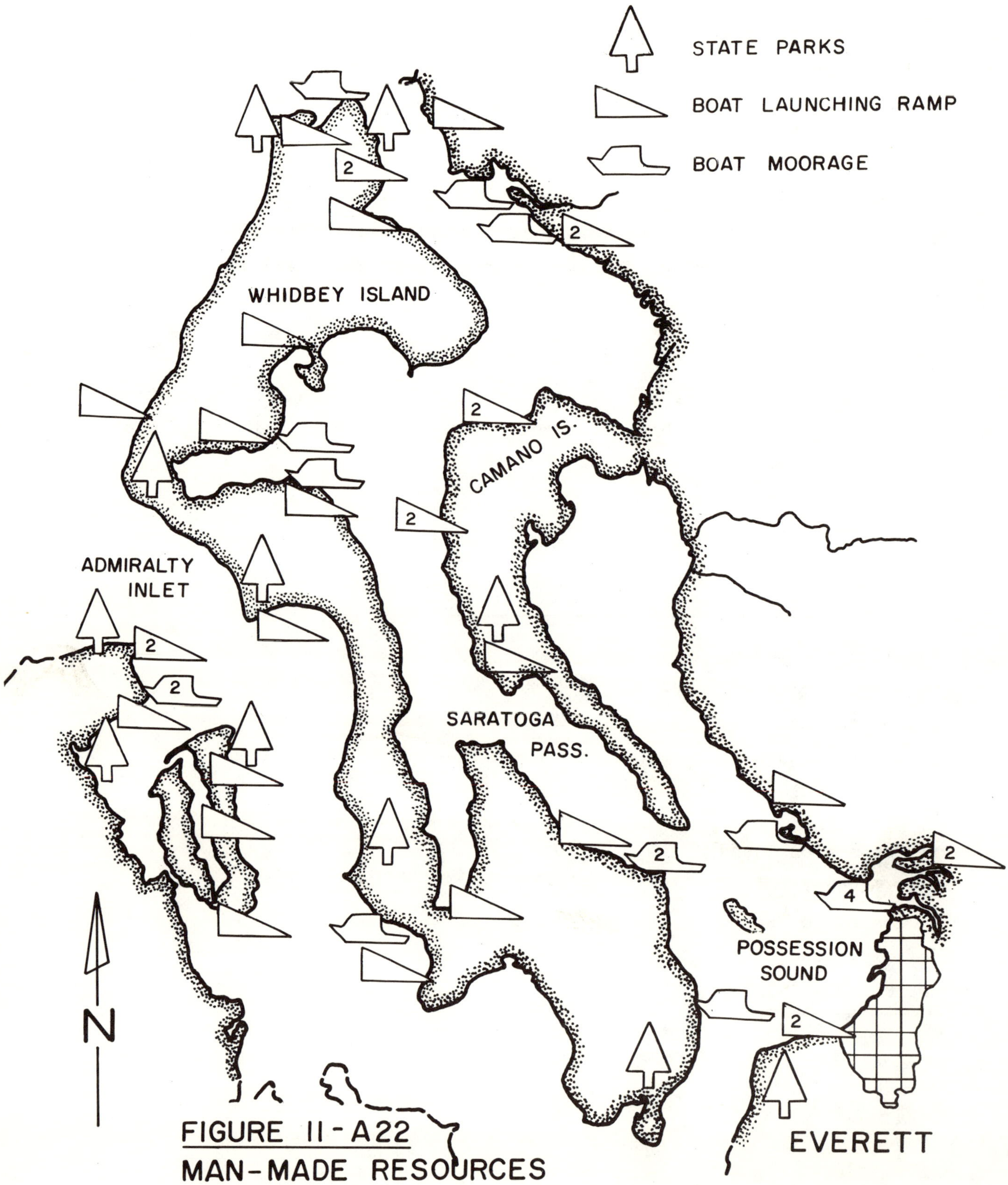

FIGURE II-A22
MAN-MADE RESOURCES
SUBZONE 3

d. Subzone 4: Northern Puget Sound

1. Geography

Northern Puget Sound includes the most populated and industrialized of the subzones in this study. On the eastern shore of northern Puget Sound a nearly continuous urban/suburban complex extends over 60 miles from Everett to Seattle to Tacoma. The shores bordering this eastern part of the subzone have a very high density of industrial, residential, and recreational facilities, many of which are of very high economic value.

The eastern shore is dominated by narrow beaches of gravel and rocks and a few sandy beaches which are heavily utilized for recreational purposes. Generally, the shores are bordered by steep banks of considerable relief or by man-made structures such as piers, but accessibility from land is quite adequate by public roads.

The western shores of this subzone are predominantly residential and recreational. As bedroom communities of Seattle, settlements near ferry services are comparatively dense in residential use, whereas the areas not directly adjacent to the ferry landings at Kingston, Winslow, and Bremerton are slightly agricultural or primarily recreational. For the most part, the character of the shoreline of the western side of this subzone is uniformly narrow rocky beaches adjacent to banks or cliffs which are of rugged relief and of limited accessibility from land.

However, in comparison to the eastern shore of this subzone, the western shore is indented with numerous bays and long sheltered inlets. Port Madison, Liberty Bay, Dyes Inlet, and Sinclair Inlet are sheltered from the main body of Puget Sound by Bainbridge and Blake Islands to the east. The shores of these long narrow bays are muddy tide flats in some places, in contrast to the steeper rocky shores of the more unprotected waters.

2. Currents

The tidal currents in this subzone are relatively weak when compared to the conditions in other subzones as shown in Figures II-A23 and II-A24. From the main body of the Sound, from Point Wells to Elliott Bay and south to Point Pully, the tidal currents are normally less than a knot during peak ebb and flood. However, constricted passages such as Agate Passage and Rich Passage may exhibit local velocities up to 6 knots.

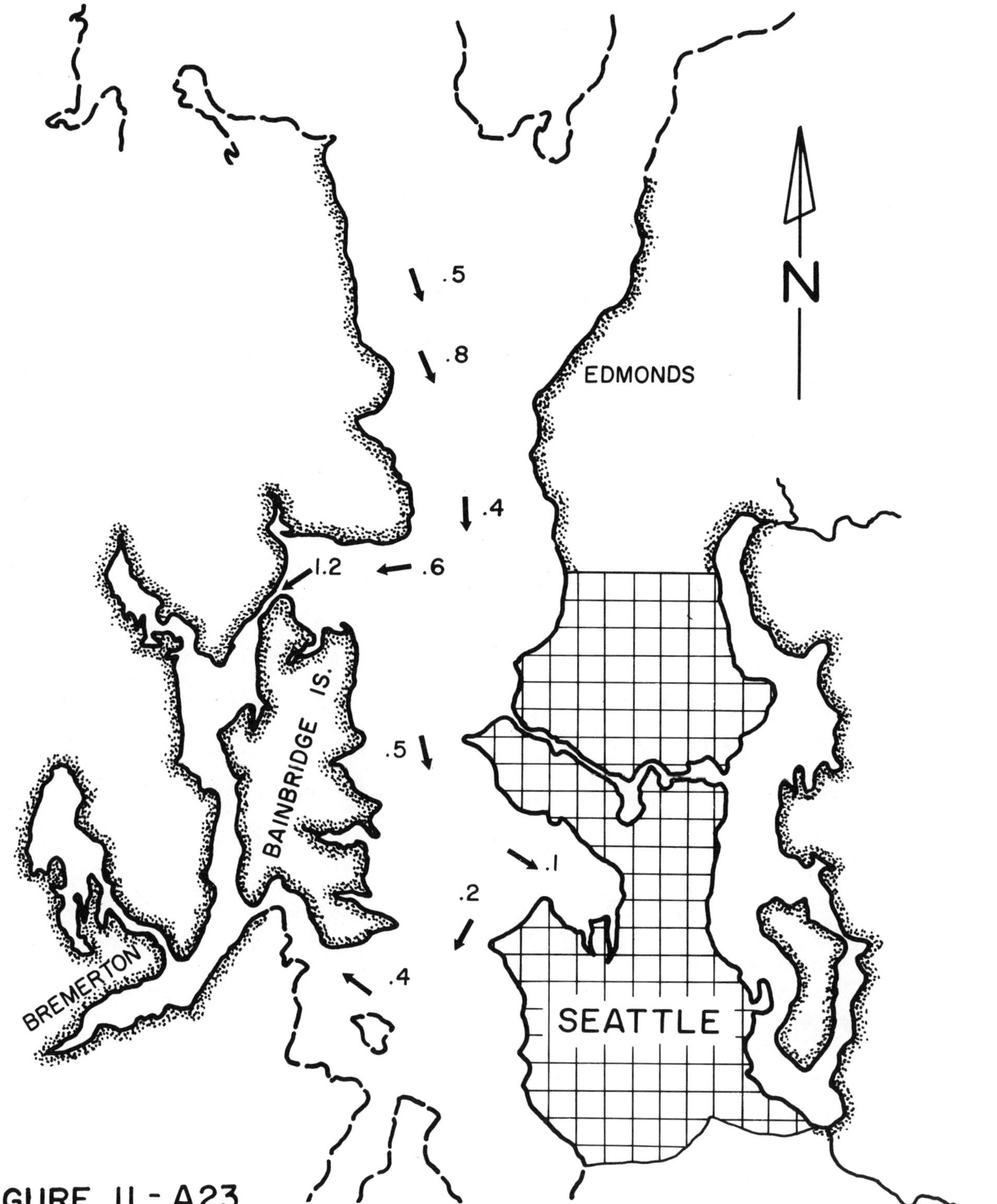

FIGURE II - A23
MAXIMUM FLOOD CURRENTS
SUBZONE 4 (IN KNOTS)

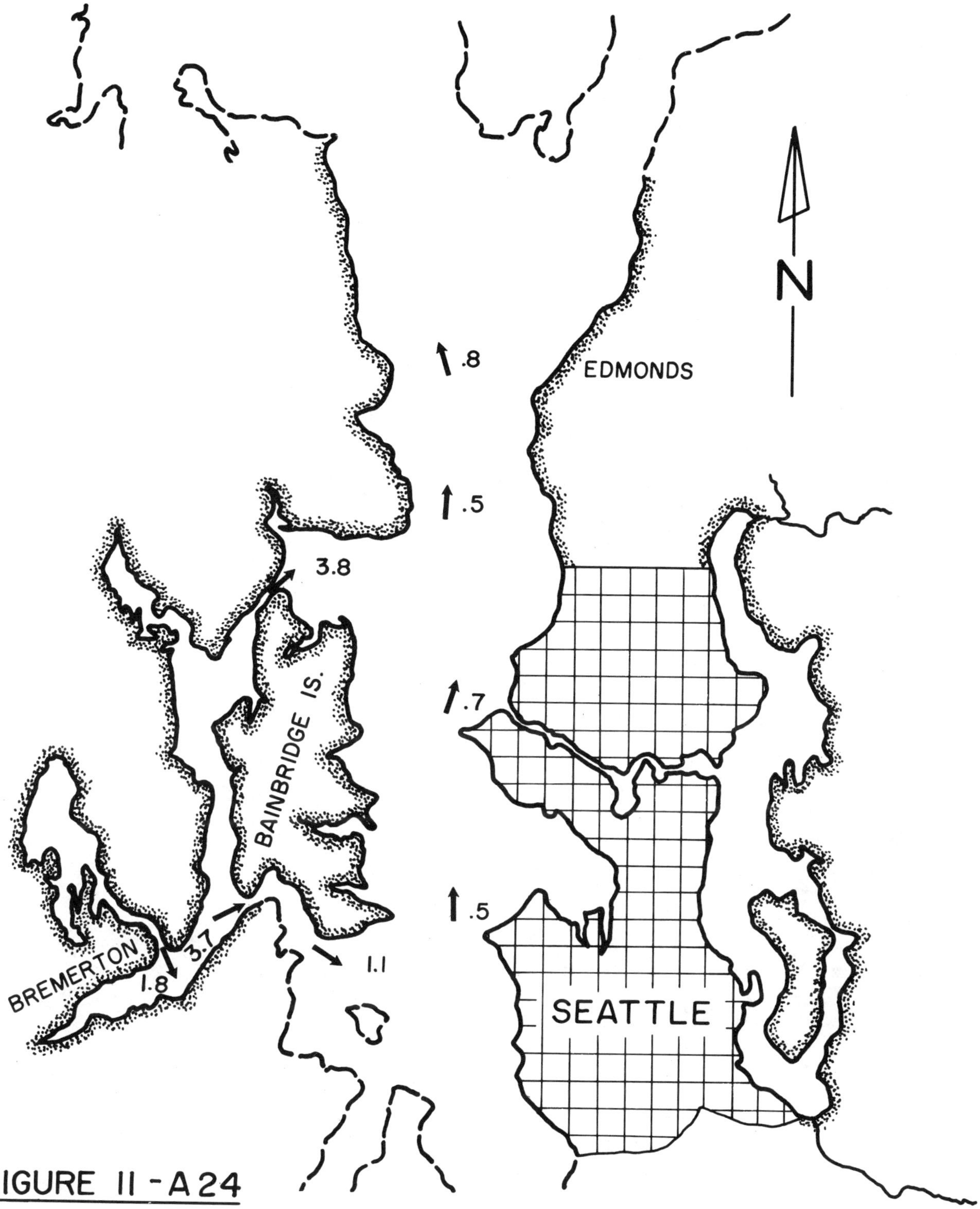

FIGURE II-A24
MAXIMUM EBB CURRENTS
SUBZONE 4 (IN KNOTS)

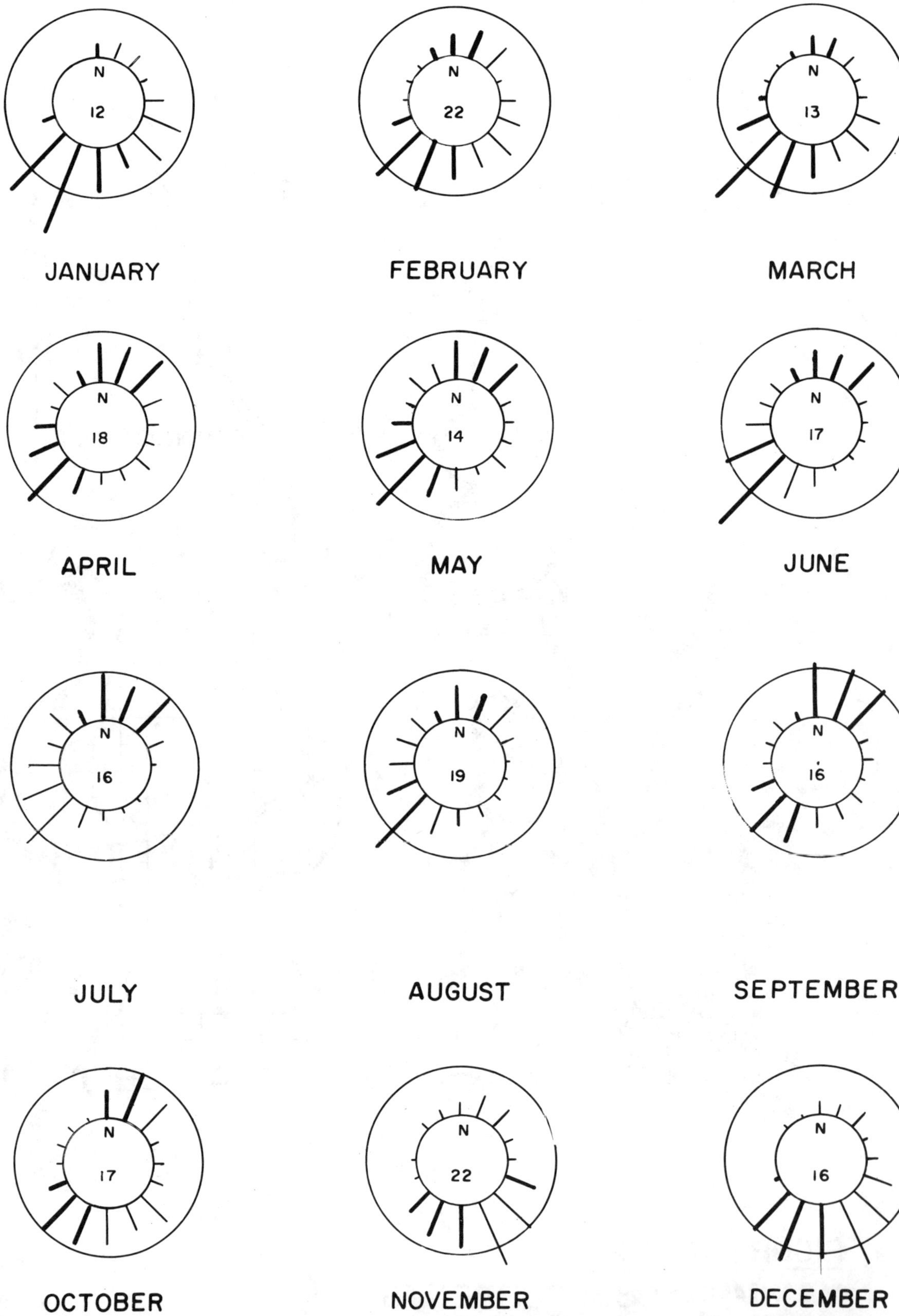

FIGURE 11-A25, WIND PATTERNS FOR SEATTLE AFTER HARRIS, 1954. WIDTH OF BAR INDICATES MONTHLY AVERAGE WIND VELOCITY, EACH RADIAL UNIT INDICATES 10 % OF TOTAL OBSERVATIONS, CENTRAL NUMERAL INDICATES PERCENTAGE OF CALM WIND 0 - 10 MPH ——, 11 - 19 MPH ——[29]

3. Winds

From the period of October to March, the prevailing winds in this subzone are southwesterly to 20 mph. From April until May, the winds tend to be south-southwesterly to 10 mph, and from June until September, the surface winds tend to be northerly to 10 mph. Figure II-A25 depicts the monthly wind pattern in Seattle.

4. Waves

Throughout the year, the limited fetch in this subzone restricts wave heights to less than two feet in height except under the stress of anomalously high winds when the wave height may be up to 6 feet locally.

5. Natural physical resources

Though this subzone is heavily populated by man, a large and varied collection of sea life exists along the shores. Figures II-A26 and II-A29 show the distribution of this life, which is described below.

i. shellfish

Crabs: Dungeness and red crabs are found throughout this subzone, although they are not abundant.

Oysters: Oysters are uncommon throughout this region. Some occurrence of Pacific oysters is in Liberty Inlet and the enclosed part of Sinclair Inlet. Olympia oysters occur near Poulsbo and Bremerton in small numbers.

Clams: Littleneck, butter, and Manila clams are prevalent around Poulsbo, in Port Madison, and Fletcher Bay. Littleneck clams are also common from Edmonds to Meadow Point and from Point Pully to Alki Point.

Other shellfish: Sparse blue mussel populations are found on the eastern shore of this subzone, while they are moderate to heavy on the western shore. Geoducks are rather sparse, while a few pink scallops exist in Elliott Bay and along the Shilshole breakwater. Spot shrimp are common in many parts of Elliott Bay, along with pink shrimp. Squid occur sporadically in many of these areas. Octopi are abundant near Vashon Island and Bainbridge Island and near most rock breakwaters.

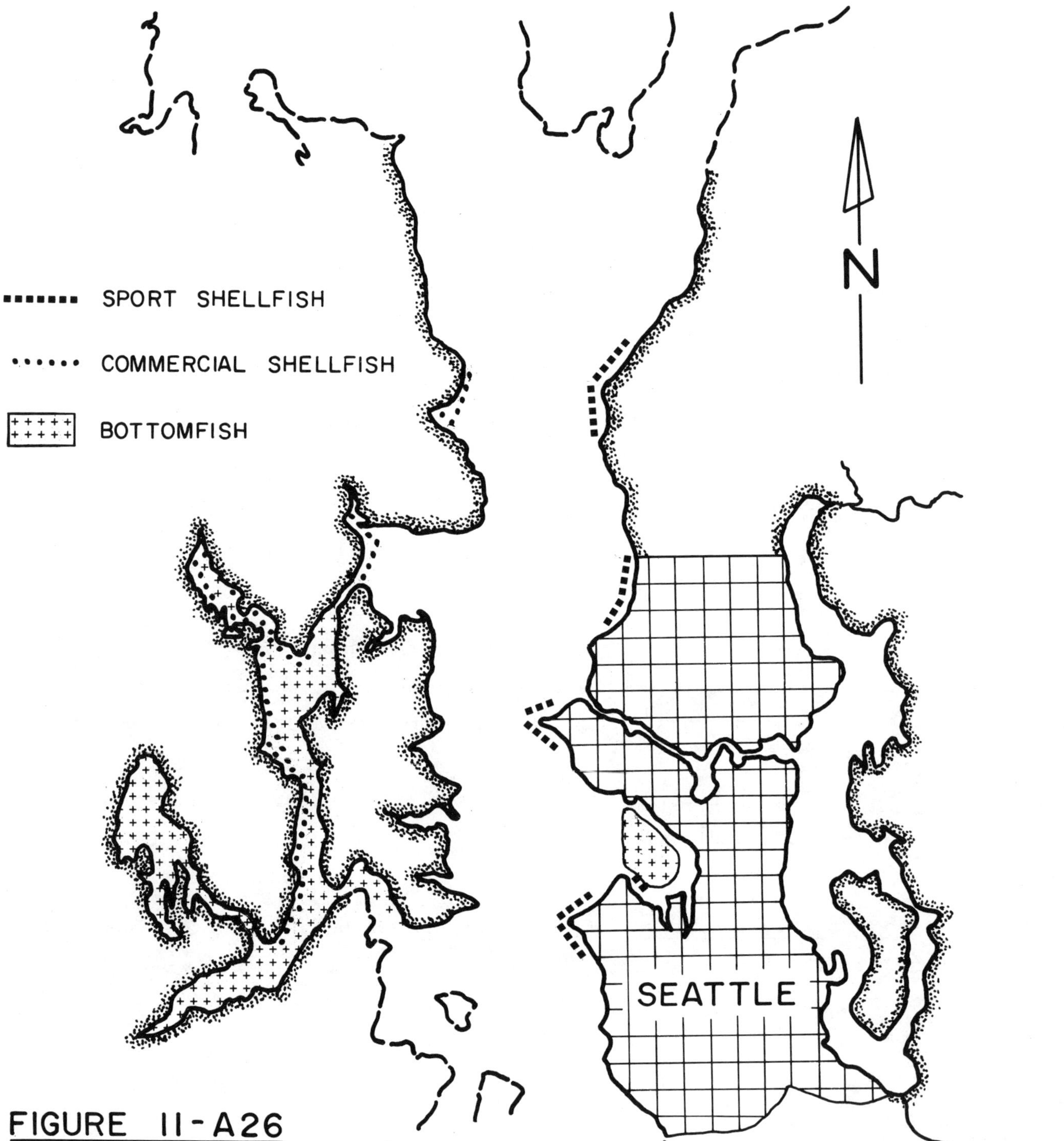

FIGURE II-A26
PHYSICAL RESOURCE GROUNDS -
SHELLFISH & BOTTOMFISH
SUBZONE 4

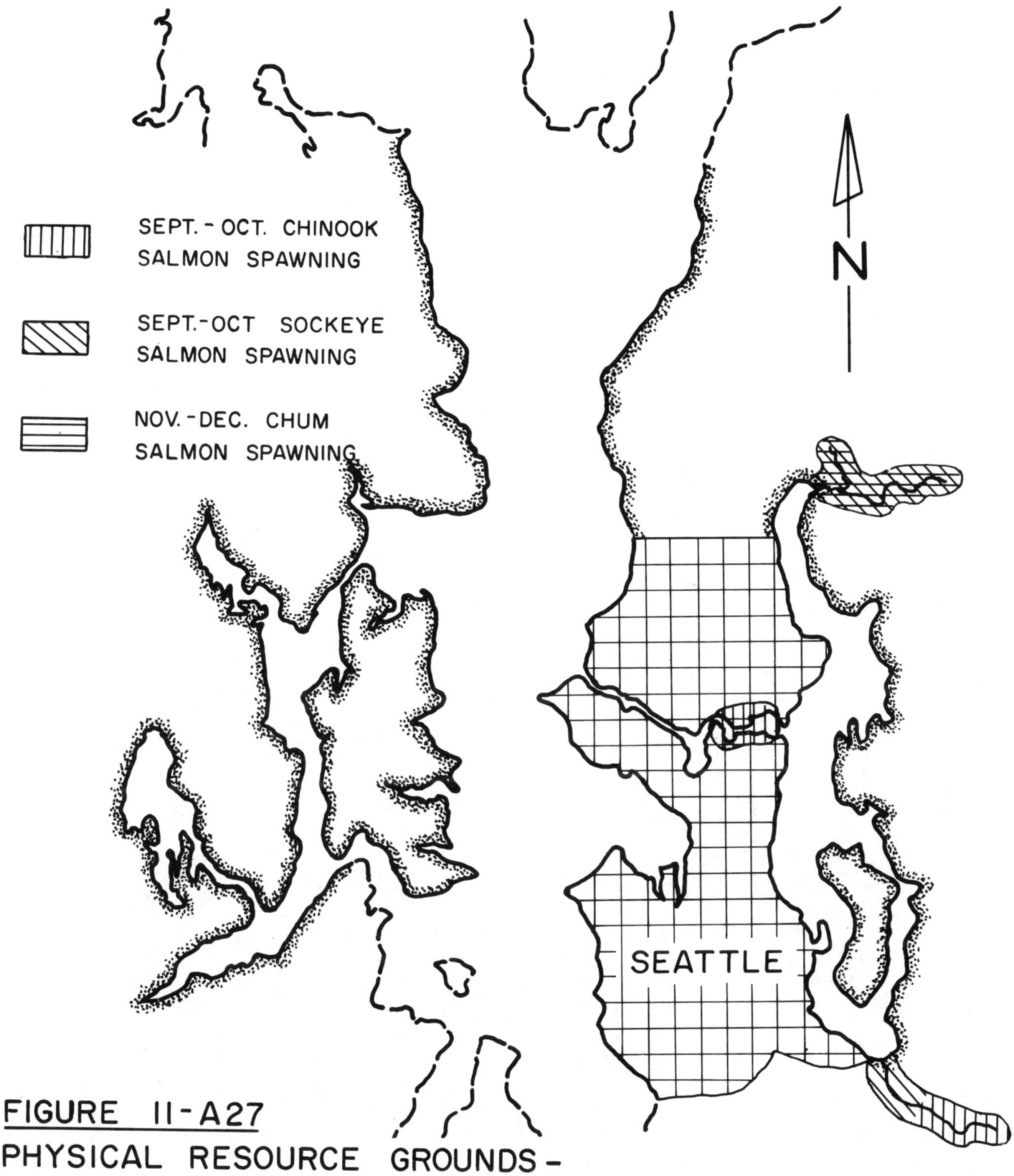

FIGURE II - A27
PHYSICAL RESOURCE GROUNDS - SALMON, SUBZONE 4

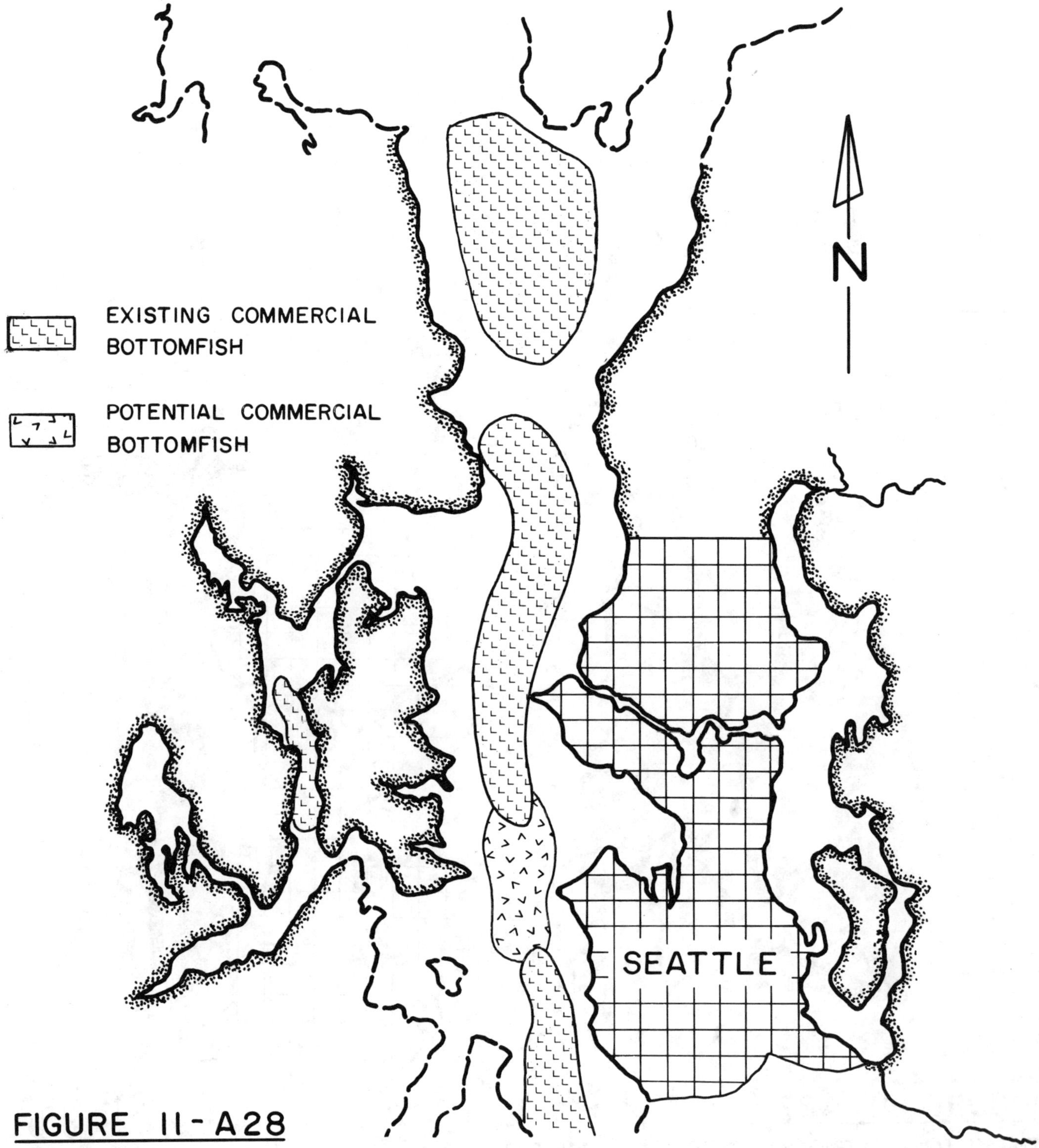

FIGURE II-A28
BOTTOMFISH GATHERINGS
SUBZONE 4

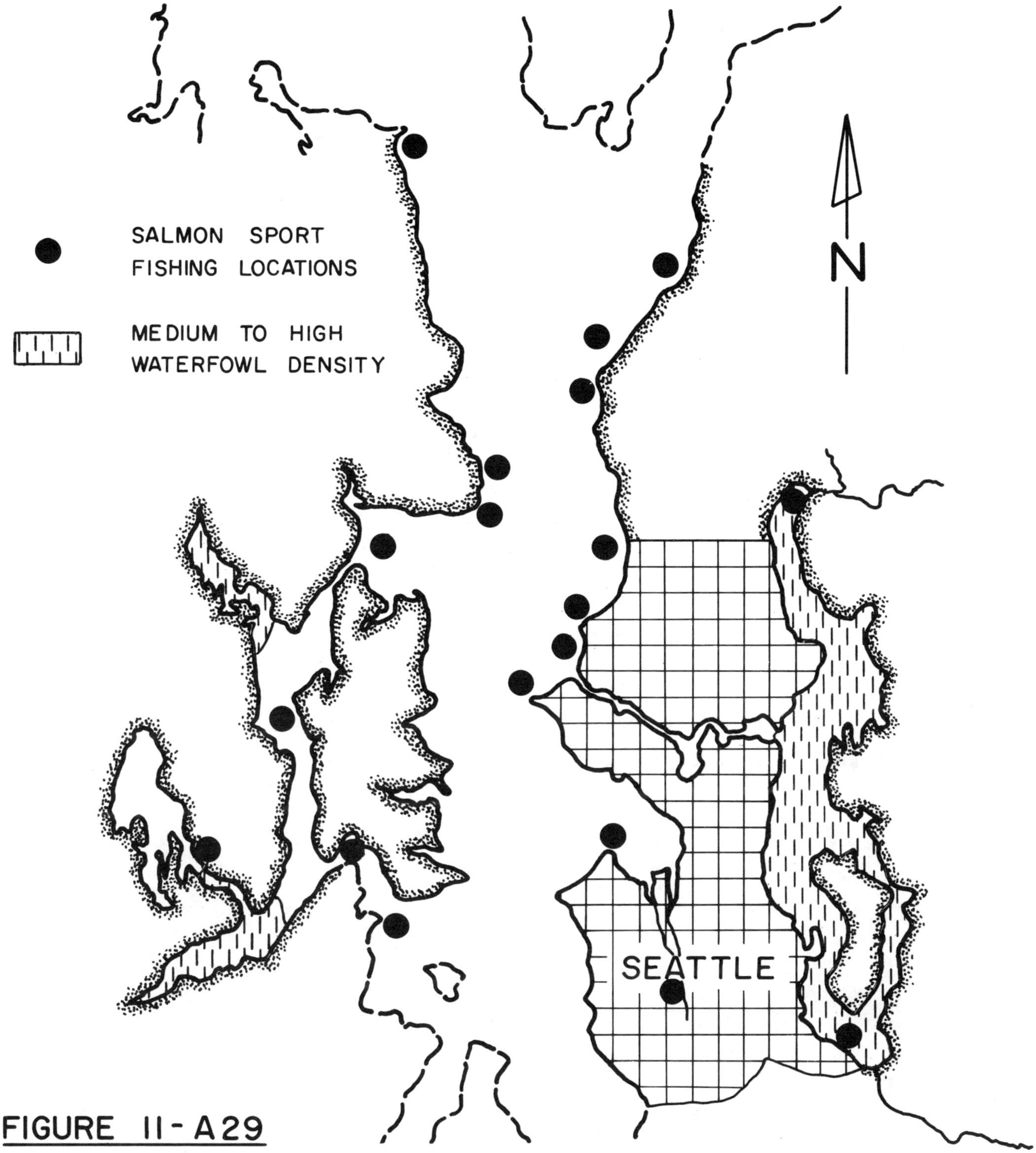

FIGURE II-A29
SALMON & WATERFOWL GATHERINGS
SUBZONE 4

ii. <u>fish</u>

Point No Point, Possession Bay, Ballard-Jefferson Head, Elliott Bay, and Manchester-Bremerton may have large numbers of salmon grounds at times, as well as Shilshole and Elliott Bays, and Richmond Beach to Meadow Point (see Figure II-A27). The major sport fishing grounds in this region include the Kingston-Port Madison area, Richmond Beach, Meadow Point, and Ballard, as shown in Figure II-A29.

iii. <u>waterfowl</u>

Waterfowl are found throughout this subzone, but are not abundant. Moderate to high waterfowl densities occur near Poulsbo and Port Orchard (see Figure II-A29).

6. Man-made physical resources

The large populace residing in this subzone and their yen for water-related recreation is borne out by the vast number of large launching ramps and rental boat moorages. Though state parks are not prevalent due to the dominance of shoreline by municipalities, public beaches are nevertheless abundant. In the Seattle area alone, there are over 20 city and county beaches, mostly in the southeast area and along the western shore of Lake Washington.

State parks, ramps, and moorages are listed below and their locations are shown in Figure II-A30.

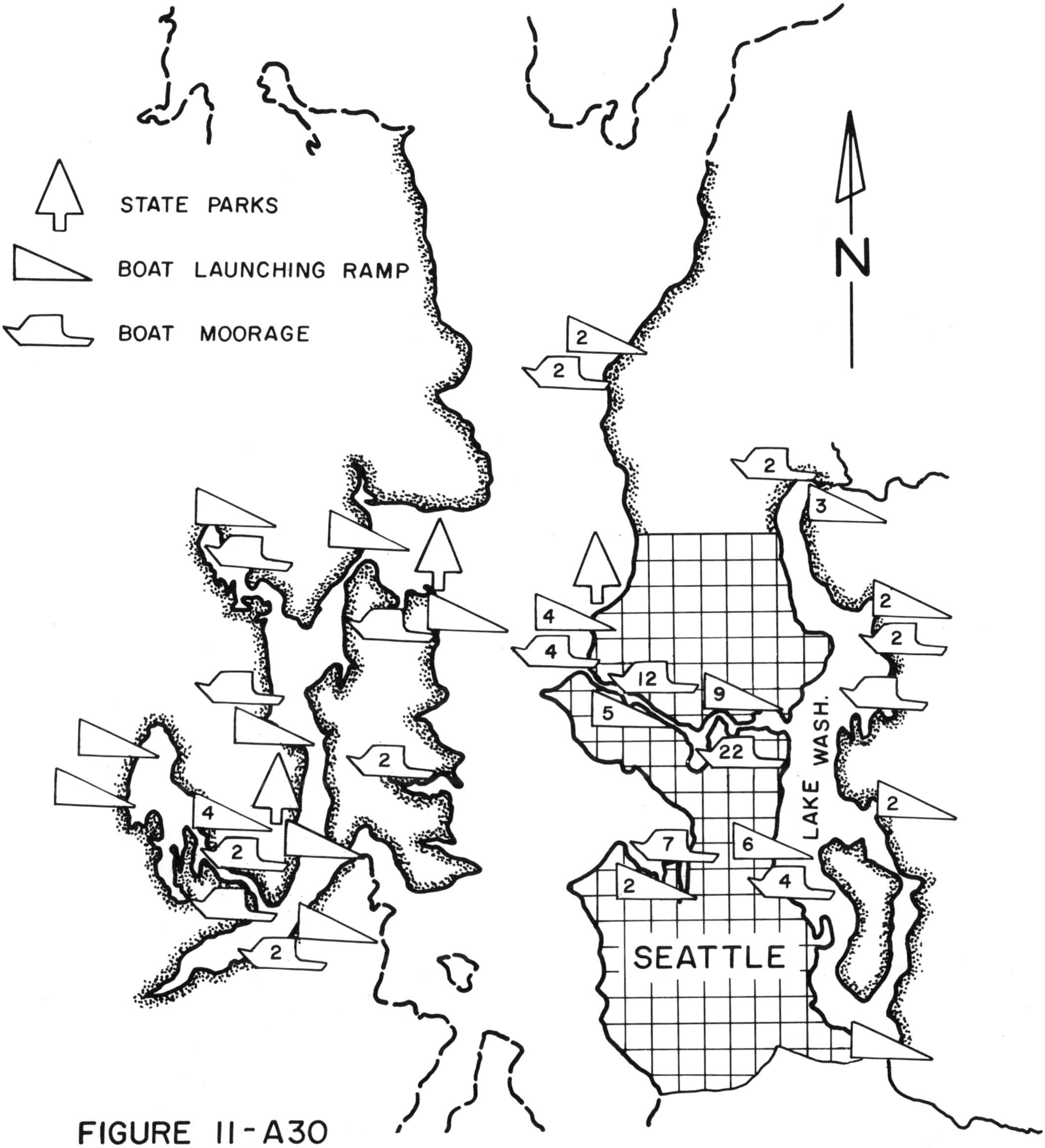

FIGURE II-A30
MAN-MADE RESOURCES
SUBZONE 4

TABLE II-A6: MAN-MADE RESOURCES, SUBZONE 4[30]

Location	State Parks	Launching Ramps	Rental Moorages
Edmonds	0	2	2
Shilshole Bay	1	4	4
Salmon Bay	0	5	12
Lake Washington Ship Canal	0	9	22
Lake Washington	0	14	9
Duwamish River	0	2	7
E. Side Bainbridge Island	2	5	4
Sinclair Inlet-Port Orchard	1	4	6
Dyes Inlet	0	6	2
SUBZONE 4 - TOTAL	4	51	68

City parks and beaches are not shown on the map.

d. Subzone 5: Southern Puget Sound

1. Geography

Southern Puget Sound narrows from a typical width of approximately 5 nautical miles in the Northern Sound to form a complex system of long passages, embayments, and inlets which are predominantly 1-2 miles in width. Although this subzone contains the industrial cities of Tacoma and Olympia, the area is dominated by recreational and residential land use.

Numerous larger islands such as Vashon, Maury, Fox, McNeil, Anderson, Hartstene, and Squaxin are of limited accessibility by public and private ferry systems. In the eastern part of the subzone, from Point Pully to Nisqually Reach, the shoreline is typically narrow and sandy with gravel and rocks in some places. Generally, the steep beaches are bordered by steep banks or cliffs of considerable relief.

To the west, long narrow inlets such as Carr, Case, Hammersley, Totten, Eld, Budd, and Henderson as well as Nisqually Flats exhibit muddy estuarine tidelands at the shallow ends. The many inlets in the western portion not only have banks adjacent to the shoreline in many cases, but also have dense vegetation and only limited public highway which constrains accessibility from land. The shoaling depths and irregular channels also hamper access of the shoreline from the water by boat in the western inlets.

Numerous small boat harbor/settlements for recreational craft are dispersed throughout the subzone.

2. Currents

Southern Puget Sound displays generally weak or variable tidal currents in most of the larger and open bodies of water. However, several very restricted passages often exhibit tidal currents of unusually high velocities due to large volumes of water which must be transported through them. Around Point Defiance and through the Tacoma Narrows, maximum tidal currents may reach velocities of up to 5 knots. Large eddies created are a menace to the navigation of smaller vessels.

South of Fox Island and at the mouths of the narrow inlets, maximum tidal currents up to 3 knots may occur locally. Weaker tidal currents in Colvos Passage during the tidal flood may become 1-1.5 knots northward during the ebb flow. Figures II-A31 and II-A32 depict the flood and ebb currents, respectively.

3. Winds

During the period from October to March, surface winds tend to be south-southwesterly with speeds to 20 mph. During April to May, prevailing winds are southerly to 20 mph, and from June to September the winds become northeasterly to 10 mph. Monthly wind patterns for Olympia are shown in Figure II-A33.

4. Waves

Throughout the year, the limited and interrupted fetch existing over the waters of the subzone as well as the moderate to light winds which prevail interact to produce waves of less than a maximum height of 4 feet from October to March. During the period from April to September, maximum wave heights are less than 2 feet.

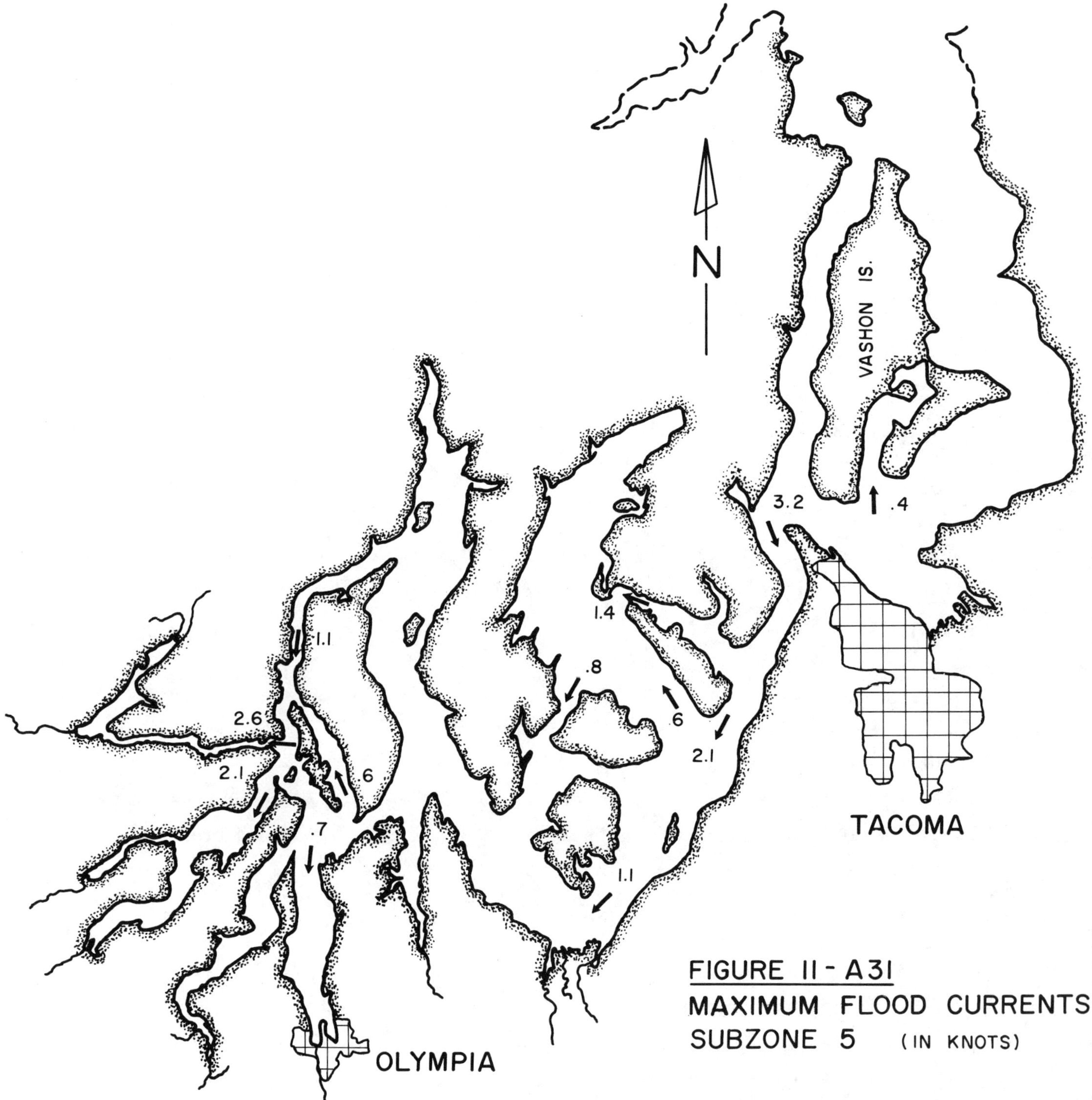

FIGURE II - A31
MAXIMUM FLOOD CURRENTS
SUBZONE 5 (IN KNOTS)

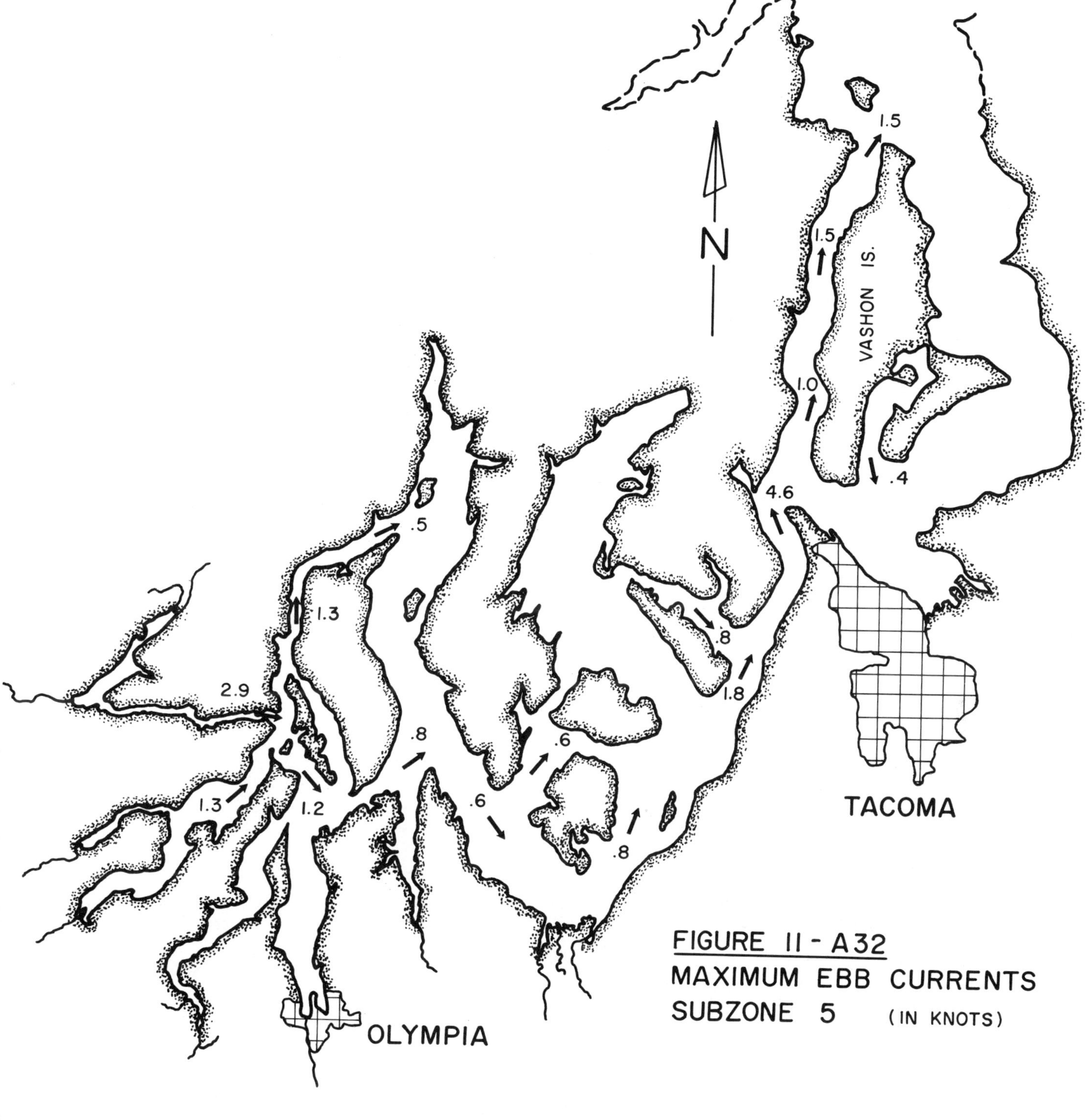

FIGURE II - A32
MAXIMUM EBB CURRENTS
SUBZONE 5 (IN KNOTS)

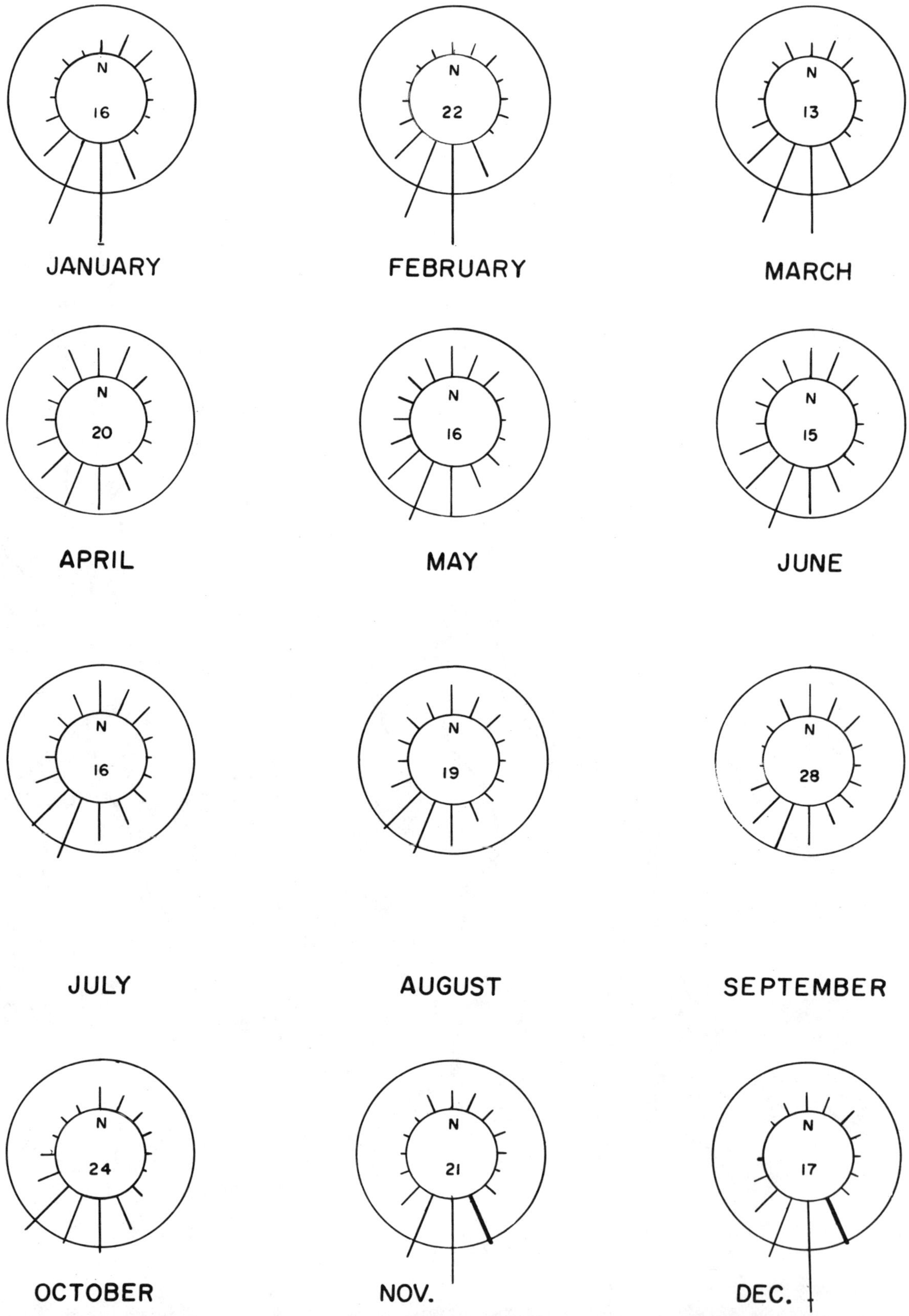

FIGURE II-A33, WIND PATTERNS FOR OLYMPIA AFTER HARRIS, 1954. WIDTH OF BAR INDICATES MONTHLY AVERAGE WIND VELOCITY, EACH RADIAL UNIT INDICATES 10 % OF TOTAL OBSERVATIONS, CENTRAL NUMERAL INDICATES PERCENTAGE OF CALM WIND. 0 - 10 MPH ——, 11 - 19 MPH ——[31]

5. Natural physical resources

The physical layout of this subzone is ideal for the large population of shellfish and marine fish which populate the shorelines and waters. The types and distribution of this sea life are detailed below, and shown in Figure II-A34 through II-A37.

i. shellfish

Crabs: Dungeness crabs are found throughout the region although nowhere in great abundance. The richest concentrations are just north of Commencement Bay (especially in eelgrass), on Nisqually Flats, and in Oro Bay. Red crabs are also common near Nisqually Flats.

Oysters: Oyster Bay (Totten Inlet) is the most productive area in Puget Sound for Pacific oysters. The species of oyster is also common in Carr, Eld, Case, Budd, Skookum, and Henderson Inlets, Pickering and Peal Passages, near and Hope Island. Also Pacific oysters are plentiful in Oakland Bay and near the Nisqually Flats. Other beds include Dyes Inlet, in Ostrich, Oyster and Mud Pass and the enclosed part of Sinclair Inlet. Native and Kumamoto oysters are found in this area, mainly in north and south Case Inlet, Vaughn, North, and Rocky Bays, along Pickering Passage, in Oakland Bay; Skookum Inlet; Oyster Bay, from Rocky Point to Mud Bay in Eld Inlet; Silver Spit and Gull Harbor (Eld Inlet); and from Chapman Bay to South Bay in Henderson Inlet. Native oysters occur in small numbers in Sinclair Inlet.

Clams: Native littleneck, butter, and Manila clams are common around Bremerton, Rolling Bay, near Vashon Island. Manila clams are very abundant in Carr, Case, Skookum, Totten, and Budd Inlets, and Oakland Bay. Butter clams exist in much the same area as the Manila, and are most prevalent in Carr and Case Inlets. Large numbers of littleneck, butter, and horse clams are present from Point Pully to Dash Point. Softshell clams are abundant in muddy bottoms through most of the area.

Other shellfish: Geoducks are common on all sandy mud flats, especially on Nisqually Flats, in north Carr, north Case, Totten, and Eld Inlets and in Oakland Bay. Piddocks are found in clay bottoms near Point Defiance. Cockles are very common throughout the many inlets at the southern

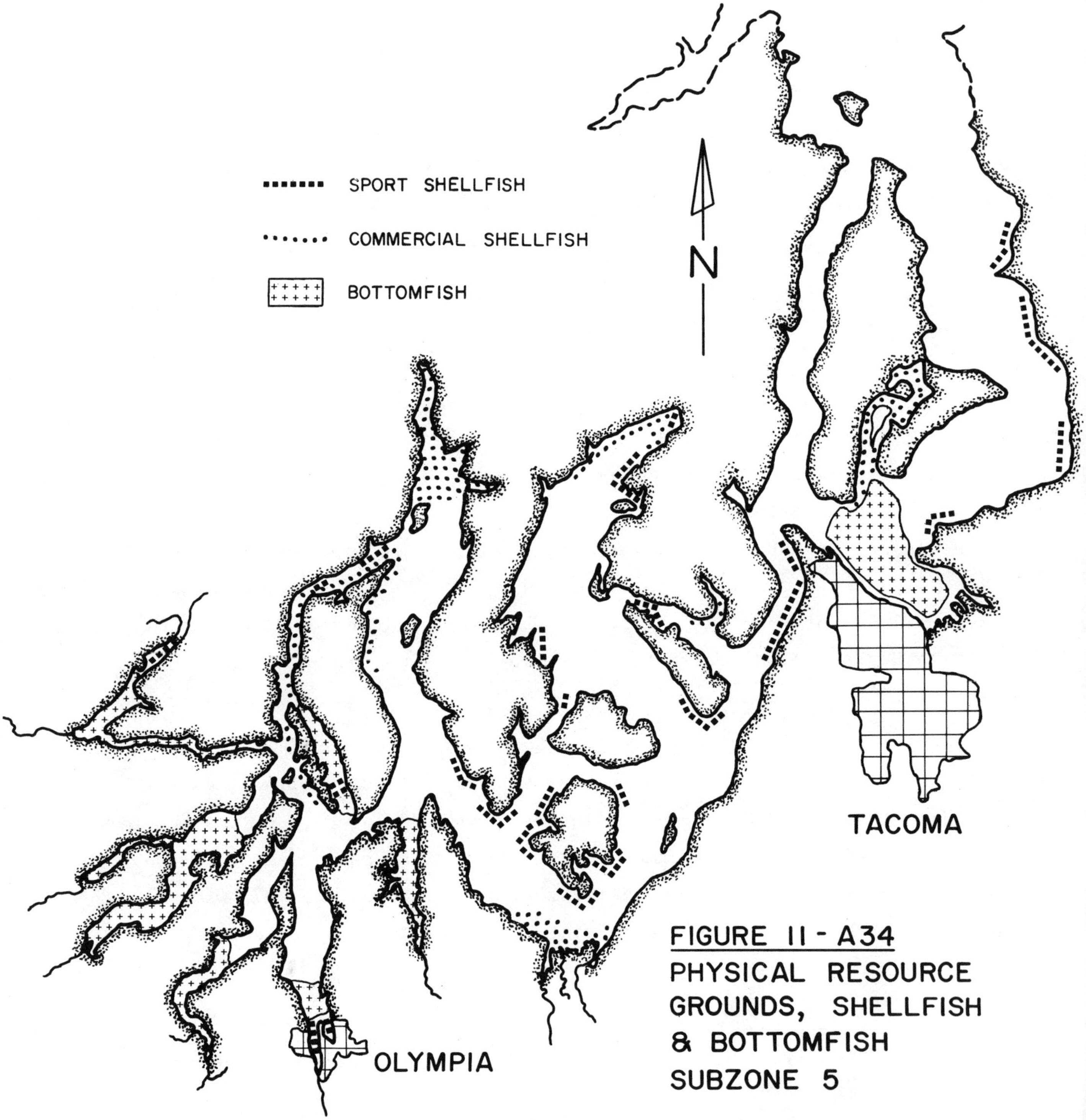

FIGURE II - A34
PHYSICAL RESOURCE GROUNDS, SHELLFISH & BOTTOMFISH
SUBZONE 5

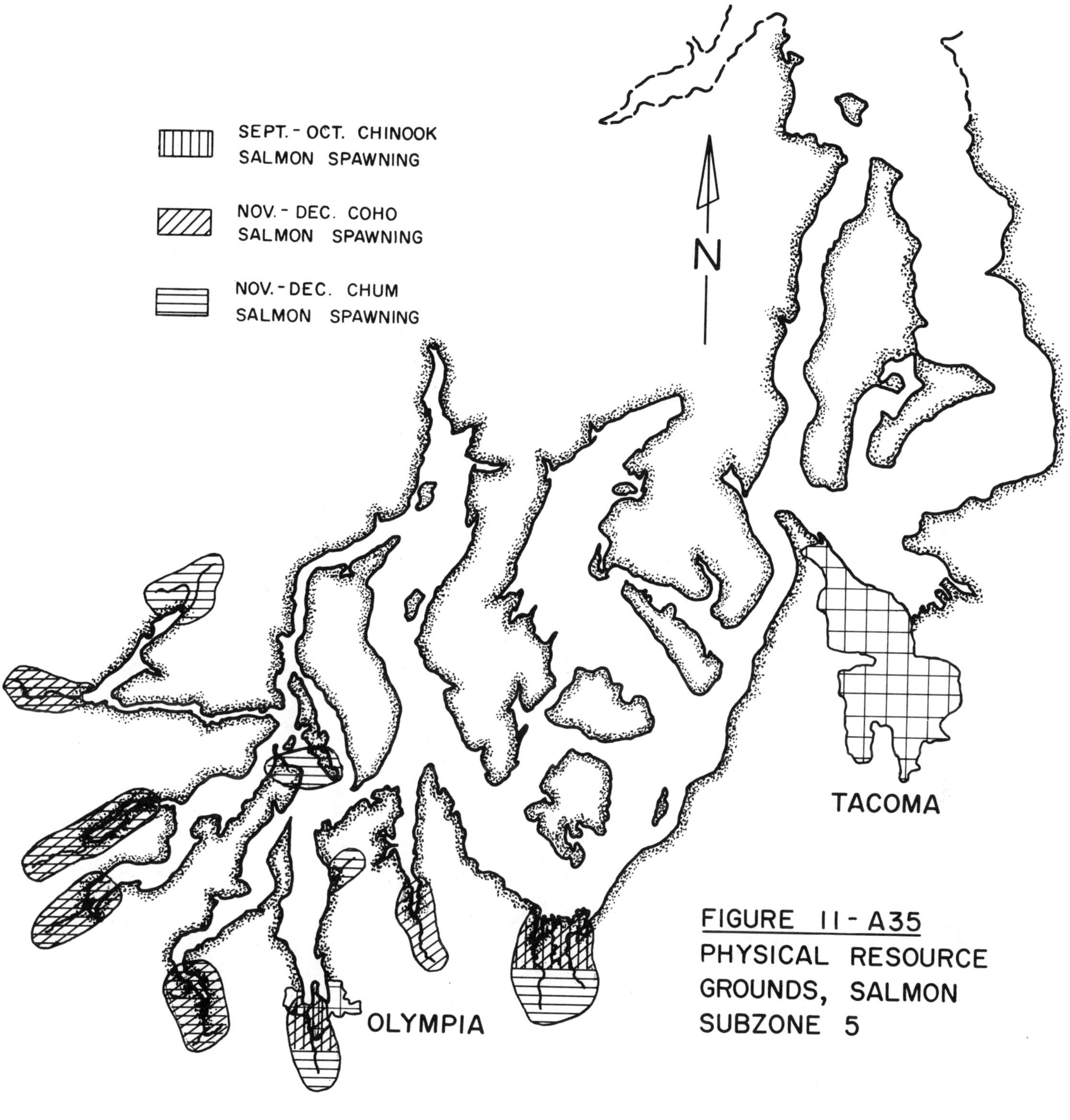

FIGURE II - A35
PHYSICAL RESOURCE GROUNDS, SALMON SUBZONE 5

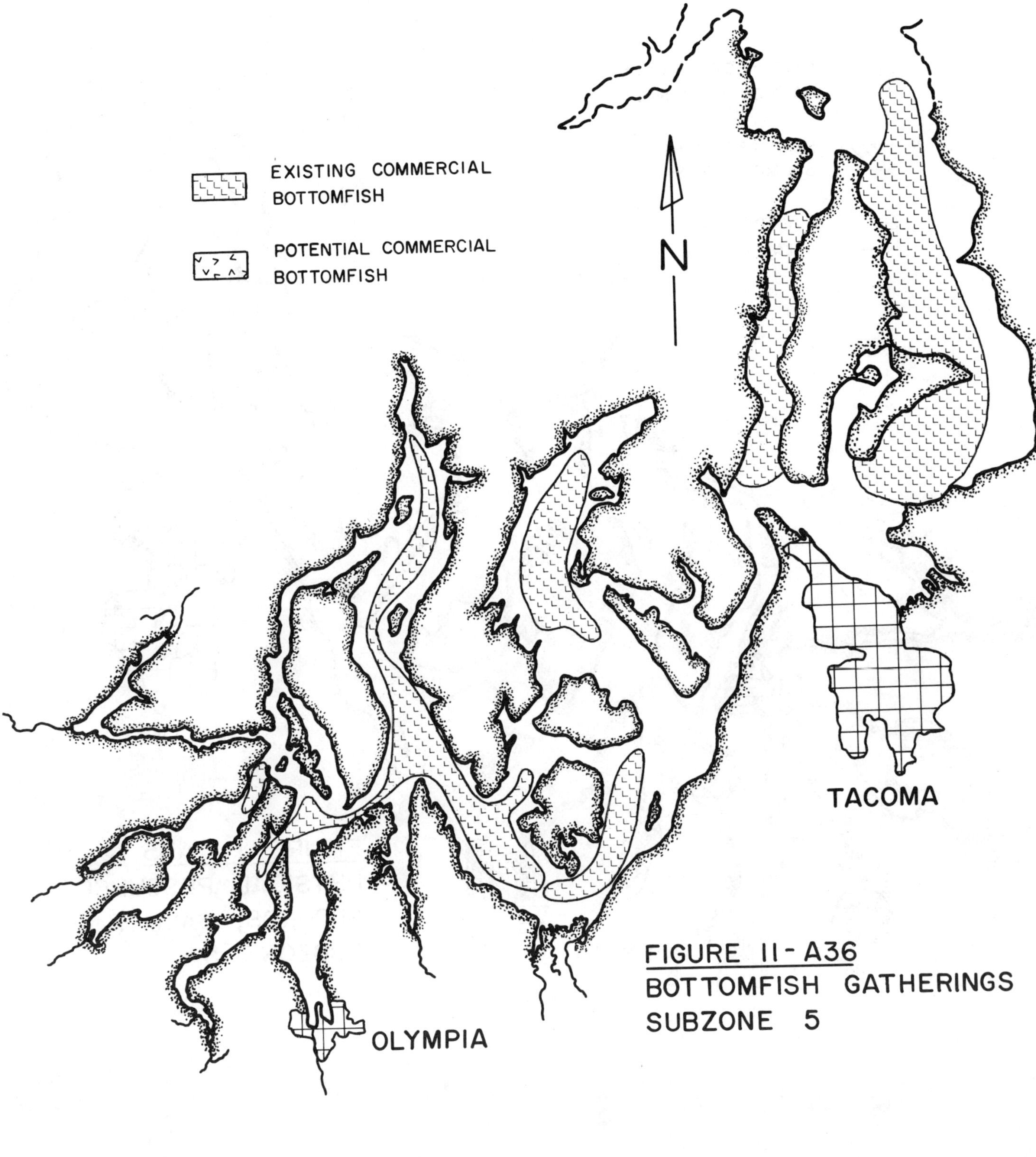

FIGURE II-A36
BOTTOMFISH GATHERINGS
SUBZONE 5

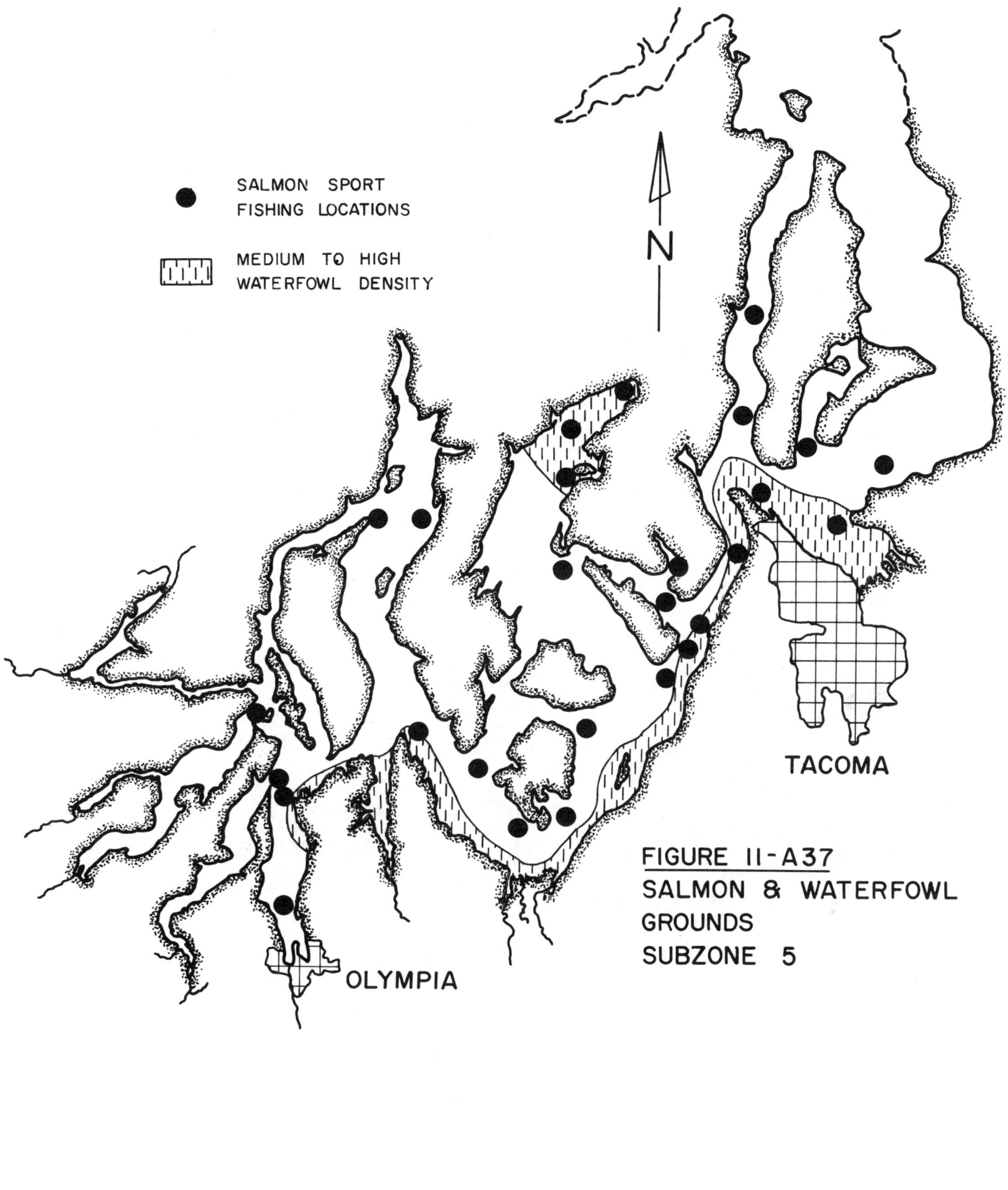

FIGURE II-A37
SALMON & WATERFOWL GROUNDS
SUBZONE 5

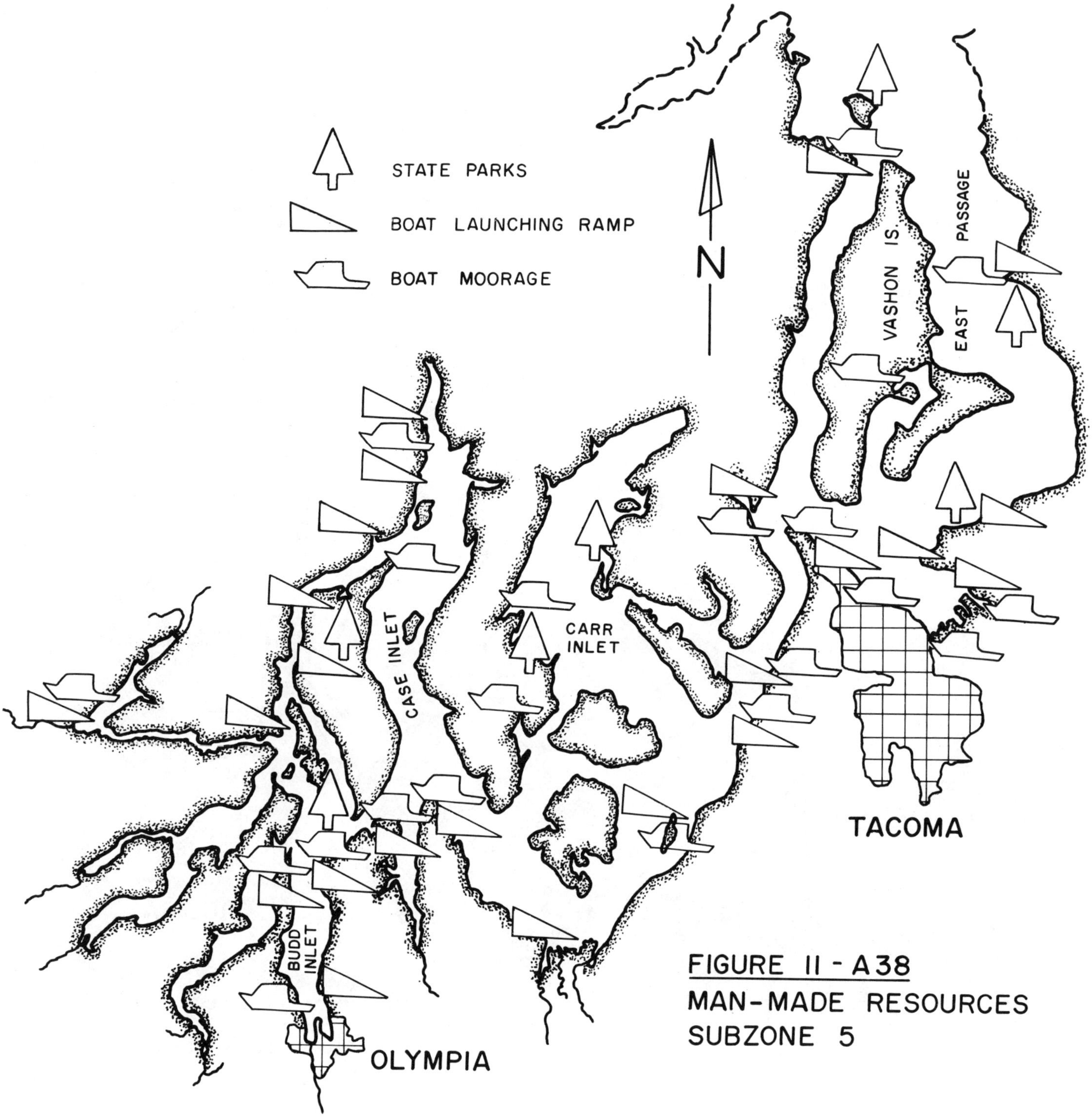

FIGURE II - A38
MAN-MADE RESOURCES
SUBZONE 5

end of the Sound, as are blue mussels, especially near Point Defiance. Four species of shrimp occur in this subzone. Pink shrimp dominate in the southernmost area, occurring in Carr Inlet, around McNeil Island, in Hale and Drayton Passages, near Anderson Island and Nisqually Beach and off of the east side of Hartstene Island. Spot shrimp dominant in East Passage and Dalco Passage. Pink scallops are common just north of Commencement Bay in water depths between 0 to 100 feet, and are found elsewhere in areas of strong currents and sandy bottoms. Large schools of squid are present in the Tacoma Narrows, Dolco Passage, Hale Passage, Wollochet Bay, Carr, Case, and Budd Inlets, and around Oakland Bay. Octopi are common around Fox and Anderson Islands, in Case and Carr Inlet, Dana Passage, Vashon Island vicinity, and around Point Definace and the Tacoma Narrows.

ii. fish

Salmon are mostly found in the waters around Point Defiance, Point Evans, Wollochet Bay, Hale Passage, and "Sand Spit" (NW Fox Island), northern tip of Carr Inlet, Johnson Point, and Steamboat Island. Figure II-A35 depicts these locations. Most commercial fishing occurs north of the Tacoma Narrows except for Case and Carr Inlets. Sport fishing locations include Dash Point, Brown's Point, Tacoma Narrows, and Point Definace (see Figure II-A37).

iii. waterfowl

Waterfowl are found in all areas of this subzone. Medium to high densities occur in Henderson Bay, Nisqually Reach, Budd Inlet and all along the east side of the Sound from Commencement Bay to Nisqually Flats, and from Nisqually Flats to Budd Inlet (see Figure II-A37).

6. Man-made physical resources

The protective inlets, plentiful in subzone 5, are ready-made for water-related recreational activities. There are six state parks, 30 launching ramps, and 39 rental boat moorages in this region of Puget Sound. Figure II-A38 shows the locations of these resources.

TABLE II-A7: MAN-MADE RESOURCES, SUBZONE 5[32]

Location	State Parks	Launching Ramps	Rental Moorages
East Passage	1	1	1
W. side Narrows	0	1	4
S. Vashon Island	0	0	3
E. side Narrows	0	4	6
Case Inlet	2	7	2
Hammersley Inlet	0	1	1
Budd Inlet	0	6	6
Nisqually Reach	0	4	1
Commencement Bay	1	6	13
Carr Inlet	2	0	2
SUBZONE 5 - TOTAL	6	30	39

f. Subzone 6: Hood Canal

1. Geography

Hood Canal is a glacially formed waterway extending about 60 miles from its juncture with Admiralty Inlet between Tola Point and Foulweather Bluff, which also marks the Canal's only outlet. The Canal has a long shoreline, 242 miles, and deep waters in excess of 600 feet in some places. Dabob Bay is the Canal's principal extension.

Many parts of the shoreline are wooded nearly to the waterline. The shores are generally steep with the eastern shoreline being gentler than the more rugged western portion. Mixed sand and mud tidal flats occur near river mouths. Elsewhere the majority of Hood Canal consists of short, relatively steep gravel beaches comprised of medium to small gravel on the upper beach and coarse to large gravel with scattered boulders on the middle and lower tidal areas.

Highway 101 runs the length of Hood Canal along the western side. Much of the shore is in private ownership and direct access is difficult except at larger towns and parks. Vehicle access to the eastern shore is quite limited with no single road spanning the full length of the Canal. The prevalent shoreline use is for single-family dwellings utilized mostly on weekends and during the summer months. The area is actually rather sparsely populated with no major resort or recreational complex in operation. Certain areas are used for commercial shellfish operations, but generally speaking, Hood Canal is primarily recreational in use.

All ports in Hood Canal can be described as minor and they include: Bangor, Seabeck, Brinnon, Pleasant Harbor, Holly, Eldon, Quilcene, Hoodsport, Union, and Port Gamble.

2. Currents

Tidal currents in Hood Canal at times attain velocities exceeding 1.5 knots. There are heavy tide rips northward of, and around, Foulweather Bluff, sufficiently heavy to be dangerous to small boats and to break up log rafts. This is most pronounced when the ebb current from the main body of Puget Sound meets that from Hood Canal off the point, and particularly so with the ebb against a strong northerly or northwesterly wind. Off Point Hannon and Hazel Point, tide rips occur at time sufficiently strong to be troublesome to tugboats with log tows. Figure II-A39 and II-A40 summarize the flood and ebb current.

3. Winds

During the period from October to May, prevailing winds are southwesterly with speeds to 20 mph. From June to September, winds tend to be northeasterly with speeds to 10 mph.

4. Waves

Due to the closed and sheltered character of Hood Canal, waves do not exceed a maximum of 6 feet and are generally much less.

5. Natural physical resources

The climate and water conditions of Hood Canal enhance the growth of natural resources, both for commercial and sport purposes. Figures II-A41 through II-A44 depict the spawning grounds. The following paragraphs discuss in detail the specific types of resources of the region.

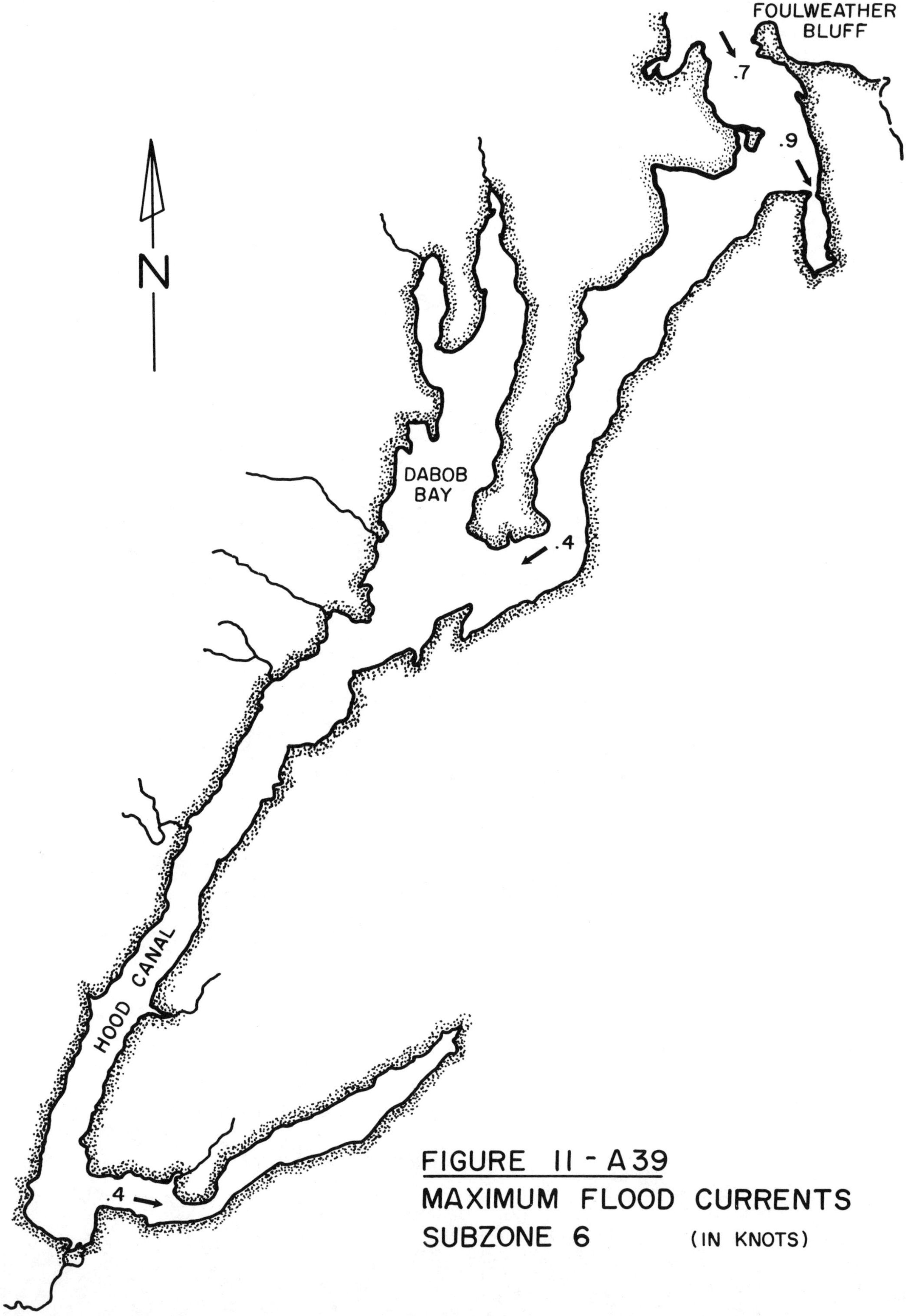

FIGURE II - A39
MAXIMUM FLOOD CURRENTS
SUBZONE 6 (IN KNOTS)

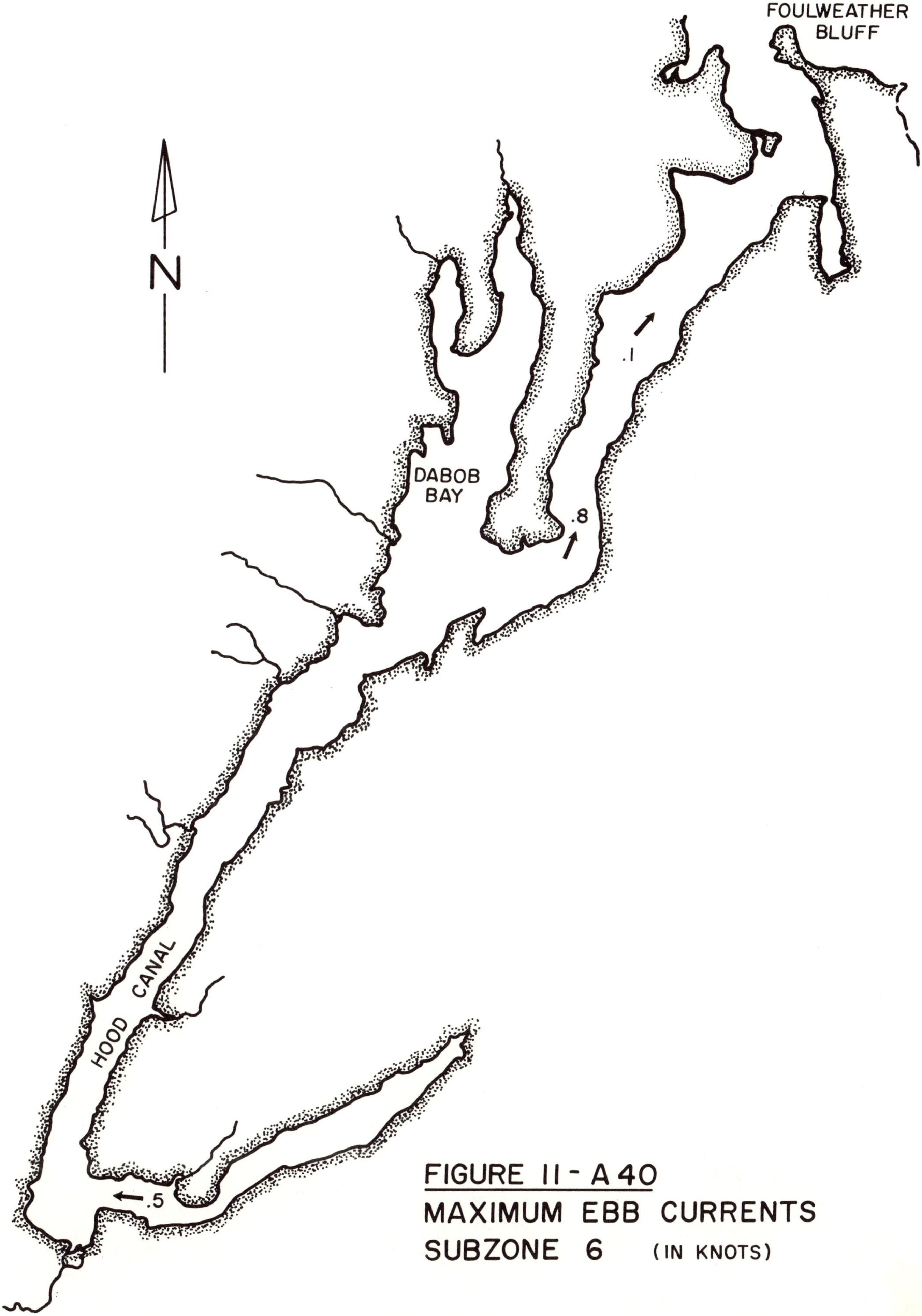

FIGURE II - A 40
MAXIMUM EBB CURRENTS
SUBZONE 6 (IN KNOTS)

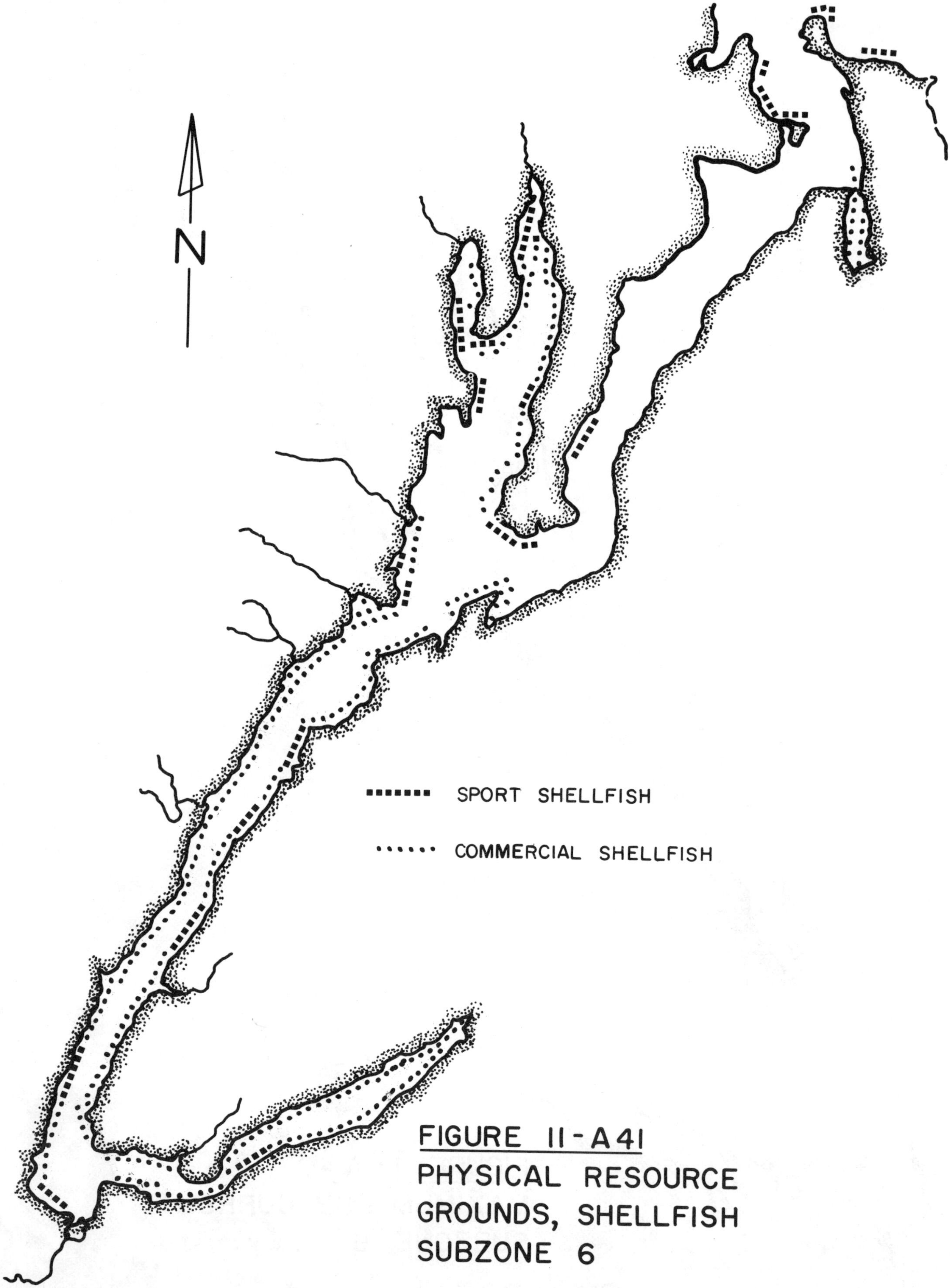

FIGURE II-A41
PHYSICAL RESOURCE GROUNDS, SHELLFISH
SUBZONE 6

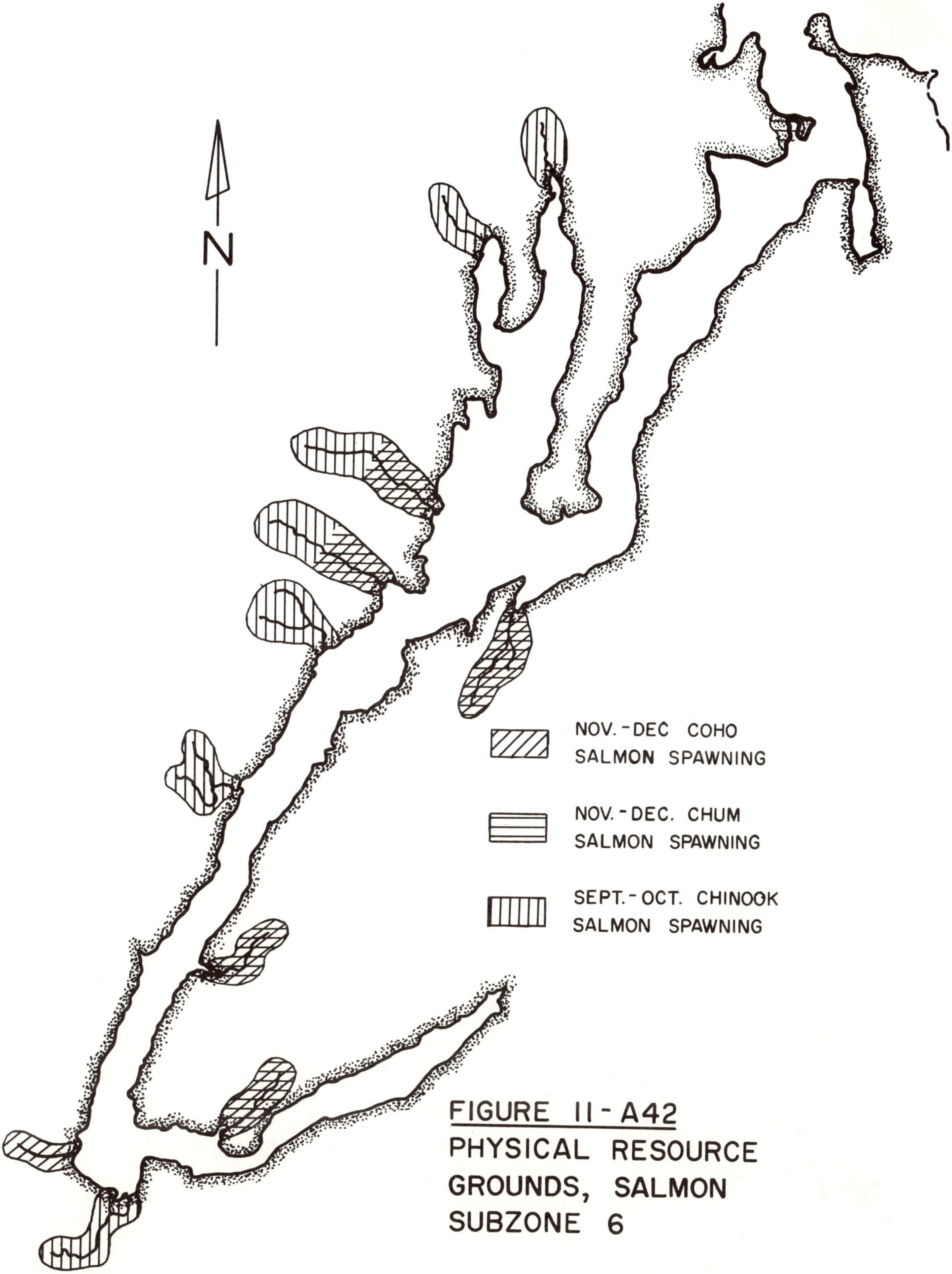

FIGURE II-A42
PHYSICAL RESOURCE
GROUNDS, SALMON
SUBZONE 6

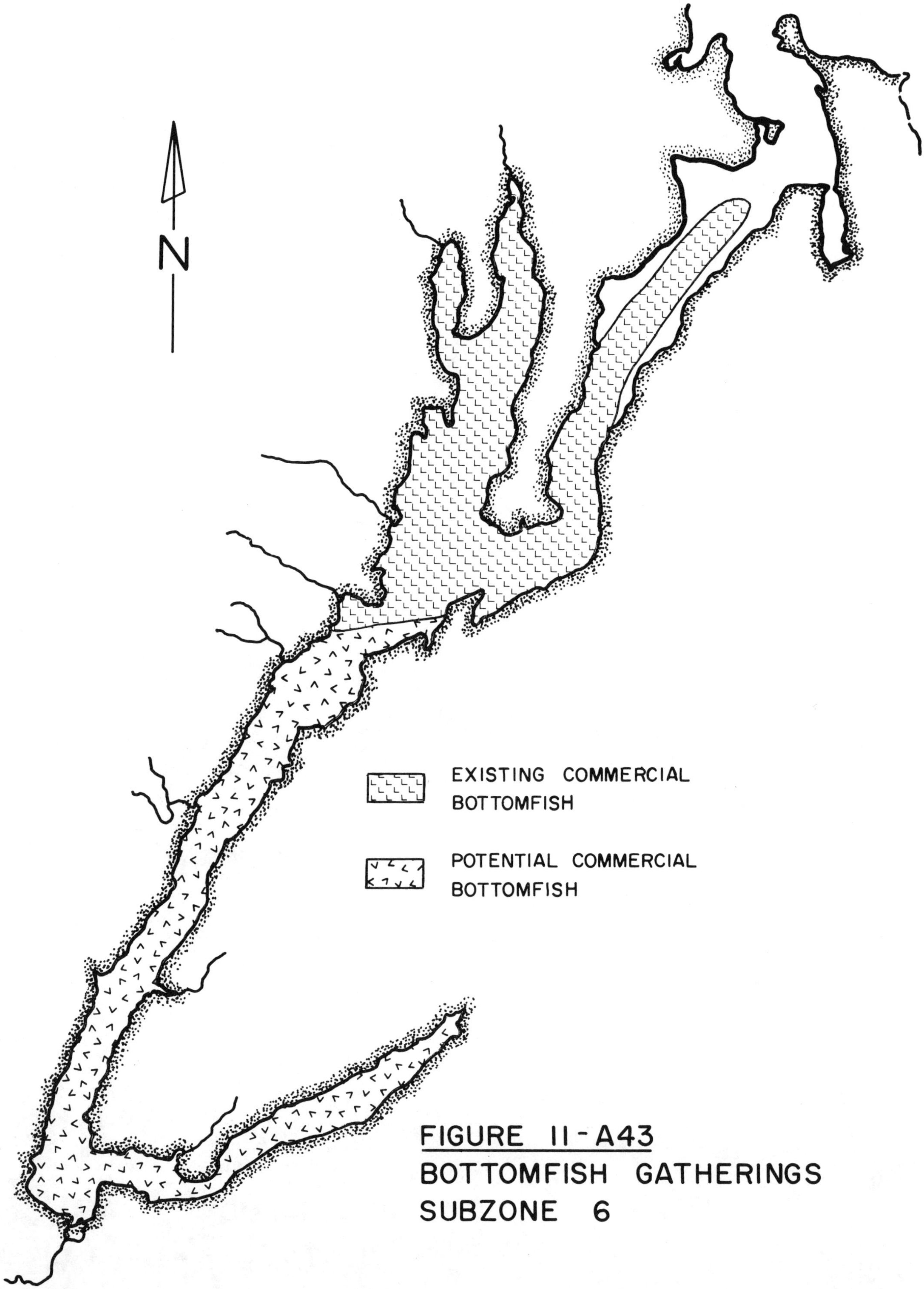

FIGURE II-A43
BOTTOMFISH GATHERINGS
SUBZONE 6

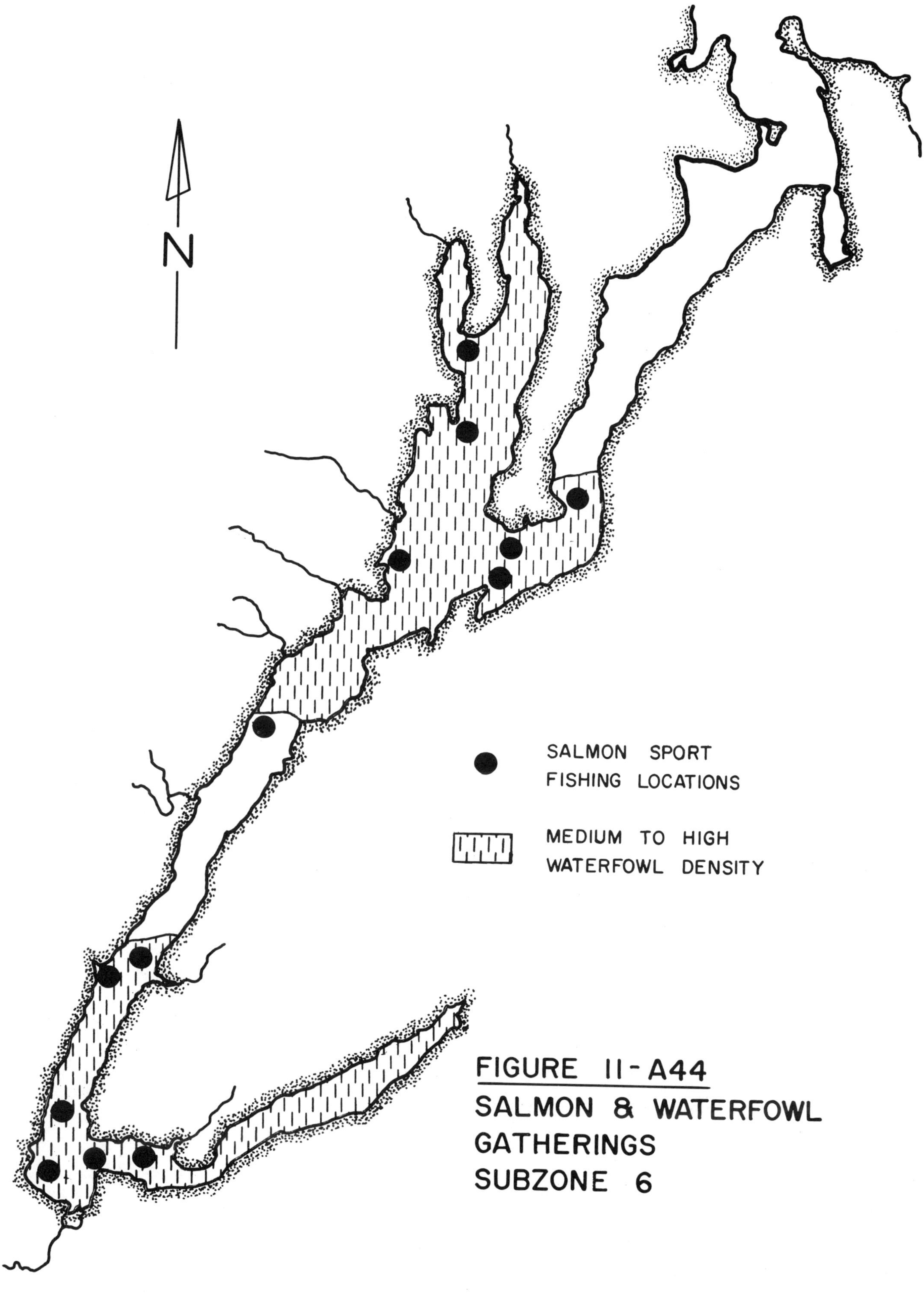

FIGURE II-A44
SALMON & WATERFOWL
GATHERINGS
SUBZONE 6

i. shellfish

Crabs: Dungeness crabs exist in shallow areas throughout the canal, but not in large numbers. Highest concentrations are at tidal flat dropoffs, and in sandy areas with eelgrass. The greatest concentration of Dungeness crabs appears to be in Quilcene Bay during the spring. Red crabs are moderately abundant from shore to about 60 feet deep throughout the canal.

Oysters: Pacific oysters are plentiful throughout the canal. Olympia oysters in limited numbers are found in the northern end of Dabob and Quilcene Bays, along Dosewallips Flats, and in Lynch Cove, along with scattered small populations throughout. Kumanoto oysters only in isolated locations.

Clams: Littleneck, horse, Manila, and butter clams exist on gravel beaches throughout the area. Softshells tend to be on the sandy mud flats near river mouth. Geoducks are in deeper waters. Clams are common throughout the area.

Shrimp: Spot shrimp are most abundant in the subzone. Also pink, coonstripe and sidestripe shrimp occur here. Shrimp are found throughout Hood Canal. The greatest quantities are found in Dabob Bay, Quilcene Bay, and off of Dewatto, Dosewallips River, Duckabush Rover, and Hoodsport. Shrimp are usually found in 30 to 60 fathoms of water.

Other invertebrates: Blue mussels are in moderate abundance especially north of Oak Head. Pink scallops occur throughout on sandy bottoms with relatively strong currents at a depth of about 40 feet. Rock scallops are found near rock outcroppings throughout the area. Squid are readily available year round. Octopi are common, especially near crab concentrations.

ii. fish

Salmon exist in significant populations the year round. During winter and spring, chinook salmon congregate in Jackson Cove, Hazel Point (late spring), Oak Head, Pleasant Harbor, Bald Point, across from Sisters Point, and east of Union. During summer and fall, salmon are found throughout the canal with concentration noted in Figure II-A42. The major

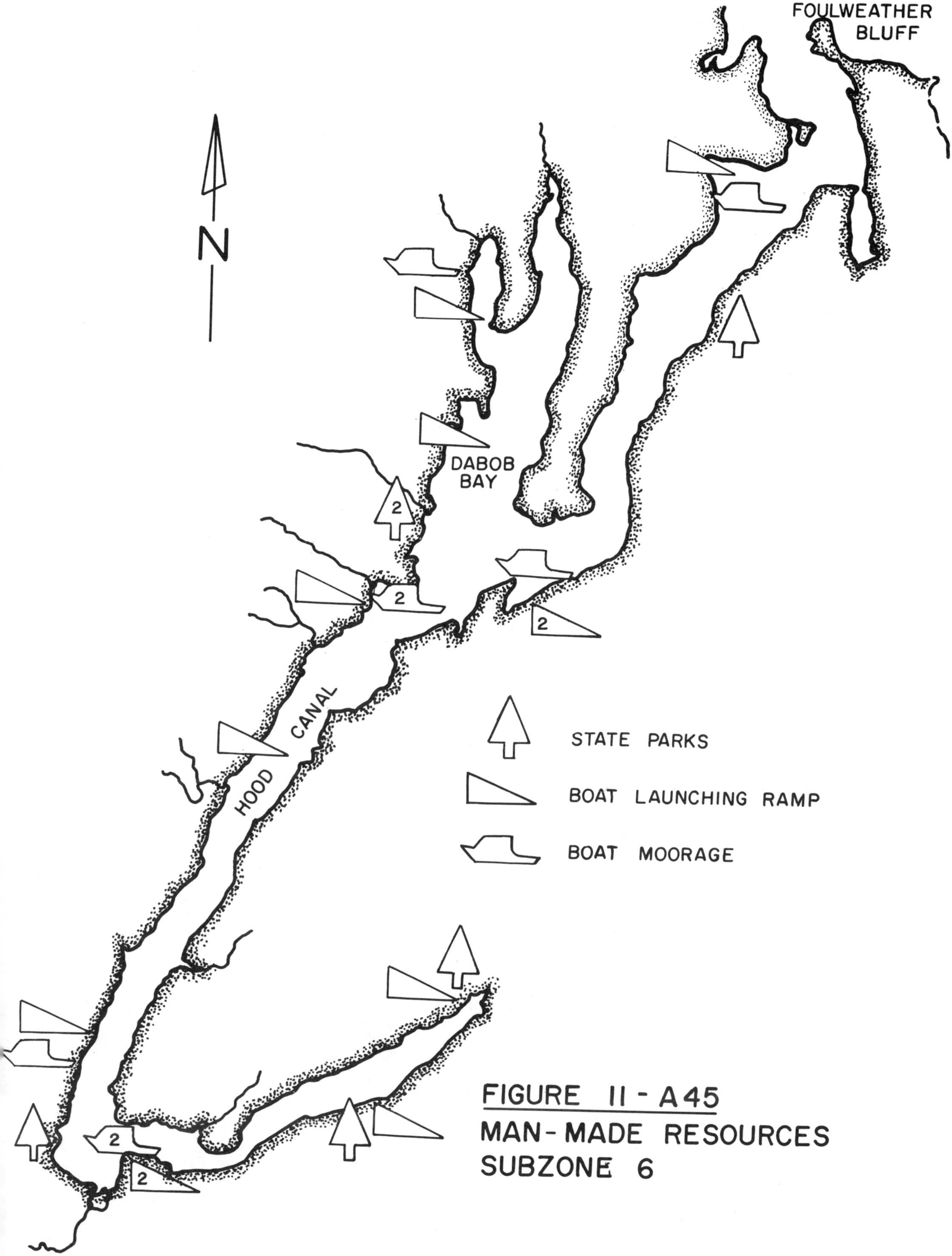

FIGURE II - A45
MAN-MADE RESOURCES
SUBZONE 6

fish grounds in this subzone are at the entrance to Hood Canal, as shown in Figure II-A44.

iii. waterfowl

Waterfowl are found throughout Hood Canal, with medium to high density found across the entrance to the canal, approximately from Port Townsend to Port Gamble. Other concentrations are found in the general vicinity of Dabob Bay and in the entire southern portion of the canal, south of Dewatto Creek (see Figure II-A44).

6. Man-made physical resources

The closed waters of Hood Canal with its long shorelines are ideal for water-related recreation. This is evidenced by the numerous state parks, launching ramps, and rental moorages along the banks of the canal. The following table lists the number of facilities, which are shown in Figure II-A45.

TABLE II-A8: MAN-MADE RESOURCES, SUBZONE 6

Location	State Parks	Launching Ramps	Rental Moorages
S. Arm-Hood Canal	3	5	2
Central Hood Canal	0	1	0
North Arm-Hood Canal	1	4	2
Dabob Bay	2	3	3
SUBZONE 6 - TOTAL	6	13	7

The general public has access to 4 launching ramps and 2 rental moorages located in this subzone.

ENDNOTES SECTION II-A

[1]Puget Sound and Approaches, A Literature Survey, University of Washington, Department of Oceanography, 1953, Volume 1.

[2]Ibid, p. 4.

[3]Ibid, p. 5.

[4]Ibid, p. 173.

[5]Ibid, p. 174.

[6]Ibid, p. 174.

[7]Ibid, p. 174.

[8]U.S. Naval Oceanographic Office, Glossary of Oceanographic Terms, Special Publication 35, 2nd edition, U.S. Government Printing Office, 1966, p. 64.

[9]Ibid, p. 165.

[10]Friebertshauser, M.S., and Duxbury, A.C., "A Water Budget Study of Puget Sound and Its Subregions," Limnology and Oceanography, in press.

[11]Ibid.

[12]Puget Sound Task Force, Puget Sound and Adjacent Waters, Appendix XI, Fish and Wildlife, p. 3-1.

[13]Ibid, p. 3-40.

[14]Ibid.

[15]Puget Sound Task Force, Puget Sound and Adjacent Waters, Appendix XIII, Water Quality Control, pp. I-45 and I-46.

[16]Puget Sound Task Force, XI, p. 3-40.

[17]77th Annual Report, Washington State Department of Fisheries, p. 138.

[18]Ibid.

[19]Groundfish Report #9, Washington State Department of Fisheries, 1970.

[20]Ibid.

[21]Puget Sound Task Force, Puget Sound and Adjacent Waters, Appendix VIII, Navigation.

[22]Harris, R. G., *Surface Winds over Puget Sound and the Straits of Juan de Fuca and their Oceanographic Effects*, University of Washington, 1954, unpublished M.S. thesis.

[23]Puget Sound Task Force, VIII, p. 11-12, 12-7.

[24]Puget Sound Task Force, *Puget Sound and Adjacent Waters Appendix X, Recreation*, 1970, p. 12-1.

[25]Harris

[26]Puget Sound Task Force, VIII, p. 3-15, 4-12, 13-3.

[27]Harris

[28]Puget Sound Task Force, VIII, p. 4-12, 6-7, 7-11.

[29]Harris

[30]Puget Sound Task Force, VIII, p. 8-20, 8-21, 11-12.

[31]Harris

[32]Puget Sound Task Force, VIII, p. 9-12, 10-7, 11-12.

[33]*Ibid*, p. 11-12.

II-B. WATERBORNE COMMERCE

1. INTRODUCTION

The primary objective in this discussion of waterborne commerce and navigational aspects of the Puget Sound region is:

1. to define and describe the relative shipping densities and traffic patterns based on statistical data, and
2. to integrate this data into such a form that it can be used as the input to a model forecasting the probability of a ship collision that could result in an oil spill.

In order to achieve this objective, the present and projected trends of several factors were evaluated. These include:

1. Total commerce in the region in tonnage.
2. Total bulk petroleum commerce in the region in tonnage.
3. Total number of vessels engaged in commerce in the region.
4. Total number of vessels engaged in bulk petroleum commerce in the region.
5. The degree of traffic attributable to shipyards and repairs.
6. The degree of traffic resulting from the Washington State ferry system.

It is hoped that from the data and discussion presented, the reader can gain further insight into his own assessment of possible dangers of oil spillage resulting from collisions between vessels engaged in commerce. The results presented should by no means be accepted as final. Continued accumulation of data to refine the results is of utmost importance to the effectiveness of assessment.

The nature of the problem at hand required many subjective evaluations, both on the part of the study group and external contributors of information, in order that some quantitative, objective evaluation could be made. The collision probability data thus must be viewed as contributing some sense of relative comparison, and not an absolute standard, for Puget Sound. However, it is shown that the results are consistent with historical data and in this sense validate the model, in terms of an absolute scale.

Present shipping trends are discussed in Part 2, while projections in shipping over the next fifty years are made in Part 3. Future trends in vessel size and capacities are examined in Part 4. A model for predicting ship collision on Puget Sound, detailed in Appendix 13, utilizes the shipping and vessel characteristic trends to estimate the number of expected vessel collisions leading to an oil spill. The results of this study are presented in the Part 5 of this section for the period 1970-1990.

For purposes of this study, the commerce data has been classified as follows:

For ports within the boundaries of the State of Washington:[1]

> Foreign Imports and Exports: traffic between United States and foreign ports, including the Canal Zone.
>
> Coastwise Receipts and shipments: domestic traffic receiving a carriage over the ocean. Traffic confined to Puget Sound is "internal" rather than "coastwise."
>
> Internal Receipts and Shipments: traffic between ports or landings wherein the entire movement takes place on inland waterways. The Puget Sound region is considered as an internal body of water.
>
> Local: movements of traffic within the confines of a port, except for car and general ferries.

For ports bordering Puget Sound in Canadian Waters,[2] the classifications are as follows:

> International Seaborne Shipping: vessels classed in foreign service, that is: when a vessel arrives from or departs for a foreign port, or a cargo is loaded for or unloaded from a foreign port. The registry is other than Canadian or British Commonwealth.
>
> Coastwise Service: vessels in Canadian or British Commonwealth registry that (1) sails between two Canadian ports, and unloads or loads no foreign freight. (2) A foreign vessel granted a waiver to engage in coasting service.

The foreign imports and exports and the coastwise receipts and shipments data for the United States has been combined to arrive at an approximate figure for traffic entering and leaving the Puget Sound region from or to some external point. The equivalent figure for the Canadian data is represented by the international seaborne shipping figures.

Traffic that is strictly internal is shown by the internal receipts and shipments classifications as well as the local classification in the U.S. data.

The equivalent internal traffic figures for Canadian traffic are represented by the coastwise service classification.

The following petroleum products were considered in the compilation of data. Specific commodities are identified using the Standard Industrial Classification (SIC) adopted in 1965 by the Department of Commerce. No equivalent system is available for the Canadian commodities.

A. United States Commodities

SIC	Bulk Petroleum Products
1311	Crude Petroleum
2911	Gasoline
2912	Jet Fuel
2913	Kerosene
2914	Distillate Fuel Oil
2915	Residual Fuel Oil
2916	Lube Oils and Grease
2917	Naphtha and Other Petroleum Solvents
2991	Petroleum Products, Not Elsewhere Classified

B. Canadian Commerce

Bulk Petroleum Products

Crude Petroleum
Gasoline
Fuel Oil
Lubricating Oils and Grease
Asphalts and Road Oils
Petroleum and Coal Products, Not Elsewhere Classified

2. PRESENT SHIPPING

The navigation resources of the Puget Sound region combine deepwater ports, easily accessible from the Pacific Ocean with a multitude of sheltered waterways. The deep waters afford passage to the largest vessels presently afloat as well as those vessels planned for the future. Table II-B1 shows that only two major ports on the Sound have limiting depths at their harbor entrance; all others are unlimited. Combined with the sheltered waters and the projected increasing size of ocean vessels, this makes the Puget Sound region a natural for expansion of seaborne commercial trade. However, a look at the present degree of commerce is necessary to establish a baseline.

TABLE II-B1: MAJOR PUGET SOUND HARBORS[3]

Major Ports	Controlling Depth at Entrance	Bulk Petroleum Terminals
Bellingham	Unlimited	11
Anacortes	50'	11
Everett	Unlimited	7
Seattle	Unlimited	52
Tacoma	Unlimited	11
Olympia	40'	5
Port Angeles	Unlimited	3

In 1967, the Puget Sound ports had steamship service from the area (U.S. ports only) with the following Lines:

TABLE II-B2: PUGET SOUND STEAMSHIP SERVICE[4]

Primary Destination	Number of Lines
Alaska	5
Intercoastal	3
Hawaii	1
Mexico and Central America	5
West Indies and Caribbean	5
South America	3
Japan, Hong Kong, Phillipines, Formosa, Korea and Okinawa	18
Indonesia, India, Pakistan, Malaysia, Vietnam and Persian Gulf	10
South Seas, Australia and New Zealand	7
Northern Europe	12
Mediterranean	6
South, West and East Africa	1

In addition to steamship service summarized above, six barge lines serve Alaska. Also, there are ten coastwise carriers of petroleum and lumber operating for their private use. Local freight in the region is handled by two lines as well as ferries. The Washington State Ferry System carries cross-sound and inter-island traffic. Three ferry lines operate between the area and British Columbia and the Alaska Ferry System operates between Puget Sound and southeast Alaska.

Ferry traffic has grown in the last several years and it appears it will continue to grow in the future. Table II-B3 shows the growth of vehicles and passengers carried by the Washington State Ferry System. A better estimate can be made by describing the total number of transits in each of the ferry lines serving the Sound. This data is represented in Table II-B4.

Information related to the amount of military traffic in the region is not available. United States Army and Navy vessels, such as transports and Navy tankers, entering or clearing without commercial cargo and foreign military and naval craft, have been excluded from all data.

TABLE II-B3: WASHINGTON STATE ANNUAL FERRY TRAFFIC[5]

Year	Vehicles (1000's)	Passengers* (1000's)
1952	2,136	4,937
1953	2,305	5,065
1954	2,356	4,994
1955	2,456	5,172
1956	2,569	5,262
1957	2,585	5,328
1958	2,681	5,443
1959	2,696	5,355
1960	2,612	5,197
1961	2,559	5,354
1962	2,616	6,737
1963	2,569	5,699
1964	2,707	5,790
1965	2,843	3,149
1966	3,149	6,297

*Exclusive of driver

TABLE II-B4: FREQUENCY OF FERRY TRANSITS[6]

Ferry Transits	Approximate Number of One Way Crossings per day
Steilacoom to Anderson and McNeil Islands	N.A.
Fauntleroy to Southworth via Vashon	60
Seattle to Bremerton	34
Seattle to Winslow	52
Edmonds to Kingston	60
Port Townsend to Keystone	16
Mukilteo to Columbia Beach	84
Anacortes to Sidney, B.C. via San Juan Islands	6
Port Angeles to Victoria, B.C.	8
Point Roberts, B.C. to Sidney, B.C.	N.A.
Seattle to Victoria, B.C.	2
Tacoma to Tahlequah	40
ESTIMATED TOTAL ONE-WAY FERRY TRANSITS PER YEAR	132,130

TABLE II-B5: MAJOR SHIPYARDS IN PUGET SOUND REGION

Duwamish Shipyard Inc.	CRD
Lake Union Drydock Co.	CRD
Lockheed Shipbuilding & Construction Co.	RD
Todd Shipyards	NCRD

where

C - Conversion

R - Repair

D - Drydock

N - New Construction

Estimated average turnaround - 4 weeks

Traffic attributable to shipyards

(4 yards x 52 weeks/year)/ 4 weeks/ship

= 52 ships/year

The figures for the amount of shipping that enter the region destined for a shipyard is, at best, only approximate. The best that can be done is to estimate the average repair turnaround time and multiply that by the number of major shipyards in the region. This estimate gives some 52 ships per year. This data and analysis can be seen in Table II-B5.

Several graphs and charts have been prepared to show the following information:

1. Estimated percentage of total commercial traffic received by each port in terms of number of vessels, Table II-B6.
2. Distribution of refined petroleum tonnage from external sources (Foreign, Coastwise, and International Seaborne Shipping) Figure II-B1.
3. Distribution of refined petroleum tonnage in internal waters of Puget Sound region (Internal and Coastwise Service) Figure II-B2.
4. Distribution of crude petroleum tonnage from external sources. (Foreign, Coastwise, and International Seaborne Shipping.) Figure II-B3.

Table II-B6 represents the distribution, by percentage, of vessel movements into and out of each of the ports in the Puget Sound region. The percentage distribution gives some indication of shipping activity in the various areas of the region. In terms of vessel activity, the Port of Seattle has approximately 40% of the total movements. The other ports vary from as low as 0.5% at New Westminister, British Columbia to 13.2% at Tacoma.

Figures II-B1 through II-B3 are approximations of the port distributions of various crude and refined oil tonnages. The percentages of tonnage per port are summarized in Table II-B7. The arrows show the direction of net commodity flow while the amount of product received at each port is shown as a percentage of the tonnage. It is interesting to note that approximately 87% of the refined oil entering and leaving the Puget Sound region terminates and originates at either Anacortes or Seattle.

In the same fashion, 66% of the internal refined oil moves within approximately the same boundaries, those consisting of Seattle on the south, Anacortes on the north, with Victoria and Port Angeles on the west.

TABLE II-B6: DISTRIBUTION OF COMMERCIAL TRAFFIC BY PORT (1967)[7]

(number of vessels)

PORT	INBOUND				OUTBOUND				CUMULATIVE			% OF SUM TOTAL
	Self Propelled		Non-Self Propelled		Self Propelled		Non-Self Propelled					
	Tank	Other	Tank	Other	Tank	Other	Tank	Other	Tank	Other	Tank & Other	
Neah Bay		8,003	8	1		8,000	8	1	16	16,005	16,021	5.2
Port Angeles	13	1,680	146	328	12	1,673	146	332	317	4,013	4,330	1.4
Port Townsend	3	4,175	85	1,674	3	4,177	81	1,679	172	11,705	11,877	3.8
Waterway		643	1	833		645		834	1	2,955	2,956	.9
Port Gamble												
Hammersley Inlet		1,411	45	284		1,411	44	284	89	3,390	3,479	1.1
Olympia		2,660	101	257		2,663	101	258	202	5,838	6,040	1.9
Tacoma	40	18,364		2,037	40	18,362	517	1,519	597	40,282	40,879	13.2
Seattle	247	53,997	1,563	5,810	247	53,995	1,564	5,810	3,621	119,612	123,233	40.0
Lake Washington Ship Canal	2	9,432	64	1,736	2	9,434	64	1,735	132	22,337	22,469	7.3
Everett		10,761	112	841		10,766	112	842	224	23,230	23,454	7.6
Anacortes	206	8,074	274	403	203	8,071	274	404	957	16,952	17,909	5.8
Bellingham		8,858	125	524		8,862	119	526	244	18,770	19,014	6.2

TABLE II-B7: DISTRIBUTION OF PETROLEUM COMMERCE, BY PORT[8]

Port	Internal Refined (%)	Foreign Refined (%)	Foreign Crude (%)
Neah Bay	.1	0	0
Port Angeles	3.0	.6	0
Port Townsend	.6	.2	0
Hammersley Inlet	.3	0	0
Olympia	1.7	0	0
Tacoma	16.5	.4	34.0
Seattle	29.0	44.0	.4
Lake Washington Ship Canal	2.7	.2	0
Everett	1.0	0	0
Anacortes	22.7	43.0	65.6
Bellingham	1.6	.1	0
Vancouver and other B.C. ports	-	11.0	0
Victoria	7.0	.5	0
TOTAL SHORT TONS	5,088,521	6,503.917	1,481,426

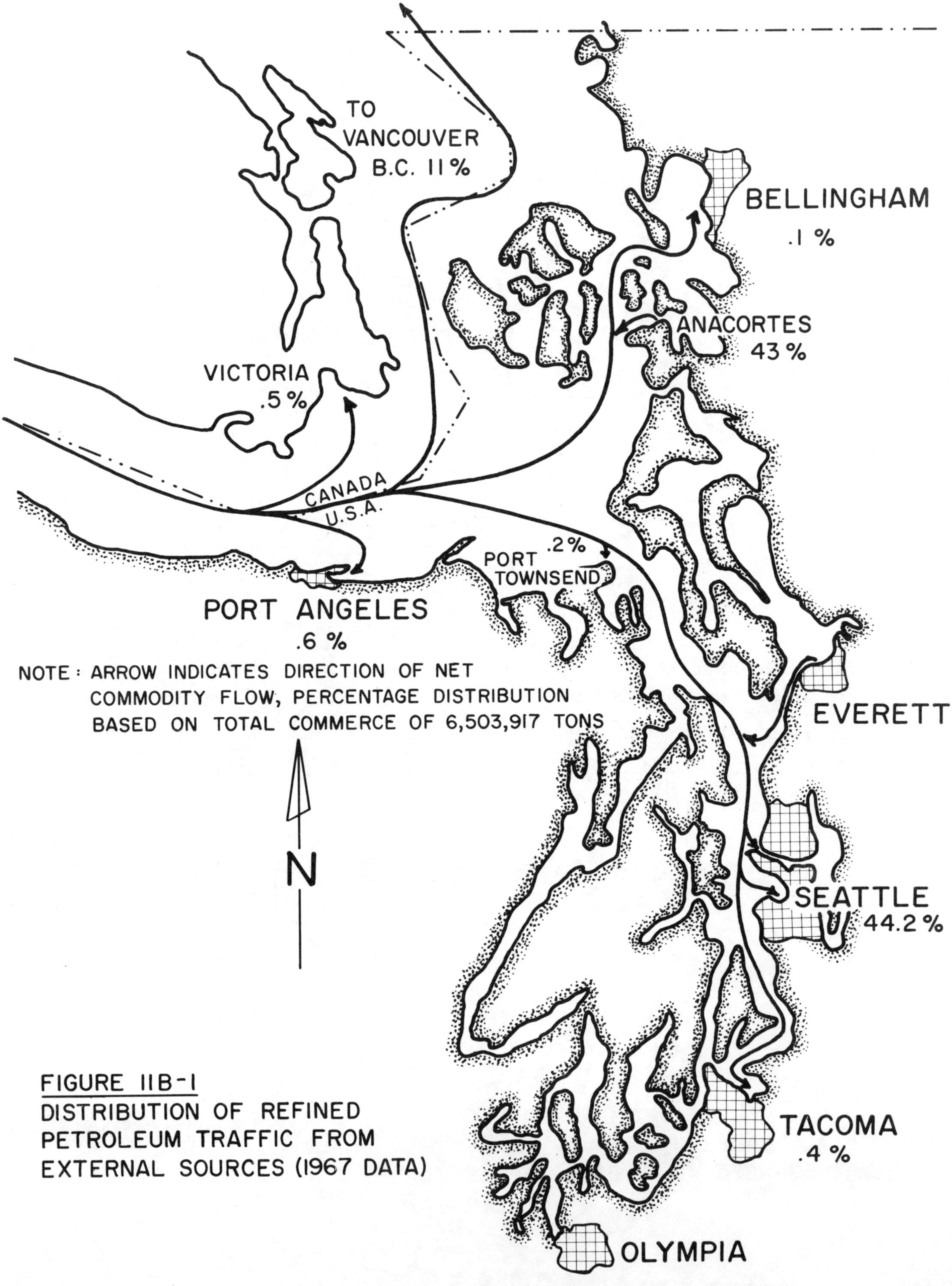

FIGURE IIB-1
DISTRIBUTION OF REFINED PETROLEUM TRAFFIC FROM EXTERNAL SOURCES (1967 DATA)

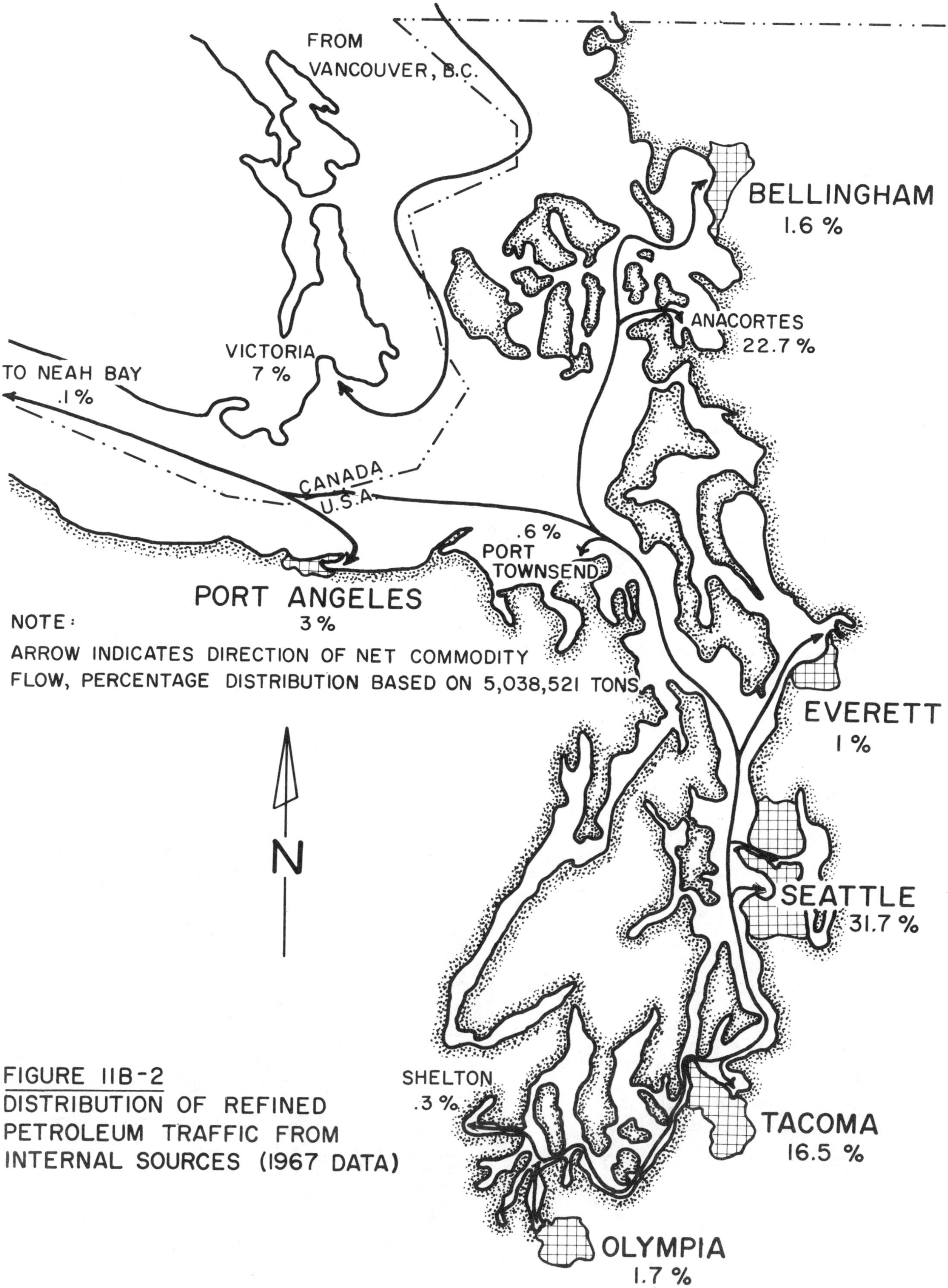

FIGURE IIB-2
DISTRIBUTION OF REFINED PETROLEUM TRAFFIC FROM INTERNAL SOURCES (1967 DATA)

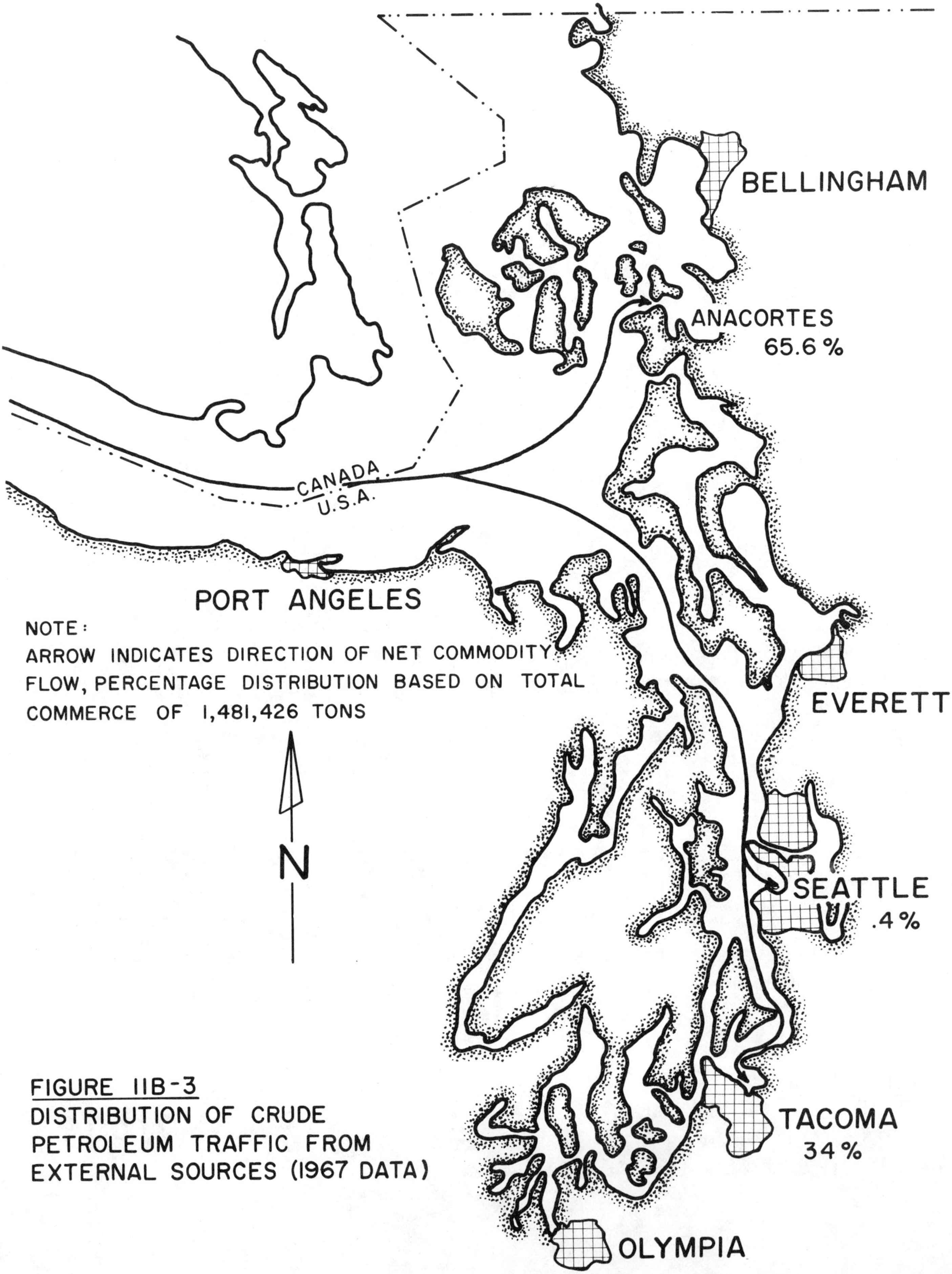

FIGURE IIB-3
DISTRIBUTION OF CRUDE PETROLEUM TRAFFIC FROM EXTERNAL SOURCES (1967 DATA)

Thus, it appears the majority of refined oil moved on Puget Sound is confined to this area with a smaller portion for external ports passing through the Strait of Juan de Fuca.

It should be noted that 16.5% of the internal movement of refined oil terminated or originated at Tacoma. This appears to be small; but when combined with the contribution of crude oil (Figure II-B3), the activity in the area becomes more significant. There is very little internal movement of crude oil, when compared to refined products. Most of the crude oil that comes to the Northwest (93%) comes via pipeline. Of the crude oil that does go by vessel, the majority of the external flow goes to Tacoma (34%) and Anacortes (65.6%).

The routes of major concern, then, with regard to the transportation of oil through these areas appear to be, in order of concern:

1. Point Wilson to Seattle
2. Strait of Juan de Fuca
3. Rosario Strait to N. Lummi Island
4. Haro Strait to Boundary Pass
5. Seattle to Tacoma

More will be said about the relative dangers involved in each of these areas in Part 5 of this section, but for now suffice it to say that any subjective evaluation of an ability to prevent or respond to an oil spill should be concentrated at present to these areas. For future planning some evaluation must be made of the impact of the transportation to Puget Sound of North Slope crude.

There is one final note to the compilation techniques employed in international shipping. There exists some duplication in terms of the Canadian and U.S. accounting of commerce. Also, there may be some question as to whether all coastwise service traffic remains internal. The degree of duplication appears to be small (66,000 tons), while the primary shipping contribution to U.S. waters is from the Port of Vancouver.

The total amount of petroleum products shipped from Vancouver to U.S. Puget Sound ports is approximately 66,000 tons. This is then the amount of petroleum product that is accounted for twice, once at Vancouver and once at the U.S. Puget Sound port. There is no data available for commodities shipped from other British Columbia ports to the Puget Sound region, however any amount there might be is felt to be very small. Thus for practical purposes the duplication of tonnage accounting can be ignored.

The total amount of petroleum commerce resulting from coastwise service at the ports bordering on the Strait of Georgia is approximately 1,730,000 tons. Of this amount, it is estimated that 354,000 tons is shipped in coastwise trade to British Columbia ports outside the Strait of Georgia area that are most easily reached through Haro Straits. Three primary ports were considered to be in this category: Victoria, Port Renfrew, and Port Alberni. Thus, the contribution to total shipping activity in the Puget Sound region from Canadian coastwise shipping activity can be considered as limited to this contribution. There are undoubtedly other unaccounted for sources of coastwise Canadian shipping that pass through the region under study. However, it is felt that the contribution from Vancouver is largely representative of the actual amount.

3. PROJECTED SHIPPING

The needs of navigation in the Puget Sound region are estimated for the years of 1980, 2000, 2020. Waterborne commerce projections are extended considering primarily historical trends and economic considerations. The trends of vessel size and draft are also considered.

Assumptions regarding the probable direction and level of national economic growth were adopted. These explicit assumptions are:

a. The Federal Government, as a matter of national policy, will actively support programs designed to stimulate economic growth.

b. Sufficient quantities of water of acceptable quality will be available through timely development to avoid being a constraint to economic growth.

c. There will be no general war nor any appreciable cessation of the cold war throughout the period to 1980. Expenditures on national security will continue to account for approximately ten percent of the GNP. After 1980, gradual disarmament will decrease the relative cost of military expenditures.

d. There will be a continued relaxation of trade tariffs and quotas and an accompanying expansion of international commerce.

e. The U.S. population will expand to:[11]

1980	259,584,000
2000	338,219,000
2020	469,126,000

f. The Federal Government will use resources energetically to promote maximum employment, production, and purchasing power. Accordingly, employment will prevail at approximately 96 percent of the total civilian labor force throughout the forecast period.

g. U.S. GNP will increase in billions of 1960 dollars to:[12]

1980	\$1,130
2000	\$2,472
2020	\$5,402

h. Development of technological process, together with expansion of workers skills and capital formation, will increase productivity per man hour approximately 2.9 percent per year.

It must be noted that these projections are to be considered as rough estimates of the future. The data base upon which these figures are founded does not have the scope and depth with which to make accurate estimates. These projections deserve more study based on further facts from a detailed examination of traffic origins and destinations, and the economic factors governing this flow of traffic. However, the necessary information with which to make this type of an evaluation is not available today, nor in the foreseeable future. Thus, the limits of these projections must be recognized for what they are and that they can be modified with updated information.

The base period from which the projections are made, although relatively short (1952-1966), does represent one in which economic conditions were rather stable without major distortions of wars and recovery from major recessions. The projections for total commerce and bulk petroleum is based on extension of trends over the past few years, without a recognition of the growth forces reasonably expected to govern the flow of commodities through Puget Sound ports. Where commodity flows can be correlated with local and national economic parameters, projections are made on this basis.

<u>FIGURE II-B4</u>

PROJECTED TOTAL WATERBORNE COMMERCE - FOREIGN & DOMESTIC

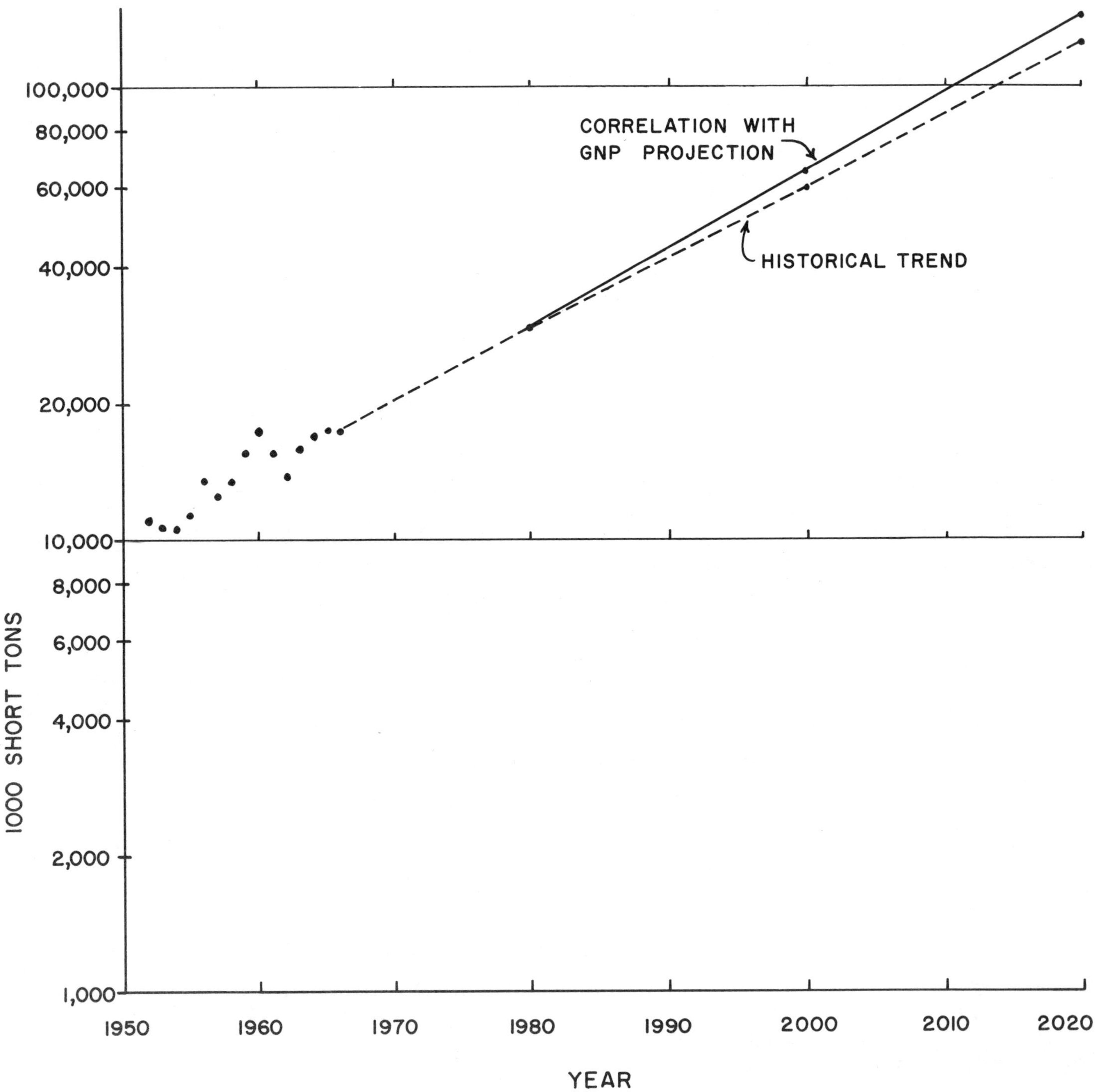

Both linear and compound growth trends have been fitted to the data for the period 1952-1966. The equation form selected was based on judgement and the degree of correlation. Time trends were developed through a linear regression model, while a logarithmic transformation of the base data was used to obtain compound growth trends.

Total waterborne commerce is projected in three general groupings; they include: (1) Foreign and Domestic Coastwise, (2) Domestic Internal, and (3) Total - the sum of the previous two.

1. Foreign and Domestic Coastwise: This is commerce with foreign nations and parts of the United States outside of the Puget Sound region. The historical record shows a progressive increase from approximately 11,000,000 short tons in 1952 to over 17,000,000 tons in 1966.[13] This is an average compound growth rate of 3.6 percent. Projecting this growth rate, waterborne commerce will reach 29,000,000 tons in 1980, 60,000,000 tons in 2000, and 123,000,000 tons in 2020. No account of the economic forces that may change in the future is included in these figures. However, to reflect these effects the historical behavior of this commerce shows a good correlation with the GNP. These projections are shown graphically in Figure II-B4.

2. Domestic Internal: This is traffic confined to the region. In the past, this tonnage has shown an erratic pattern rising from 20,000,000 short tons in 1952 to a peak of 27,000,000 tons in 1959 and thereafter stabilizing near 25,000,000 tons through 1966.[14] The average growth rate has been 1.9 percent. Assuming this to continue, annual tonnages will be 34,000,000 by 1980; 49,000,000 by 2000; and 71,000,000 by 2020, as illustrated in Figure II-B5.

3. Total: The total commerce in the region has increased from 31,000,000 short tons in 1952 to almost 42,000,000 tons in 1966.[15] Rapid growth occurred in part of this period and has stabilized since 1960. The historical trend growth is calculated at 2.5 percent. As shown graphically in Figure II-B6, the projected annual total commerce will be 62,600,00 tons by 1980; 103,000,000 tons by 2000;

FIGURE II-B5

PROJECTED TOTAL WATERBORNE COMMERCE - INTERNAL

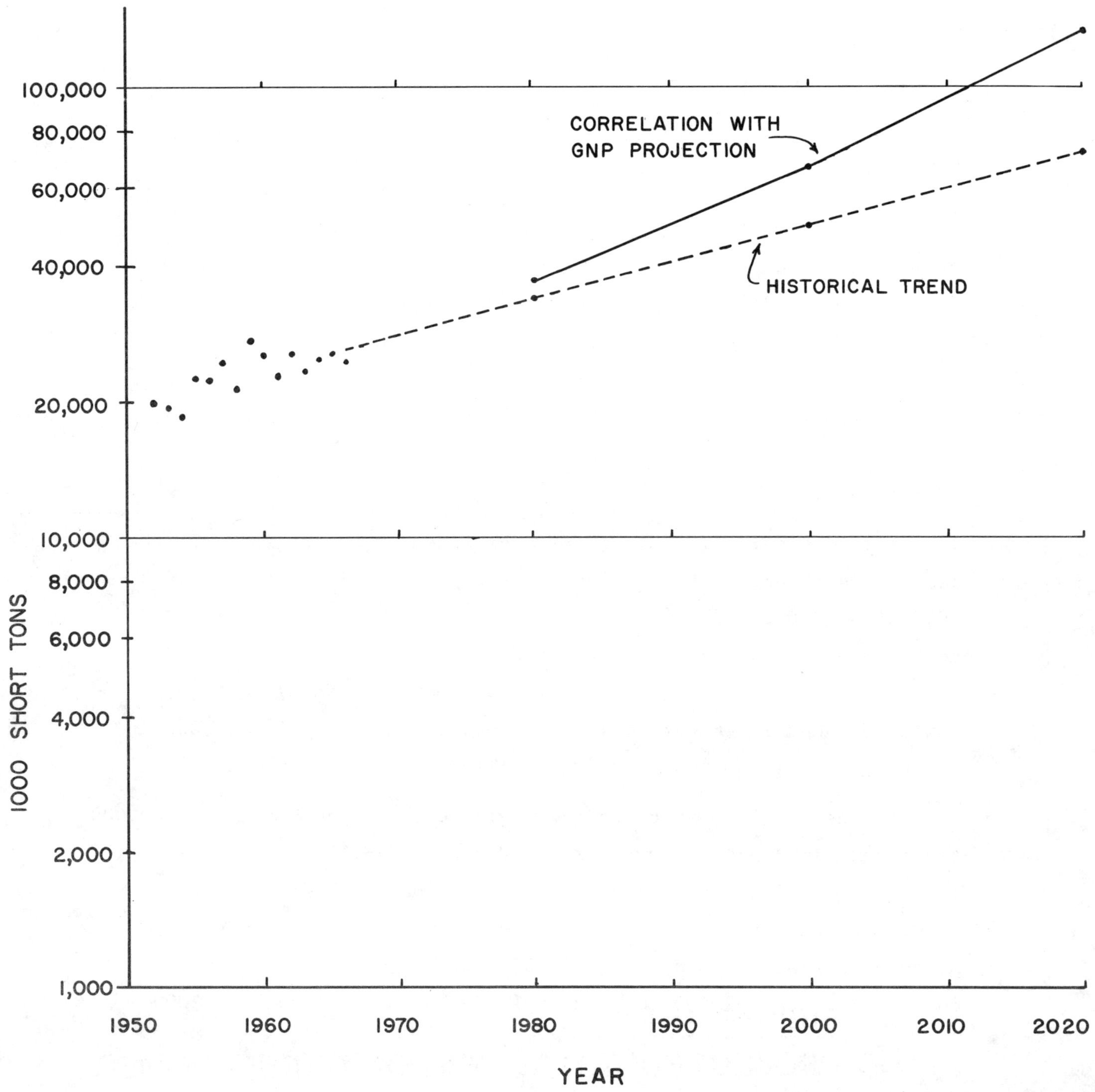

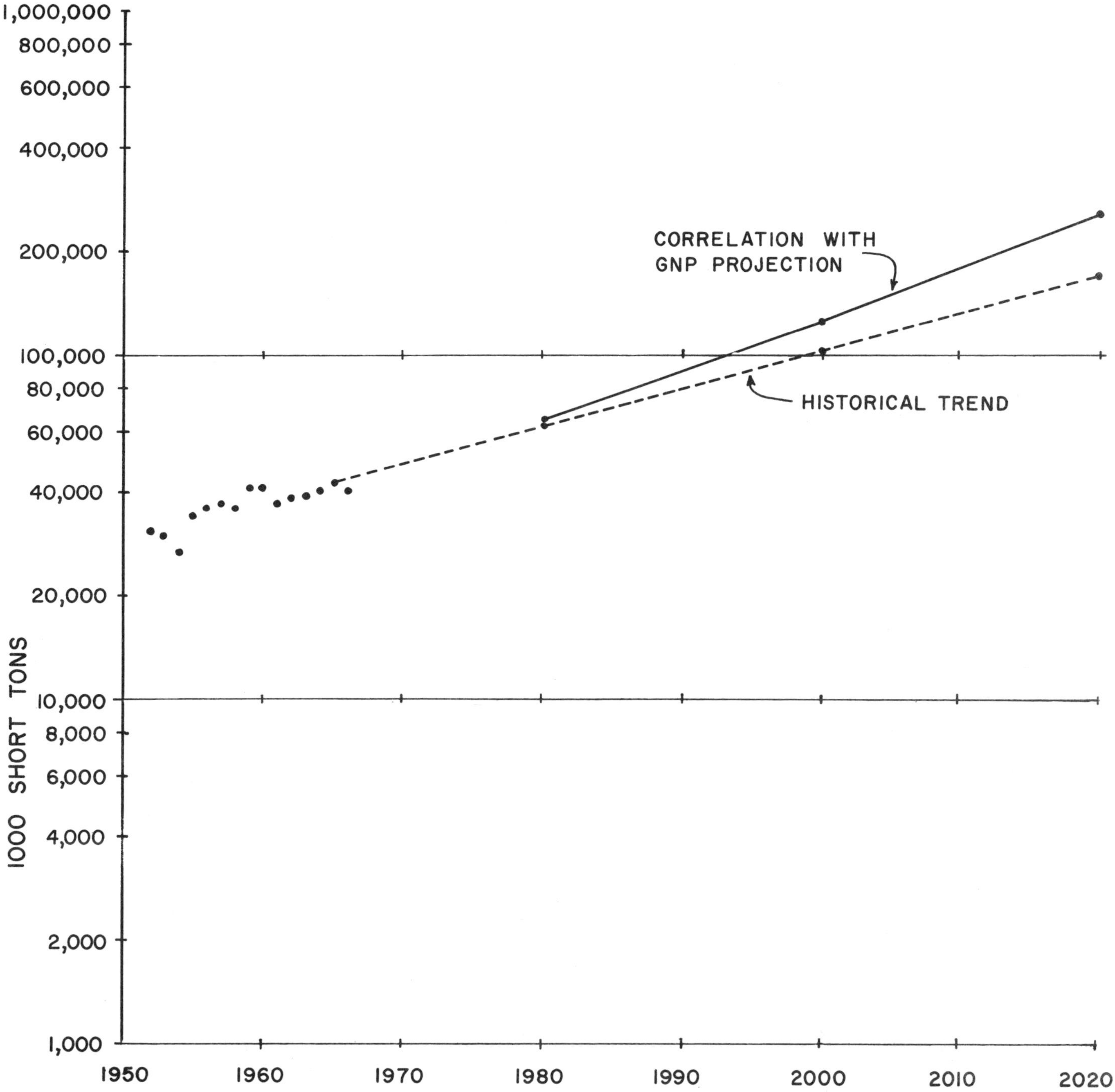

FIGURE II-B6
PROJECTED TOTAL WATERBORNE COMMERCE

and 170,200,000 tons by 2020. Recognizing economic forces, the historical trends are compared to GNP and are found to have a reasonable correlation. The projections based on the GNP index show somewhat higher forecasted totals. These are: 65,500,000 tons by 1980; 126,300,000 tons by 2000; and 259,100,000 tons by 2020.

Bulk petroleum projections are also classified in the same three general groupings as the total waterborne commerce projections.

In 1963, the petroleum refinery capacity satisfied about 46 percent of the consumption of petroleum in the region, the capacity of the refineries being 165,100 barrels per calendar day (b/cd) of which 157,700 b/cd were produced by refineries in the Puget Sound region.[16] The refinery complex has expanded from a base of two refineries in 1954 producing 9,500 b/cd in 1964. The location of the new sites has been influenced by available deepwater ports which permit the loading and unloading of large tankers. Secondly, the geographical location of the region is convenient to foreign imports of crude oil as well as the dispatching of the refined product to major Northwest markets. Finally, there is dependable inexpensive electric power readily available. The electrical energy requirements to refine a barrel of crude are approximately three kilowatt hours.

The continued future growth of the petroleum industry depends on the current search for oil reserves in the coastal waters of Oregon and Washington, and the distribution method employed for the proven crude oil reserves estimated at 9.6 billion barrels in the North Slope fields in Alaska.[17] Present plans for marketing the oil includes a pipeline from the fields to Valdez and then by tanker to Puget Sound and parts of California. Other plans include a pipeline across Canada and the possibility of a pipeline from the Puget Sound region to Chicago exists. Presently being developed is the Atlantic Richfield refinery at Cherry Point. Its capacity will be in the neighborhood of 100,000 b/cd, bringing the total refinery capacity of the Puget Sound region to 272,000 b/cd by 1980.

There remains, at this time, uncertainty as to how the Alaskan oil will be distributed. The pipeline to Valdez has yet to receive government approval. There is increasing speculation of a cross-Canada pipeline that would completely bypass the Northwest markets. On the other hand, plans for the new refineries are being continued and orders have been placed by Atlantic Richfield for oil tankers to carry oil from Valdez, presumably to Northwest markets.

The implementation of the Valdez plan could mean the Puget Sound region would experience a large growth in refineries and that the region would become a net exporter of refined products. However, the uncertainty that remains and the decisions yet to be made preclude anything but a rough approximation of future petroleum commerce recognizing the probability of future changes.

Foreign and Domestic Coastwise Petroleum Commerce: This commerce has grown from 6,000,000 short tons in 1952 to a high of 10,000,000 tons in 1960.[18] The growth rate based on these trends is 3.2 percent. Future annual commerce based on this growth rate will be 13,900,000 tons by 1980; 26,200,000 tons by 2000; and 49,000,000 tons by 2020. See Figure II-B7.

Domestic Internal Petroleum Commerce: Refined petroleum moves into the Puget Sound markets by tanker, barge, tank car, tank truck, and pipeline. The volume of this commerce increased from about 2,300,000 short tons in 1952 to 9,800,000 in 1963 and then decreased to 6,300,000 tons in 1966.[19] The trend prior to 1963 indicates as annual growth rate of 475,800 tons per year; thus projecting 16,000,000 tons by 1980; 25,600,000 tons by 2000; and 35,100,000 tons by 2020, as shown in Figure II-B8. However, the construction of pipelines connecting major centers has altered this trend since 1963 as revealed by the decreasing tonnages. Most likely, future tonnages of waterborne petroleum commerce will be represented by some random variation around 6,300,000 tons. This variation will depend primarily on pipeline capacities, location of consumers and transportation cost structures.

Total Bulk Petroleum Commerce: The best estimates support a continued growth of the petroleum industry in this region. The development of the Alaska fields will, in all probability, have a major impact on this area. Although the projections herein are based on many factors that are by no means stable, the forecast can be considered to be a reasonable estimate of future commerce. As shown in Figure II-B9, the total projections for 1980 are 20,200,000 short tons; 32,500,000 tons by 2000; and 55,300,000 tons by 2020.

FIGURE II-B7

PROJECTED BULK PETROLEUM - FOREIGN & DOMESTIC, COASTWISE

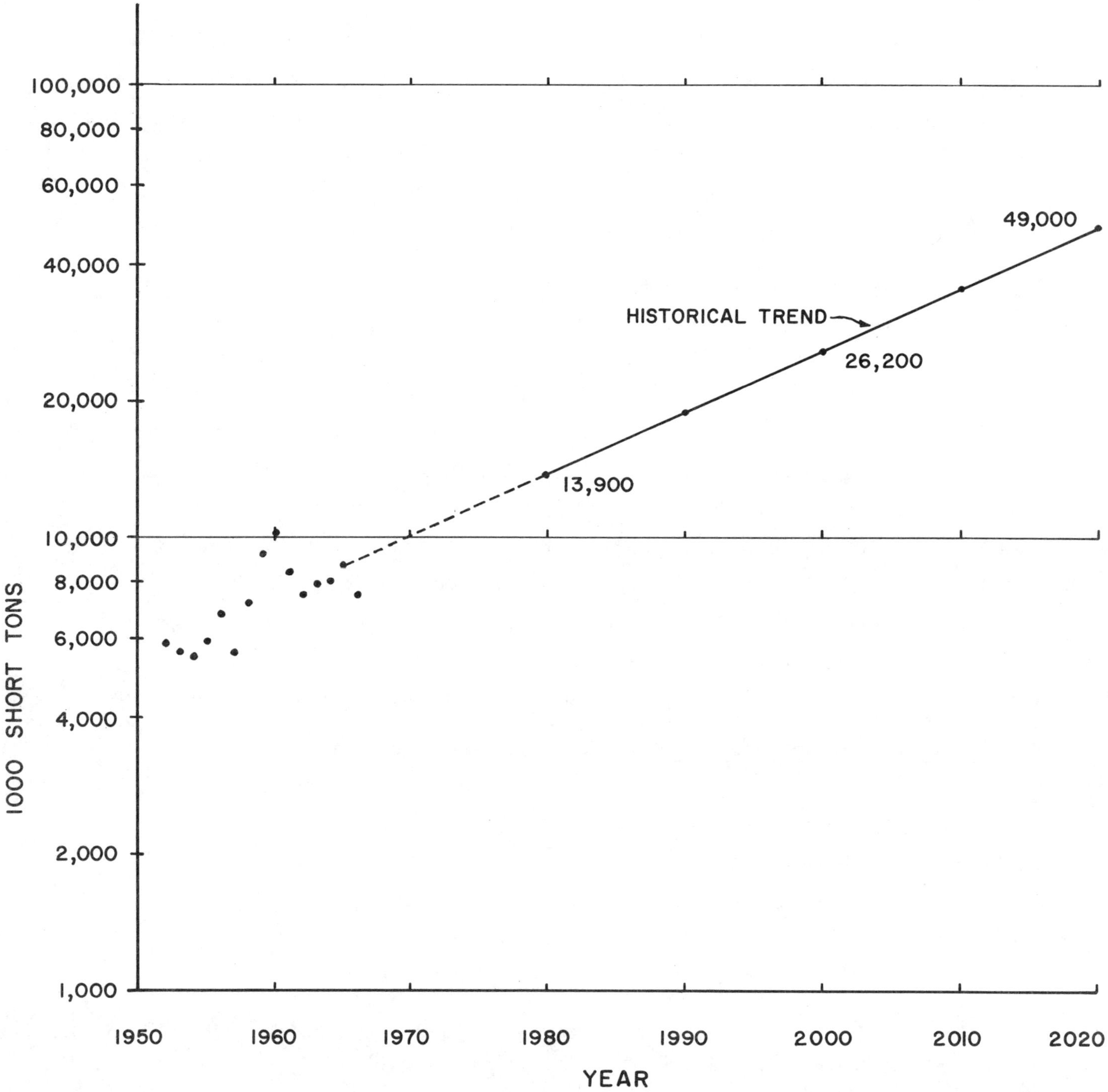

FIGURE II-B8

PROJECTED BULK PETROLEUM - DOMESTIC INTERNAL

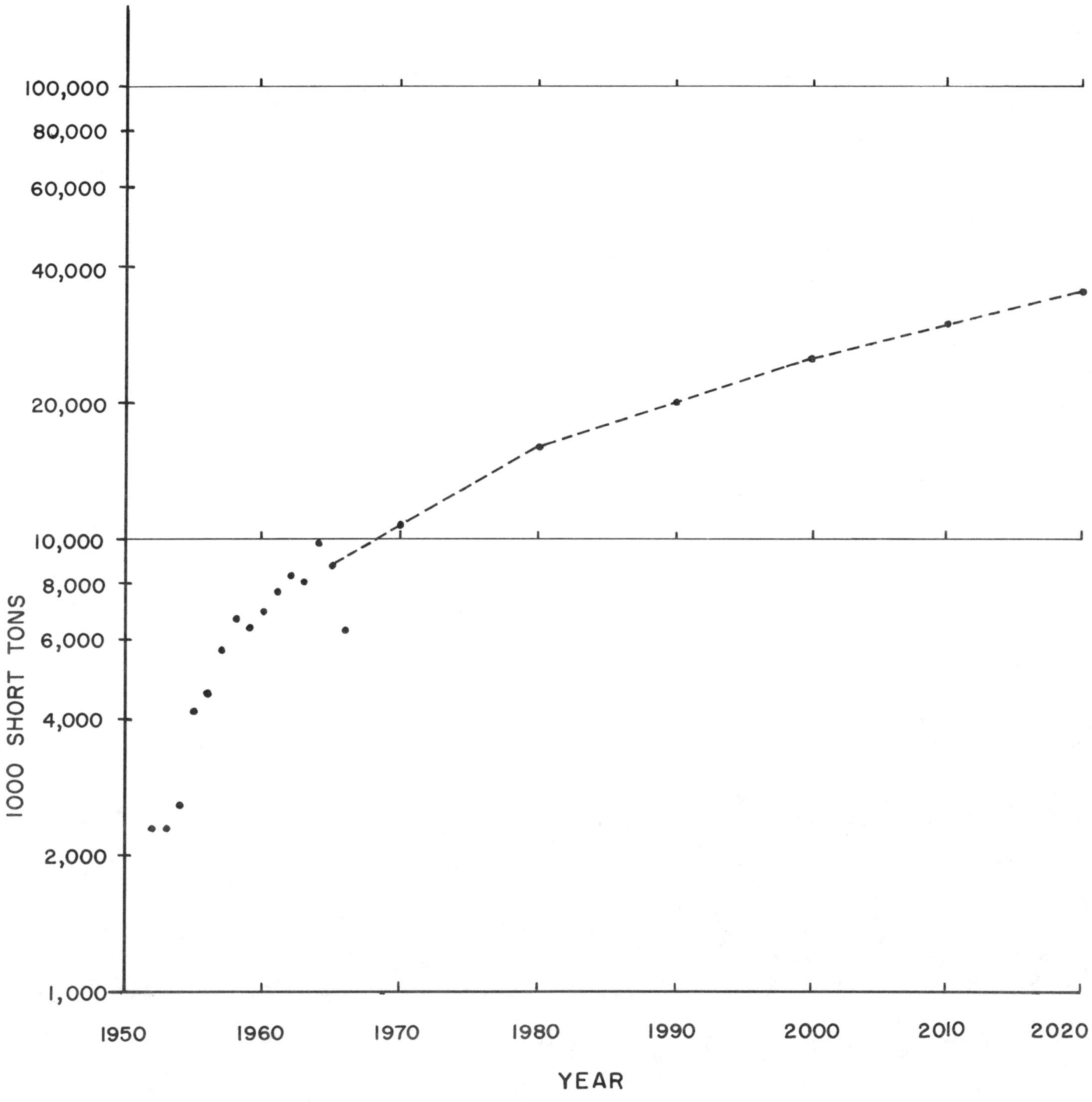

FIGURE II-B9

PROJECTED BULK PETROLEUM - TOTAL COMMERCE

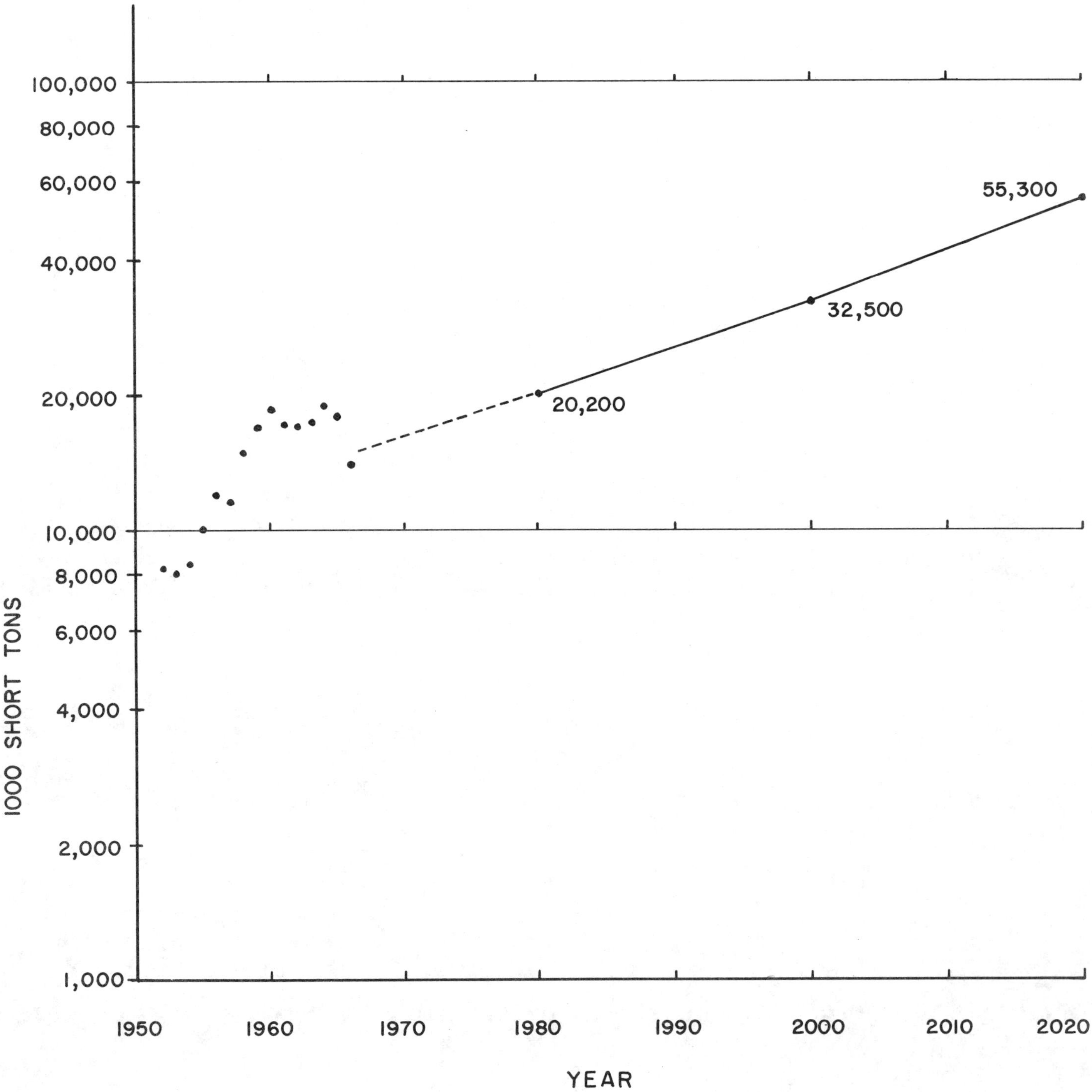

4. FUTURE OF WATERBORNE VESSELS

The trend in vessel construction is toward larger, more efficient sizes. The increased sizes are particularly evident in the bulk carriers and tankers. As the Puget Sound region is physically well suited for larger vessels, it can most likely expect to receive its share of them in the future.

Vessels are classified in three categories: freighters, tankers, and bulk carriers. Vessel trends within the Puget Sound area are based on U.S. Maritime Commission[20] estimates and other unpublished information from various government agencies.

Freighters. Freighters are defined as ocean-going ships, including container ships that carry general cargo. The number of freighters in the world merchant fleet has increased from 10,782 in 1956 to 11,611 in 1966. The number of freighters in the 10,000-10,999 dead weight ton class decreased by 1,090 vessels while the number of vessels in classes 11,000 dwt and greater increased.

The average size of freighters has increased from 7,600 dwt in 1956 to 7,800 dwt in 1966.[21] This is projected to an average size of about 10,000 dwt by 2020. The number of freighters in service and their size projections are plotted in Figures II-B10 and II-B11, respectively.

Tankers. The tanker fleet has increased from 2,834 in 1956 to 3,654 in 1966. During this same period, the number of tankers under 17,000 dwt decreased and those over 17,000 dwt increased.

Tanker size has increased from a maximum size of 35,500 dwt in 1954 to 205,000 dwt in 1966. Currently, 312,000 dwt tankers are in service with 500,000 dwt tankers scheduled for operation before 1975.[22] Studies are being made for 1,000,000 dwt tankers with a draft of 104 feet. Under present technology, this is about the maximum practical size.

Average sizes have increased from 15,100 dwt in 1956 to 26,900 dwt in 1966. This trend projected will result in the average tanker size of 100,000 dwt with an average draft of 48 feet by 2020. Current tanker fleet numbers and projected sizes are shown in Figures II-B12 and II-B13, respectively.

FIGURE II-B10

FREIGHTER SERVICE TREND - BY SIZE

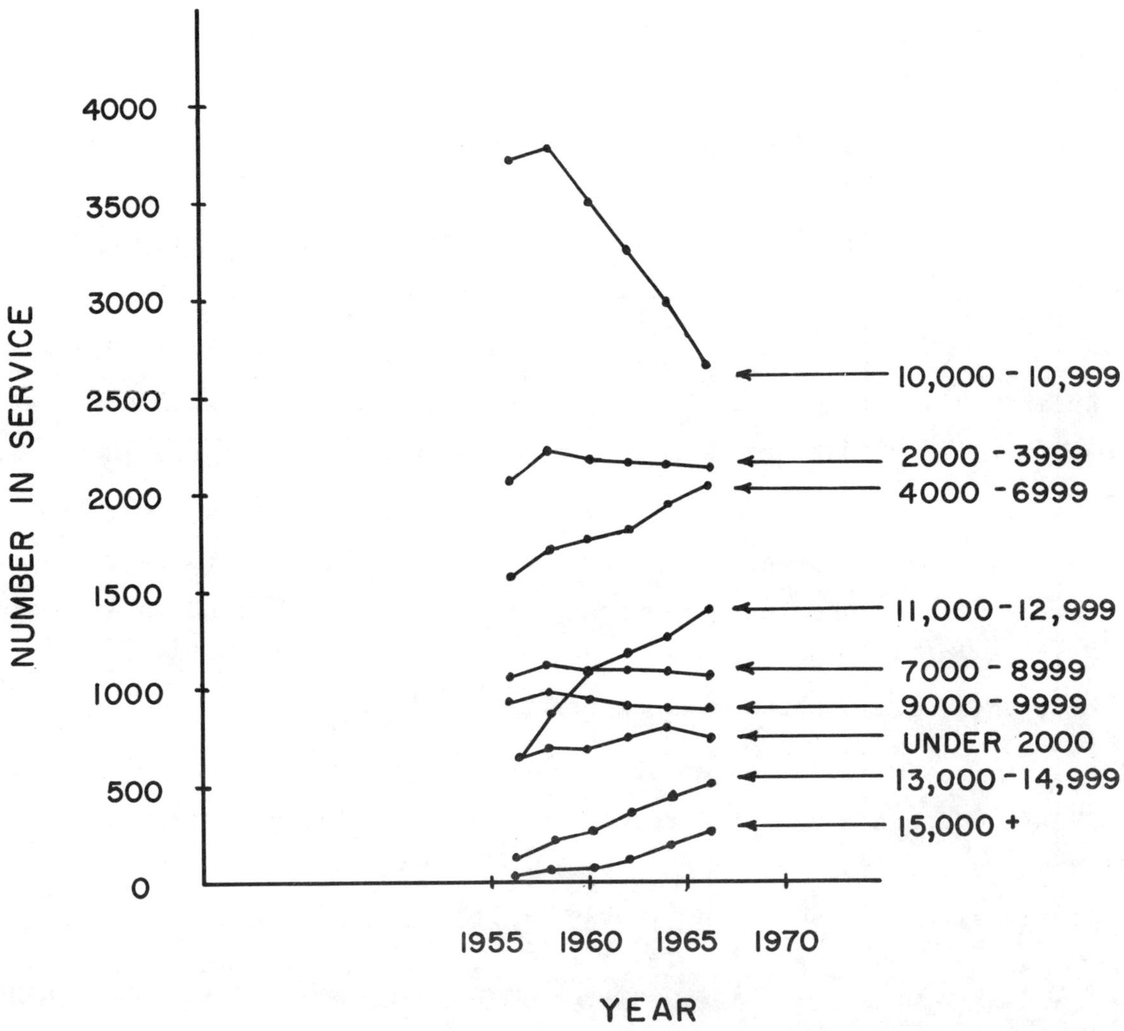

FIGURE II-B11

PROJECTION OF FUTURE FREIGHTER SIZE

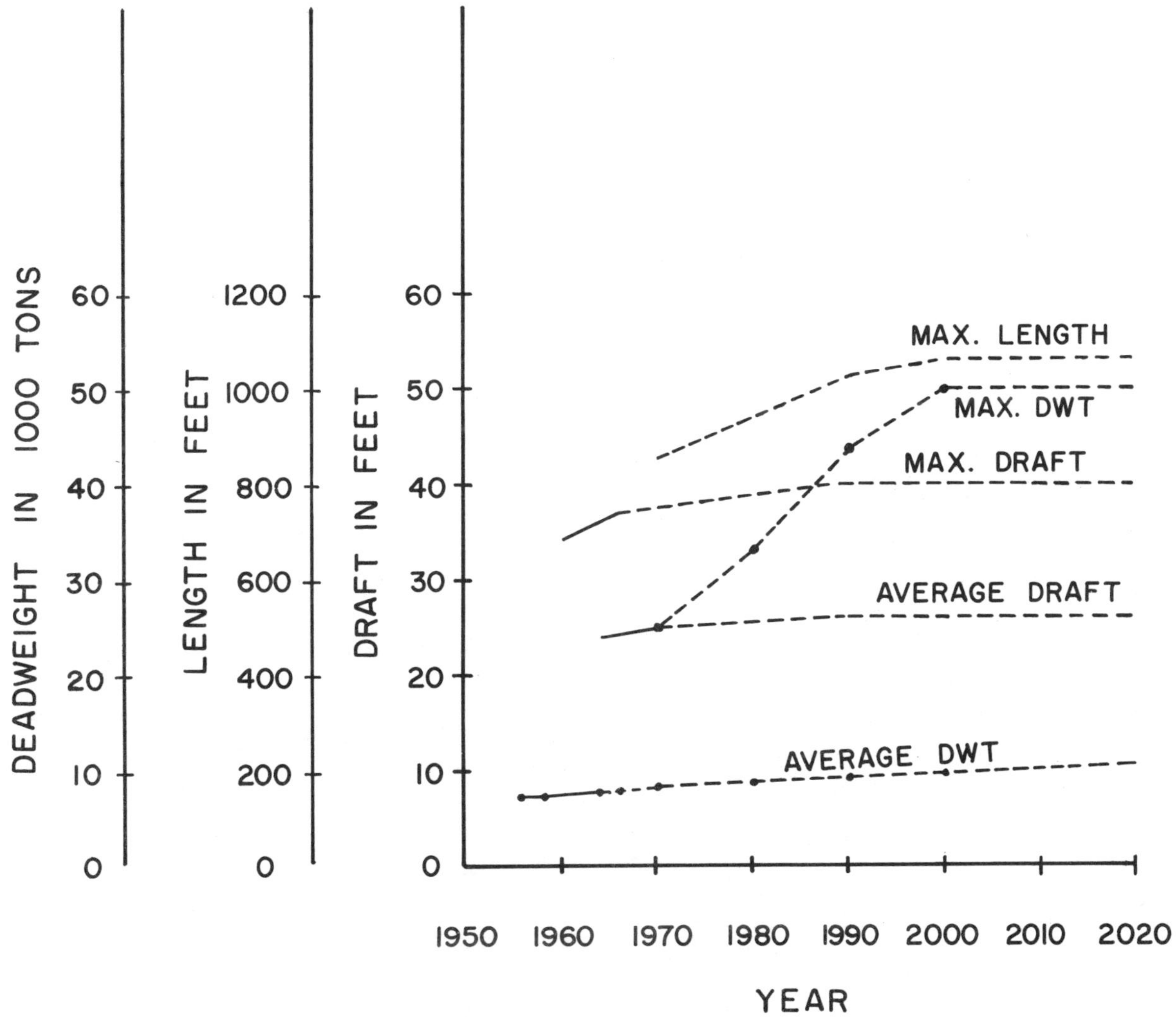

FIGURE IIB-12

TANKER SERVICE TREND - BY SIZE

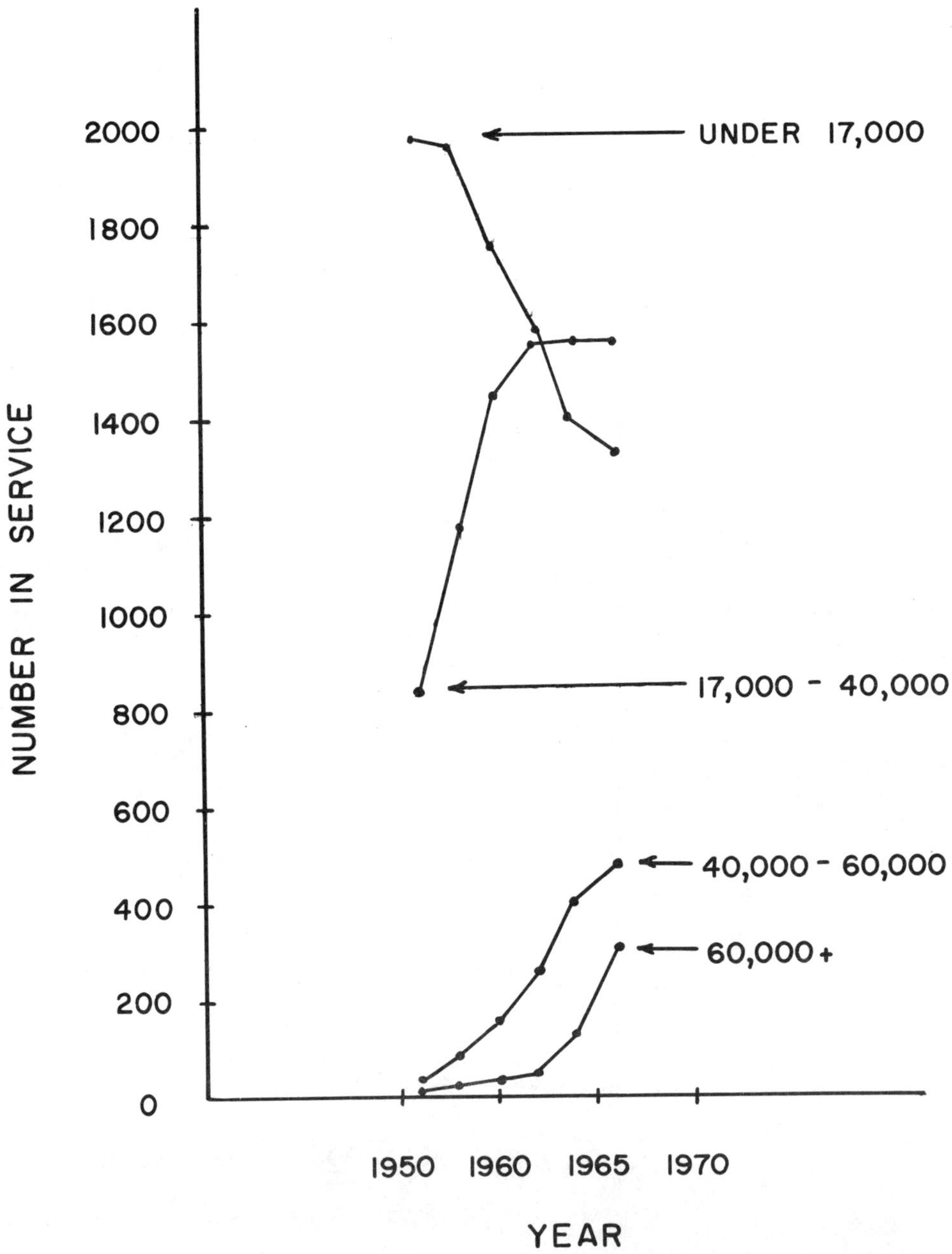

FIGURE II-B13

PROJECTION OF FUTURE TANKER SIZE

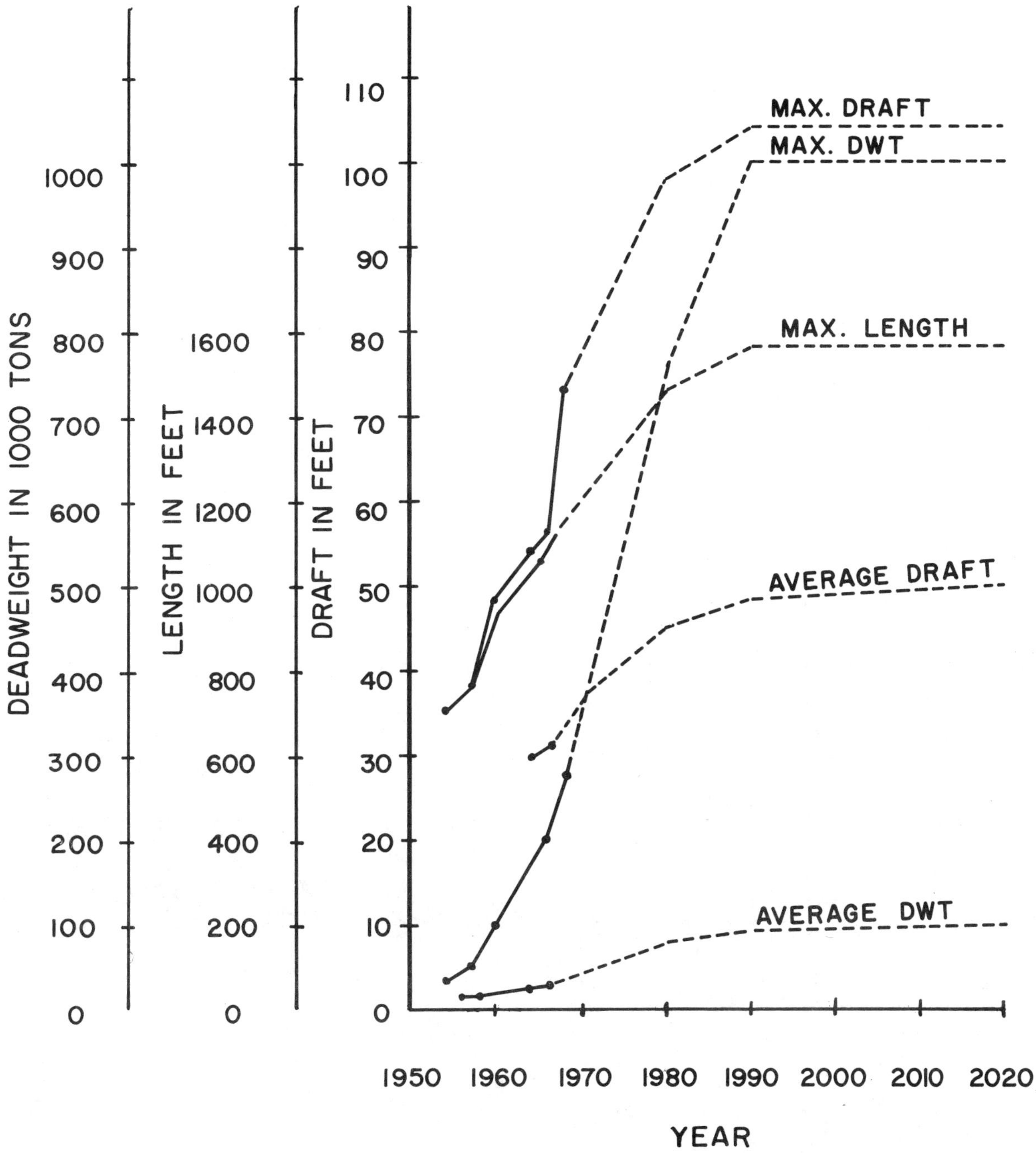

FIGURE II-B14

BULK CARRIER SERVICE

TREND - BY SIZE

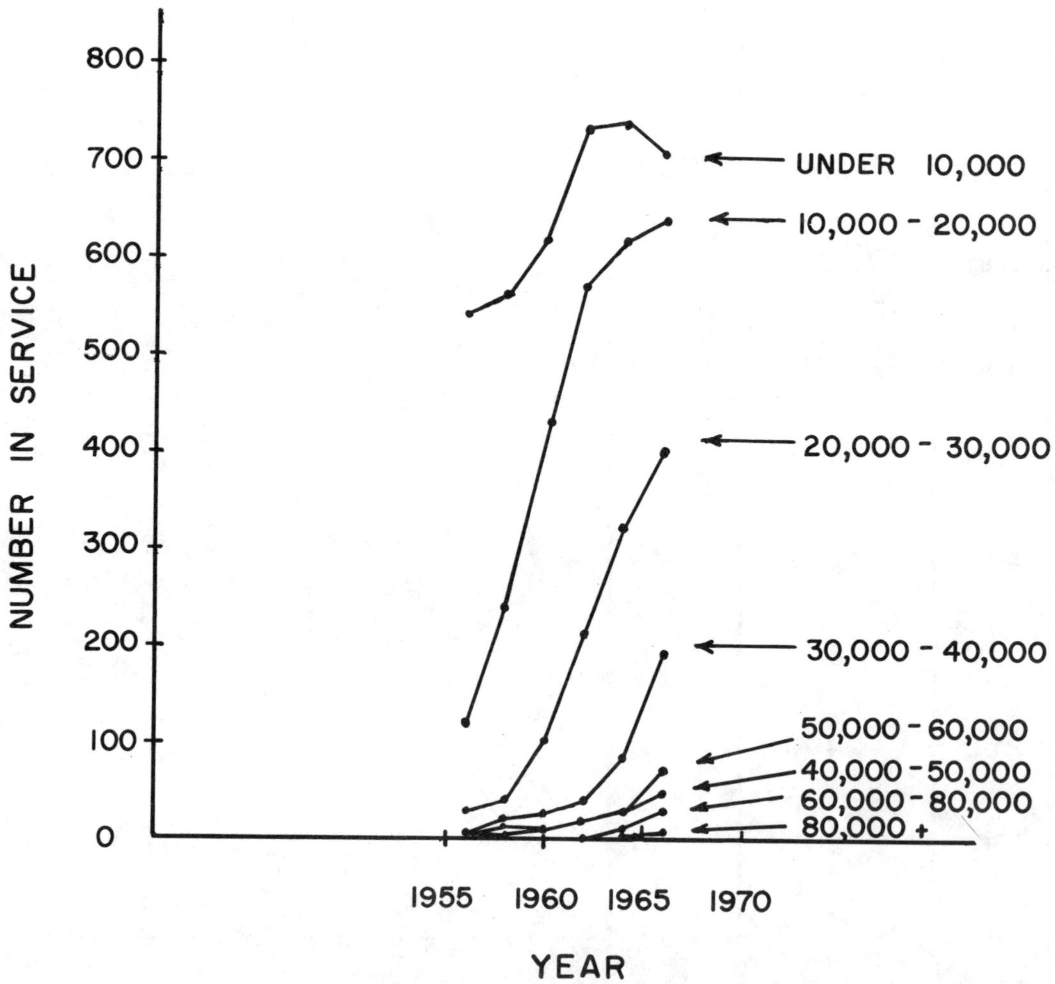

Bulk Carriers. These vessels include ore carriers, ore/oil carriers, and coal carriers. From 1956 to 1966, their number increased from 704 to 2,104. All sizes of these vessels have increased in number as shown in Figure II-B14, except for those under 10,000 dwt.

Bulk carriers could be built as large as tankers but because of the type of service and the physical limitation of the majority of ports, their practical maximum size is about 400,000 dwt. The projected average deadweight in 2020 is approximately 30,000 dwt as shown in Figure II-B15.

5. PROBABILITY OF A SHIP COLLISION

a. Introduction

Any solution to problems regarding the probability of a ship collision on Puget Sound will, at best, be approximate. The nature of the input data, the scope of the problem, the multitude of environmental variables - all these serve to limit the degree of accuracy of the solution. The answers presented must be recognized for what they are - the product of a mathematical model constructed with only a few vital parameters, subjectively evaluated and selected and founded on simplifying assumptions. Nevertheless, the solutions arrived at do represent an order of magnitude accuracy, and allow one to assess the relative probability densities in various areas of Puget Sound.

b. Model Description

The Puget Sound region has been broken into five separate channels as shown on Figure II-B16 and defined in Table II-B8. For the most part, the channels are defined to emphasize geographical constrictions in the area and to consider the primary shipping routes within the area. By applying this physical data plus vessel movement data for each channel to the prediction model some assessment of the relative danger of a shipping collision leading to a possible oil spill within the region can be assessed.

Complete development of the model and the solutions are included in Appendix 13 and will not be repeated here. Only the results of the model are shown.[23]

FIGURE II-B15

PROJECTION OF FUTURE BULK CARRIER SIZES

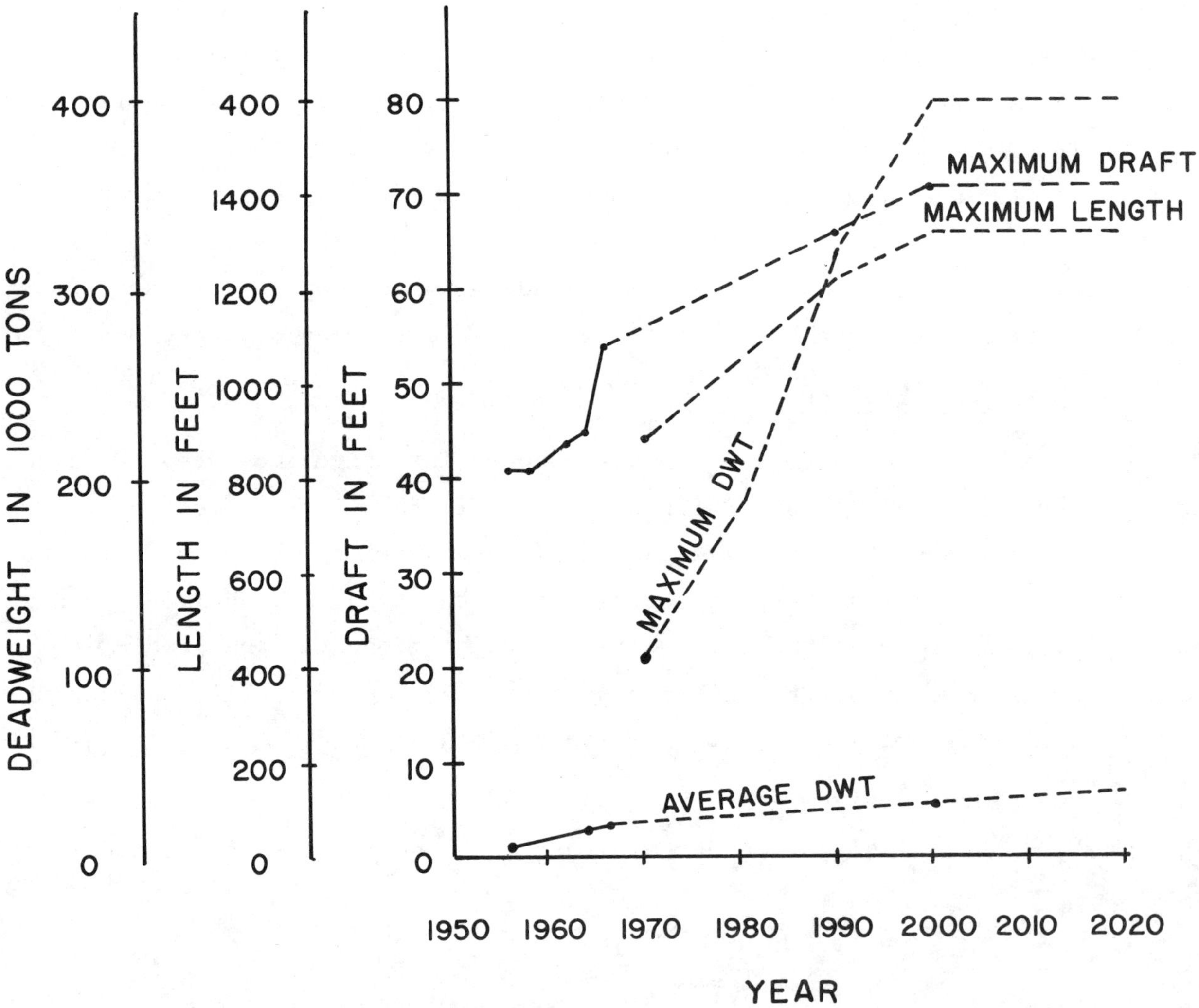

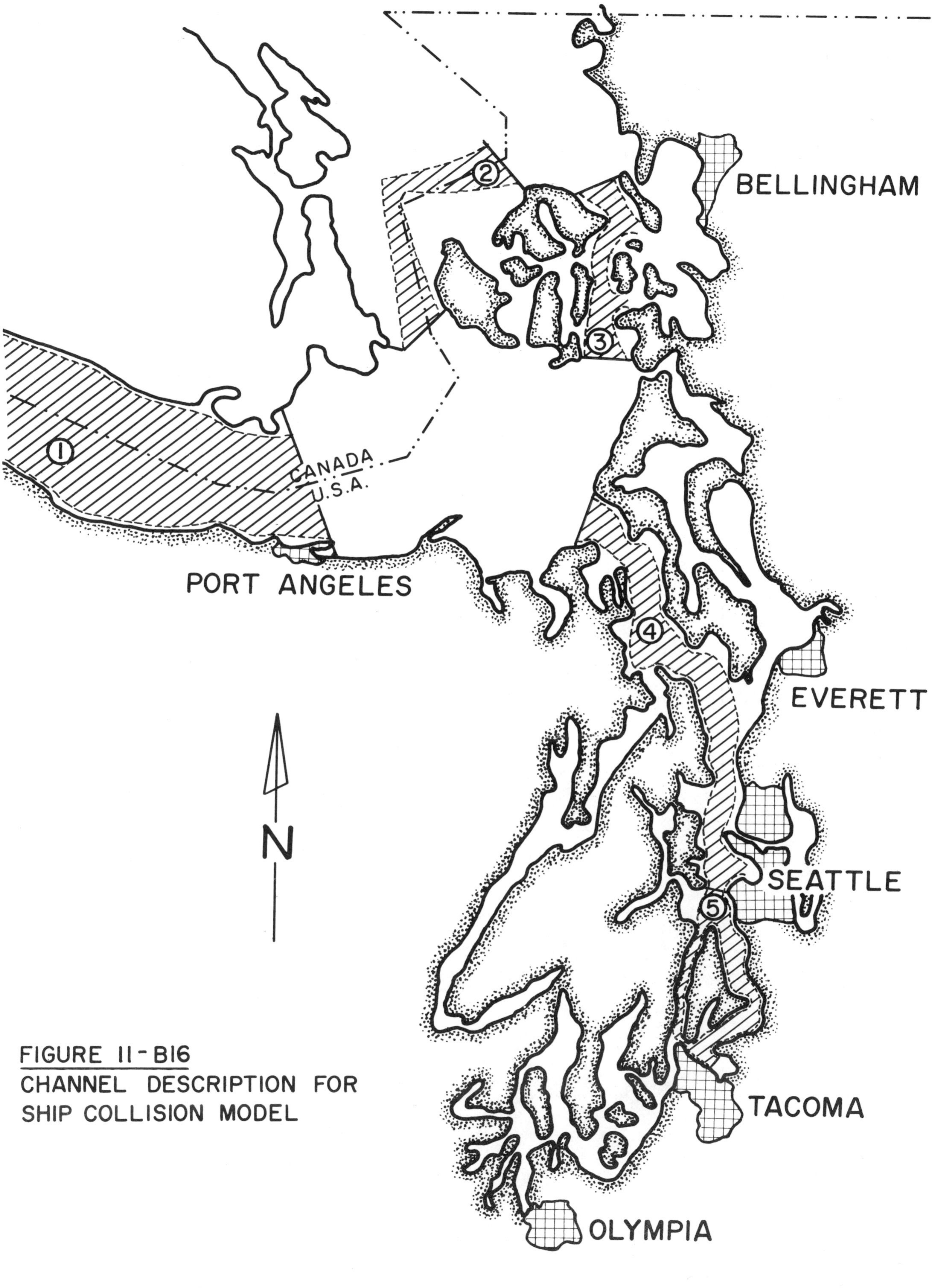

FIGURE II-B16
CHANNEL DESCRIPTION FOR SHIP COLLISION MODEL

FIGURE II-B17

PROJECTED VESSEL COLLISION WITH OIL SPILLS (FIVE CHANNEL SUMMARY FOR PUGET SOUND)

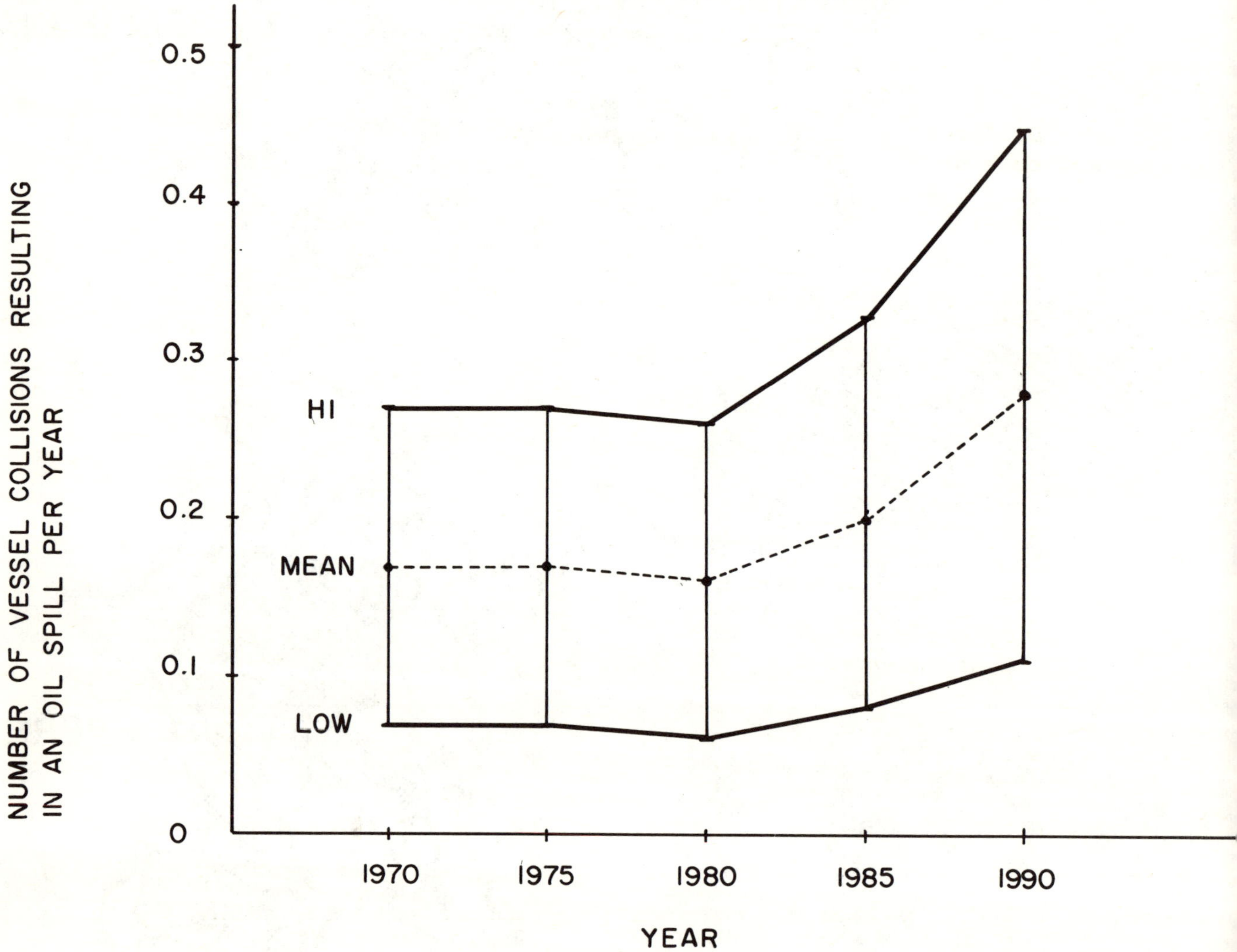

TABLE II-B8: CHANNEL DEFINITION FOR SHIP COLLISION MODEL

	Approximate Average Width (miles)	Approximate Average Length (miles)
Channel 1		
Cape Flattery-Bannilla Pt. to Dungeness Spit - Trail Islands	12.2	65.0
Channel 2		
Discovery Island - Cattle Pt. to East Pt. - Alden Pt.	4.2	28.9
Channel 3		
Pt. Colville - West Pt. to Lawrence Pt. - Village Pt.	3.19	18.0
Channel 4		
Pt. Wilson - Admiralty Head to Elliott Bay	3.45	39.0
Channel 5		
Restoration Pt.-Alki Pt. to Commencement Bay (Two Passages)		
Colvos Passage	1.01	17.3
East Passage	2.53	23.5

c. Model Results

The total number of collisions per year in Puget Sound resulting in an oil spill is shown in Figure II-B17. These figures were obtained by summing the annual expected values for the individual channels. Also, adding the expected value for any group of years will give the expected value for that period.

The expected number of oil spills in Puget Sound over the next ten years, 1970-1980, resulting from a vessel collision is predicted to range from a low of 0.67 to a high of 2.69 vessel collisions. For the period 1980-1990 the corresponding values are 1.07 and 3.56. These figures are obtained by summation of channel values. Table II-B9 shows, for the individual channels, the expected number of collisions resulting in oil spill:

TABLE II-B9: NUMBER OF COLLISIONS LEADING TO AN OIL SPILL (PER TEN YEAR PERIOD)

Channel number		Oil Spill Collisions Per Ten Year Period (1970-1980)	Oil Spill Collisions Per Ten Year Period (1980-1990)
1	Lo estimate	.152	.185
	Hi estimate	.594	.749
2	Lo estimate	.074	.093
	Hi estimate	.300	.374
3	Lo estimate	.074	.093
	Hi estimate	.300	.374
4	Lo estimate	.300	.375
	Hi estimate	1.200	1.500
5	Lo estimate	.074	.093
	Hi estimate	.300	.374

Clearly Channel 4 appears to have the highest chance of incurring an oil spill as the result of a vessel collision, much more so than the other channels as defined.

In Figure II-B16, two bodies of water on the Sound have been excluded from this model. These are the large bodies of water just east of the Strait of Juan de Fuca and the Strait of Georgia. The aggregate pattern of ships moving in these areas exhibits more of a random behavior than that shown in the designated channels. Thus, the Sperry model was felt not to be valid in analyzing these bodies. Instead, a free gas analogy model was applied. Unfortunately, well documented data on movements and collisions in these areas was unavailable. This, plus the fact that the predicted results for these bodies came from a different model, invalidate comparing the numerical values with those presented above. For the interested reader, this second model and its resulting probabilities are presented in Appendix 13.

ENDNOTES SECTION II-B

[1]For further explanation of these definitions, the reader is referred to: Waterborne Commerce of the United States, Part 4 - Waterways and Harbors, Pacific Coast, Alaska, and Hawaii, U.S. Army Corps of Engineers.

[2]For a further explanation of these definitions, the reader is referred to Dominion Bureau of Statictics, Shipping Report, 1967.

[3]Puget Sound Task Force, Puget Sound and Adjacent Waters, Appendix VIII, Navigation, Puget Sound Task Force, 1970, p. 1-13.

[4]Ibid, p. 2-4.

[5]Economic Evaluation, Kitsap Penninsula, Bainbridge Island Connector Bridge Routes, January 1969, Washington State Department of Highways.

[6]Figures based on summer 1970 ferry schedule, Washington State Ferries.

[7]Waterborne Commerce of the United States, Calendar year 1967, Part 4, U.S. Army Corps of Engineers.

[8]Ibid.

[9]Shipping Report, 1967, Part IV, Dominion Bureau of Statistics, Table 23.

[10]Ibid, Table 24.

[11]Puget Sound Task Force, VIII, p. 2-27.

[12]Ibid.

[13]Ibid., p. 2-28.

[14]Ibid.

[15]Ibid.

[16]Pacific Northwest Economic Base Study for Power Markets, U.S. Department of Interior, Vol. II., Part 11c, Petroleum, 1966.

[17]This value has been calculated by the American Petroleum Institute as their estimate, dated January 1971, it was quoted by Mr. C. O. Nickle of Calgary, Alberta in a speech at the Pacific Northwest Trade Association Convention, Vancouver, B.C. April 29, 1971.

[18]Puget Sound Task Force, VIII, p. 2-32.

[19]Puget Sound Task Force, VIII, p. 2-33.

[20]U.S. Maritime Commission, "A Statistical Analysis of the World's Merchant Fleets", 1956, 1958, 1960, 1962, 1964, 1966.

[21]Puget Sound Task Force, VIII, p. 2-43.

[22]*Ibid*.

[23]The model is based on analysis developed by Sperry Piedmont Company and reported in *Final Report, Lookout Assist Device Feasibility Studies*, Vol. I. August 1965, Maritime Administration Contract Ma-3374.

II-C. OIL HANDLING FACILITIES

1. INTRODUCTION

An extensive network of oil handling facilities has developed, over the years, to receive, transfer and distribute the enormous volume of petroleum products transported on Puget Sound. This section presents an overview on the operations of these facilities. To obtain first hand information, several members of the study team visited a number of locations. Comparative analysis of oil spill prevention and control techniques at these plants is deferred until Section III-C, Part 4, with detailed discussion on each site documented in Appendix 1. A location study of the significant oil handling facilities in and around Puget Sound is also included in this section to point out the relative geographic concentrations of oil operations.

Oil handling facilities are of two main classes, distribution centers and refineries, although some could perform both functions. The products they handle range from heavy fuel oil to diesel fuel, aviation fuel, and the various grades of gasoline. Appendix 14 includes a glossary of terms on the various petroleum products. A distribution center receives the refined products by barge, pipeline, railroad tank car, or tanker and stores the products in their storage tanks. Then, the products are distributed to various dealers and customers, either by pipeline (for example, there is a pipeline going from the Shell Oil Company's Harbor Island facility to Seattle-Tacoma International Airport) or by directly fueling ships at the distribution facility. As an example, all the Washington State Ferries are fueled at Shell's Harbor Island facility, and most other distribution facilities can bunker tugs. Two other methods of distributing the oil are by truck and by barge. One other type of facility is the marine service station, which fuels small pleasure boats.

For purposes of this study, the only difference between a refinery and a distribution facility is that the refinery only receives crude oil, and a distribution center receives refined petroleum products. Most of the oil handling facilities in Puget Sound are distribution centers.

2. DESCRIPTION OF DISTRIBUTION CENTER

The most noticeable feature of an oil handling facility is the tank farm, an area within a large retaining wall which houses the oil storage tanks. Typically, these tank farms can contain from 1 to 60 tanks, with a capacity of from 20,000 to 2,000,000 barrels per farm. The retaining walls are necessary to contain any spill or leakage from a tank. Pipelines feed from the tank, through the retaining wall, to the pier. Normally, there is a valve on the pipeline at the tank, and a manifold network of pipelines leading from the tank valve to the retaining wall. The manifold network of pipelines makes it possible to pipe one product per pipeline from the tank farm to the pier. Within the tank farm there is usually an oil-water separator system (a gravity separator). This is a cement- or steel-lined pit which is filled with waste oil, or filled by the storm drain system. The water, being heavier, sinks to the bottom of the pit and is passed through some oil absorption material on its way to the sound. The oil is skimmed off the top and either burned at the facility, sent back to the refinery, or sold to an oil reclaiming organization.

Another major feature of an oil handling facility is the pier, with its array of pipelines and valves. These pipelines, one per product, have a number of different valves on it. There usually is a valve located along the pipeline as it runs to the pier; this is to prevent oil spillage in the event of damage to the pier. At the end of the pipeline there is another valve called a header or riser valve. At this valve the oil transfer hose is connected. There can also be check valves (one way flow valves) in the pipeline and relief valves (to account for the effect of thermal expansion) around the other valves. Beneath the riser or header valves is a trough to catch oil dripping from the hose connection. Some facilities use drip pans or buckets for this purpose.

Since oil handling facilities transfer oil by means other than barge or tanker, there are fillstands for loading trucks, and railroad tank car handling areas. To keep these oil transfer areas supplied with oil, additional pipelines are needed, and to keep the oil flowing there must be pumps located along the pipeline. The pumps are either electric motor operated or steam powered. At or near the pumps there is normally another set of valves. Some valves are motor controlled which will allow the operator to close the valve very quickly. Also, there are remote control valves, which can be closed from a place other than at the valve. Some facilities have storage tanks uphill from the pier, making it possible to use gravity flow when filling a barge.

3. POTENTIAL ORIGINS OF OIL SPILLS

At first glance, the oil transfer procedure may appear quite simple: just connect up the hoses, open a few valves, and turn on the pumps. But, to do this safely and economically requires a rather rigid set of instructions which must be followed; otherwise, the direct consequences of an oil spill must be confronted. Five potential spill areas have been identified. These areas are:

1. The vessel receiving or discharging oil.
2. The transfer connections:
   a. the barge valves,
   b. the hoses,
   c. the header or riser valve.
3. The pier:
   a. the pipeline on the pier,
   b. the valves (relief, check or bleeder).
4. The pipeline between the pier and the tank.
5. The tank farm:
   a. the retaining wall,
   b. the pumps, and
   c. the manifold and valves.

To activate a comparative analysis of different oil transfer operations, eleven locations were visited. The selection basis was a combination of randomness and convenience. The sites visited were:

1. ARCO, Pier 11, Harbor Island.
2. ARCO Refinery, Cherry Point.
3. Manchester Naval Installation.
4. Mobil Oil Refinery, Ferndale.
5. Phillips Petroleum, Pier 33, Seattle.
6. Shell Oil Company, Pier 19, Harbor Island.
7. Shell Oil Refinery, Anacortes.
8. Standard Oil of California, Pier 32, Seattle.

9. Standard Oil of California, Point Wells, Richmond Beach.
10. Union Oil Dock, Pier 71, Elliott Bay.
11. Union Oil Company Pier, Edmonds.

The procedures at most of the facilities are very similar, and oil facilities have a lot of operations in common. Some of the similarities found within Puget Sound's oil industry are:

1. All facilities handle a variety of products and have their own special color code in addition to the universal color code. For instance, red painted valves are used for gasoline, but to distinguish between the various grades of gasoline the facilities have made slight changes in color to these valves on gasoline pipelines.

2. Night operations are common practice so the piers are lighted. These lights are the primary visual aid used at night to detect an oil spill, though none of them hang out over the water. No other means are available to detect an oil spill.

3. No weather problems occur except occasional adverse current conditions during docking.

4. The larger facilities have the capability to clean out ship's ballast tanks and to purge their fuel tanks, plus the ability to store the waste products. What to do with waste oil remains a problem at most facilities.

5. All facilities have similar valving systems:
    a. a tank valve,
    b. a header or riser valve,
    c. at least one more valve in between, and
    d. few remote controlled valves.

6. Hoses are used to make the connection between the barge valve and the header or riser valve.

7. Oil is transferred by pumping it, although gravity flow is possible.

8. Fuel lines are left full to counter the problem of pipe corrosion.

9. Few check-off sheets are used.

10. The requirements set forth by the Coast Guard are well followed.

Only two facilities in Puget Sound have the capability to do a reasonable job of cleaning up an oil spill. Both are located on Harbor Island in Seattle. The first is the Puget Sound Tug and Barge Company with their Marine Oil Pick-Up Service (MOPS). The second is the Shell Oil Company facility.

MOPS has made available their oil spill response plan to Puget Sound's oil industry,[1] while the Shell oil spill plan is designed to clean up a small spill on their dock, or resulting from damage to one of their trucks when on the road. Their oil spill plan is an adequate initial step. They have, on hand and readily deployable, the basic equipment needed, including 300 feet of an oil containment boom on their pier which can be deployed through rollers at the end of their pier, and a 16-foot boat to handle the boom and their absorbent material. Their plan to clean up any spill on the road is to rapidly deploy three pallets of cleanup equipment to the spill site.

4. GEOGRAPHIC DISTRIBUTION OF OIL HANDLING FACILITIES

There are 77 bulk oil handling facilities in and around the waters of Puget Sound based on a study by the Army Corps of Engineers.[2,3,4] These facilities receive and ship various petroleum products in addition to bunkering larger and smaller vessels. They are primarily located in the metropolitan areas of Bellingham, Everett, Seattle, Tacoma, and Olympia. Exception to the case are the Sandy Point, Cherry Point and March Point areas. Fueling of small vessels, especially pleasure boats is also available at marine service stations. There are over 100 such stations scattered throughout Puget Sound.[5]

Tables II-C1 through II-C6 list the oil handling facilities in each of the six subzones of Puget Sound, as defined in Section II-A of this report. For each subzone, the number of facilities, storage capacity, and important operation are drawn. These are pictorially represented in Figure II-C1 through II-C6. Marine service stations are shown separately on both the tables and figures.

Details of military and other federal government oil handling facilities were not available for this report. But, one publication listed five naval facilities in Puget Sound and these are noted on the figures. They are located in subzones 4, 5, and 6.[6]

TABLE II-C1: LOCATION OF OIL HANDLING FACILITIES - SUBZONE 1[2,3,4]

	Facilities Receiving Bulk Petroleum Products	Facilities Bunkering Vessels	Facilities with Barges for Bunkering Vessels	Capacity of Barges (BBL)	Facilities Fueling Small Vessels	Facilities that Consume all of Received Products	Total Capacity of Facility (BBL)	# of Marine Service Stations 5
Port Angeles	11	0	0	-	2	3	274,156	2
Neah Bay/Seiku	0	0	0	-	0	0	-	7
TOTAL	11	0	0	-	2	3	274,156	9

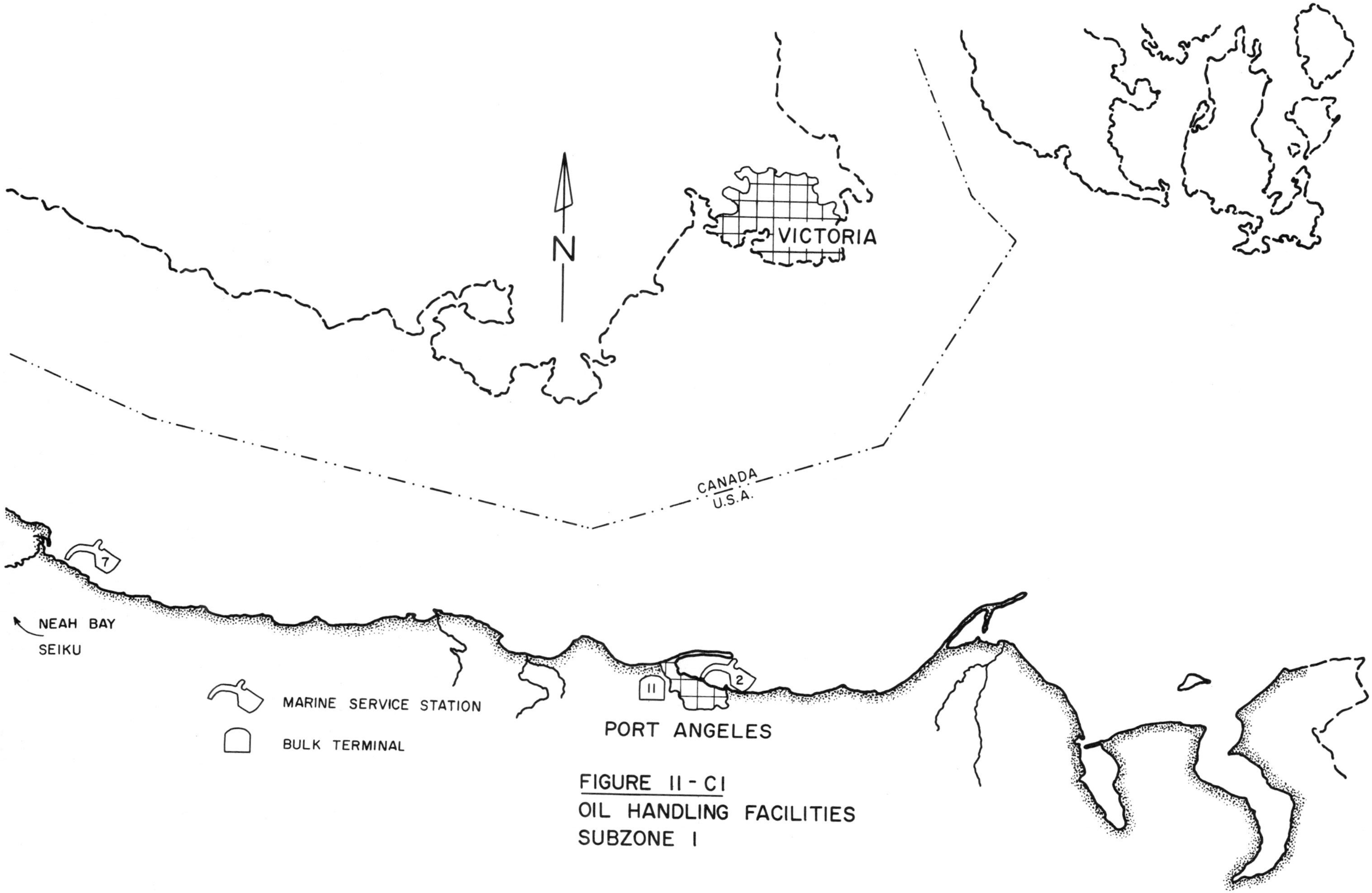

FIGURE II - C1
OIL HANDLING FACILITIES
SUBZONE I

TABLE II-C2: LOCATION OF OIL HANDLING FACILITIES - SUBZONE 2[2,3,4]

	Facilities Receiving Bulk Petroleum Products	Facilities Bunkering Vessels	Facilities with Barges for Bunkering Vessels	Capacity of Barges (BBL)	Facilities Fueling Small Vessels	Facilities that Consume all of Received Products	Total Capacity of Facility (BBL)	# of Marine Service Stations[5]
Birch Bay	0	0	0	-	0	0	-	1
Boundary Pass	0	0	0	-	0	0	-	1
Haro Strait	0	0	0	-	0	0	-	1
Strait of Georgia	1	1	0	-	0	0	2,370,000	4
Bellingham Bay	8	0	0	-	2	1	120,285	1
Guemes Channel/ Padilla Bay	5	2	0	-	3	0	6,621,300	3
Hale Passage	0	0	0	-	0	0	-	1
San Juan Channel	0	0	0	-	0	0	-	3
East Sound- Harney Channel	0	0	0	-	0	0	-	1
Wifidalgo Island	0	0	0	-	0	0	-	1
TOTAL	14	3	0	-	5	1	9,111,585	17

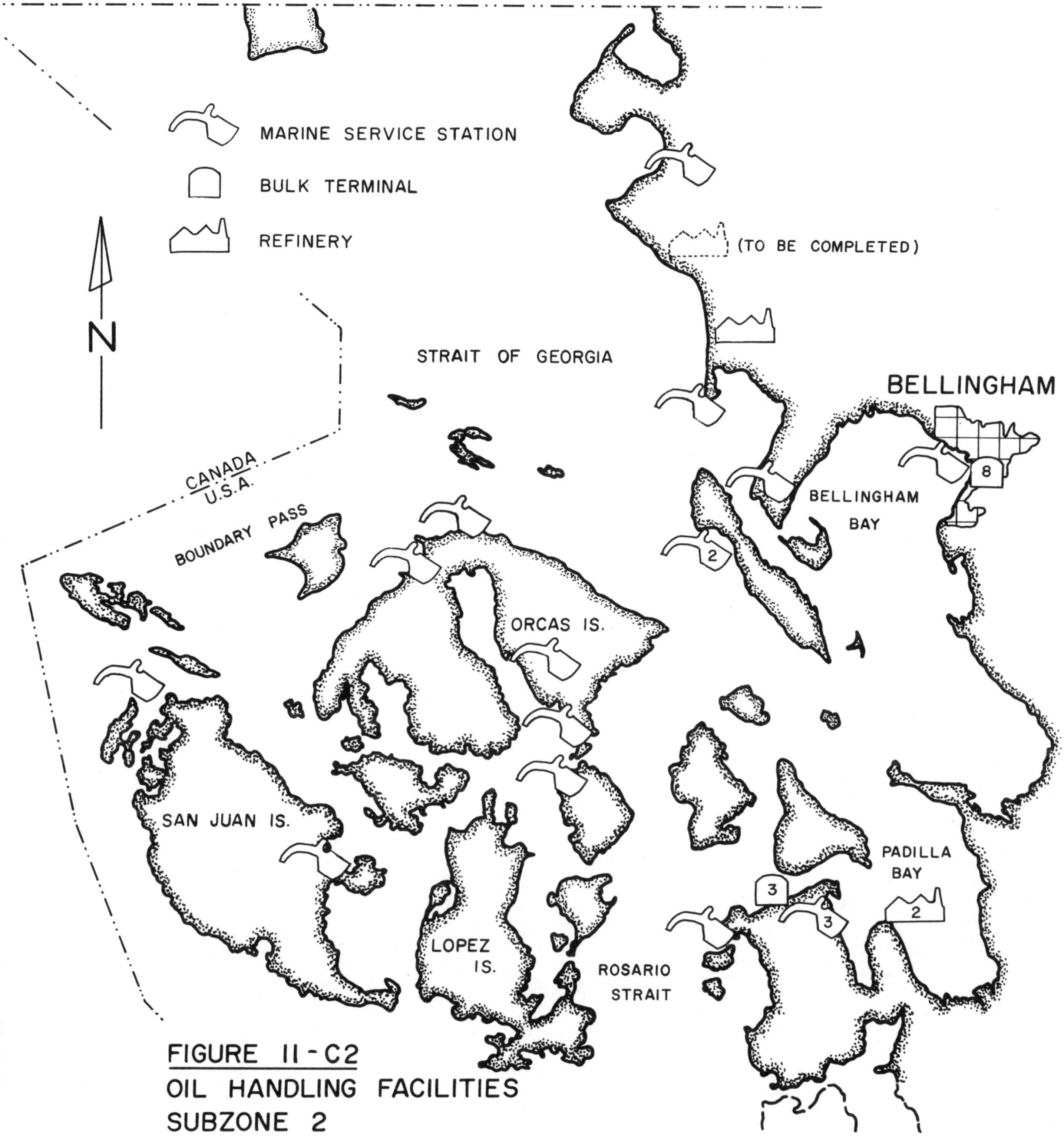

FIGURE II - C2
OIL HANDLING FACILITIES
SUBZONE 2

TABLE II-C3: LOCATION OF OIL HANDLING FACILITIES - SUBZONE 3[2,3,4]

	Facilities Receiving Bulk Petroleum Products	Facilities Bunkering Vessels	Facilities with Barges for Bunkering Vessels	Capacity of Barges (BBL)	Facilities Fueling Small Vessels	Facilities that Consume all of Received Products	Total Capacity of Facility (BBL)	# of Marine Service Stations[5]
Skagit Bay	0	0	0	-	0	0	-	2
Port Gardner	4	1	0	-	0	3	105,112	2
Deception Pass	0	0	0	-	0	0	-	1
Admiralty Inlet - East Side	0	0	0	-	0	0	-	1
Saratoga Passage	0	0	0	-	0	0	-	4
Possession Sound	0	0	0	-	0	0	-	3
Port Townsend	0	0	0	-	0	0	-	2
TOTAL	4	1	0	-	0	3	105,112	15

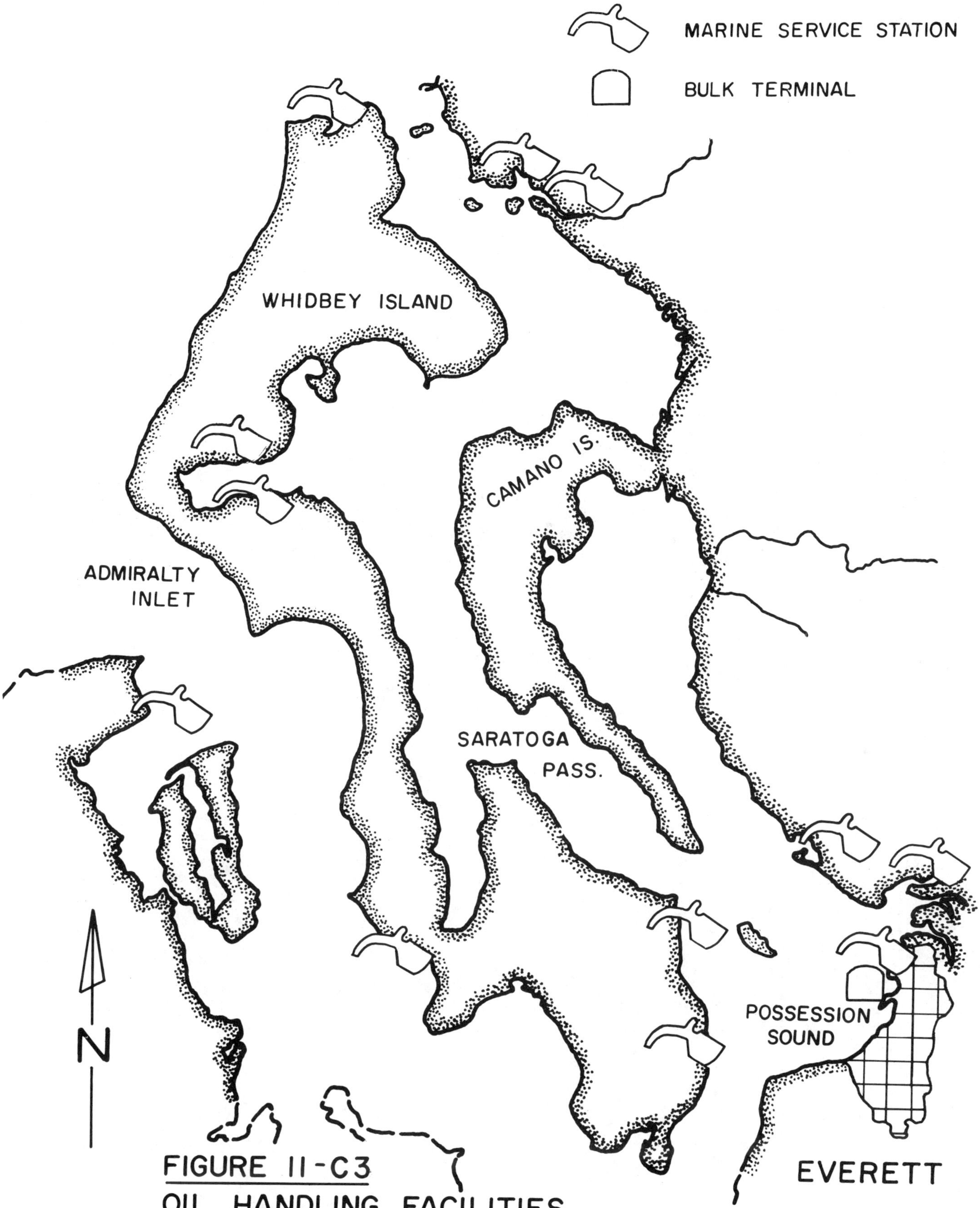

FIGURE II-C3
OIL HANDLING FACILITIES
SUBZONE 3

TABLE II-C4: LOCATION OF OIL HANDLING FACILITIES - SUBZONE 4[2,3,4]

	Facilities Receiving Bulk Petroleum Products	Facilities Bunkering Vessels	Facilities with Barges for Bunkering Vessels	Capacity of Barges (BBL)	Facilities Fueling Small Vessels	Facilities that Consume all of Received Products	Total Capacity of Facility (BBL)	# of Marine Service Stations[5]
Edmonds	3	2	2	18,870	0	1	2,890,600	2
Elliott Bay	1	1	1	9,435	0	0	139,400	0
Shilshole Bay	0	0	0	-	0	0	-	2
Salmon Bay	8	0	1	9,435	6	1	271,878	3
Lake Washington Ship Canal	2	0	0	-	1	1	62,655	6
W. Waterway	2	1	1	9,435	0	1	677,800	2
E. Waterway	6	5	5	47,175	1	0	2,790,829	1
Lake Washington	4	0	0	-	0	2	138,476	4
Sinclair Inlet	0	0	0	-	0	0	-	4
Dyes Inlet	0	0	0	-	0	0	-	1
E. Bainbridge Island	0	0	0	-	0	0	-	2
TOTAL	26	9	10	94,350	8	6	6,971,638	27

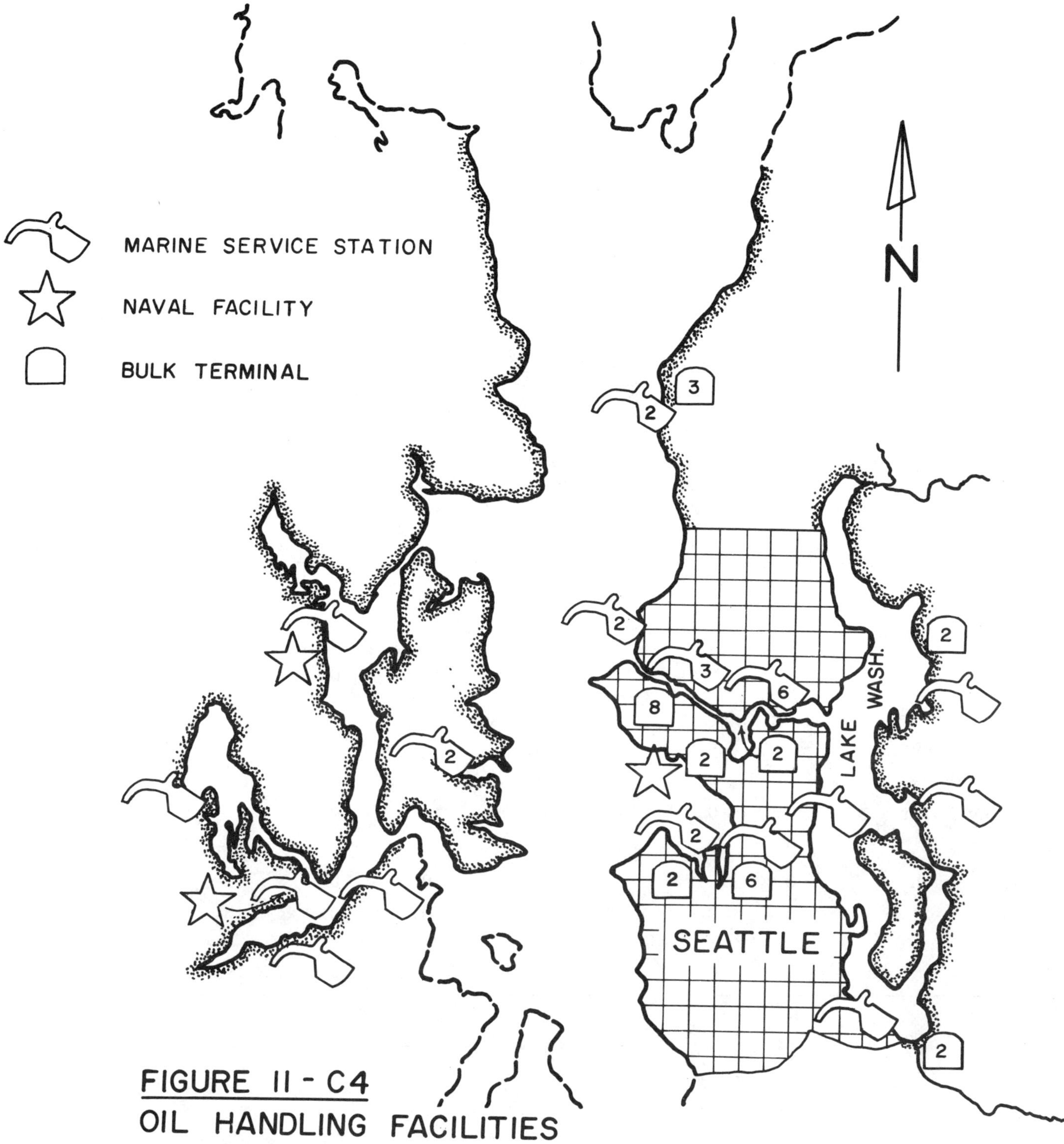

FIGURE II - C4
OIL HANDLING FACILITIES
SUBZONE 4

TABLE II-C5: LOCATION OF OIL HANDLING FACILITIES - SUBZONE 5[2,3,4]

	Facilities Receiving Bulk Petroleum Products	Facilities Bunkering Vessels	Facilities with Barges for Bunkering Vessels	Capacity of Barges (BBL)	Facilities Fueling Small Vessels	Facilities that Consume all of Received Products	Total Capacity of Facility (BBL)	# of Marine Service Stations[5]
Commencement Bay	16	0	3	26,700	4	6	2,697,339	10
Budd Inlet	6	0	0	-	3	0	135,943	7
E. Passage	0	0	0	-	0	0	-	1
S. Vashon Island	0	0	0	-	0	0	-	1
W. Side, Narrows	0	0	0	-	0	0	-	4
E. Side, Narrows	0	0	0	-	0	0	-	5
Hammersley Inlet	0	0	0	-	0	0	-	1
Case Inlet	0	0	0	-	0	0	-	2
Nisqually Reach	0	0	0	-	0	0	-	1
Carr Inlet	0	0	0	-	0	0	-	1
TOTAL	22	0	3	26,700	7	6	2,833,282	33

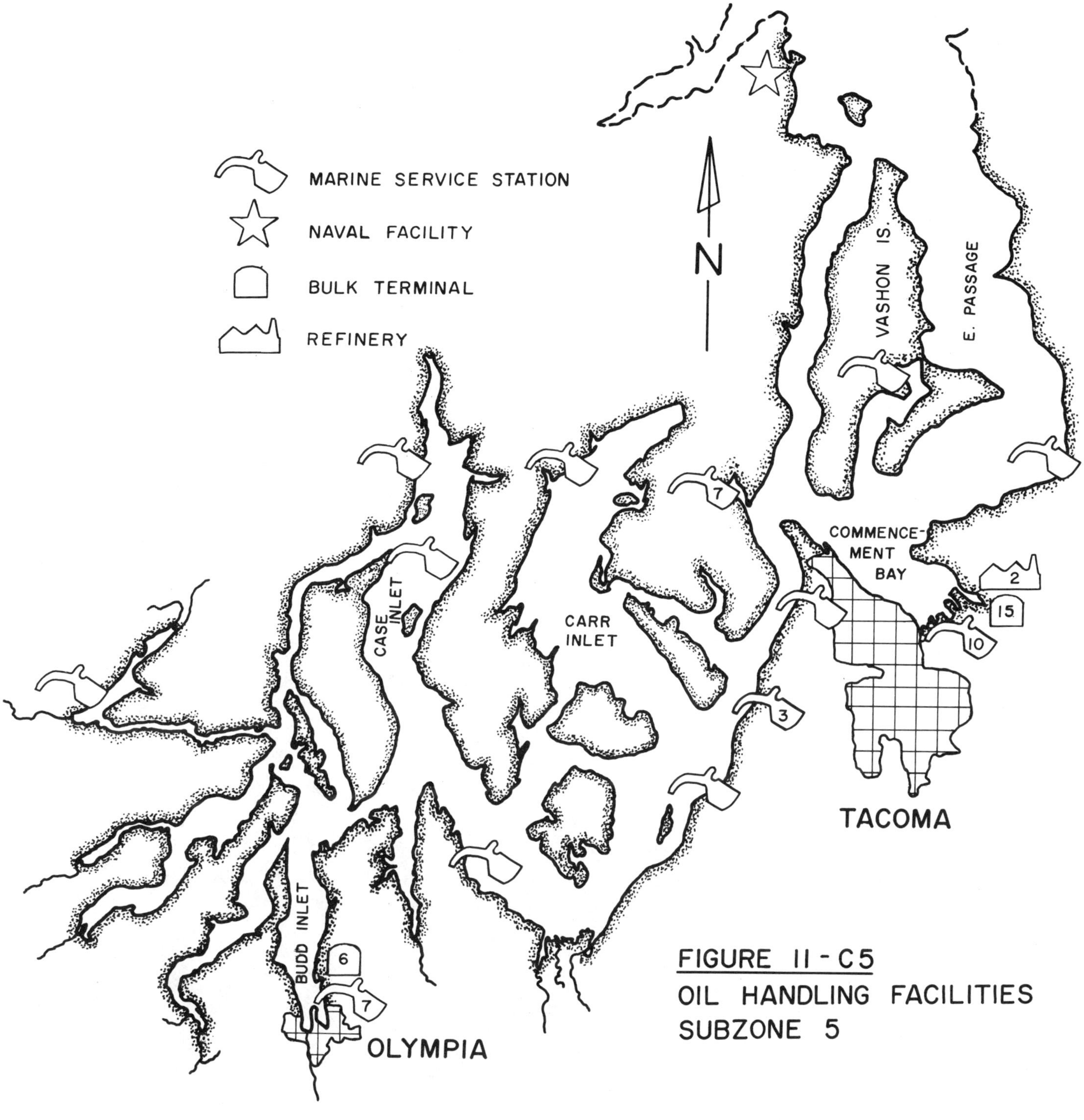

FIGURE II - C5
OIL HANDLING FACILITIES
SUBZONE 5

TABLE II-C6: OIL HANDLING FACILITIES - SUBZONE 6[5]
(All Marine Service Stations)

S. Arm Hood Canal	3
Central Hood Canal	1
N. Arm Hood Canal	2
Dabob Bay	2
TOTAL	8

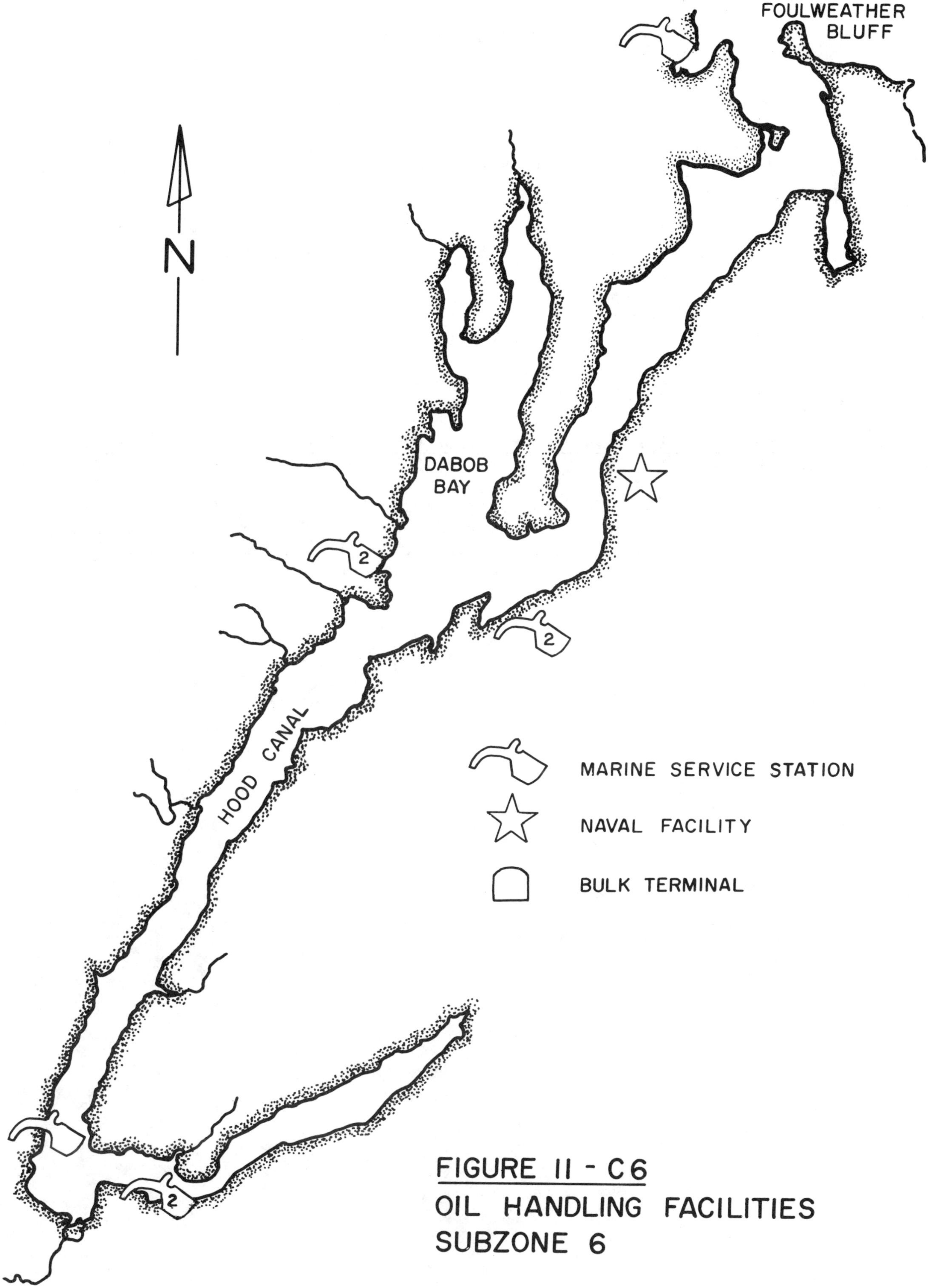

FIGURE II - C6
OIL HANDLING FACILITIES
SUBZONE 6

ENDNOTES SECTION II-C

[1]See Section III-B Part 2, for a critique on the MOPS plan.

[2]Corps of Engineers, The Ports of Port Angeles, Anacortes, Everett, and Bellingham, Washington, Port Series #37, Part 2, U.S. Government Printing Office, 1965, pp. 18, 40, 73, 107.

[3]Corps of Engineers, U.S. Army, The Ports of Tacoma, Grays Harbor, and Olympia, Washington, Port Series #35, part 2, U.S. Government Printing Office, 1964, pp. 40-41, 76, 96.

[4]Corps of Engineers, U.S. Army, The Port of Seattle, Washington, Port Series #36, Part 2, U.S. Government Office, 1964, pp. 93-95.

[5]Puget Sound Task Force, Pleasure Boating Study, Puget Sound and Adjacent Waters, 1968, pp. 26-33.

[6]Study of Equipment and Methods for Removing Oil from Harbor Waters, Battelle Memorial Institute, August 26, 1969, report no. CR.70.001, p. A-12.

[7]Port Series #35, p. 52, and Port Series #36, p. 119. For the Seattle area, the number of facilities having barges for bunkering vessels totaled six, and each barge had a different capacity. Since the deployment of these barges was unknown, the total storage capacity of the six barges was proportionally allotted to each facility.

II-D. ECONOMIC EVALUATION OF PUGET SOUND RESOURCES

1. VALUATION OF RESOURCES

Adverse effects of oil spills and runoffs of oil products into Puget Sound and adjacent waters can be considered from two points of view:

1. the adverse effect of oil to the natural environment, and
2. the adverse effects of legislation aimed at protecting the natural environment, i.e., adverse effects to industry and the Washington State economy.

The first is the most widely considered because of present day interest in the national and international ecology. The latter is not as widely espoused because of present attitudes toward business and the attendant stigma of modern capitalism. In considering the adverse effects and the welfare of the state, as well as the environment, it is important that one have an understanding of valuation pitfalls and inadequacies. It is also important that a realistic consideration be made of the needs of people as well as the needs of the natural environment.

One of the biggest pitfalls that people from both camps fall into is cardinal evaluations. No matter what research is used in analysis, no matter how sophisticated the techniques of measurement, no matter how many experts attest to the accuracy, there can be no finite evaluation for the following reasons:

1. The frame of reference used for evaluation is crucial in determining the relevance and accuracy of the evaluation. Those in industry will have varying viewpoints as they try to evaluate either from past profits or from future profits and the optimism or lack of optimism in predicting the demands of the future. Those in regulatory agencies will differ with all others as they either use figures, value to future generations, priority determinations based on expected, societal attitudes, their personal priorities and/or opinions. In turn, the citizens valuation of their ecology will also be dependent on what they like and what they use most frequently and/or believe is of value based on their personal background and subcultural heritage.

2. Because of changing dollar values, inflation, the dynamics of price to ever changing supply and demand relationships and technology changes that effect costs, durability, and aesthetic appeal of goods, it is extremely difficult to determine the replacement costs and sentimental value of existing assets and resources. The task of determining such a value for a given point in time would require Bayesian techniques on an individual and organizational basis over a vast range of considerations. The cost of such information far exceeds its value to society, using state-of-the-art statistical techniques.

3. In determining tomorrow's values, the most society has to go on are today's techniques, today's value, today's priorities, today's societal attitudes, and today's beliefs as to what tomorrow's society will be like. To say the least, we do not have the tools nor the talent for such a prognosis. Important here is the understanding that the accuracy of predictions decreases at an increasing rate as the analyst moves through time with his projections.

4. Though many valuations impress the observer and overwhelm him with complex and sophisticated data and manipulation techniques, it is questionable as to the increased accuracy of such techniques and their margin of increased benefit. Many can make a strong argument that the increased cost related to the increased accuracy is a misallocation of resources.

The above reasons are cited so that the reader might get a feel for the importance of using ordinal concepts rather than cardinal concepts when dealing with value and so that care will be taken not to believe, as the gospel, the finite figures in dollars or quantities that imply accuracy.

In trying to live up to present day society's responsibility to the future generations, it is important that leaders recognize the needs of people and the burden of the future that will be placed on societal resources. Recognizing that power is derived from the constituents in our society, one must consider expected population increases and attendant demands. Table II-D1 compares the population of 1960 in the Puget Sound and adjacent waters region

TABLE II-D1

POPULATION TRENDS AND FORECASTS PUGET SOUND AREA: 1960 and 1980[1]

Area	1960 Enumerated Population	1980* Forecast Population
STATE	2,853,814	4,086,385
Puget Basin	1,768,117	2,582,910
King-Kitsap-Pierce-Snohomish	1,512,979	2,231,967
Economic Area 1	128,477	163,383
Clallam	30,022	40,398
Grays Harbor	54,465	69,230
Jefferson	9,639	11,562
Mason	16,251	21,048
Pacific	14,674	17,066
Wahkiakum	3,573	4,141
Economic Area 2	144,177	197,742
Island	19,638	35,414
San Juan	2,872	3,145
Skagit	51,350	69,871
Whatcom	70,317	89,347
Economic Area 3 (Kitsap)	84,176	110,405
Economic Area A (King)	935,014	1,355,544
Economic Area B (Pierce)	321,590	467,683
Economic Area E	172,199	299,174

*Forecasts from the source were stated as a high, medium, and low figure. For ease of handling only medium forecast figures were extracted.

with expectation for 1980. The important observation here is not to the exact numbers but to the relative change and to what areas this change will take the greatest toll. In essence, the population of the Puget Sound region is to almost double in the twenty year period. This means that consumer demands for durable and non-durable products will also double. This means that demands for natural resources by residents will also double. Instead of a little less than one million jobs in the region, there will need to be about 1.8 million jobs. The condition of jobs is closely related to demand for natural resources in the area of recreation because ongoing trends indicate that the work week will continue to decrease as disposable income increases; a condition that has grave implications for camping, fishing, hunting, scuba diving, sailing, motor boating, bicycling, and the myriad of other activities that require products and a scenic environment rich in facilities and natural life forms.

Senator Warren G. Magnuson, Washington Senator and Chairman of the Senate Committee on Commerce, has placed a valuation on Puget Sound of $500 million.[2] Just as accurately, one could claim that because 50% of the state's economic activity takes place in this region and because the Port of Seattle enhances the value of the region to industry, the value can be placed at 40-50% of the state's GNP, which was $16 billion in 1968. Here then is a valuation that ranges between $6.4 and $8.0 billion. Both valuations may be correct. Interested parties may disagree on the exact figures but there is one item all will most likely agree upon; the Puget Sound and adjacent waters region has an astronomical value.

2. MAJOR RESOURCE GROUPS

In the present debate there appear to be three major groups trying to protect their individual interests and their responsibility to society as they see it:

1. the fisheries industry and those private and government groups involved or dependent on the industry,
2. the tourist and conservation groups of both private citizens and government agencies, and
3. the manufacturing industry (primarily the oil industry) and those dependent on such an industry.

At present, it appears that these groups of concerned parties have taken absolute stands of having things completely their way or else. This approach seems almost ludicrous when one considers the educational and cultural level of our segment of civilization, except for the fact that each group is made up of people, most of whom live in a very narrow world. As an aside, it might be worthwhile if each and every group took on an advocacy position of their opponents and attempted to learn, understand, and defend the position they had previously opposed with vigor and enthusiasm.

The fishing industry people have more than just economic value on their side; they also have the societal attitude of approval and support for the harvesting functions of our society. It is unique that as the job quantities decrease and output increases in the fishing and farming industries and the political base diminishes in shear numbers, the political influence remains constant or increases. The Puget Sound region has greater importance to the national fishing and shellfish industry than it does to the Washington State economy in a purely commercial dollar sense. In the sport fishing area, the impact is greater as this aspect is closely tied with tourism. Tables II-D2 through II-D7 give examples of evaluations of the commercial and sport fishing industry as well as a breakdown of the catch value and processed value of the 1968-69 commercial activities for the State of Washington as well as a breakdown of the Puget Sound catch and its contents. As one can readily see, the value is not small and the intent here is not to belittle an industry but to bring in as pragmatic an approach as possible to the tasks that face all of us. Without a doubt, the Puget Sound region is one of the richest fishing grounds in the nation and for that consideration alone, prime importance must be given to the national and international responsibility of the local citizenry to preserve such a resource.

Tourism, on the other hand, is not an established industry in the sense that fishing is because this industry is an outgrowth of shorter work weeks and higher disposable incomes. Tourism is not only an ally of the fishing industry but also the conservationists, sportsmen, and environmentalists; a diversified group to say the least. While tourism is presently the fourth largest industry in the State of Washington and the Pacific Northwest, it is surpassed only by food, manufacturing, and forest products; it is a growing

TABLE II-D2: SUMMARY OF SALMON SPORT FISHING VALUES[3]

Recreational Salmon Fisheries

a.	Cross annual expenditures by sport salmon fisherman	$40,000,000
	1. Direct fishing costs . . . $20,000,000	
	2. Indirect fishing costs . . $20,000,000	
b.	Addition to State incomes by salmon fishing expenditures of out-of-town anglers	5,000,000
c.	Net annual economic benefits	65,000,000

Non-salmon Marine Recreational Fisheries

a.	Shellfish-minimum annual net economic benefit	1,560,745
b.	Finfish-minimum annual net economic benefit	203,140
	Total	$1,763,885

Capital Investment in Sport Salmon Fishing

a.	Investment in charter boat and salmon charter facilities	$24,000,000
b.	Investment by fisherman (Gross annual expenditures includes depreciation costs, etc., on boats and equipment)	
	1. Annual expenditures on salt water fishing equipment	12,630,000
	2. Investment in private boats by sport salmon fisherman prorated against use for salmon fishing	102,950,000[A]
		$139,950,000

[A]Based on 106,000 Coast Guard registered vessels 16-23 feet and employing 10 h.p. engines or over. The assumption is that all such vessels are used in sport salmon fishing and that part of their estimated value can be prorated to salmon fishing.

TABLE II-D3: SUMMARY OF COMMERCIAL FOOD FISH VALUES[4]

Commercial Valuation - 1966	
Wholesale Value - Washington Landings	$40,709,568
Value added to imported products	10,923,143
Total	$51,632,711
Disposition of Wholesale Receipts	
Value of catch to fisherman	$21,169,988
Wages and salaries of processors and handlers . . .	14,150,000
Taxes (other than income taxes)	1,350,000
Other expenses, income taxes and returns to owners	14,850,000
Total	$51,519,988
Net addition to state incomes - 1966	$46,180,000
Net economic yield - 1966	$10,000,000
Capitalized value of income to state residents from commercial fishing	$877,420,000
Capital Investments in Commercial Fishing Industry	
Vessels and gear	$92,000,000
Plants - equipment and inventory	32,000,000
Oyster lands	15,000,000
Total	$139,000,000

NOTE: Factors are not additive. Annual values relate to wholesale value, net additions to state income and net economic yield.

TABLE II-D4: PUGET SOUND FISHERY VALUES[5]

Commercial and Sport

	Direct		Indirect	
	Commercial	Sport	Commercial	Sport
Salmon:				
Chinook	$99,456	-	$314,530	-
Coho	470,307	-	215,557	-
Pink (odd yr)	753,085	-	765,546	-
Sockeye	68,340	-	248,986	-
	$1,391,188	$15,742,048	$1,544,619	$6,640,676
Other Anadromous Fish	-	3,053,800	-	2,296,450
Freshwater Fish	-	1,367,130	-	1,820,100
Shellfish:				
Clams	$31,812	-	$240,385	-
Oysters	304,972	-	4,344,670	-
Crabs	48,731	-	552,947	-
Shrimp	4,791	-	24,912	-
Octopus	197	-	10,548	-
Squid	8	-	1,071	-
	$390,511	$42,750	$5,174,533	$86,700
Other Fish:				
English Sole	$96,778	-	$136,130	-
Rock Sole	1,582	-	2,896	-
Flounder	8,700	-	29,538	-
Other Flatfish	196	-	4,881	-
Pacific Cod	7,527	-	171,531	-
Lingcod	760	-	26,401	-
Rockfish	1,406	-	5,567	-
Smelt	10,543	-	11,355	-
Other Food Fish	4,467	-	11,766	-
	$131,959	$55,755	$399,966	$14,745
Herring/Reduction	$50,927	-	$113,696	-
Scrap Fish/Animal Food	116,809	-	119,285	-
	$167,736	-	$232,981	-
TOTAL - ALL FISH	$2,081,392	$20,261,483	$7,352,099	$10,858,671
BY AREA	$22,342,875		$18,210,770	
PUGET SOUND	$40,553,645			

TABLE II-D5: PUGET SOUND COMMERCIAL FISH LANDINGS - 1969 (lbs.)[6,7]

Species	North Puget Snd.	Bellingham Bay	Central Puget Snd.	Port Susan	South Puget Snd.	Strait of Juan de Fuca	Hood Canal	TOTAL
English Sole	431,326	7,121	1,058,397	250	1,660	133,689	369	1,632,812
Rock Sole	19,013	1,174	2,683	-	72	16,549	590	40,081
Petrale Sole	180	-	-	-	-	2,499	-	2,679
Douck Sole	765	-	22,385	-	-	-	-	23,150
Starry Flounder	385,378	31,550	101,166	-	22,870	13,409	-	554,373
Sand Sole	6,539	3,426	482	-	13,007	10,981	-	34,435
True Cod	843,517	39,188	41,455	34	-	701,449	945	1,626,588
Ling Cod	71,949	1,321	497	-	-	15,602	-	89,369
Rockfish	2,065	61	5,696	1,100	2,381	43,227	121	53,551
Misc.	12,556	-	3,264	-	1,700	11,763	-	30,383
Food Fish Subtotal	1,773,288	83,841	1,236,025	1,384	381,997	949,168	2,025	4,087,421
Reduction	1,630,039	1,323	701,100	7,134,135	-	949,168	-	9,519,996
Animal Food	803,579	47,527	1,059,681	513,043	-	53,399	7,483	2,899,115
TOTAL	4,206,906	132,691	2,996,806	7,648,562	423,687	85,805		16,506,532

In 1969, 22,385,814 pounds of salmon were caught commercially in Puget Sound, worth $8,209,018 to the fisherman.

TABLE II-D6: SPECIES COMPOSITION OF 1969 REDUCTION AND ANIMAL FOOD LANDINGS - (lbs.)[8]

REDUCTION

Species	North Puget Snd.	Bellingham Bay	Central Puget Snd.	Port Susan	South Puget Snd.	Strait of Juan de Fuca	Hood Canal	TOTAL
Turbot	13,000							13,000
Unspecified	1,209,679					49,047		1,258,726
Pacific Hake	35,115		497,700	7,134,135				7,666,950
Ratfish	174,563		165,180			4,352		344,095
Dog Fish	189,267	1,323	38,220					228,810
Mud Shark	500							500
Skate	7,915							7,915
TOTAL	1,630,039	1,323	701,100	7,134,135	---	53,399	--	9,519,996

TABLE II-D6: SPECIES COMPOSITION OF 1969 REDUCTION AND ANIMAL FOOD LANDINGS - (lbs.)[8] (Cont'd)

ANIMAL FOOD

Species	North Puget Snd.	Bellingham Bay	Central Puget Snd.	Port Susan	South Puget Snd.	Strait of Juan de Fuca	Hood Canal	TOTAL
Butter Sole	6,164							6,164
English Sole	240		464,069	371	381,864	12,710	1,211	860,465
Starry Flounder	313							313
Turbot	11,907	8,922				1,964		22,793
Unspecified	568,512	37,754	4,550			11,459		622,275
Tom Cod	36,995	851						37,846
Pollock	179,448							179,448
Pacific Hake			106,000	512,100				618,100
Ratfish			426,063	572		54,270	5,325	486,230
Sculpin			800					800
Dogfish			57,999			5,402	947	64,348
Skate			200					200
Rockfish					27			27
Sea Perch					106			106
TOTAL	803,579	47,527	1,059,681	513,043	381,997	85,805	7,483	2,899,115

TABLE II-D7: ECONOMIC VALUE TO COMMERCIAL FISHERMAN - 1969[9]
(ex vessel)

Species	Pounds Caught	Value/Pounds ($)	Total Value ($)
English Sole	1,632,812	.09	147,000
Rock Sole	40,081	.08	3,200
Petrale Sole	2,679	.125	334
Dover Sole	23,150	.073	1,710
Starry Flounder	554,373	.06	33,200
Sand Sole	34,435	.1250	4,320
True Cod	1,626,588	.065	106,000
Ling Cod	89,369	.065	8,790
Rock Fish	53,551	.065	3,490
Misc. Species	30,383	.06	1,970
Food Fish - TOTAL	4,087,421	---	310,000
Reduction	9,519,996	7.06*	383,000
Animal Food	2,899,115	7.06*	106,000
TOTAL	16,506,532	---	$800,000

*assume .04 as average

Processed value of food fish is approximately 3 times the ex vessel price.

industry. The tourist industry is growing at an increasing rate while industries such as lumber and wood products are decreasing in magnitudes of jobs and output, as shown in Table II-D8. The industry, according to the Puget Sound Task Force Report, will probably move to the number three position in the Pacific Northwest by 1985 and the number one position by the year 2000. In 1968 there were 9,100,000 travelers (Table II-D9). According to a 1969 Battelle Northwest study, there were 9.6 million travelers who expended some $375 million in the state in that year, of which 50% was in the Puget Sound region.[12] By the year 1980, there is an expectation of 15 million visitors expending $614 million (see Table II-D10) and affecting the job situation by an increase of some 24,000 jobs, up to 64,000 jobs (direct, indirect, or induced) for a gross output of $1 billion. Other facts that give the industry potential muscle in decision making on matters such as the oil problem are that every dollar spent by a traveler generates two dollars of business activity through the multiplier effect, and that vacationers and visitors of friends and relatives make up the largest percentage of non-resident travelers. According to the Task Force Report, 30% of the state's residents participate in sport fishing.

Table II-D10 has a chart of non-resident traveler expenditures. It might be interesting to note that vacation expenditures are growing at an increasing rate. Thus, from an industry point of view, local resources are extremely valuable to this generation and more so to future generations because it is dependent not on production but on degrees of primitiveness and preservation of the Puget Sound in its entirety, from Admiralty Head through the San Juan Islands, rich in life forms and extremely sensitive to contamination by waste and exploitation down to Jones Beach outside of Olympia. It should also be noted that a complement to the tourist industry is the motor and sail boating industry. In the Puget Sound, 180,000 of the 223,000 boats owned in the state are owned by residents of the Puget Sound and adjacent waters area and that 30% of those residents outside the area bring their craft into the area to reap the benefits of Puget Sound.[14]

The third category is industry. Whatever people would like to see, the truth of the matter is that Puget Sound is a natural deep water port and the economy of the state is intricately tied to this condition. Table II-D11

TABLE II-D8: TOTAL EMPLOYMENT AND OUTPUT BY INDUSTRY (1963 and projected 1980) PUGET SOUND AREA[10]

(Employment in 1000's; sales in millions of dollars)

Industry	Employment 1963	Employment 1980	Output 1963	Output 1980
Agri., For., Fish., and Mining	23.7	18.2	196.7	261.1
Food and Kindred Products	15.9	19.5	698.5	1,240.9
Lumber and Wood Products	19.7	8.3	413.7	371.3
Paper and Allied Products	9.4	14.7	349.2	683.1
Chemicals	2.3	1.9	70.4	138.6
Petroleum Refining	1.2	1.3	255.9	511.8
Stone, Clay and Glass	3.8	5.0	92.4	172.5
Primary Metals	4.1	7.3	118.8	518.6
Other Non-durable Mfrs.	15.1	19.7	168.7	344.3
Other Durable Mfrs.	86.2	175.7	1,815.9	5,460.7
Trans., Comm., and P.V.	40.2	36.2	615.6	1,192.8
Whsle. and Retail Trade	140.0	202.6	1,250.3	2,269.4
Services	144.0	230.1	1,149.5	2,185.9
Construction	41.2	54.5	673.8	1,359.7
Government	115.8	178.1	--	--
TOTAL	662.6	973.1	7,869.4	16,710.7

TABLE II-D9: PROFILE OF THE TRAVEL ACTIVITY WASHINGTON 1968[11]

Number of Travelers	9,100,000
Jobs associated with out-of-state travelers	38,000
Expenditures	
Dollars spent by travelers	$323,000,000
Dollars spent per traveler	$35
Amount invested in new facilities (18 month period)	$16 to $18 million

TABLE II-D10: DIRECT DOLLAR EXPENDITURES OF NON-RESIDENT TRAVELERS IN WASHINGTON STATE[13]

By Kind of Purchase and Type of Traveler 1963, 1967, 1980

(thousands of dollars)

Kind of Purchase	Total			Vacations		
	1963	1967	1980	1963	1967	1980
Food	$19,934	$24,807	$50,839	$6,481	$8,065	$16,528
Personal Service	71,768	89,311	183,033	26,582	33,080	67,794
Transportation	13,747	17,107	35,059	5,385	6,701	13,733
Communications	819	1,019	2,088	291	362	742
Petroleum Products	16,347	20,343	41,691	7,029	8,747	17,926
Shopping - Textiles	120	150	307	43	53	109
Apparel	8,474	10,546	21,613	3,236	4,027	8,253
Furniture	3,370	4,194	8,596	1,284	1,598	3,275
Other	602	749	1,535	180	224	458
Trade	82,649	102,853	210,786	26,642	33,155	67,947
Imports	22,919	28,522	58,543	8,458	10,526	21,572
TOTAL	$240,750	$299,600	$614,000	$85,611	$106,538	$218,338
PERCENT (approx)	100%	100%	100%	31%	35%	36%

TABLE II-D11: LARGE WATER USING MANUFACTURING INDUSTRIES, PROJECTED OUTPUT TIED DIRECTLY AND INDIRECTLY TO FINAL MARKETS - PUGET SOUND AREA, 1980 - (MILLIONS OF 1963 DOLLARS)[15]

Industry	Total Output 1963	Total Output 1980	Sales to Consump-tion	Sales to State & Local Gov'ts.	Sales to Invest-ment	Sales to Govern-ment	Sales to Exports
Food and Kindred Products	698.5	1,240.9	595.8	50.3	0.2	17.7	576.8
Lumber and Wood Products	413.7	371.3	21.2	24.1	59.0	6.8	260.4
Paper and Allied Products	349.2	683.1	52.1	9.3	3.1	1.8	616.9
Chemicals	70.4	138.6	17.1	5.9	7.5	2.1	105.9
Petroleum	255.9	511.8	223.0	29.5	16.5	11.3	231.5
Stone, Clay and Glass	92.4	172.5	27.1	32.8	57.9	4.3	50.5
Primary Metals	118.8	518.6	6.7	13.6	28.0	7.8	462.8

TABLE II-D12: ESTIMATED VALUE OF SCUBA DIVING IN PUGET SOUND[16]

Cost of Equipment for Average Sport Diver	
Tank	$90
Regulator	80
Backpack	15
Weight Belt	15
Knife	10
Watch	15
Wet Suit	70
Fins	15
Mask	10
Divers Flag	10
Life Jacket	20
Spear Gun	30
Depth Gauge	10
Compass	5
	$470

Average number of dives per year	10
Average life expectancy of equipment (a diver who dives more frequently than 10 dives a year may get as many as 250 dives from his equipment, on the average)	10
Each dive costs in equipment	$4.70
Cost of air fill	1.50
Cost of certification ($45 divided by 100 dives)	.45
Cost of transportation to diving area; car, boat, charter, ferry, accommodations, etc.	3.35
Total Estimated Cost per Dive	$10.00

Total estimated number of divers in the Puget Sound Area	20,000
Total estimated use-cost for scuba diving annually	$2,000,000

shows the large water-using manufacturing industries of which the petroleum industry is one. Unless someone can stop the world now, stifle the demands of consumers, and stabilize to the point of rigidity all of our economic functions, the population and its geographical location, there can be no denying the fact that in the period from 1960 to 1980 (Figure 1), the population is going to double in the Puget Sound region. It is a reality that the environmental issue of oil and water pollution is one of being able to afford the cost. As soon as large numbers of jobs are threatened, as soon as the ramifications of stifling legislation, should any be passed, are felt, the citizenry affected will most likely react in a way beneficial to them as individuals and possible injurious to society as a whole, a situation analogous to the paradox of thrift touted in economics. Sixty-three percent of the 1,500,000 jobs in the state are held by the eight counties of Snohomish, King, Pierce, Kitsap, Whatcom, Skagit, Island, and San Juan. This area holds 61% of the state's population and receives about 61% of the state's personal income. Applying these percentages to the employment and output data in Table II-D8 and the large water-using industry data of Table II-D11, one can again see the astronomical value of the Puget Sound region from an industrial and purely economical point of view. Tourism is not going to become number one industry without a fight. Existing industries, managers, and investors will fight hard for the allocation of resources and legislation to protect their present position and their future. This is only rightly so, for their responsibilities of their position dictate such response and effort.

3. VALUATION OF RESOURCE CONSERVATION

To further complicate the task of valuation and determination of economic and societal adverse effects, one can look at the other two data sources: scuba diving and duck hunting, shown in Table II-D12 and II-D13, respectively. These two serve as an example of how society has only scratched the surface of evaluation and measurement of adverse effects. Just how much, in the aggregate, is a trip below the water's surface? How much value or cost lies in shooting a Canadian Honker from the sky? How are the costs and benefits to individuals related to a society? Is it worth $10 or so per day for every day of outdoor rest a city dweller is able to obtain? Is it worth $10 per

TABLE II-D13: ESTIMATED WATERFOWL HUNTER USE IN PUGET SOUND[17] (1961-1965)

Puget Sound Basins	Hunter-days	Estimated Value $5/hunter-day
DIRECT		
Whidbey-Camano	12,600	
Snohomish	20,500	
Seattle area	13,400	
Puyallup	9,000	
Total	55,500	$277,500
INDIRECT		
Skagit-Samish	65,000	
Stillaguamish	17,200	
Nisqually-Deschutes	14,000	
West Sound	20,000	
Total	116,200	$581,000

day to maintain the environment and facilities so that the travelers, city folk, and tourist can utilize the asset four or five times a year? Is the society of the Puget Sound region sincerely convinced of the importance of maintaining the natural conditions or are they only interested in the idea as long as they are still able to enjoy (1) synthetics made from oil, (2) autos, (3) boats that pollute as viciously as any spill, only slower, and (4) use oil, plus make a high disposable income with a shorter work week due to industrialization, "progress," and technology.

4. RECOMMENDATIONS

The attempt here is to be positive and yet realistic to the fact that any solution, which will benefit society today and tomorrow while minimizing adverse effects, must be an optimization. The situation is already out of the control of this region alone. Other national factors come to bear and influence the growth and operations of the ecology and economic environment in the Puget Sound region. So many of the resources that the region has are irreplaceable that due care must be taken by all concerned; by industry to eliminate the onus of reprisal legislation, by tourism and conservationism in coordinating expansion, growth and utilization, and by fisheries in cooperating rather than blocking. Everyone must give up something in this endeavor. The alternative to each giving up a little part is that we all lose.

ENDNOTES SECTION II-D

[1]Washington State Census Board, Population Forecast - State of Washington 1965 to 1985, Department of Commerce and Economic Development, 1966, pp. 62-63.

[2]In a letter dated August 7, 1970, addressed to the group that prepared the 'Environmental Statement on the Proposed Leasing of Puget Sound Shorelines and Beds of Navigable Waters for Oil and Gas Exploration.'

[3]Washington Department of Fisheries, Plan for Washington State Food Fisheries, Washington Department of Fisheries, 1970, p. 3.

[4]Ibid, p. 13.

[5]C. W. Dolan, T. F. Savage, and R. C. Clark, Jr., Environmental Statement on the Proposed Leasing of Puget Sound Shorelands and Beds of Navigable Waters for Oil and Gas Exploration, The Mountaineers - Puget Sound Group of Sierra Club - Friends of the Earth - Interested Private Citizens, 1970, p. 12.

[6]Washington Department of Fisheries, Year 1969 - 79th Annual Report, State of Washington, 1970, p. 78.

[7]D. Donato, "The 1969 Washington Trawl Landings by Pacific Marine Fisheries Commission and State Bottomfish Statistical Areas," Groundfish Data Report, Series #9, Washington State Department of Fisheries, 1970.

[8]Ibid.

[9]Ibid.

[10]Puget Sound Task Force, Puget Sound and Adjacent Waters: Apendix IV, Economic Environment, 1970, p. D-43.

[11]D. L. Henricksen and R. E. Hughs, Jr., Travel Activity in Washington State, Department of Commerce and Economic Development, Business and Economic Research Division, 1969, p. 2.

[12]Battelle Northwest, Impact of Non-Resident Travel on the Washington State Economy, Department of Commerce and Economic Development, 1969.

[13]Ibid., pp. 5-8, 5-9.

[14]Puget Sound Task Force, Pleasure Boating Study, Puget Sound and Adjacent Waters, 1968, p. 4.

[15]Puget Sound Task Force, IV, pp. D-54, D-55.

[16]Dolan, et al., p. 61.

[17]Ibid., p. 15.

II-E. INDUSTRY EFFORTS IN PROTECTING RESOURCES

1. INTRODUCTION

The previous sections have described the economic value question of the resources in and around Puget Sound, and signified the locations of priority areas with respect to consequences of their exposure to environmental degradation. This section examines the efforts on the part of the oil industry, current and recent past, to minimize the consequences of an oil spill.

Prior to the Alaskan North Slope discoveries in 1968, there was little public concern for oil spill prevention efforts in Puget Sound. Not one bill relating to water pollution by oil was introduced before the State legislature in the 1965 and 1967 sessions.[1] There had never been a local oil spill to cause any alarm. On the West Coast, protection of California waters dominated the efforts of industry. The Santa Barbara channel spill of 1969 heightened industry's efforts in this region. Water conservation is a long run effort, requiring investment in plant and equipment as well as training of personnel. It does not happen overnight. Industry has done a credible job at Santa Barbara, but it has taken them one serious accident and many futile dollars to reach this end. The possibility of transporting North Slope oil to Puget Sound, plus the California oil spills, have awakened the concern of both the public and industry in the Pacific Northwest.

The public has, all of a sudden, become fully conscious of the economic value of Puget Sound and its vulnerability to destruction, having taken it for granted such a long time. This economic value relates more to the man-made resources, such as leisure time activities, tourism, and commercial fishing, and is quantified for the short run in terms such as yearly revenue or profits. What has been missing is the detailed analysis of the long run economic value and consequences of irreplaceable resources such as shellfish and plant life. The public has, through the political system, demanded increased efforts in water conservation,[2] but it should also strive to define the economic benefits of such efforts.

Industry, on the other hand, has been critically examining current and past performances in local water conservation.[3] The results of these surveys are not available for analysis, but an evaluation made for this report has

TABLE II-E1: QUESTIONNAIRE ON OIL SPILL PREVENTION

1. What efforts are you currently expending on oil prevention in the Puget Sound and surrounding waters?
2. What are the dollar costs of such efforts? (Please correlate cost to type of effort, if possible, and state figures as annualized costs.)
3. How long has your firm been engaged in such efforts?
4. What is your estimated cost for the previous efforts?
5. Do you foresee any dramatic change in the level of your effort in the near future? (Next three years in terms of dollars.)
6. How are the dollar costs of your efforts recovered? (Who bears the burden and how?)
7. If you have a model for assessing burden of costs, please provide a brief description.
8. Is your firm currently being assessed for oil spill prevention efforts by other groups or organizations? If so, what is the annualized cost to you?

revealed various inadequacies in Puget Sound.[4] These include lack of up-to-date cleanup devices, and fragmented techniques and procedures. As an initial step to improve the situation, ten oil firms located in the Puget Sound area have formed a cooperative to update efforts.

2. INDUSTRY'S WATER CONSERVATION EXPENDITURES IN PUGET SOUND

There are four basic functions in the processing of oil. These are:

1. the production of crude oil
2. refining of the oil
3. shipping of the refined products
4. marketing of the products

It is in the first three areas that the oil spillage problem exists. As far as Puget Sound is concerned, the two major functions are refining and transporting petroleum products. Thus, industry's efforts, ideally, are geared to:

1. the prevention of oil spills
2. effectively containing and cleaning up an oil spill

These efforts should apply to both refining and transportation. An attempt was made to analyze industry's effort.

The first step in the analysis was to determine possible sources of information on oil spill prevention efforts. In Puget Sound, the refining segment is dominated by three firms: Mobil, Texaco, and Shell (ARCO at Cherry Point is not yet operational). The transporting of oil in the Sound is divided between the major oil producing firms and the hauling service companies. Cursory examinations of the situation revealed that the cost of any effort with respect to oil spills is eventually reflected in the selling price of the product. Therefore, it was concluded that the oil producing firms would be the proper source for information.

A questionnaire shown in Table II-E1 was prepared outlining specific topics of interest. This document was administered on a trial basis to four firms whose local representative firms were contacted, with follow-up letters to corporate headquarters offices of two out of the four firms. The results of the initial survey included four non-quantitative responses with

TABLE II-E2: COMPARISON OF SALES & VOLUME FOR 3 OIL FIRMS[8]
(1967 Figures)

SALE OF PETROLEUM PRODUCTS (Refined and Crude):

Firm	Barrels	%
Mobil	526,300,000	30.3
Texaco	704,085,000	40.5
Shell	504,045,000	29.2
TOTAL	1,734,430,000	100.0

SALES DOLLARS:

Firm	Dollars (10^6)	%
Mobil	5,771.8	39.6
Texaco	5,121.4	35.2
Shell	3,657.7	25.2
TOTAL	14,550.9	100.0

PRODUCTION OF REFINED PRODUCTS:

Firm	Barrels
Mobil	503,100,000
Texaco	549,102,000
Shell	300,910,000
TOTAL	1,353,112,000

no mention of dollar costs and two non-responses. The general tone of the responses was that the requested information was not readily available.

Further literature search revealed that the American Petroleum Institute (API) had published a report on industry expenditures for environmental protection.[5] In the end, this was the primary source of industry data utilized. The API report defines expenditures incurred by the 45 largest oil producing firms in the United States. Of those listed, Mobil ranked as second largest, Texaco was third, and Shell ranked sixth.[6] The basis for ranking was sales dollars. The 1967 compilation on sales was the most up-to-date information available.

Utilizing Moody's industrial manual,[7] it was ascertained that the sales dollar correlated closely with the sales volume of petroleum products, including crude and refined (see Table II-E2). From the 1967 figures, the total sales in petroleum products by Mobil, Texaco, and Shell was approximately 1,734,430,000 barrels. Also shown in the same table is the production of refined products for the three firms. The total was 1,353,112,000 barrels in 1967.

The 1963 refining data for the Puget Sound region revealed that 47,300,000 barrels were produced,[9] primarily by the three firms previously mentioned. Taking the United States refined output annual growth rate of 3%,[10] an output of 53,400,000 barrels were refined in the Puget Sound region in 1967. This amount represents 3.94% of the total refinery output of Mobil, Texaco, and Shell.

How are these figures translated to costs? Moody's showed that oil company expenses for cost of goods sold and administration ranged between 68% and 78% of sales. Thus, it can be concluded, from the available data, that cost also has a direct relationship with production. (Table II-E2 demonstrated that sales and production were related.) Therefore, one can assume that the three firms in question incurred 23.4% of total expenditures of the oil industry since their combined sales equaled 23.4%.

Since all oil producing firms have comparable backgrounds and orientations, each firm should incur approximately the same percentage of its expenditures for environmental conservation efforts. Thus, these three

TABLE II-E3: INDUSTRY EXPENDITURES FOR OIL SPILL PREVENTION IN PUGET SOUND[11]

	1967	1970
CAPITAL EXPENDITURES:		
Transportation	$8,500	$50,000
Marketing	20,000	97,800
Refining	335,000	785,000
OPERATION AND MAINTENANCE:		
Transportation	11,500	34,450
Marketing	6,160	29,700
Refining	178,000	272,000
ADMINISTRATIVE EXPENDITURES:		
Transportation	845	1,390
Marketing	2,660	7,230
Refining	43,200	91,000
RESEARCH AND DEVELOPMENT:	-	-
TOTAL	$605,865	$1,368,570

firms should incur 23.4% of each of the various categories of conservation expenditures defined in the API report. Moreover, conservation efforts for the Puget Sound region should be 3.94% of the combined efforts of Mobil, Texaco, and Shell, or 0.837% of the total industry cost for appropriate categories listed in the API report. Since little, if any, crude oil production exists in the Puget Sound region, the percentage figure applies only to transporting, manufacturing and marketing.

The expenditures in the API report are divided between air and water conservation. Since this report is concerned with the latter topic, it is these costs which are relevant. Table II-E3 shows a table of the appropriate expenditures. Since the report gives no detailed breakdown of the costs, the only meaningful measure is the examination of total dollar expenditures, \$605,865 for 1967 and \$1,368,570 for 1970. At this juncture, it is not possible to verify or dispute these figures, since the data is not available, as previously noted. Also, it is not known what portion of this expenditure was earmarked for advertising of water conservation efforts.

Therefore, one can hypothesize:

1. that \$1.4 million expended by the oil industry is adequate to protect the Puget Sound waters, or
2. that industry's efforts, at whatever the true dollar level, are adequate to protect Puget Sound.

3. EFFECTIVENESS OF OIL INDUSTRY'S CONSERVATION EXPENDITURES

To analyze the effectiveness of the expenditures and test the above hypotheses, water conservation techniques at two refineries, namely Shell and Mobil, were evaluated. Though no dollar costs were revealed, on-site inspection of equipment and techniques were provided. Details of these findings are presented as part of Appendix 1. Highlights are presented here in Table II-E4. Though some water conservation efforts are evident, they are a long way from being effective. The most obvious shortcoming is that one refinery does not have adequate knowledge of the water conditions around its facility to predict the behavior of a potential oil spill. The other is associated with the stockpiling of oil spill cleanup equipment. Neither facility had any estimate as to the cleanup capability of the equipment on hand.

TABLE II-E4: EVALUATION OF REFINERY CONSERVATION TECHNIQUES

	Shell-Anacortes	Mobil-Ferndale
Provisions for:		
Purging Fuel Tanks	yes	-
Ship Ballast Treatment	yes	yes (tanks)
Oil Spill on Dock	no	no
Oil Spill on Grounds	yes	yes
Waste Oil Recovery	yes[1]	yes[2]
Oil Spill Booms	no	no[2]
Oil Detection Equipment	no	no
Oil Spill Cleanup Materials	no[3]	no[3]
Analysis of Water Behavior in Vicinity of Facility	yes	no

1. Drip pans, holding tanks, gravity separator, API separator, bio-treatment.
2. A pre-1960 cork boom. Not adequate to cope with water conditions of the area.
3. Some straw and absorbent, but not readily deployable nor is exact quantity known.

The results of the above evaluation can lead to two conclusions. First, assuming that $1.4 million was actually spent in 1970, $1.4 million expended by the oil industry is inadequate to protect the Puget Sound waters. The second conclusion is that the oil industry has shortchanged the Puget Sound area in its water conservation efforts. It would appear that the latter seems more plausible at this time, again owing to the lack of data to test the first hypothesis.

The picture is not all bleak, however. Earlier this year, ten oil firms formed a cooperative organization to "...protect the Sound - and themselves - from damage by oil spills."[13] While the oil industry has made some progress, it does not have the required resources to develop and produce integrated prevention and cleanup systems. For an individual oil firm to acquire funds for oil spill prevention, it must rely on recovery of such costs from its profits.[14] At present, there is no evidence the firms are incurring such expenditures and passing on these expenses to the consumer. Thus, it must rely on outside specialists to fill this void. To function effectively, these outside contractors must be aware of the industry's specific needs, and of the industry's own progress in the area of oil spill prevention and cleanup.

Information on industry efforts in water conservation is not readily accessible. Usefulness of the available data is, at most, marginal. Above and beyond this, industry has not critically examined serving the public interest in water conservation.[15] It appears that industry has not evaluated nor heeded to conservation criteria applicable to specific facility locations, but relied on generalities. Techniques and hardware effective for cleanup of an oil spill in the Louisiana Gulf may not be applicable in Puget Sound. Also, the priorities in resource protection standards in the gulf region may differ drastically from those in the Pacific Northwest. Industry must become aware of such criteria, and if such criteria do not exist, it must demand that society establish these.

In order to determine the marginal benefit of current and future efforts in preserving the environment, the oil industry must be candid with organizations external to itself as to its efforts and related costs, and the perceived benefits.

4. RECOMMENDATIONS

As a starter, all industry efforts should somehow be translated to dollars, since this is the common denominator of most economic analysis. Also, since the value of the dollar is time dependent, it is important to specify the type of dollar expenditure defined by a given investment. The common financial reporting period is one year; hence, equivalent annual costs for capital investments is a proper statistic. This recommendation applies specifically to the API report.

The point to be made here is this: the oil industry should strive to improve intra- and inter-communications, and should strive to eliminate ambiguity in what they report. Then, the oil industry can begin working hand-in-hand with other fields of expertise in coping with the water conservation problem.

While it is true that the industry has not been thoroughly effective in its water conservation efforts, they should not have to shoulder the entire burden. In the free enterprise system, the objectives of the private firm are profit maximization, and compatibility with society's welfare. While the first is easy to quantify, the second is, and will be, a difficult topic to pinpoint and quantify. Five years ago, the public placed no demand on industry in regard to oil pollution in Puget Sound. With the recent public clamor over this issue, industry has begun to respond. In order for both industry and the public to exploit the Puget Sound environment in harmony, each must understand the other's goals and values. Where conflicts in goals arise, tradeoff analysis may be necessary, as situations in the States of Delaware and Maine have demonstrated. Each may have to give up a little to preserve the water quality standard and the standards of the free enterprise system. Ultimately, such efforts will allow the oil industry to serve the public interest equitably.

ENDNOTES SECTION II-E

[1]Legislative Digest and History of Bills of Senate and House of Representatives for Washington State, 39th Session, 1965, and 40th Session, 1967.

[2]The 41st Session (1969), saw three bills introduced which dealt with oil pollution of water. The first half of the 42nd Session (1971), has already witnessed six such bills before the legislature.

[3]American Petroleum Institute, Report On Air and Water Conservation Expenditures of the Petroleum Industry, 1966-1970, 1971, Crossley Surveys, Inc. A questionnaire is contained in Appendix II of this document.

[4]The evaluation of oil handling facilities is presented in Section III-C part 4, and Appendix 1 of this report.

[5]See footnote 3.

[6]Newsfront, 25,000 Leading U.S. Corporations, 1970, see SIC 29, 13.

[7]Moody's Investor Service, Moody's Industrial Manual, 1970.

[8]Ibid., pp. 2658, 2636, 3064.

[9]Based on 157,000 b/cd. as reported in Pacific Northwest Economic Base Study for Power Markets, U.S. Department of Interior, Vol. II, Part 11c, Petroleum, 1966. This figure multiplied by 300 days to get 47,3000,000 barrels.

[10]Based on data found on page 656 of Statistical Abstract of the United States, 1970, U.S. Department of Commerce.

[11]Data shown is 0.0837% of related data shown on pp. 11, 14, and 17 of the API report.

[12]Based on analysis of available on-site equipment, condition of facility, and discussion with personnel.

[13]Page D., "Oil Companies Unite to Contain Possible Spills," Seattle Post Intelligencer, April 17, 1971.

[14]Conversation with Mr. Arnold Hinkley, Chief Engineer and Operations Manager, Northwest Region, Western Operations, Standrad Oil of California, April 27, 1971.

[15]The terms 'public interest' as used here refers to the aggregate of all self-interest groups outside the oil industry.

[16]The State of Delaware has enacted a law to ban oil refineries and other heavy industries from locating new plants along its 125 mile long ocean, bay, and river coastline. Thus, that State has placed high social value on conservation. The industrial growth in the affected localities, Delaware Bay, carries the second highest waterborne petroleum traffic in the nation based on number of barrels handled. (Sources: Time Magazine, July 12, 1971, p. 51; and Dillingham Report.) A corollary effect of the state's actions was to stop progress on the development of an offshore terminal for oil transfers involving supertankers. Source: Business Week, July 2, 1971, pp. 20-21.

[17] Maine's Environmental Improvement Commission turned down a proposal for refinery construction on Penobscot Bay. The reason given was that the refinery lacked financial and technical ability to meet pollution standards, and would not 'fit harmoniously into the natural environment' of the area. (Source; Time Magazine, August 2, 1971, p. 47). There were two categories of tradeoffs in this issue. The first involved additions to the labor force versus the ecological and scenic effects of a refinery. The second tradeoff concerned increased shipping traffic from the refinery interfering with commercial fishing and lobstering in the bay. (Source: Wall Street Journal, August 10, 1971, p. 1).

SECTION III

PREVENTION AND CONTROL OF OIL SPILLS ON PUGET SOUND

This segment of the report is divided into three parts - identification of oil spill sources, containment and cleanup of these spills, and prevention of such oil spills. Section III-A categorizes potential sources of spills, quantities involved, and associated probabilities. Section III-B analyzes current state-of-the-art oil spill response capabilities, culminating in a recommended inter-agency proposal. The section also delineates orders of priority in resource conservation, proposed containment and cleanup techniques for specific locales, and the required organizational cooperation.

Section III-C delves into the topic of oil spill prevention. Areas covered include tankers design and operation, traffic control, oil detectors and oil handling facilities. Again, state-of-the-art techniques and equipment, and evaluation and their applicability to Puget Sound are discussed.

III-A. IDENTIFICATION OF SOURCES OF OIL SPILLS

1. INTRODUCTION

The purpose of this section of the study is:

1. to identify the sources of potential oil discharge into Puget Sound, and
2. to establish a ranking of sources to indicate which of these require the greatest priority for preventive attention.

Before a listing of all sources can be analyzed and ranked, an objective scheme of criteria must be chosen to evaluate the impact of the various types of spills.

Ideally, a strictly quantitative analysis would employ the following relationship to establish priorities:

Ranking = Probable Frequency of Spill x Probable Quantity of Spill x Weighting Factor

where

Weighting Factor is a function of (1) oil type spilled and (2) the maximum concentrations that result (time dependent effects). (For example, the weighting factor would allow for possible damage variation between the case where a specific quantity of a certain oil type is spilled instantaneously and the case where the same quantity is spilled over a longer period of time.)

At the present time, data is insufficient to allow an accurate estimate of statistically probable spill frequencies and quantities. While some extensive data is available for the statistical analysis of the probability of waterborne traffic collision or grounding, this data does not directly relate to oil spill frequency or quantity in a particular body of water, or part thereof. An analysis of probable waterborne accidents involving commercial vessels is described in Part 5 of Section II-B and Appendix 13 of this report. The data for shipping is probably more complete than for any of the other potential sources of oil discharge, but it needs further refinement for a strictly objective quantitative analysis. In addition, a useful estimate of the weighting factor is dependent upon research actively in progress by various elements of industry, government, and private institutions.[1] Also, the data assessing oil spill effects is not yet conclusive.

Even though a strictly quantitative statistical analysis of probable spill frequency and quantity is not possible at this time, a complete listing of sources ranked in a subjective manner can be useful for the direction of immediate preventive and control measures analysis. It must be stressed that the resultant ranking of sources is preliminary, and based upon incomplete data. Its purpose and intent is not to suggest that it has been proved that any source assigned a high priority has been or will damage Puget Sound resources, but that each is a threat to these resources. It is hoped that the discussion of the requirements for a strictly objective quantitative analysis will indicate the type of data that should be obtained for more refined statistical studies in the future.

The following categories of oil discharge have been identified on Puget Sound, and each category will be examined in the following sections to obtain a ranking for preventive priority.

a. Waterborne traffic (large vessels).
b. Waterborne traffic (intermediate size vessels).
c. Waterborne traffic (small vessels).
d. Vessel/shore transfer facilities.
e. Vessel/vessel transfer operations.
f. Shore storage and transportation facilities.
g. Municipal, industrial, private waste disposal.
h. Submarine oil drilling.

Within each category, the parameters examined include source of spill, quantity of oil spilled, and the probable frequency of occurrence.

2. SOURCES OF OIL SPILLS

Category (a): Waterborne Traffic (Large Vessels)

1. Sources:
    a. Bulk oil carriers (including tankers, barges, oilers).
        i. Structural failure of tank spaces due to grounding, collision, explosion and fire, foundering.
        ii. Intentional pumping of tank compartments or bilge spaces in case of emergency.

b. Non-petroleum cargo vessels.

i. Structural failure due to collision, grounding, fire and explosion, foundering.

ii. Pumping bilge spaces.

2. Quantities:

Considerable quantities of various types of petroleum are transported over the waters of Puget Sound. Table III-A1 summarizes the annual volume of each major type of product from 1965-1970. The sum total of refined products represents approximately 42% of the total amount transported by all modes, while the total for crude oil represents about 7% for all modes.[2] The bulk of the crude oil coming into the Sound travels by the Trans-Canada pipeline. Most of the crude oil that is shipped by water is exported from Puget Sound.

Gasoline represents the largest portion of the petroleum products shipped in the Sound, averaging 4.6 million short tons over the six years analyzed. Most of the gasoline is shipped within the Sound from refineries to population centers. The majority of gasoline is moved by small vessels or barges.

Distillate oil is also shipped in large quantities, as noted in the table. Although a sizable portion of the distillate fuel transported to Puget Sound markets, significant amounts are also shipped outside to other states and countries. Residual fuel oil, kerosene and jet fuel are transported primarily to Puget Sound markets.

3. Probable Frequency:

When dealing with large vessels (tankers, barges, and freighters), an estimate should be made of their potential for an oil spill of a given size. A quantitative analysis of the loss of a certain amount of oil over a given time based on probabilities is useful in answering the problem of the relative size of different sources of oil pollution.

To encompass the many assumptions and uncertainties inherent in a statistical prediction of this nature, and to form a basis for further discussion of potential oil spill situations, the following procedure is outlined to arrive at an order-of-magnitude estimate for oil potentially spilled by a collision involving a large vessel:

TABLE III-A1: ANNUAL PUGET SOUND OIL TRANSPORT VOLUME[3]

(in 1000 short tons)

Year	Total	Crude	Gasoline	Distillate Fuel Oil	Residual Fuel Oil	Lube Oil and Grease	Jet Fuel	Other
1965	15,783	1,163	6,646	5,014	1,442	101	186	1,231
1966	12,348	671	4,638	3,655	1,837	73	189	1,285
1967	12,528	1,481	3,905	4,195	1,365	70	299	1,212
1968	14,750	2,496	4,271	4,207	2,106	119	773	778
1969	12,580	1,619	3,635	3,932	2,009	264	1,054	517
1970	12,891	1,899	3,871	3,930	1,701	226	899	365

1. Determine the probability of ship collisions in Puget Sound - this has been done for various areas of the Sound using traffic and historical collision data.

2. Compute probabilities as applied to commercial vessels only - given the probabilities in step 1. Proportion these according to the population ratio of commercial to total ship traffic.

3. Distribute this commercial probability among ship type using frequency of port call data.[4]

4. Compute an average size for each ship type according to number of barrels of oil carried - for freighters and barges compute an average size from Waterborne Commerce Data - for tankers use size estimates by port of call as reported in Puget Sound Task Force Report.[5]

5. Estimate an average extent of damage by vessel type for each collision by using IMCO casualty card file[6] for tankers and by making further assumptions for barges and freighters. Since partial damage to a ship can cause a major oil spill, these data must not be overlooked. The damage card data from IMCO can therefore give an estimate of the type and heavy damage included in the partial loss column. (If the ratio of partial to total loss is considered for the entire fleet due to stranding and contact damage to be about 50 to 1, and the ratio reported in the IMCO data cards to be about 10 to 1, then these cards must reflect heavier than average partial loss incidents.)

6. Combining all of the above data properly, an estimated potential spill size due to the chance of a collision on Puget Sound may be determined.

7. Check this number with one based on world fleet casualty statistics prorated to tonnage of oil spilled on the Sound.[7]

Following this elaborate procedure, an order-of-magnitude estimate of the potential amount of oil that would be spilled on the Sound from ship collision can be estimated. This estimate may be unwarranted if the oil spill hazard from other situations is neglected. The tendency is to consider ship collisions as the most probable source of ship damage leading to an oil spill. However, the recent oil spill data suggests that collision is not the only significant source of waterborne oil spills.[8]

Information from Lloyd's Registry shows that, up to 1966, the total damage to oil tankers could be credited to four main causes of relatively equal magnitude - wrecked, foundered, burned, and lost by collision. Further information from the Liverpool Underwriter's Assn. indicates that partial damage to ships in the world fleet can be equally attributed to weather damage, collisions, and "other" causes. The concept of a traffic advisory system is directed toward only part of these potential hazards. Land based radar and bridge-to-bridge communication among all vessels in the Sound could minimize the possibility of collisions, and groundings. But, older and less seaworthy ships are subject to foundering. Also, tankers are subject to fire and explosion damage. The latter two potential hazards could not be minimized by a traffic advisory system. Therefore, collision predictions are not the only factor to be considered with waterborne traffic oil spills.

As mentioned earlier this data is based upon the total number of ships in commission and is not truly accurate. The partial loss data compiled by Liverpool Underwriters is valuable in that it does include all ship damage. This basis could possibly be adjusted with IMCO damage card data to give a reasonable picture of heavy damage which could lead to oil spills. The question naturally arises in this context as to whether a fire or explosion causing partial damage to the ship would lead to as large an oil spill as a partially ripped bottom. Even if these details could be decided on some arbitrary scale, the Liverpool data is not presented so as to discriminate vessel size. The original files would have to be consulted for monthly accident reports for further analysis.

The total loss data analyzed by Beers[9] present some interesting trends. Among his conclusions are:

1. Small ships historically have a high casualty rate as part of a pattern dictated by coastal trade; its trend has not worsened.

2. Collisions show increased losses for oil tankers which are largely due to fire consequent upon collision.

3. "Burnt" losses have increased for dry cargo ships, but mostly in port; the worst trend of all is for oil tankers burned in port.

These conclusions are interesting but not at all inclusive. Total losses average around 0.5% (by number) of the world fleet per year, whereas partial losses are about 35% (by number) per year. Added to this is the fact that fires and explosions tend to weigh heavier in the total loss column. These combined facts indicate that great caution must be employed when extrapolation of damage data to potential for oil spills is attempted.

The main result drawn from this discussion is that the threat of oil spills can be equally distributed among weather damage, collisions, fires and explosions, contact damage, and "other" casualties. With this in mind, any estimate of oil lost from ship collision must be amplified by an arbitrary factor of at least two. Considering that the above data is for world shipping trends, with only limited input of Puget Sound data, an estimate of oil spill potential is very preliminary, at best.

Category (b): Waterborne Traffic (Intermediate Size Vessels)

Category (b) also includes tugs, fishing vessels, larger recreational "inboard" boats.

1. Sources:
   a. Bilge pumping.
   b. Engine operation: vessels that allow machinery exhaust to discharge through cooling water effluent have potential to release oily water depending upon condition of machinery.

2. Quantities:

Difficult to determine. Probably less than 1 gallon per boat per month. However, there were approximately 200 tug boats, 1500 fishing vessels, and over 20,000 inboard pleasure craft on Puget Sound in 1965.[10]

The following data is available on quantities:[11]

Registered inboard boats - 18,200

Mean annual fuel consuption, inboard boats - 536.9 gal/boat which includes diesel and gasoline (approximately 95% of inboards use gasoline, 5% use diesel).

3. Probable Frequency:
   a. Intermittent for bilge pumping.
   b. Continuous for all boats in operation with motors operating.

Category (c): Waterborne Traffic (Small Vessels)

Category (c) also includes outboard boats with two cycle engines.

1. Sources:
   a. Leakage and spillage from portable tanks.
   b. Inherent inefficiency of two cycle engines passes oil with cooling water and exhaust.

2. Quantities:

Depending upon condition of motor, up to 1/2 pint of lubricating oil per 1 gallon of fuel might be lost during operation.

In 1965, there were 94,400 outboard boats on Puget Sound of which the vast majority were two cycle. Mean annual fuel consumption 189.5 gallons of fuel per boat.

3. Probable Frequency:

Continuous with motor operation.

However, boating is seasonal.

Boating Activity - % Total Boats per Month:[13]

January	37%
February	38
March	41
April	57
May	83
June	97
July	100
August	99
September	97
October	69
November	41
December	33

Category (d): Vessel/Shore Transfer Facilities

1. Sources:[14]
   a. Bulk Transfer.
      i. Leaking transfer equipment (i.e. hoses, Chiksans) during operation.
      ii. Leakage during removal and handling of transfer equipment.
      iii. Improper filling procedure.
   b. Unit Fueling Facilities.
      i. Leaking hoses in operation.
      ii. Removal and handling of transfer equipment.
      iii. Overfilling of tanks.

2. Quantities:
   a. Bulk Transfer.

      Large diameter hoses and large flow rates; quantity depends upon nature of conduit failure and procedure before and after transfer.

      There are 77 piers on Puget Sound for the receipt of oil, and 13 for shipment.

   b. Unit Fueling Facilities.

      There are 30 fueling piers on Puget Sound for large vessels with large flow rate transfer systems.

      There are over 100 marine service stations on Puget Sound for smaller vessels.[16] Quantity depends upon nature of conduit failure. Up to several gallons of refined oil may be spilled per fueling event due to technique of "topping off" tanks by filling until fuel pours out air vents.

3. Probable Frequency:
   a. Bulk Transfer.

      Some dripping may occur during hose or conduit handling, but quantitative data unavailable.

   b. Unit Fueling.

      Dripping before and after each fueling possible, but no quantitative data present.

      "Topping off" overfilling spillage may occur each fueling event.

Category (e): Vessel/Vessel Transfer Operations

1. Sources:

Fueling or bunkering larger vessels by barges, oilers, or lighters create the same problems as vessel/shore transfer operations.

2. Quantities:

There are ten facilities in the Seattle area and three in Tacoma that fuel vessels from barges.[17] Total capacity for these barges is 104,050 barrels.

3. Probable Frequency:

Details on frequencies not available at publication time.

Category (f): Shore Storage and Transfer of Oil and Products

1. Sources:
   a. Tank farms (refineries, storage).
      i. Structural failures.
   b. Tank trucks.
      i. Structural failure due to wreckage near shoreline or near sewer intake.
      ii. Overfilling, emptying procedures.
   c. Railroad Tank Cars.
      i. Structural failure due to accident or derailment near shoreline or sewer system.
      ii. Overfilling, emptying procedures.
   d. Pipelines.
      i. Structural failure due to excessive pressure or geologic event (i.e. earthquake, foundation or slope failure).

2. Quantities:
   a. Tank farms.
      Dependent upon nature of failure and safety cofferdam effectiveness.
   b. Tank Trucks.
      Up to 10,000 gallons/truck if entire cargo lost.
   c. Up to 10,000 gallons/car if total load discharged.
   d. Pipeline.
      Large volume pipeline (Olympic or Trans-Canada) may be considerable depending upon nature of failure.

Smaller Tributary Lines.

Dependent upon flow rates and types of failures.

3. Probable Frequency:

Details of analysis were not available at publication time.

Category (g): Municipal, Industrial, Private Waste Disposal

1. Sources:
   a. Industrial.
      i. Refineries - "Oily water" effluence, ballast transfer from tankers, barges.
      ii. Shipyards - accidental spillage during repair and maintenance of vessels.
   b. Municipal.

      Addition of waste oil to sewer system.
   c. Private.

      Illegal disposal of waste oil in fill areas or water.

2. Quantities:
   a. Industrial.

      Industrial firms are regulated to approximately 100 parts per million oil discharge into the sewer system or waterways so that effluence is limited to several barrels per day, at most.[18] This effluence includes lubricating and quenching oils, plus oil-water emulsions. A 1969 Seattle area study indicated that of the estimated 1.3 million gallons of industrial waste oil, approximately one million gallons were reclaimed.[19] The remaining 300,000 gallons were unaccounted.

      Data on ship and tank cleanings reveals between 2-2.6 million gallons of oily wastes are accumulated annually, of which approximately 60% is oil.[20]

      No data was available on quantities of oil handled or discharged from shipyards, but their effluence has been handled by tank or ship cleaning firms.[21]

b. Municipal.

The quantities of oil wastes added into sewer systems is uncertain, but as much as two million gallons per year are discharged in King and Snohomish counties alone. The percent of oil in this waste is unknown. Also unknown is the amount of this oil that reaches Puget Sound. Though no written standard exists a maximum allowable discharge 10 ppm into Puget Sound has been used by the State Department of Ecology in relation to the Seattle Metro system.[22]

c. Private.

The quantities of waste oil discharged from private sources are primarily through boating and automotive uses. The former has been covered in Category (c). Seven independent studies have been conducted on estimating the amount of oil waste in King, Snohomish, and Pierce counties. Since these encompass the bulk of the populous in the Puget Sound region, the numbers are indicative of the magnitude of the oil waste situation. The quantities estimated range from one million to 6.6 million gallons of waste motor oil annually, with a mean of 4.9 million gallons.[23] Of this amount, approximately 2.5 million gallons are collected and re-refined. The remainder is unaccounted for.[24]

3. Probable Frequency:

a. Industrial.

i. Refineries - Effluence continuously.

ii. Shipyard data unavailable.

b. Municipal.

Uncertain, probably daily.

c. Private.

Uncertain, probably daily.

Category (h): Submarine Drilling Operations

1. Sources:

   a. Exploration Drilling.

      i. Oily drilling mud leakage.

      ii. Leakage through equipment failure.

      iii. Leakage through faults in strata.

      iv. Structural failure due to collision because of navigational hazard.

   b. Production Drilling and Operations.

      i. Conduit leakage.

      ii. Handling after production.

      iii. Leakage through geological failures (faults).

      iv. Structural failure due to collision because of interference with navigation.

2. Quantities:

   Pollution due to escaping drilling mud difficult to evaluate - probably slight.

   For equipment failure due to collision, operational defect,or geological failure, quantity of crude oil could be essentially unlimited. (Eighty percent of submarine lands in Central Puget Sound that might be available for drilling in the future is in water deeper than 60 feet. The technological problems in repairing well heads in depths up to 600 feet are considerable. Therefore, any accident on a drilling platform would be difficult to control.[25])

3. Probable Frequency:

   a. For drilling mud, continuous during drilling operation.

   b. From operational failure, insufficient data; from collision due to interference to navigation, insufficient data.

      From geological failure, preliminary evidence[26] indicates that recently active fault zones are present under Puget Sound and that these faults might be a problem to submarine drilling.

3. SUMMARY

In 1970, there were 242 oil spills reported in the State of Washington.[27] The usual method of detection was visual, and most spills were reported by other than the responsible party. Thus, it was difficult to assess the amount of oil spilled or the cause. Table III-A2 lists these spills by source and size. In 42 percent of the spills, the source was unknown, and for about 90 percent of the spills, the quantity spilled was not known. In order to improve statistical analysis of oil spill probabilities, more sophisticated detection and reporting techniques are necessary.

Table III-A3 shows the geographical distribution of the 1970 reported oil spills. The subzones have been described in Section II-A. Almost 80% of the reported spills occurred in subzones 4 and 5, encompassing the Seattle, Tacoma, and Olympia metropolis, and Bremerton. As shown in Section II-B and II-C, these are also the areas of heavily concentrated waterborne commerce and oil transfer facilities.

While no compilations are available on quantities of oil spilled either in the State or nation,[28,29] a study was made on estimated oil discharged into the world's waters in 1969.[30] Table II-A4 summarizes this study. The applicability of this data to Puget Sound is questionable for several reasons. First, the worldwide data is for all areas, coastal and inland. Second, the method of analysis is not known.

Data is available on the quantity of oil spill for a given port, Milford Haven, S.W. Wales, Great Britain, which is an arm enclosing 70 miles of coastline. It is primarily an oil port, with three refineries and one tank farm. Surveillance consists of continuous patrolling through day and night, with 20-30 inspections daily.[31] The oil spill data has been accumulated since 1963.[32] Table III-A5 presents a summary of the oil spilled in quantity and percent of total transference in that port. Table III-A6 lists the sources and sizes of the oil spills.

The number of incidents per year has increased some, though the percentage of tankers involved in oil spills has slightly decreased. It is interesting to note that, even with such detailed information, respected scientists and

TABLE III-A2: REPORTED OIL SPILLS IN WASHINGTON STATE, 1970[27]

Source	Number of Spills Reported	Quantity of Spill Known	Quantity of Spill Unknown	Known Quantity (Gal) <100	Known Quantity (Gal) 100-1000	Known Quantity (Gal) >1000	Unknown Quantity Min	Unknown Quantity Mod	Unknown Quantity Ext
a. Waterborne Traffic Large Vessels	21	1	20	1	0	0	12	6	2
b. Waterborne Traffic Intermediate Size Vessels	3	0	3	0	0	0	3	0	0
c. Waterborne Traffic Small Vessel	4	1	3	0	1	0	2	1	0
d. Vessel/Shore Transfers	11	9	2	5	1	3	0	2	0
e. Vessel/Vessel Transfers	0	0	0	0	0	0	0	0	0
f. Shore Storage	18	5	13	2	0	3	8	5	0
g. Municipal	8	0	8	0	0	0	8	0	0
h. Industrial	61	5	56	2	2	1	37	15	4
i. Private	15	1	14	1	0	0	11	3	0
j. Unknown	101	4	97	2	2	0	74	15	8
TOTAL	242	26	216	13	6	7	155	47	14

TABLE III-A3: GEOGRAPHIC DISTRIBUTION OF 1970 REPORTED OIL SPILLS IN WASHINGTON STATE[27]

Subzone	Total	Known (Gallons) <100	100-1000	>1000	Unknown Min	Mod	Ext
1	5	0	0	0	2	2	1
2	10	1	2	0	4	1	2
3	7	0	0	2	3	2	0
4	154	9	5	1	102	32	5
5	42	1	0	1	33	5	2
6	5	1	0	0	2	1	1
Other	19	1	1	1	11	3	2
TOTAL	242	13	8	5	157	46	13

TABLE III-A4: ESTIMATE FOR QUANTITIES OF OIL DISCHARGED INTO WORLD WATERS, 1969[30]

(gallons per year)

Source	Quantities		Percent of Total
1. Tankers			
a. Normal operation using control measures	9,040,000		
b. Normal operation not using control measures	150,000,000	159,040,000	10.7
2. Other Ships (Bilges, etc.)		150,000,000	10.1
3. Offshore Production (Normal oprations)		30,000,000	2.0
4. Accidental Spills			
a. Ships	30,000,000		
b. Non-ships	30,000,000	60,000,000	4.0
5. Refineries and Petrochemicals		90,400,000	6.0
5. Industrial and Automotive			
a. Highway vehicle spentoils	542,000,000		
b. Industrial plus other vehicles	452,000,000	994,000,000	67.2
GRAND TOTAL		1,483,440,000	100.0

TABLE III-A5: QUANTITIES OF OIL SPILLED AT MILFORD HAVEN[32]

Year	Total Quality Handled 10^6 Tons	Quality Spilled-Tons	% of Total Spilled
1963	13.0	9.8	0.00008
1964	17.7	8.8	0.00005
1965	24.9	35.8	0.00014
1966	28.9	30.5	0.00011
1967	28.2	267.4	0.00095
1968	30.0	13.6	0.00006
1969	39.9	16.5	0.00005
1970	41.3	14.9	0.00004
Yearly Average	28.0	49.7	0.00018

1 short ton = 273 gallons

TABLE III-A6: FREQUENCY AND SOURCE OF OIL SPILLS AT MILFORD HAVEN[32]

Year	Number of Tankers	Number of Spill Incidents	Probability of Spill %	Source or Incident								Size of Incident		
				In Passage or at Mooring	Loading	Discharging	Ballast	Deballast	Bunking	Misc.	Jetties	<80 Gal.	80-160 Gal.	>160 Gal.
1963	1,236	28	2.3%	5	1	5	1	2	1	5	8	18	5	5
1964	1,392	34	2.4%	0	8	3	2	3	3	7	8	17	13	2
1965	1,985	83	4.2	0	11	9	6	3	13	17	24	44	21	18
1966	2,378	72	3.0	2	17	13	3	0	8	14	15	38	21	13
1967	2,680	50	1.9	1	10	17	3	1	4	9	5	27	16	7
1968	2,669	52	1.9	2	9	10	4	0	9	6	12	27	24	1
1969	3,226	58	1.8	3	11	22	0	4	9	3	6	36	17	5
1970	3,440	55	1.6	3	15	13	4	3	8	3	6	35	17	3

engineers often misquote the data in their publications.[33] Therefore, improvement in accumulating data on oil spills in Puget Sound should be well documented and made available to interested parties in such a manner as to eliminate ambiguity.

4. CONCLUSIONS

As indicated in the introduction, quantitative data are insufficient to allow a strictly objective ranking of oil discharge sources on Puget Sound. However, several sources appear to be worthy of close attention because of their very large handling volumes or their very large potential frequency of spillage. Several classes of sources will be relegated to a lesser priority because existing data is not yet conclusive. Therefore, the following listing of priority is suggested as a guide for interim analyses of preventive methods and as an indication for areas of further study where data is inadequate.

First Priority:

(a) Waterborne Traffic (Large Vessels).
(d) Vessel/Shore Transfer Facilities.
(g) Municipal, Industrial, Private Waste Disposal.

Second Priority:

(c) Waterborne Traffic (Small Vessels).
(e) Vessel/Vessel Transfer Operations.
(h) Submarine Oil Drilling (if undertaken).

Third Priority:

(f) Shore Storage and Transportation Facilities.
(b) Waterborne Traffic (Intermediate Size Vessels).

The above priority ranking suggests a method of allocating the scarce resources of manpower and technology to cope with oil spills. Because of time constraints, the study team had to realistically narrow its field of investigation. This report concentrates on waterborne traffic and vessel/shore transfer facilities, though parts of the conclusions and recommendation apply equally well to large and small vessels, and vessel/vessel transfer operation categories.

Municipal, industrial, and private waste disposals are recognized as a high priority item, and much has been documented on methods to alleviate this situation. An oil disposal facility has been proposed for the Seattle area.[34] This facility would handle private, municipal, ship cleaning, and industrial oil wastes. As of this writing, no further progress has been made on the project.

Processing plants for handling private oil wastes are presently in operation in various locations around Puget Sound.[35] A plan to collect and recycle crankcase oil directly from private auto owners was initiated by Superior Refineries, in April, 1971.[36] Evidence has shown that response to this program is negligible.[37]

Recycling of marine waste oil is performed by Chemical Processors at their Pier 91 Seattle plant. This outfit processed about 300,000 gallons of marine oil in the first month of operation. The firm has formulated plans, through a third party, to handle the waste oil from the prospective tankers from Alaska.[38]

Submarine oil drilling has been relegated to second priority only because there currently is no major oil drilling operation on Puget Sound. The American Petroleum Institute chaired a 1970 conference relating to oil drilling and has published a document with numerous articles on preventive techniques in this field.[39] In addition, other publications have documented technological advances on the same topic.[40] This report does not attempt to criticize the published information; neither does it delve into the analysis of drilling operations.

Shore storage and tranfer of oil products are covered indirectly in this report. The recommendations for vessel/shore transfer facilities are applicable in most cases. One area in need of further investigation is the probable frequency of shoreside spills. There is presently no mandatory reporting procedure of shoreside spills. In fact, there are no known statistics on the number of shoreside oil transfer operations in the Puget Sound area.

Improvements in data collection techniques will enhance the quantitative definitions of sources, probability and amounts of oil spills. It is projected that these new informations will reinforce this priority rankings assigned by the study team.

ENDNOTES SECTION III-A

[1]Included are Shell Oil Company, the National Marine Fisheries Service, and the American Petroleum Institute.

[2]Dolan, Savage, Clark, Environmental Statement on the Proposed Leasing of Puget Sound Shorelands and Beds of Navigable Waters for Oil and Gas Exploration, 1970, p. 3.

[3]Data for 1965-1969 provided by annual reports, Waterborne Commerce, Part 4, by U.S. Army Corps of Engineers. 1970 data supplied by a galley proof of that year's **report**.

[4]Ibid.

[5]Puget Sound Task Force, Puget Sound and Adjacent Waters, Appendix VIII, Navigation, 1970.

[6]IMCO sub-committee on Ship Design and Equipment, 5th Session, Grounding Casualty Statistic, May 8, 1970.

[7]IMCO sub-committee on Ship Design & Equipment, Statistic on Grounding and Stranding Damage, May 7, 1970. Lloyd's Registery keeps detailed information on total losses from collision, whereas Liverpool Underwriters Association keeps data on partial losses that could cause total oil spills. Local tonnage data for Puget Sound can be extracted from either the Corps of Engineers reports or the Puget Sound Task Force Report.

[8]System Study of Oil Spill Cleanup Procedures, Volume 1; Analysis of Oil Spill and Control Materials, Dillingham Corporation, p. 10.

[9]Beers, W. J., "Analysis of World Merchant Ship Losses", a paper presented at Royal Institution of Naval Architects, London, March 28, 1968.

[10]The tug population was obtained from estimates by Foss Launch and Tug, and Puget Sound Tug and Barge Companies. The fishing fleet and pleasure boats population are found in the Puget Sound Task Force Report, Appendices XI and VIII, respectively.

[11]Puget Sound Task Force, Pleasure Boat Study, Puget Sound and Adjacent Waters, 1968, pp. 4, 20.

[12]Ibid. p. 4.

[13]Ibid. p. 18.

[14]See section II-C; III-C, part 4; and Appendix 1 of this report for detailed analysis.

[15]Data supplied in Corps of Engineers Port Series reports, also presented in Tables II-C1 thru II-C6 of this report.

[16]See Tables II-C1 thru II-C6 and Figures II-C1 thru II-C6.

[17]See Tables II-C4 and II-C5.

[18]Standard used by State Department of Ecology and City of Seattle with respect to hexane solubles.

[19]Seattle Area Oil Waste Disposal Facility Study, Cornell, Howard, Hayes, and Merryfield, 1969, p. 12.

[20]Ibid, p. 11.

[21]Ibid., p. 11.

[22]Conversation with Mr. A. Poole, engineer for Metro, August 16, 1971.

[23]Cornell, et al, p. 9.

[24]Ibid. p. 8.

[25]Dolan, p. 4, p. 24.

[26]Sylvester R. E., The Determination of Active Fault Zones in Puget Sound, Washington, by Means of Contours Seismic Profiling, unpublished masters thesis, University of Washington, 1971.

[27]Monthly reports of reported oil spills for 1970, provided by Water Pollution Control Commission. A minimum reported spill occurs when the investigator sees no oil at the site, or where light oil is. A moderate spill is one which requires some cleanup, usually involving a heavier oil. An extensive spill is one where the investigator cannot estimate the area affected by ground surveillance.

[28]Hollis M. Dole, Assistant Secretary of Interior, stated on p. 38 of the April 27, 1970 issue of Oil and Gas Journal, that in 1968, 714 oil spills occurred in U.S. waters, of which 2 came from oil wells, 347 from docked vessels, and 300 from shore facilities.

[29]A letter dated July 22, 1971, from Mr. George Moeins,of the Division of Oil and Hazardous Materials, Environmental Protection Agency, to the editor of this report indicated that records of oil spills are stored on computers for the agency's internal use.

[30]The source of this data is the December 1970 issue of Oil and Hazardous Materials Research Newsletter, Volume 3, Federal Water Quality Administration. The original table was in metric tons. To obtain gallons, the conversion formula, 1 metric tone = 301 gallons, was used.

[31]Dudley, Captain G., "The Probelms of Oil Pollution in a Major Port", Biological Effects of Oil Pollution on Littoral Communities, edited by J. D. Carthy and D. R. Arthur, June, 1968, p. 22.

[32]"Written Submission by L. R. Breynon for Alaska Pipeline Service Company Report on Oil Spill Statistics at Milford Haven. Mr. Breynon is head of Environmental Studies in the Technical Development Division of British Petroleum.

[33]D.R. Arthur, "The Biological Problem of Littoral Pollution by Oil and Emulsiforms -- A Summing Up " in Carthy and Arthur. He states that in 1966, 29,000 million kilograms of oil were handled at Milford Haven, with 29,000 kilograms spilled, (1 kg = .905 short tons = 1 metric ton). He also presents a relative figure of 0.01% for spillage amount. The two quantitives quoted correspond closely to the Breynon table. But, many an eminent scientist and engineer has used the misquoted percentage figure. Until now, it is a common household word to many oil spill analyses.

[34]Cornell, et al.

[35]Among these are Superior Refineries in Woodinville, and National Oil Refining in Tacoma.

[36]Article in April 25, 1971 issue of Seattle Times.

[37]Conversation with personnel of Superior Refineries in August 13, 1971.

[38]Conversation with Mr. Kent Clark, vice president of Chemical Processors.

[39]1969 Offshore Technology Conference, Preprints, Volume I & II, American Petroleum Institute.

[40]Ocean Industry, June 1970, pp. 50-51.

[41]In a conversation on November 12, 1971 with Mr. John C. Doolittle of the Washington State Oil Spill Cooperative, the following information was obtained:

> "In the last quarter of 1971 the Washington State Oil Spill Cooperative (WSOSC) has progressed toward their intended goal. They have committed $100,000 toward the purchase of 1000 feet of off-shore Bennett boom and a 40-foot long Huskey SBG floating skimmer. This equipment is to be stored in the Port of Bellingham with the skimmer arriving in early November and the boom in late December. The Coop has also retained for three months a technical and managerial advisor/coordinator to provide the WSOSC membership with a three year illustrative budget and to draft an oil spill response plan. The contingency plan draft will be completed by late December. The Coop is currently considering hiring a full time manager to formalize the organizational plans currently being developed and to coordinate the activities in the oil spill response plan."

III-B. CONTROL OF OIL SPILLS

The cost-effectiveness of controlling (contain and cleanup) an oil spill is dependent on the operational response, capability, resource protection priority, and the implementation of reliable equipment. This section examines, in detail, each of these factors, evaluating the current state-of-the-art and recommending improvements where appropriate. The environmental resource data of Section II-A, the trends in oil handling on Puget Sound as discussed in II-B, II-C, and III-A, along with the economic evaluation, form an integral background of the forthcoming discussions. In addition, the critiques of equipment in Appendices 5 and 6 have influenced the study team's choice of integrated containment cleanup system.

Part 1 of this section discusses the organizational and operational aspects of a response plan. A system of resource protection priorities is presented in Part 2, while configuration for containment and cleanup devices are proposed in Part 3.

1. OPERATIONAL RESPONSE PLAN

a. Introduction

Up to the present time the response to major oil spills and the means directed at containing and cleaning them up have been inadequate indeed. Each succeeding spill revealed new techniques and methods of combatting the spread of oil, but because of the sheer magnitude of a large spill the present overall capability has not advanced significantly.

The Puget Sound region, ecologically one of the cleanest waters in the world, has only now begun to face up to the problems inherent with the transportation of oil. The historical indifference toward solving the problem can be credited primarily to a lack of significant oil shipment on Puget Sound. This coupled with low numbers of oil spill incidents has led many people to believe that a 'real' problem does not exist in the regional waters. With the discovery of the North Slope crude oil and its possible transhipment to markets via Puget Sound waters the problem has been brought close to home. The probable future impact of the discovery will be to double the crude oil movement within the regional waters as well as an increase in refined product shipment due to growth in refinery capacity. Perhaps more important, at least half of this oil will move through narrow channels, marked with navigational hazards, and will be carried by vessels exceeding 100,000 dwt capacity.

The standard answer for what should be done about oil spill control today is to entrust it to technology. Supposedly, this mystical word will permit industry to extract 20,000 barrels of oil from Puget Sound waters when a cargo tank of a large tanker is split open from grounding or collision. In spite of the many advances in oil spill recovery techniques, the resources at command are pitifully ineffective when compared to the requirement dictated by a spill of any significant size. Claims about future capabilities of various cleanup and containment devices do not alter today's cleanup ability. Instead of relying on these promises for the future it is necessary that the most efficient utilization of present scarce resources be undertaken in the event of a crisis. To do so requires the formulation of an organization to control and put to work the ability these scarce resources possess to combat an oil spill. Such plans do exist, but the question that must be asked concerns the ability to respond operationally to a spill of large magnitude and duration. Specifically, are the present contingency plans sufficient, does the skilled manpower in this region to make the plans operational exist, what what would be the costs of such an undertaking? These are only a few of the many questions that must be answered before the effectiveness of a response plan can be evaluated. All along it must be kept in mind that present response systems are the result of a reaction to crises that are tempered with little or no experience.

Furthermore, no two oil spills behave the same because of the multitude of variables involved. A spill on Puget Sound waters will react differently and cause different problems than a spill on some other water. Due to this characteristic, a more specific perspective is taken of the present response capability in the region. The following discussion is not to judge the acceptability, but to evaluate the response system that is the present *modus operandi* for dealing with oil spills. In so doing the major attributes of an effective response system are presented. The present structure and capability of oil response readiness on Puget Sound is looked at. Finally, some recommendations are made with the major effort directed toward enabling oil spill response on Puget Sound to be an operational reality.

b. Major Prerequisites for an Effective Response System

An organization designed to respond quickly and efficiently to periodic random happenings is not to be considered as any ordinary day-to-day organization. This peculiarity becomes especially evident when one considers that the organization is staffed with non-professional, for the most part, and volunteer people. The various component parts of the organization have different interests and motivating pressures in spite of a single objective; and the organization, by necessity, operates in an environment where speed and accuracy of decisions are of utmost importance. The design of an organization to respond effectively to the crisis of an oil spill obviously is difficult; however, not impossible. The plan discussed in this section is patterned after the industry response plan proposed by the Dillingham Environmental Company in 1970.[1] The ideas and recommendations are to some extent, utilized and made specific for the characteristics in the Puget Sound region. As well they have formed the framework with which the existing response capability is measured and evaluated.

1. Component Parts

Basically a response plan must specify the following categories to be effective:

a. Organization: A formal organization structure with the lines of authority clearly delineated.

b. Operations: The detailed procedural steps need to be developed.

c. Logistics: Must include all the necessary contractural pre-arrangements for the procurement of resources and transportation to the site of the spill in the shortest time possible.

d. Preparedness: The entire response plan must be operational prior to the spill event, with regard to the specific functions that need be performed, and at a level commensurate with the financial capability of the relevant sectors in the Puget Sound region.

2. Objective and Attributes Required

The primary objective of the organization is to contain and clean up oil on water at the least economic cost and in time to minimize the damage inflicted upon the environment. In achieving this primary goal there are any number of attributes the organizational response plan needs to possess. The primary characteristics are discussed below:

a. The plan, because of legal constraints (i.e., liability of the offender), must be voluntary on the part of industry. That is, if industry does not agree to accept the provisions therein it cannot be forced to do so. If agreement is not reached, effectiveness is decreased. A company deciding to act alone in relieving its liability for an oil spill it created can do so - response plan or not, so long as this duty is discharged without jeopardizing to interests of the public at large.

b. The myriad of problems created by all but the smallest oil spills today dictate a relatively large, though not necessarily rigid, organization to solve these problems.

c. Because of financial cost considerations, many key positions need to be filled with the assignment of temporary people in the local response area. These would most likely be selected from interested government agencies and participating private industry. It is felt that the incremental cost attached to a decreased organizational efficiency is much less than the loss of incremental benefits of staffing the organization with full-time personnel. It would be expected, however, that the response organization be professionally staffed to the point where the incremental cost of the partially staffed organization is equal to the loss of incremental benefit as a result of that action. A benefit to the industry point-of-view in this idea of temporary staffing is that ultimately industry maintains more control over the operation, and hence its financial liability.

d. For total response planning, an independent communications network complete with the appropriate central command and field communications centers is of utmost importance. The need for effective communications cannot be overstressed in light of the problems encountered as a result of poor communications in past spill incidents.[2] It does not appear unrealistic to procure, through a one time expenditure, the necessary communications equipment and design into this primary system an interface of periphery secondary systems, comprised of all interested parties in the region, rather than relying on a patchwork of temporary and unreliable crisis-constructed systems.

e. The requirement of rapid response to a spill incident dictates the need for a continuously operational alert system that not only will bring the response effort into full operation quickly, but that will also confirm the degree of response required. The recent Anacortes spill showed the great need for an accurate appraisal of the spill situation so as to intelligently activate the degree of response required.[3]

f. The authority and responsibility relationships between cooperative parties for the initiation of containment and removal operations needs to be clearly delineated before the fact as extensive cash flow expenditures become ongoing with each decision. In a parallel vein, the specific conditions under which the public interest is no longer being represented by the offending polluter in his cleanup operation needs to be spelled out so as to relieve the on-scene commander (OSC) and the respective government agency (USCG) from possible future litigation and the assumption of the costs of the cleanup. These authorities, responsibilities, and takeover conditions must be well defined, clearly stated, and thoroughly understood by all the interested parties in this response region.

g. Because of basic financial constraints, the capability existing today must be utilized to its full extent. This requires pre-planning, which in effect this proposed response plan is intended to do. Nowhere should a present effective response plan be downgraded in view of "future developments" of the technical capability to remove oil spills from water.

h. Arrangements must be made to stockpile all the necessary supplies, and materials required to combat a spill. The deployment of such apparatus to the site of the spill also needs to be pre-arranged. The distribution of the supplies and materials within the region is subject to many factors, primarily being the relative assessed hazard of a spill in each and every subzone within the region. In any regard, at least a minimum level is required to provide for an initial response. (More details on this are in Part 2 of this section.)

i. To maintain the operational preparedness required, there is a need for some type of ongoing formal training of regional personnel who will fill the various organization billets in the event of a response. This is necessary because of the many parties that are integrated into the response organization. It is necessary that all concerned receive the same interpretations and gain a common understanding of the organization in which they are participating.

c. Present Status of Puget Sound Response Capability

For a response plan to be truly operational the organizational attributes discussed above are of vital importance. Of overriding consideration in the development of this plan is the fact that it must be operational much the same as the municipal fire department; and secondly, along these same lines, that it evoke an immediate response action. Thus, the organization must be flexible yet authoritative, large but not cumbersome. This is especially a requirement for the Puget Sound region, as all its waters are closely bounded by land rich in man-made and natural resources. Along with the waterway itself, all are of vital necessity to the future economic development of the region. The need for more than a paper plan is thus magnified because of these special considerations.

The organizational response capability now present in this region lies with two basic plans. The first, and perhaps the most general, is the contingency plan of the U. S. Coast Guard.[4] This plan addresses itself primarily to the response capability of the various government agencies. The procedures outlined adequately to interlock a local regional plan into a national response capability and dictate under what circumstances the local capability can be expanded to receive aid outside the immediate response area. It constitutes a viable framework for those individuals operating at the upper level of responsibility. The guidelines of what action to take are fairly definite down to the level of the sub-regional response center (SRC). Beyond that level, however, the plan loses its effectiveness, the very level in the total organizational response structure where the actual work of removing the oil from the water takes place.

The second formal plan is the joint venture of Puget Sound Tug & Barge Company and Pac-Mar Services, Inc., known as Marine Oil Pick Up Service (MOPS).[5] This plan in essence is not a response plan at all, but a prospectus offering of the service capability of MOPS. MOPS is available as a third party contractor to individuals or firms that bear the responsibility of oil cleanup but do not have the expertise nor the resources with which to accomplish such a task (which virtually includes all oil handling facilities on the Sound). The MOPS plan is perhaps the closest thing this region has to an operationally ready response capability. Although not spelled out on paper per se, the firm appears to have the capability of operating effectively once called into action. This is due to their experience and expertise in this field.[6]

Supplementing the MOPS operational capability are the response plans of the various individual companies involved in the business of oil transportation, receiving, refining, etc. This study group is aware of the plans of 11 such facilities, however, it is noted in another section of this report of the general inadequacy of the response capability of these firms.[7] For the most part these eleven oil handling facilities conformed to the minimum requirements of the Coast Guard Standard for oil handling facilities.[8] However, their ability to respond to a spill situation, should one occur, is questionable, as shown by the results of the survey and discussion with the operating personnel of these facilities.[9] Regardless of the fact that the sum total of these fragmented response capabilities may be considered significant, their total contribution is minimal unless they become integrated through some collective agreements into an overall organization response plan.

Major oil refiners and handlers, in and around Puget Sound, have formulated an organization to provide response capability for oil spill cleanup. This group, the Washington Oil Spill Cooperative, is made up of ten private firms, and was created in April, 1971. After several months of negotiating financial and political questions with their respective parent companies, each firm has begun to provide the required resources to fulfill the cooperative's objectives. These goals include the assessment of the required response to an oil spill, evaluation of current capability, stockpiling of equipment and materials, and training of personnel. At publication time of this report, no progress report of the cooperative's efforts was available.*

---

*See Endnote 41 this section.

To summarize the preceding discussion, two response plans are of any significance in terms of their ability to cope with larger spills. They are the previously mentioned Coast Guard contingency plan and the MOPS Plan. The Coast Guard plan is well organized insofar as its ability to develop an effective governmental response and provide direction from the higher level of responsibility. It also provides technical and factual information as to the various oil spill combatant methods and the availability of equipment and materials for the man in the field, but does not provide the necessary direction to ensure the necessary attainment of optimal operational results. The MOPS plan contains a degree of this capability but is of value only when, as a third party, they are contracted to remove the spill. Clearly what is needed are some guidelines for the personnel directing the cleanup operation as to what is necessary for a complete and thorough job. A discussion of these requirements follows in the development of an integrated operational plan below.

d. Proposed Integrated Overall Operational Response Plan for Puget Sound

Portions of existing response plans are included in an overall response framework to show what areas presently appear adequate, as well as to point out areas where potential deficiencies exist. From an elaboration of these points, recommendations are made that, if adopted, would give Puget Sound a much improved capability to respond to an oil spill over what presently exists.

The total response model is represented by Figure III-B1, a block diagram of the major functions necessary for effective spill response and their inter-relationships.

1. Participants

Present support capability is both operational in nature as well as technical in terms of advisory staff. A scientific advisory group has been formulated in the Coast Guard plan.[10] Their function is to offer information and supporting data as to the appropriate plan of action that should be taken by the sub-regional response team (SRT). This support includes, but is not limited to, technical expertise in plans of containment,

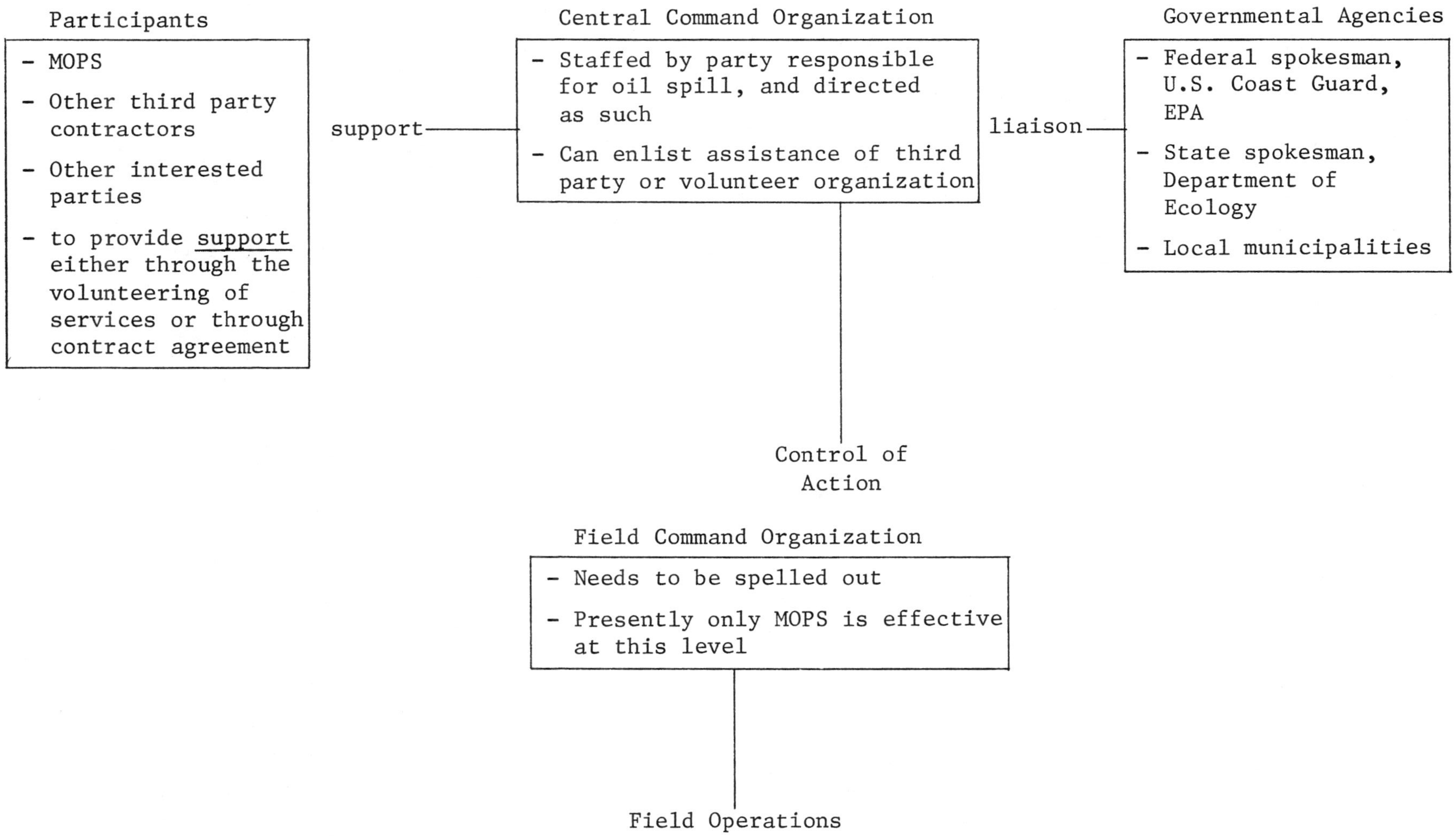

FIGURE III-B1: PROPOSED RESPONSE ORGANIZATION

cleanup and environmental restoration. A second support advisory group consists of members of the industrial community that can offer expertise on operations and physical peculiarities of onshore and offshore structures, water-front facilities, ships, barges and other sources. Both groups are to meet and support the SRT in the event of a moderate or major spill (a spill greater than 100 gallons in internal waters).

Operational support can be provided by the MOPS organization on a contract basis, or by any number of interested third parties who may volunteer or contract their services to the party responsible for the spill. The main strength in this operational support capability appears to be MOPS. Other companies do not have an adequate or immediate response capability other than perhaps management expertise and labor resources to effectively cope with a spill. At present even this limited capability is questioned because of their apparent lack of a totally integrated readiness plan.

It appears then that the Central Command responsibilities will be supported primarily by MOPS should a spill occur in the near future. It is doubtful, though, under the circumstances, that the responsible party would turn control of the cleanup operation over to MOPS unless the necessary agreements had been made previously before the fact.

2. Central Command

The proposed Central Command will be staffed by personnel of the responsible party supported in key areas by MOPS personnel. Presently at issue in this type of arrangement is the ability of the responsible party personnel to provide a level of expertise and management knowledge so as to best cope with the spill. The public interest is legally represented by the Coast Guard On-Scene Commander (OSC) who has discretionary power to assume command when he deems the efforts of the responsible party are not sufficient. However, there are no established guidelines through which this authority may be exercised. Herein lies the problem that needs to be resolved through the establishment of these guidelines.

It is anticipated that MOPS can make a significant contribution to the overall Central Command function primarily due to their past experience in dealing with oil spills. In the event of a spill all possible efforts should be expended to place some of this expertise in the position where it can do the most benefit regardless of the contractual arrangements implicit in the relationship.

The Central Command Center is the hub of control and, as such, it must contain the necessary equipment and supplies. It does not appear of benefit, in Puget Sound, to duplicate this center to assure more readiness. One center centrally located, preferably adjacent to the primary staging area, is sufficient. At present there are two Central Command Centers. The MOPS organization has such a resource located at their present base of operations.[11] The Coast Guard has provided a center for the federal response effort at the 13th CGD Regional Response Center (RRC) and for the OSC at the Coast Guard Stations which is designated as the Sub Regional Response Center (SRC).[12]

Both centers have shortcomings. The MOPS center is privately owned and not available for use unless under contract. The RRC provided by the Coast Guard is available only to federal officials. What is needed is a center completely furnished with the necessary materials that could be made available to the party responsible for cleanup with no questions asked.[13] Funding for such a center could come through the appropriate state agency. The amount will be miniscule compared to the total amount involved in a cleanup operation and would be in the best interests of all segments of society.

3. Governmental Agencies

The primary liaison exists between the various federal and state agencies and the Central Command organization. This liaison exists and is necessary because of the Federal and State statutory responsibilities for response to pollution incidents.[14] The primary federal responsibility is to ensure that the party responsible for the oil discharge is taking adequate action to remove the pollutant or adequately mitigate its environmental effects. As pointed out previously some policy guidelines need to be developed in order to carry this out. So long as effective action is being taken, the federal activity is primarily to observe and monitor the progress, providing advice and counsel as required. Where proper action is not taken by the responsible party, further federal response needs to be initiated. Those federal agencies possessing the necessary facilities and resources useful in combatting a spill can be called into action consistent with their operational requirements and within the limits of existing statutory authority. The exact nature of these federal response capabilities in this region are spelled out explicitly.[15]

State response takes place through the Department of Ecology, Water Pollution Control Commission, whose present director is Mr. John A. Biggs. State law requires the party owning or having control over oil that is spilled into state waters to be responsible for the immediate removal of the oil. The state has further statutory authority to assume the responsibility of cleanup where the party at fault fails in his obligation to do so as required by law, and where the public interest is at stake. All expenses incurred are subject to reimbursement by the responsible party.[16]

The capability of the various governmental agencies to fulfill their responsibilities appears adequate for the present. This is not to say they should not continually evaluate these capabilities. At the writing of this report, the study group had no direct knowledge of a state response contingency plan. A plan apparently was being formulated however, as reported by Mr. Harry Tracy of the Department of Ecology.[17] It is recommended that the state make public any plan they may now have under consideration at the earliest possible date. The Federal response plan has been critiqued and is included in another section of this report.[18] The substance of the criticism revolved around the following points:

a. definition of spill classification
b. response communications capability
c. establishment of a baseline data bank
d. priority of critical water area classifications
e. utilization of the Scientific Advisory Group
f. training of response personnel
g. adequate listing of equipment and services resources

No further comment is made on these points in this section.

4. Field Command

Up to this point all the various sections of the overall response plan discussed have dealt with the necessary support and liaison functions as well as the upper level of responsibility in directing a cleanup operation. This last section is categorized as perhaps the most important, for it is under this organization that the actual cleanup will be accomplished. (A subsection of the Field Command is Field Operations, which is discussed under a separate heading.) Unfortunately, of all the response capability in Puget Sound to date, this function is perhaps the most overlooked and the weakest. The Coast Guard contingency plan provides very little information to those directing a cleanup operation in the field. Critical water areas are briefly enumerated, some obvious cleanup and disposal techniques are discussed, and a questionable listing of resources is provided.

Therefore, the effort here is to outline a general plan of attack to provide operational direction in combatting an oil spill to those less than completely familiar with technicalities required. It is highly recommended that a plan of this nature, or some derivative thereof, be disseminated among all interested parties to be used by their personnel as needed. The main advantage of such an arrangement is the ease of coordination of field activity with the other aspects of the response plan already fairly well established and understood.

At present, the best answer to a field command organization is perhaps the MOPS organization. They possess the necessary resources, personnel and expertise to sustain large scale field operations. There are other private parties available but not of the necessary size. However, as stated before, these organizations are of value only if their services are contracted for. There needs to be an operational plan that provides the specifics as to the ordering of equipment, mobilizing labor forces and other resources, and the coordination of field efforts with the other sections of the overall response plan.

Such a model plan has been developed by the Dillingham Corporation. The substance of this plan is adopted for utilization in this report with the recommendation that it be expanded upon and a major effort be directed toward the establishment as the standard for Puget Sound. The following discussion presents the basics of the plan.

Once a spill situation occurs certain factors become of utmost importance. These include:

- operations, which can be broken into four phases:
  a. activation of response plan
  b. initiation of offshore control activities
  c. initiation of shoreside protection and cleanup activities
  d. termination of operations and demobilization of equipment and labor resources
- logistics
- preparedness

A brief comment is made about each in turn.

The key to conducting efficient operations is how the response plan is activated. This in turn depends primarily on the initial assessment of the size and nature of the oil spill. Because of the importance of this aspect of the response it needs to be made by an individual in the upper levels of responsibility in the response organization - most likely the general manager. It would also be advisable that he be accompanied by representatives of the State and Federal interests. To ease the uncertainty in the making of this decision several questions need to be asked. Below are a series that, if correctly answered, can provide a reliable reference point for commencing effective cleanup operations.

1. What is the nature of the material spilled? (Chemical qualities.)
2. What is the exact location of the origin of the spill?
3. What has been the quantity of material spilled and is the spill continuing?
4. How much time has elapsed from the initial spill and to what degree has the material spread?
5. What is the source of the spill and the quantity of material remaining in the source?

6. What is the local situation (weather, degree to what action has been taken, etc.)?
7. Is the capability of the responsible party for controlling the situation adequate?

All logistical support needs to be pre-arranged for the entire region in order to ensure that the necessary resources are available at the least economic cost. The responsibility for maintaining a logistical arrangement is primarily a preparedness function discussed in the next paragraph. The responsibility for maintaining logistical readiness should include:

1. All oil spill control equipment and material purchased.
2. The inventory, storage, and maintenance of this equipment and material.
3. All contractual arrangements for other equipment, personnel and facilities.
4. All transportation arrangements.
5. Maintenance of up-to-date equipment and material specifications.

Complete records of all logistical resources need to be maintained including periodic (3-6 month interval) inventory updating of all available resources.

For any response plan to be wholly effective the preparedness function becomes important. Full preparedness by necessity should include the following specifications:

1. Organizational and maintenance of an oil spill alert system.
2. Maintenance of an up-to-date roster of response plan personnel.
3. Maintenance and periodic operation of the region's Central Command Center.
4. Maintenance of an up-to-date inventory of all available oil spill control equipment as well as periodic inspections as to their serviceability.
5. Training in control techniques to all parties who may be expected to participate in an oil spill incident.
6. The planning and initiation of mobilization exercises designed to simulate an actual oil spill incident, with a resultant evaluation of the experience.
7. Provision of up-to-date information to the various governmental authorities concerning the readiness of the response organization.

Such specific preparedness factors could be managed through a cooperative effort on the part of interested parties in this region. This effort could either be in the form of lending management personnel for short intervals of time on some rotation basis, or the pooling of funds contracting a third party to perform the task. Regardless of the method employed, the establishment of the preparedness responsibility is of utmost importance. However, it should be kept in mind that one-time expenditures for adequate equipment and materials should not be sacrificed at the expense of a rather large and elaborate preparedness staff incurring annual costs.

5. Field Operations

Once the response plan has been actuated, the Central Command Center has been established, liaison with the governmental agencies has commenced, and the initial survey of the spill has been taken, the business of cleaning up the spill begins. Generally, for a spill of any size, operations should be conducted round-the clock until the operations are terminated. The conductance of field operations is described primarily by a listing of the task requirements for any one phase of the operation. A complete discussion of the task requirements is presented in the Dillingham Report.[19] Lack of time and space prevent a complete adaptation of these task descriptions to Puget Sound in this report.

It is advisable, though, that these ideas be adapted to this region and incorporated into a type of field operation handbook that could as well become a part of the Coast Guard regional contingency plan. The task descriptions in the Dillingham report incorporate much of what has been learned on previous spills. The four stages of the operational activities as previously listed are now expanded into a sequence of required tasks. Specific reference to each task is made to the Dillingham report, and noted after each item:

Task Descriptions

1. <u>Mobilize for Control Action</u>

a. Establish and man the field command center (1e).

b. Establish communications between the field and central command centers (1f).

c. Commence documentation of the oil spill incident (1g).

d. Commence an environmental analysis ex post for the spill. (It is assumed an ex ante analysis has already established some type of baseline.) (1h)

e. Commence the plotting and prediction of oil slick movements (1i).

f. Formulate the initial plan for control action (1j).

g. Alert the logistical network for the possible movement of control resources (1k).

h. Make final considerations as to the initiation of a control action (1*l*).

2. Initiation of Offshore Control Activity

a. Implement the selected offshore control plan (2a).

b. Implement the logistical system of pre-arranged resources (2b).

c. Submit periodic control status reports to all interested parties (2c).

d. Begin public information disclosures (2d).

e. Monitor and cooperate if necessary in the salvage of the source of the oil spill (2e).

3. Initiation of Shore Side Protection and Cleanup Activities

a. Implement the selected Shore Side Control plan (3a).

b. Maintain liaison and appropriate communications as above in 2b, 2c, 2d;(3b).

c. Begin to dispose of the oil residue and rubbish (3c).

d. Rehabilitate waterfowl and other threatened wildlife (3d).

e. Provide the necessary assistance in settlement of claims resulting from the oil spill (3e).

4. Termination of Operations

a. Evaluate the requirements for continued control action (4a).

b. Begin to recover the control equipment (4b).

c. Make a professional assessment of the effects of the oil spill (4c).

d. Prepare a final report on the spill incident with recommendations for future incidents (4d).

e. Summary and Recommendations for an Integrated Operational Response System

The Puget Sound region at present has a response organization sufficient at the upper levels of response management, primarily in the liaison function of the federal agencies. No plan has been developed by the State of Washington to date, so the exact position they would occupy in the overall response is not completely certain. The major problem in the response capability appears to be with the actual control operations. What is required in this area is to advance a standard operational plan that can be maintained through the cooperative efforts of all the interested parties. The major points of such a plan have been discussed in this section of the report.

The following list of recommendations should carefully be considered in a true assessment of this area response capability. At the risk of some repetition, it is the consensus of the study group that the following factors need be considered in order to improve the Puget Sound oil spill response capability:

1. A baseline of relevant data needs to be collected and stored in such a form that rapid retrieval will permit maximum use. It is recommended that further work be done on the storage of information by digital computer.

2. That training of individuals most likely to become involved in the cleanup of a spill be instituted to provide knowledgeable people experienced in the techniques of oil spill control.

3. That an integrated independent communications network be established that can be placed in the field and made operational within a matter of hours.

4. A detailed up-to-date listing of all available equipment, material, and labor resources be maintained through the cooperative efforts of interested parties. An effort needs to be directed at the establishment of minimum standards for equipment and materials. For example, straw mulchers that break straw too finely are of little value, as there is a optimal size straw for soaking oil.[25]

5. Previous arrangements need to be made for all resources including firm cost estimates and prior contract agreements as well as the means to transport the resources to the spill site.

6. A comprehensive analysis and identification of high priority areas within the region is required. Before this can be accomplished, it is necessary that some agreement be reached as to the definition of what constitutes high priority (see part 2 of this Section).

7. Both during and after the spill incident an evaluation of the environmental impact of the spill needs to be conducted. This should be a part of the pre-arranged part of the response plan and conducted by local disinterested third parties so as to introduce minimum bias into the results. The results of such an evaluation need to be made automatically available to the appropriate agencies.[20]

8. A capability needs to be developed for all-weather surveillance of the movement of the oil slick.

9. Mobilization exercises need to be conducted to test the readiness of the response organization as well as to simulate and evaluate the response system.

10. A minimum level of effectiveness of a response organization needs to be defined so as to set some standard by which a response effort can be gauged.

11. The field command response effort needs to be further defined. One such method is suggested in this report.

12. Specific policy guidelines need to be established to delineate the extent of the authority of the on-scene commander over the responsible party. For instance, under what circumstances does the OSC have the authority to make the decision that an inadequate effort is being made to cleanup the spill.

2. RESOURCE PROTECTION

a. Introduction

The purpose of this section is to integrate the data on the environment and resources of Puget Sound[1] with the evaluation of containment and cleanup devices[2] into the oil spill response plan. Several important qualifications are mentioned here prior to the discussion. These include:

1. Complete and detailed information on the environment, physical resources, and effects of cleanup material on these resources has not been accumulated.
2. Capability of containment and cleanup devices under adverse meteorological and hydrographical conditions is still under investigation.[3]
3. Statistics on oil spills in Puget Sound are not comprehensive enough to allow accurate predictions on frequency and locations of future spills.[4]
4. Valuation of the physical environment and its inherent resources are presently conducted under various criteria, which frequently are in conflict with each other.[5]

Even with these constraints, the consensus of opinion dictates that some assignment of priority rankings is necessary to protect societies resources with a limited arsenal of equipment and manpower. The alternatives to this approach are:

1. Expend sufficient manpower and equipment to completely protect all resources in the Puget Sound area.
2. Expend no effort and "let nature take its course."

Neither one of these choices would minimize the cost to society. The former would impose extremely burdensome capital outlays without commensurate benefits, while the latter would impose potentially severe losses to society today, and even harsher penalties on successive generations. Thus, some basis must be established for selective protection of resources. Three such bases are proposed in this report.

b. Valuation of Resources

The methods of determining the value of resources in Puget Sound include:

1. renewability,
2. economics, and
3. arbitration.

Each method will be discussed separately.

Renewability: The first method compares resources which are renewable (man-made) against those non-renewable (natural). It would evaluate certain natural resources such as a biological community of particles types of organisms as being more important than any other man-made facilities such as a marina, regardless of financial value. Simply stated, this method assigns priority to those resources that would require the longest period of time to be replaced (if ever) regardless of cost. It is assumed that man-made resources (real property) can be readily replaced or repaired.

Economics: The second method, economics, is perhaps the most accepted and widely used method of resource comparison. However, the parameters of this system are often very difficult to evaluate and only a partial comparison may result. Factors abetting this difficulty include the evaluator's viewpoint, and the intrinsic worth versus cash outlay question for a given commodity. For example, the financial value of a man-made facility such as a yacht basin can be accurately determined by an estimate of current replacement value, but what value does man assign to the convenience of easy access to waterborne recreation? It is assumed here that questions of partial replacement and/or primitive assessments due to litigation are de-emphasized. On the other hand, the economic value of commercial species of salmon or oysters is difficult to determine because of the ambiguity of determining the causes of catch variables (no absolute measure of standing stock and the uncertainty about the period of time necessary for stocks to return to pre-spill levels). Furthermore, the assessment of the relative economic value of sport fishing and purely aesthetic resources are only crudely suggested by estimates of tourist and recreational expenditures. Therefore, local man-made resources of readily assignable value tend to become the most highly valued by this system of comparison and because of the difficulties inherent in assigning a financial value upon natural resources.

Arbitration: The third method, arbitration, acknowledges that the relative value of certain resources are presently indeterminate or hardly known. For example, the paucity of conclusive data concerning the importance of non-economic plant and animal species to the sustenance of economic species, via the complex food web, precludes the dismissal of the importance of the non-economic species *a priori*. Furthermore, the problem of evaluating the recreational and aesthetic values of non-commercial biota and other natural resources (such as beaches) makes a strictly economic mode of comparison seem inadequate. Unfortunately, the arbitrary criteria for evaluation are most susceptible to dispute by parties of differing interests and are not as practical as the first two methods of comparison.

For the purposes of this section, the relative value of subzone resources subject to oil contamination will be determined by both economic and non-economic criteria. Since it is the intent of this section to provide potentially useful information for rational emergency decisions for resource protection, both options for strictly economic and non-renewable resource protection will be presented, and the decision regarding the relative value of each criterion will have to be determined by the emergency response decision maker. Within the limits of the response capability that may be allocated, it is hoped that both economic and non-economic criteria will be given priority.

It should be emphasized that the philosophy of this section is to present an interpretive synthesis of available data for an interim contingency plan for emergency action on Puget Sound and adjacent waters. The recommendations contained herein are not to be construed as final nor absolute; the recommendations are offered in good faith to (1) improve the data base from which emergency decisions can be made, (2) emphasize topics for further analysis, and (3) stimulate discussion and pre-emergency planning.

On the other hand, the priorities documented in this report are the best available today. In case of an oil spill tomorrow, they should be integrated into the response plan. Continual refinement of the values of resources by accumulation and evaluation of information is essential to its effectiveness.

c. Resource Protection Priorities

As indicated in the introduction, the relative uncertainty involving the frequency and distribution of future oil spills on Puget Sound and adjacent waters seriously hinders the possibility of making very specific recommendations for the location of a minimum emergency response facility (-ies). In addition to the great variability in environmental conditions (currents, winds etc.), the character of the particular type of petroleum product that is spilled determines to a large degree the kind of containment and cleanup response that is most appropriate. For example, the more viscous crude or bunker C materials spread less rapidly and interact with water very differently than the lighter refined fractions such as gasoline.

In view of the uncertainties involved, and the number of special cases that may be considered (i.e. crude oil spilled in the turbulent currents of Admiralty Inlet versus the spillage of refined products in the relatively calm waters of Elliott Bay), it is impossible to prepare for all possible cases of emergencies due to spillage in all areas.

However, it is believed that a conservative minimum response capability might be developed to optimize the procurement and placement of emergency resources within practical financial constraints. The recommendations for each subzone which follow will be concerned with a suggested minimum response capability as well as the possible placement of additional emergency equipment and manpower if it becomes available.

Generally, response time in an accidental spillage is critical because the successful containment and cleanup of spilled oil depends upon keeping the oil film as concentrated as possible; as the spilled oil spreads and the film becomes thinner with time, possible resource contamination increases and cleanup potential decreases. Containment barrier systems may be satisfactorily effective in certain cases both to contain an oil slick and to protect special resources as a defensive measure. A distinction should be made between a capability to contain versus a capability to protect. The quantity and character of an "offensive" containment barrier required to retard and redirect oil spreading are dependent upon the environmental conditions and the necessary amount of equipment is difficult to predict for any spill size, even if the spill size can be anticipated. On the other hand, the "defensive" deployment of containment barriers to isolate especially valuable resources may be ascertained by an analysis of regional geography, and minimum quantities can be specified within certain limits.

Because of the cost and variable effectiveness of containment devices, it is not practical to make a recommendation for a minimum quantity of containment barrier that should be maintained in each area on the basis of resource occurrence alone. However, the effectiveness of sorbents both in the removal of oil from water and in the mitigation of damage to shorelines indicates that it would be prudent to maintain at least the availability of a certain minimum quantity of sorbent material accessible to every area that might be jeopardized on Puget Sound. A subjective examination of past spills, oil spreading characteristics, and costs suggests that an arbitrary minimum response capability for all of Puget Sound and adjacent waters should be the availability of enough sorbent material for 5000 gallons of oil that may be deployed anywhere in the region within a maximum of 4 hours after notification.

This minimum response capability will probably require the local storage of some sorbent material in several subzones. With this minimum response, cleanup, and protective measures can begin within the critical first 4 hours which may be supplemented later by additional emergency resources from central points in the region.

Specialized optional recommendations for the availability and deployment of emergency resources will be summarized in the following pages for each subzone.

Subzone 1: Strait of Juan de Fuca

Environmental Conditions:[6] Open water conditions, locally swift tidal currents, wind waves and swells from Pacific Ocean as well as wind waves from prevailing winds in Strait.

Prevailing waves and winds probably exceed the operating limits of commercially procurable barriers most of the time.

Shore Type: Predominantly coarse; rocks and boulders form narrow beaches bordered by banks and cliffs of variable relief. Access to eastern portion of shoreline is fair from public highways. West of Port Angeles, shoreline access from land is only intermittent. Logs and beach drift are common and may hinder shoreline operations.

Possible Sources of Spills: Fuel or cargo of vessels operating in the Strait of Juan de Fuca. Oil spilled in neighboring subzones and Canadian waters.

Special Resources:[7] Port Angeles - Inside Ediz Hook recreational and industrial property such as boats, piers, marinas, may be protectively boomed.

Dungeness Bay - Commercial and sport shellfish are prevalent in intertidal and sublittoral zones. Protective booms may be of marginal effectiveness, but deflecting booms may be most effective due to the tidal currents to divert oil slicks from bay entrance. Generous application of sorbent material on tideland shoreline and on nearshore water surface may mitigate shoreline damage.

Sequim and Discovery Bays - Commercial and sport shellfish. Dynamic or deflection booms may be marginally effective to reduce encroachment of oil slicks. Tidal currents at constricted entrances may significantly reduce the effectiveness of fixed containment booms. Sorbents applied to shoreline would probably reduce damage.

Port Townsend - Piers and moorage facilities; booms probably ineffective except very locally due to waves and currents.

<u>Subzone 2: San Juan Islands/Bellingham</u>

Environmental Conditions:[8] Open water conditions in S.W. quadrant and Rosario Strait. Milder conditions (currents and waves) prevail in San Juan Islands and in the vicinity of Anacortes and Bellingham, but tidal currents may be locally swift at constricted passages. Containment booms would probably be marginally effective except at passages where tidal currents would reduce effectiveness. Dynamic or deflection booms may reduce damage to large areas if placed near eastern passages of San Juan Islands.

Shore Types: Predominantly rocky with some sand and gravel intermittent in San Juan Islands and North of Anacortes. Eastern Padilla, Samish, and Bellingham Bays are mainly estuarine tidelands with broad tidal flats. Access to Bellingham and Anacortes areas is adequate by public roads. Shorelines of all islands of limited access by public and private ferry service. Logs and beach drift are common.

Possible Sources of Spills: Loading and handling accidents at Anacortes, Ferndale, Cherry Point; waterborne traffic in Rosario, Haro, and Georgia Straits. Refined products are primarily transported by water.

Special Resources:[9] All islands of San Juan Archipelago display valuable shellfish resources and the area may be partially protected by dynamic or deflection booms. If spill is from Rosario Strait or eastward, dynamic booms may be attempted at Obstruction and Peavine Passes, Thatcher and Lopez Passes, and San Juan Channel with marginal success predicted.

Anacortes and Bellingham - Widespread property resources (piers, vessels, small boat harbors) might be boomed with marginal success except in Guemes Channel where tidal currents would restrict the effectiveness. Generous application of sorbents to shoreline would probably reduce damage.

NOTE: Because net water circulation in Guemes Channel is to the west, this tends to flush material into open waters; therefore, spilled refined products should not be dispersed mechanically. Such dispersal would probably tend to increase local damage by incorporating the contaminant into the water column. Shoreline sorbent application is probably the best approach for refined product spillage.

Lummi Island/Lummi Bay - Aquaculture experiment and biological resources may be partially protected by dynamic booming and the application of sorbents to the shoreline.

NOTE: During months of peak flow, the suspended sediment transport of the Nooksack River may alter the density of the fresh water near its plume. This increase in density of the surface water as well as the effects due to particulate surface adhesion may significantly affect the behavior of spilled oil and may cause some sinking of surface slicks.

Subzone 3: Whidbey Island/Everett

Environmental Conditions:[10] Western portion of subzone has open water conditions which significantly restrict emergency containment and clean up operations. With the exception of Deception Pass and its adjacent area, the waters east of Whidbey Island are relatively sheltered and tidal currents, winds, and waves may be often mild enough to use containment barriers with marginal success. Admiralty Inlet to the S.W. of Whidbey Island is probably not amenable to containment operations due to tidal turbulence and swells.

Shore Types: Predominantly sandy and rocky on western Whidbey Island with high banks. Eastern Whidbey Island, Camano and mainland shorelines are primarily narrow and rocky with banks of variable relief. Accessibility to shore from land is limited from public highways. Large logs and beach drift are common on beaches and may hinder shoreline operations.

Possible Sources: Western side of Whidbey Island - vessels in Rosario Strait, Eastern Strait of Juan de Fuca.
Eastern Whidbey Island and waters adjacent to Camano Island are not commonly used by larger vessels so only significant source would be oily material spilled outside the subzone entering by Deception Pass or Possession Sound.

Special Resources:[11] Although the waters surrounding Whidbey Island represent rich areas of recreational and commercial fishing, the prevailing environmental conditions probably preclude the practical use of anything except shoreline sorbents except in the sheltered waters to the east of the Island. All waters to the east of Whidbey Island are valuable recreational fish and shellfish areas.

The Skagit and Snohomish Rivers seasonally support the spawning of large numbers of salmon which must transit salt waters.

The waters of Skagit Bay, Saratoga Passage, Port Susan, and Possession Sound are locally amenable to containment and protective booming, but the ecologically productive tidelands and shorelines are generally all of equal value and any emergency response should endeavor to protect the entire area as best as possible from the encroachment of oil spilled in other zones.

The Port of Everett exhibits considerable property resources that may be locally protected by booms, but the effectiveness of barriers may not be great due to tidal currents and wind waves.

NOTE: The seasonal flow volume and sediment transport of the Skagit and Snohomish Rivers[12] (greatest in June and December) may affect the behavior of oil spilled near the plumes of the rivers. Density and particulate surface effects may induce sinking.

Edmonds has valuable property resources in the form of piers and small boat facilities and should be amenable to protective booming under most conditions.

Subzone 4: Northern Puget Sound

Environmental Conditions:[13] Generally the conditions of Northern Puget Sound are favorable such that containment and protective booming is possible with marginal success. Except in the constricted passages around Bainbridge Island, tidal currents for the subzone are normally weak and wind waves are often insignificant (although they may become significant during exceptionally windy periods).

Shore Types: On the eastern side of Northern Puget Sound, the shores tend to be sandy and rocky with adjacent banks where man-made structures do no intervene. Access to the shoreline from public roads is quite good along the entire eastern shoreline.

On the western side of Northern Puget Sound, numerous inlets and constructed embayments indent the normal rocky shoreline. With the exception of the heads of the numerous inlets which commonly contain estuarine tideland and mudflat areas of sensitive ecological and recreational value, the rocky beaches are generally narrow with adjacent banks of variable relief. Access to the shore from land is generally fair from public roads in this part of the subzone.

Possible Sources: All types (vessels, handling, municipal, industrial sources).

Special Resources:[14] Eastern Shore - All man-made structures and recreational facilities along the shoreline are vulnerable. Of particular value are industrial and commercial piers and small boat marinas.

Biological resources such as salmon spawning sites are seasonally important.

Western Shore - Inlets and bays to the west of Bainbridge Island and adjacent to Bremerton support valuable recreational and commercial fish and shellfish resources.

NOTE: Although dynamic booming is not practical in the narrow Rich and Agate Passages during maximum tidal currents, some deflection booming attempts might be marginally successful near specific inlets. Generous application of shore sorbents is the most effective method with existing technology to protect the valuable inlets.

Bremerton and Winslow are sites of valuable recreational, residential, and public property that may be partially protected by local booming and sorbent application.

Numerous other small settlements of private recreational and residential use are dispersed throughout the subzone and attempts should be made to include them in any program of protective response.

Subzone 5: Southern Puget Sound

Environmental Conditions:[15] Generally weak tidal currents and insignificant wind waves prevail except in constricted passages (especially the Tacoma Narrows) and during exceptional wind conditions. Booming may be marginally effective under most conditions except in the turbulent areas mentioned.

Shore Types: Rocky, narrow beaches with adjacent banks of variable height prevail; at the heads of inlets in the western portion, estuarine tidelands are present.

Accessibility to the shore from land is generally quite poor from public roads and may be difficult from most of the islands.

Possible Sources: Small to intermediate size vessels, handling accidents near Tacoma, spills from other zones.

Special Resources:[16] With the exception of Tacoma and Olympia, Southern Puget Sound is primarily recreational and residential with sport and commercial shellfish concentrations. Salmon enroute to spawning sites may be very important seasonally.

Property resources at Tacoma and Olympia may be protected with reasonable success by protective booming.

The biologically productive western inlets probably will be less amenable to booming because of locally rapid tidal currents; however, deflection booms and cleanup operations with sorbents may mitigate damage.

Nisqually Reach and Nisqually Delta are valuable ecological tideland preserves that may be marginally protected by booming or the application of shore sorbents if they are available.

In addition, numerous small settlements of recreational nature consisting of small boat facilities may be protected from damage by the use of protective booming.

Subzone 6: Hood Canal

Environmental Conditions: Generally weak tidal currents and insignificant waves. Booming possible with marginal success.

Shore Types: Rock and gravel, usually narrow beaches with banks of variable height. Access from land to shoreline is intermittent by public highway and is generally quite poor.

Possible Sources of Spills: Large vessel accident in Admiralty Inlet, or spilled material from Puget Sound passing through Admiralty Inlet.

Special Resources:[17] All of Hood Canal is rich in recreational and commercial shellfish resources, particularly Dabob Bay. Deflection and dynamic booming may be attempted at Foulweather Bluff in order to reduce the encroachment of spilled oil. Although there are numerous small recreational and residential settlements, all areas on the Canal are of almost equal high value. Sorbents on the shoreline as a defensive measure would probably mitigate damage.

d. Summary and Conclusions

The recommended minimum response capability of enough sorbent material for a 5000 gallon spill with appropriate deployment resources to be available to any site in the Puget Sound region within 4 hours after notification is probably the minimum preparation for improving Puget Sound oil spill response capabilities. The four hour maximum response time for all areas would probably require storage centers in subzones 1, 2, 4.

The optional "offensive" (containment) and "defensive" (protective) barriers suggested in the subzone presentations would probably require thousands of feet of very expensive equipment if duplication is not reduced by effective coordination of the emergency response resources. Therefore, it is recommended as an optional minimum response capability (if it is decided that conditions warrant it and that funds should be allocated) that 10,000 feet of high performance booming material be available for deployment within 12 hours of notification to any point in the area bounded by Point Pully, Port Angeles, and Bellingham (subzones 1, 2, 3, 4). This quantity of boom suggested was determined from an analysis of environmental conditions and geography (for example, the entrance to Dungeness Bay is on the order of several miles wide) as well as an examination of the behavior of past spills (as at Anacortes, April 26, 1971). Because of environmental conditions and possible spill frequency estimates, 5,000 feet of high performance boom might be stored at subzones 2 and 4 respectively. As an additional option, enough absorbent material for a response to 50,000 gallons should be available at the sites of the booming devices for deployment to any site in the region within 12 hours after notification.

In conclusion, it is emphasized that the recommendations made herein are subjective interpretations of all available data offered in good faith to present an optional plan for improving the emergency response capability for an oil spill on Puget Sound. It is not claimed that the recommended quantities or suggested deployment sites will be totally effective in the case of a spill. However, this information is based on the best data available at this time and is offered to those who must consider the appropriate levels of preparation for an oil spill emergency as an interim reference point from which further emergency planning will be developed. The purposes of this section will be well served if the presentation of this interim contingency plan stimulates active discussion and constructive criticism by the public, government, and industry to creat an optimum emergency response capability for oil spills in the Puget Sound region.

3. INTEGRATED CONTAINMENT CLEANUP SYSTEMS

The initiation of both the offshore and shoreside cleanup operations is somewhat a function of the resources available at the time of spill. However, assuming the necessary advance preparations are made (i.e., adequate response plan) then the material and equipment resources do not need to be the significant constraint. The following contains the results of an evaluation of the various methods that can be employed in the actual containment and cleanup. The systems suggested are the result of extensive evaluation of many techniques and are believed to provide the most effective control in the Puget Sound region. It is suggested that a discussion of this nature be a part of the field operations handbook, previously recommended in Part 1 of Section III-B1.

Integrated containment-cleanup systems are discussed here, along with deployment techniques. Evaluation of cleanup devices has shown that sorbents, which are spread in the water, and subsequently recovered, are effective in removing oil from the water.[1] This material, combined with various recovery devices form the basis of the cleanup systems evaluated. Additions of containment booms to this system give the user an integrated oil spill control capability.[2] The different cleanup methods are:

a. sorbent and suction pumps;

b. sorbent and conveyor;

c. sorbent and manual retrieval;

d. sorbent and clamshell.

Based on recent past experiences, straw has proven to be a cost-effective material for absorbing oil.[3] Thus, the following discussions utilize this type of sorbent with recovery devices. Dispersants and other similar channel cleanup materials are not considered here because of the recommendation of the State Department of Ecology.[4]

a. Evaluation of Sorbent Cleanup Systems

The straw manual retrieval system (c) has been the most successful of the sorbent cleanup systems utilized in actual oil spills. Its value has not been exceeded when coping with spills that have spread out into small patches. There are several ways of depositing the straw onto the spilled oil. These include:

1. drop by helicopter,
2. mulcher mounted on a barge,
3. spread by men in small boats,
4. dropped from large barges onto concentrated oil slicks.

Essentially the straw is spread on the various oil slicks and is then picked up by men in small boats with pitchforks. The straw can be delivered to the oil spill site by barge. A large barge, with a capacity of 300 tons of hay and equipped with mulchers (as used by the highway department) can be towed into position down current of the oil spill. These mulchers have a capacity of delivering two tons of straw per hour. Such a rig could stay near the oil spill site for several days before exhausting its supply and would therefore be very effective in dealing with the oil spill at its early stages. The straw absorbs about five times its weight in oil, so that one barge could effectively treat 1200-1500 tons of oil if properly mixed.

Manual methods by which the straw-oil mixture can be picked up include the following:

a. Manual retrieval by men with modified pitchforks on small boats (such as purse seiners, flatties or Boston Whalers).
b. Purse seiners or other fishing boats could be hired to pick up the straw-oil mixture with modified nets. These nets should be specifically constructed for oil-straw recovery and could be owned by the oil cooperative organization in the Puget Sound area. The nets could be stored at strategic locations on Puget Sound to minimize the response time to reach the oil spill site. Construction of nets for oil-straw removal should be as follows: The nets should be approximately 20-30 feet deep and double reinforced for the top 5 feet. A batten and flotation arrangement should be constructed so that the net has at least one foot of freeboard. Preliminary models should be tested to determine the best anchoring system when the boats are towing the net to optimize the pickup

effectiveness of the oil-straw mixture. Along with the purse seiners should be barges and other equipment which could pickup the straw corraled by the purse seiners.

Method 2 can be implemented by clam shell operations. A clam shell would be positioned on top of a barge located downwind from the mulching operations. The purse seiners could then drift downwind with their catch and deliver the oil-straw mixture to the clam shell. The straw-oil mixture could then be herded close to the barge so that it could be picked up more efficiently by the clam shell. Such a clam shell was used at the San Francisco oil spill and was recorded as having picked up the most volume of oil-straw mixture of any system in operation at that time.[5] The disadvantages of clam shells are: its size and weight which limit its maneuverability; it cannot operate in rough sea states and can only work in areas where the oil-straw concentration is large.

Sorbent in conjunction with oleophilic conveyors or other mechanized equipment has had some success. These systems have picked up moderate amounts of oil during actual oil spill conditions. The main problems of such mechanized systems seem to be the high rate of equipment failure, their ineffectiveness in rough weather, clogging by floating debris.

Sorbents in conjunction with vacuum or other type of suction pumps have also been used at oil spill locations. Their effectiveness is rather limited to locations that have high concentrations of the mixture. (This could be accomplished by the use of small craft and booms to herd the mixture to the vacuum pump.) Other limitations of the vacuum system are that it can only operate in relatively calm water and is very easily clogged by debris.

b. Evaluation of Oil Spill Treatment Methods

At the present time there is no physical means possible to completely enclose an oil spill in open waters except under calm water conditions. Oil is driven to the downwind/downcurrent side of a containment boom and through various physical forces passes beneath the boom by the flow of water past the boom.[6] All that can be hoped for is some type of temporary containment or deflecting booms to be employed in the waters of Puget Sound. The following discussion treats two aspects of the oil spill control problem.

1. The treatment of oil spills in open water using booms to deflect the spread of oil, that is to exercise some control as to where it goes.
2. The treatment of oil spills close to shorelines primarily using booms to direct oil onto beaches for water cleanup.

1. Treatment of Oil Spill in Open Waters

Channeling of oil slicks leaving the vessel - This suggested use of booms to channel oil downwind of the source is recommended in all spill situations to define the oil control area and should be implemented as a first step if control action is undertaken. The only limitation in the use of booms in this matter would be where weather conditions are so severe that the boom-towing vessels are unable to maintain their positions relative to the source. Considerations in the use of booms to channel the oil are as follows:

a. The boom walls or channel walls should be independent of each other to limit the coordination required by the marine units positioning the walls.

b. The boom walls should be dynamic or under tow rather than static or at anchor to permit adjusting and repositioning the wall to suit the rate of oil release, the oil drift direction and velocity, weather conditions, and general operations in the area.

c. The upwind end of the boom or bow of the towing vessel should be sufficiently downwind of the course to avoid interference with salvage operations. At the same time, the upwind end of the boom wall should be close enough to the source to be effective in limiting the spread of the oil.

d. The upwind ends of the boom walls should be as close together as the situation permits, but should be spread sufficiently to enclose all the oil drifting downwind from the source.

e. The downwind ends of the boom walls may be permitted to position themselves in which case the channel area will be rectangular or they may be pulled toward the center of the channel to concentrate the oil. The objective would be to close the boom walls on the downwind side as much as possible to concentrate the oil while not losing significant quantities of oil over or under the walls. It is recommended that a minimum of 2000 feet of heavy duty boom (see Appendix 5) be available for use by the regional cooperative response organization in the Puget Sound area in implementing this control action.

f. The boom-towing vessels should be equipped with radar to permit positioning and station keeping at night and should be manned and provisioned to remain on station for several days at a time before being relieved. The boom itself should be equipped with lights and radar reflectors so that its location can be readily determined in low visibility conditions, and at night.[7]

The treatment of the oil in the channel should involve the use and retrieval of sorbents in the channel area and downwind or downcurrent from it (see Figure III-B2). The purse seine, and clam shell methods have been discussed previously.

A mechanical cleanup method, the boom-skimmer system can also be an integral part of this treatment. The collection of oil spilled from ships or barges can also be handled by a sweeping boom and skimmer retrieval operation (see Figure III-B3). Such a system is available complete from Bennett Pollution Controls Ltd.[8] This system is very effective since their boom is specifically designed for sweeping operations. The following paragraphs are excerpts from their oil pollution cleanup system manual.

> The two 500 ft. lengths of heavy duty boom are held to the skimmer (at a distance of less than 3 ft.) by means of two rigid arms, fitted with universal joints at both ends. The resulting gap between the skimmer and boom is sealed with a small section of boom. The first 100 ft. of each boom section (starting from the skimmer) are held together by cross cables at the 10, 30, 50, 70 and 100 ft. Each joint to the boom is made by a wire bridle, distributing the load across the boom. The cross cables are of a pre-determined length to maintain an angle of 30° maximum.
>
> The towing plane at the end of each 500 ft. boom section consists of a vertical steel pipe, attached to each longitudinal boom cable, and is headed by a torpedo shaped float.
>
> The attachment to the tugs (1) and (2) from the towing plane is via 2" nylon rope, approximately 100 ft. long. This is the safety link in the system and the rope will break if the tow tugs accidentally move in opposite directions.
>
> The angle of two depends on the current and a table is given on Figure III-B3.

FIGURE III-B2

CHANNEL CONTAINMENT & CLEANUP SYSTEM

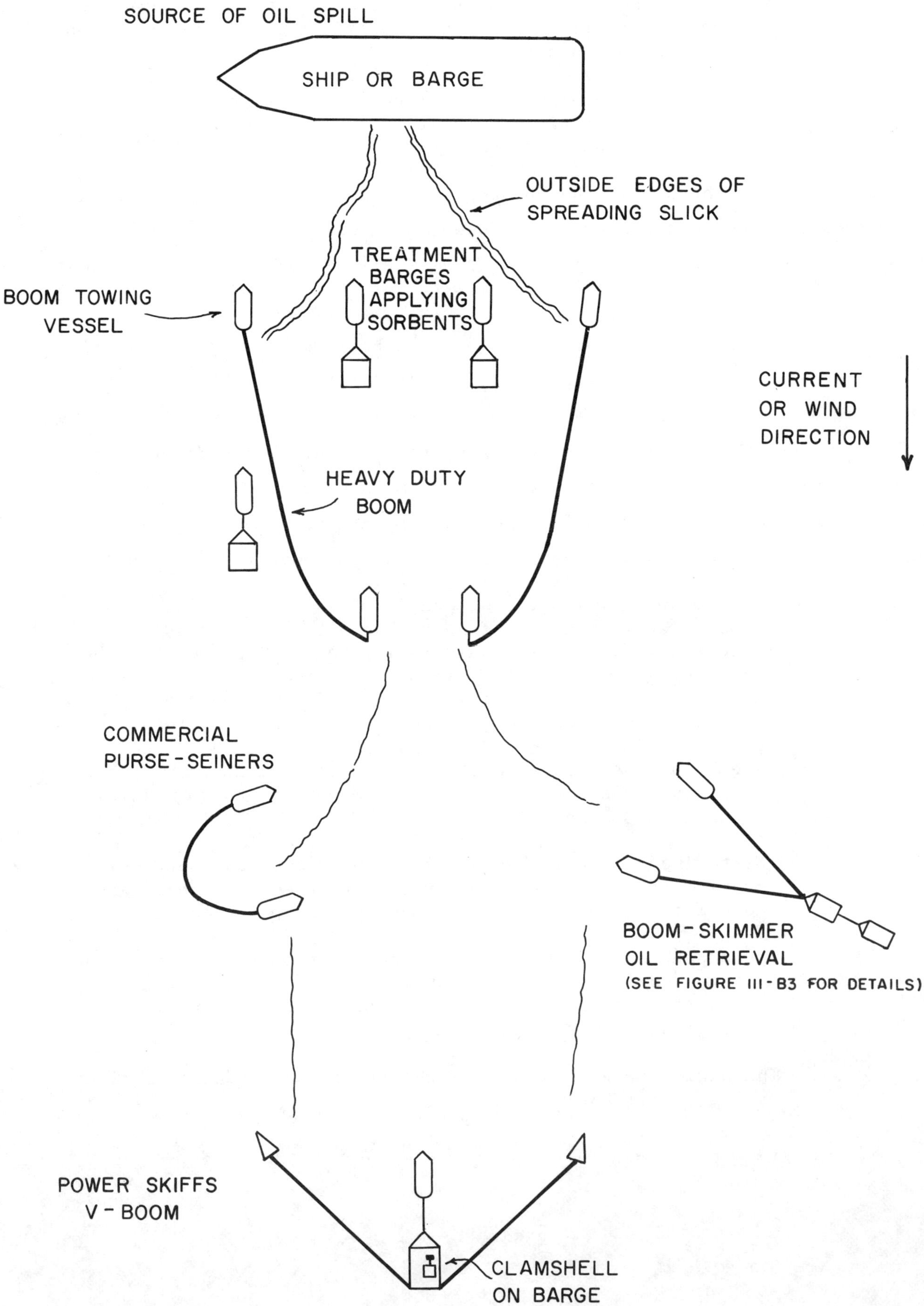

FIGURE III-B3

BOOM-SKIMMER SYSTEM

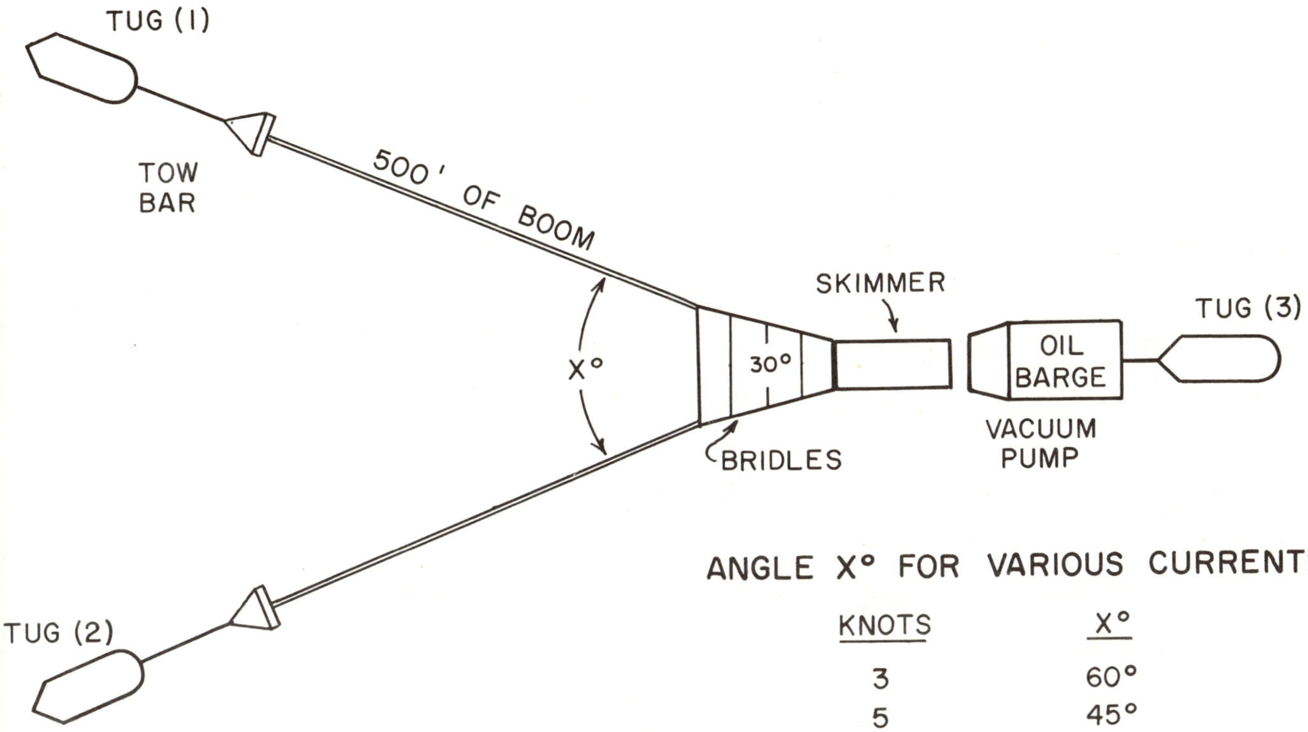

ANGLE X° FOR VARIOUS CURRENTS

KNOTS	X°
3	60°
5	45°
8	30°

No. 3 tug is the control tug, maintaining the position of the oil barge and skimmer, keeping tension on the booms and directing tugs (1) and (2).

The total recovery unit is assembled in a harbor or protected bay and towed to the oil spill area 5-8 knots, depending on sea conditions. In this operation no. 3 tug becomes the lead tug and no.'s 1 and 2 trail, keeping the booms straight and as close together as possible. The limitation on this system is the skimmer barge, which becomes inefficient in seas in excess of 4 feet, due to its pitching which allows oil to pass underneath. With the inclusion of an oilevator instead of skimmer (an absorbent belt machine), it is considered possible to maintain a 90% efficiency up to 6-foot sea conditions.

A simple v-boom arrangement and two skiffs shown in Figure II-B2. can be used to collect oil-straw mixtures and bring the mixture to either the clam shell operation or to suitable beach areas where the mixture can be easily collected and disposed. All recovery systems discussed earlier are applicable here.

2. Treatment of Oil Spills Close to Shorelines

If floating oil is approaching the shoreline in large quantities, or if the source of the spill is located close to the shore so that adequate control of the oil as it is released is not possible, it must be assumed that a significant quantity of oil will actually reach the shore. The control approach recommended is therefore one of protecting those areas where oil pollution will create the most damage or be the most difficult to cleanup, while sacrificing sections of the shoreline to use as collection points or catch basins and directing the floating oil toward these points. In many respects the strategy is to bring the oil ashore as rapidly as possible where positive control methods may be applied and to limit the opportunity for the oil to refloat or drift along a shore, contacting miles of shoreline in the process.

There are several considerations in the application of this control approach.

First, the topography and uses of the threatened shoreline must be reviewed and an early determination made as to what areas should be protected. Inlets to bays, harbors, marinas or other inland bodies of water should receive primary protection to prevent oil from entering these areas. Secondary concern should be directed toward protecting areas where damage to the ecology may be incurred such as shellfish beds, nesting grounds and saltwater marshes. Finally, those areas which are most difficult to cleanup or have large included shorelines such as rocky promontories, or irregular coasts should be shielded where possible.

Booms should be used to direct the incoming oil away from inlets, around promontories, or to bring drifting oil ashore into open beach areas. As in the case at sea, booms should not be expected to contain the oil or prevent its passage by positioning directly in the path of the drifting oil.

However, booms can be expected to be effective in redirecting the oil as is illustrated in Figure III-B4. Booms used in this manner must be attended at all times, since they must be continually positioned to suit changes in the wind and sea conditions and in the drift direction of the oil. Booms at the sea end of inlets should be backed up by secondary booms spanning the inlet, with floating sorbents on the water to pickup whatever oil passes the seaward boom. Aerial reconnaissance from helicopters will greatly assist in planning the strategic location of booms and in adjusting and positioning.

Booms can be used effectively to control the passage of oil. Since booms are ineffective when currents are large, they cannot be used to contain oil by positioning directly in the path of the drifting oil. Therefore booms should be used to deflect oil to specified areas by positioning them at a relative angle to the uncoming oil. (See Figure III-B4.) When booms are used to block inlets, they should be backed up by secondary and even tertiary sets of booms. Sorbents can also be used in between the booms to catch any oil that leaks through.

> "Temporary earthen dikes may be constructed where the situation permits to provide a solid barrier to floating oil. Dikes should be considered for small inlets and entrances on quiet water where the temporary closure will not cause excessive inconvenience and the construction and removal of the dike will not be a major inconvenience."[9]

When oil threatens a shoreline, the most effective and immediate method of protecting and catching the oil is to use sorbents. These sorbents can be applied to the oil before it reaches the beach by the use of small boats. The beach can be further protected by applying sorbents directly from the beach to the water. If the oil extends over a large portion of a beach, booms can be used to concentrate the oil mixture. This is done by putting booms out from the beach which can deflect the oil-straw mixture to those areas where the oil can be collected.

Mechanical devices can also be used to treat the oil if the weather and sea state is calm enough to permit operations. In general, however, where the oil is wide spread or thin and the slick is moving fairly rapidly, a defensive approach of directing the oil with booms rather than attempting to collect it will probably be more successful.

FIGURE III-B4, DEFLECTION OF OIL SLICK WITH BOOMS

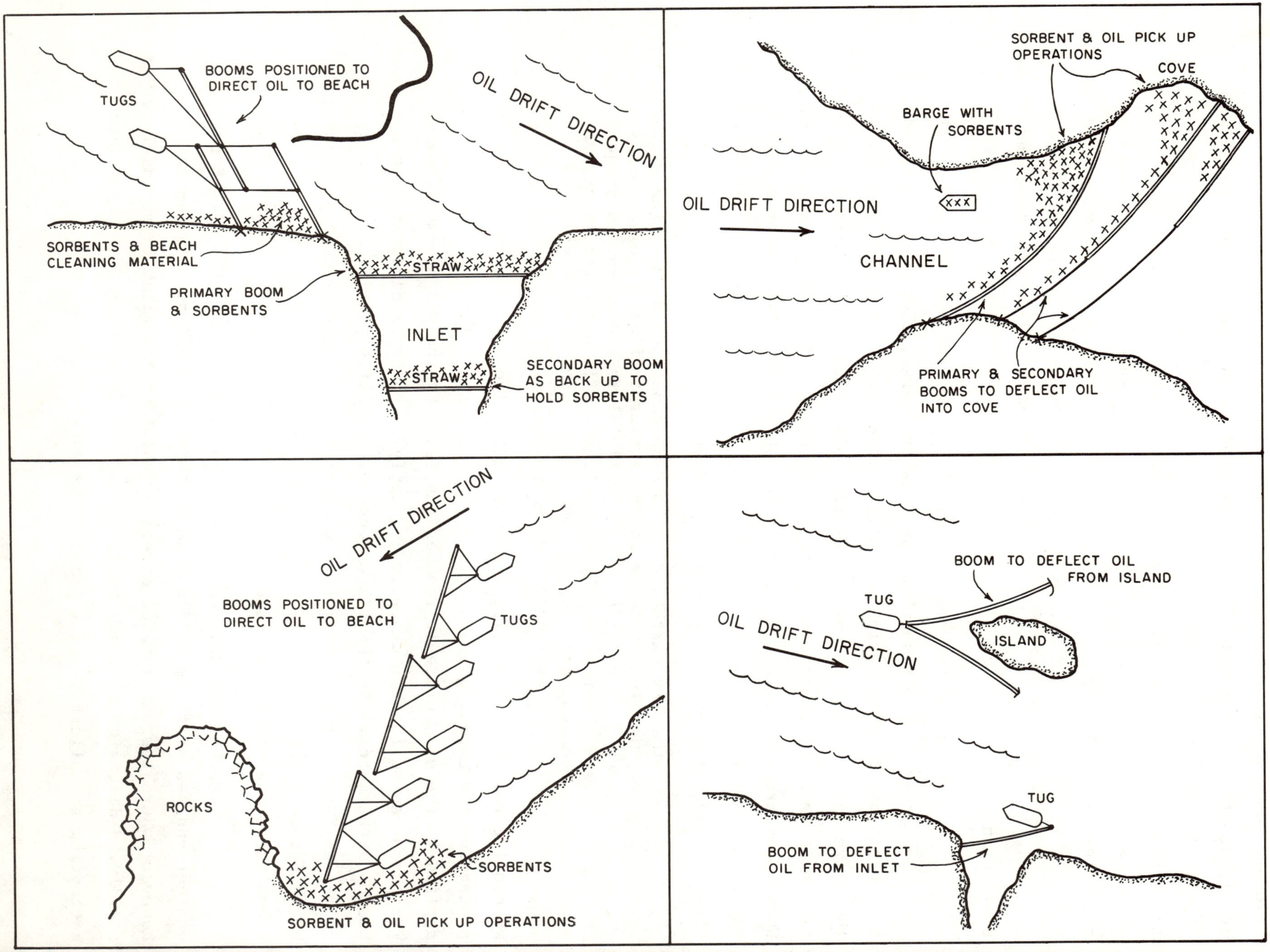

Many boom systems are available that can be used to deflect oil or collect oil on beach areas. The main requirement is that the booms have to be effective in currents. Such booms are:

a. Heavy duty or medium duty booms with long skirts (at least 24" of draft).

b. Booms made with fir trees connected end to end. These booms can be improved by using seine net that goes down at least 4 fathoms.

c. Double row of log booms with straw in between. These can also be improved by the addition of seine net that can trap the sorbent-oil mixture. (This is a specially favorable system since log booms are easily available in large quantities in the Puget Sound area.)

c. Summary

Since no cleanup system is the cure-all for oil spills, a thorough treatment of oil spills will require the use of most systems mentioned. A combination of systems will probably be most effective for a particular situation, whether it is in open water or close to shore. These systems should be deployed to the areas where they can be used most efficiently. This will require studying the local geographical conditions, weather conditions, before the deployment of the required systems. A cursory survey has been conducted in the region in an effort to determine the actual ability to cope with a spill situation. The data collection was geared toward controlling an oil spill in Subzone 2,[10] though two regions were examined - (1) the Anacortes-Bellingham area (Subzone 2,) and (2) the rest of Puget Sound and the Strait of Juan de Fuca. The specific deficiencies are reported as follows:

Subzone 2: In the immediate area, a maximum of three helicopters are available; there is only one source of contractual manpower. Vacuum trucks in the area have a limited capacity and no proven marine effectiveness. Subzone 2 has only 1300 feet of oil boom; this is only effective for harbor spills. These could do no more than protect 1 or 2 calm bays. There are no waterborne skimming units in the region, and the capacity to improvise is limited in comparison with San Francisco. There are no oily waste dumping sites in the zone. The nearest dumps are in

Marysville and Port Townsend. Subzone 2 has no heavy cleanup specialists. Perhaps most significantly, there are no standby service boats in the vicinity. A small number of tugs can be reasonably depended on in 2-3 hours, but nothing is ready to go. The region has no high speed craft capable of moving men and equipment; the existing ferry fleet offers the only means of effective transport to the San Juan Islands. No appropriate craft are available for putting equipment on isolated beaches.

Rest of Puget Sound: Many of the same criticisms apply. An adequate number of service boats exist, but there are not any high speed craft suitable for the purpose. The area has only 1700 feet of boom; the boom can be used for harbors, etc. There are 4 helicopter outfits, and only 2 skimmers, based on available information. The capacity to turn skimmers out of shops on short notice is greater than in Subzone 2. Personnel problems are present in both regions; there are no numbers of trained people; there are no groups ready to respond in short order as there are for other types of disaster.

The thrust of this cursory review suggests that the ability to respond quickly to a spill situation is severely restricted by a lack of immediately available material and equipment resources. The general public in this region should not be lulled into a false sense of security by the claims of a "paper" readiness.

An optimal integration of equipment requires a continual updating of testing and evaluation by private and public agencies. An effort should be undertaken to establish minimum standards for equipment and materials. For example, straw mulchers that break straw too finely are of little value, as there is an optimal size for straw that soaks oil.[11]

[1] Systems Study of Oil Spill Cleanup Procedures, Dillingham Environmental Company, February 1970.

[2] Anacortes Oil Spill, April 26, 1971; San Francisco Bay Spill, January 18, 1971.

[3] Original estimate was 5,000 gallons. Not until 40 hours after the spill was the final estimate of 5,000 barrels confirmed. See Appendix 12 for further discussion.

[4] "Puget Sound Oil Spill Response Plan," formulated by Puget Sound Tug and Barge Co. & Pac-Mar Services, Inc.

[5] R. W. Roe, "Report on Barge 17 Oil Spill," Texaco Refinery Dock, Anacortes, Washington, April 26, 1971, United Transportation Company.

[6] Section III-C, Part 4, and Appendix 1 of this report present detailed findings of these facilities.

[7] 33 CFR 126, revised September, 1970.

[8] See appendix 1 of this report.

[9] Seattle Coastal Region Oil and Hazardous Materials Pollution Contingency Plans, U.S. Coast Guard, dated December 1, 1971, Annex XX, Appendix 1, Table F.

[10] 3406 - 11th Ave. S.W., Seattle, and Terminal 105. On the Duwamish River.

[11] 13th Coast Guard District Headquarters are at 618 2nd Avenue, Seattle, Captain of the Port office is at Pier 90, Seattle.

[12] Dillingham, Volume II, p. 26.

[13] See Appendices 2 & 3 of this report for critiques on state and federal laws.

[14] Coast Guard Plan, Annex XX, Appendix I, Table C.

[15] RCW 90.48.325, 90.48.330, 90.48.335 of Washington State Statutes.

[16] Conversation with Harry Tracy, biologist for Department of Ecology at the University of Washington, February 11, 1971. Also stated in Department of Ecology internal memorandum, dated May 10, 1971 from H.B. Tracy to J. Behlke.

[17] See Appendix 7 of this report for a critique of the Coast Guard regional contingency plan for Puget Sound.

[18] Dillingham Report, Vol. II, pp. 43-110.

[19] The effectiveness of the present arrangement, under contract no. 68-01-0017, wherein EPA has designated Texas Instruments, Inc. to assess damages in Puget Sound waters in questionable. TI's required response time by contract is 48 hours after notification by EPA. The firm's distant location plus their unfamiliarity in the Puget Sound area are just two factors involved. For a news media evaluation, the reader is referred to the Seattle Post Intelligencer, August 15, 1971, page 1. A critique of TI's performance at the Anacortes spill is documented on Appendix 12 of this report.

ENDNOTES SECTION III-B, Part 2

[1]See section II-A of this report.

[2]See Appendices 5 & 6 of this report.

[3]Undersea Technology, July 1971, Vol. 12, no. 7, p. 8.

[4]See Section III-A & appendix 13 of this report for statistics & potential sources of oil spill.

[5]See section II-D of this report.

[6]See Section II-A; part 3, for environmental description of subzone and appendix 5 for boom evaluation.

[7]See section II-A, part 3 for discussion of man-made and natural resources of Subzone 1.

[8]See section II-A, part 3, for environmental description of Subzone 2, and appendix 5 for boom evaluation.

[9]See section II-A, part 3 for discussion of man-made and natural resources of subzones 2.

[10]See section II-A, part 3, for environmental description of subzone 3.

[11]See section II-A, part 3, for discussion of man-made and natural resources of subzone 3.

[12]Friebertshauser, M. A. and A. C. Duxbury, "A Water Budget Study of Puget Sound and Its Subregions," Limnology to Oceanography, in press.

[13]See section II-A, part 3, for environmental description of subzone 4.

[14]See Section II-A, part 3, for discussion of man-made and natural resource for subzone 4.

[15]See Section II-A, part 3, for environmental description of subzone 5.

[16]See section II-A, part 3, for discussion of man-made and natural resource of subzone 5.

[17]See section II-A, part 3, for discussions of man-made and natural resources of subzone 6.

ENDNOTES SECTION III-B, Part 3

[1]See appendix 6 of this report.

[2]See appendix 5 for evaluation of containment devices.

[3]Santa Barbara Spill, January 1969; and San Francisco Spill, January, 1971.

[4]Laws and Oil Spill Emergency Procedures, Department of Ecology, p. 10.

[5]Conversation with Jules Mayer of Standard Oil Company of California at University of Washington, February 23, 1971.

[6]See Appendix 10 of this report for a theoretical analysis of this phenomenon.

[7]This suggested set of recommendations has been excerpted from Dillingham, Volume II, pp. 69-70.

[8]See Appendix 5 of this report for evaluation of the Bennett Boom.

[9]Dillingham, Volume II, p. 88.

[10]See Section II-A, part 2, for subzone boundaries.

[11]Coast Guard Plan, Annex XV, 2800, Technical Information, p. XV-14.

III-C. PREVENTION OF OIL SPILLS

The ideal solution to the oil-on-water problem is to prevent oil discharges from occurring, but the efforts expended in this area are subject to local, State, Federal, and international regulations. This section examines four areas of prevention; namely: tankers, traffic control systems, oil detection, and oil handling facilities. Each is discussed with respect to regulations and state-of-the-art techniques.

1. OIL TRANSPORTATION ON WATER

One of the primary sources of oil spills is waterborne traffic; in particular that associated with large vessels. Efforts undertaken to prevent spills from this source occur on three fronts with regard to tankers: policy, design, and operations. Because of the nature of waterborne commerce, policy for such commerce must begin in the international arena. The implications of such policy on Puget Sound, filtered through the federal, state, and local authorities is presented in this section.

These implications take their form in regulations governing the design and operation of tankers. Both revolutionary techniques and modification of current procedures are discussed, along with their inherent benefits and disadvantages. Finally, a list of long run and short run proposals is recommended for Puget Sound.

a. Policy

The problem of oil pollution of the seas has been recognized for some time. As early as 1922 the U. S. Congress proposed an international conference on oil pollution. Studies that were carried out led to a National Oil Pollution Act in 1924. Later, the U.S. convened an Intergovernmental Conference of Maritime Nations in 1926. About the only result from this conference was an agreement among several ship owners to refrain from discharging oil within 50 miles of a coastline. After this nothing was effectively accomplished until the 1954 Convention for Prevention of Pollution of the Sea by Oil. Interim attempts included the 1936 League of Nations proposal to consider the problem (this conference never took place). The United Nations in 1949 accumulated views on the problem from various governments, and in 1953, the Economic and Social Council of the U. N. attempted to establish a council of experts to evaluate the problem. This was postponed until the 1954 conference.

The thrust of the 1954 conference was to establish zones where the discharge of oil or mixtures[1] of oil and water was prohibited. They also initiated a procedure whereby an oil handling log book was to be maintained onboard the ship. The convention adopted a requirement that by 1957 all signature governments should establish adequate oil disposal facilities in their ports. They also stipulated that one year after coming into effect, all ships of signature nations be fitted such that all bilge discharges be eased through oil water separators. The major weakness of the convention is that it did not specify uniform penalties or means of enforcement. The U. S. did not ratify the convention until 1961 in time to join the 1962 Conference.

The 1962 Conference aimed to ammend the 1954 Conference resolutions in the following ways:[2]

1. Extend the prohibited zones further to sea.
2. More clearly define the oil handling log.
3. Allow direct pumping of bilge waste overboard so long as the oil mixtures set by the 1954 Convention were not exceeded.
4. Remove the signature governments obligation to establish oily water reception facilities and instead require them to promote such facilities.

These amendments effectively reduced the restrictions previously imposed and brought the convention resolutions more in line with economic reality. Since 1962 the policy making pace has quickened. The growth of tankers during this period coupled with the greater demand for petroleum products and a new awareness of our ecological system has made the problem more severe than ever. In 1969, the 1962 Convention resolutions were once again modified in the following manner:[3]

1. Elimination of free dumping zones where the discharge of oil is not regulated with the exception that a vessel passing through such an area may have an instantaneous rate of discharge of oil content not to exceed 60 liters per mile.
2. Extend the prohibition of oil discharge close to coastlines to all ships besides tankers.
3. Limit the total discharge from a tanker to 1/15,000 its total cargo carrying capacity and require any discharge be more than 50 miles off shore.
4. Remove the exemption set in 1954 of discharge of oily residue sledge.

The 1970's opened with increased pressure to prevent the spilling of oil. In November of 1970, Secretary of Transportation John Volpe proposed before the NATO Conference on Challenges of Modern Society a complete halt to all discharge of oil into the seas by tankers and other vessels before the end of the decade.[4] The convention adopted his resolution and agreements among the NATO nations to achieve this goal include:

1. Early ratification of the 1969 amendments to the Convention for Preventing of Pollution of the Sea by Oil.
2. Support and aceeleration of the work being done by international organizations toward the development of increased prevention measures, particularily IMCO.
3. Improvement of all oil waste reception facilities.

These resolutions, then, are the sum of the efforts with regard to state-of-the-art in oil pollution prevention. Legal interpretation of international controls for oil pollution is presented in Appendix 4.

Efforts are being directed on many varied levels toward the achievement of the overall goal. The problem is of such a nature that the major efforts will continue to be expended on the international level. The significant domestic advances in oil pollution prevention maritime activities will be based on the resolutions derived from international agreement. In the United States, the Coast Guard is vested with the authority for control and regulation of the technical aspects of the maritime industry. On the international level, the Intergovernmental Marine Consultative Organization (IMCO) appears to be the most productive forum in dealing with the technical aspects of the pollution problem. To clarify the functions of both IMCO and the Coast Guard with regard to the preventative aspects of pollution a brief discussion follows.

1. Coast Guard

The Coast Guard is responsible for minimizing the loss of life, property damage and personal injury that may result from marine transportation activities. Their focus on the accomplishment of this goal is through two primary preventative programs.

a. Development and enforcement of vessel design and construction standards, and
b. licensing of personnel for maintenance of technical operating competence.

Both of these programs are managed under the Merchant Marine Safety Division of which there are five departments. Merchant Vessel Personnel (MVP) exercises control over licensing function. Merchant Marine Technical (MMT), Merchant Vessel Inspection (MVI), Merchant Vessel Documentation (MVD), and Hazardous Materials (MHM) exercise control over the maintenance and development of the technical standards function. The operational aspects of the preventative program are carried out at a local port level through the officer in charge of Marine Inspection Offices (OCMI). Branches of (MMT) have been established in four cities to provide staff support in the form of technical guidance to the OCMI's in their geographical area.

All standards with regard to vessel design, operation, documentation, or personnel competence are published in Chapter I of Title 46 of the Code of Federal Regulations (CFR). These regulations are a part of the U. S. Code (USC), the authority of which have been granted by Congress. These regulations for the most part provide the minimum standards for safe design and operation of vessels.

A listing of vessel standards that fall under the authority of the Coast Guard are cited below:

Seven subchapters prescribe the vessel design and construction standards.

Reference	Subchapter	Title
46CFR24-26	C	Uninspected Vessels
	D	Tank Vessels
70-78	H	Passenger Vessels
90-98	I	Cargo & Miscellaneous Vessels
	R	Nautical Schools
175-187	T	Small Passenger Vessels
188-196	U	Oceanographic Vessels

Three subchapters prescribe rules for system and specific equipment utilized on vessels.

Reference	Subchapter	Title
46CFR 50-63	F	Marine Engineering
110-113	J	Electrical Engineering
	Q	Equipment Specifications

Four subchapters prescribe operating limitations or conditions for various types of vessels.

Reference	Subchapter	Title
46CFR 42-46	E	Load Lines
	M	Bulk Grain Cargos
146-149	N	Dangerous Cargos
157	P	Manning

Five subchapters prescribe various administrative, personnel and investigative procedures.

Subchapter	Title
A	Procedures applicable to the Public
B	Merchant Marine Officers
K	Casualty Investigation, Suspension and Revocation Proceedings
L	Overtime Services
S	Numbering, Undocumented Vessels, Boating Accident Reports and Statistics

As stated previously, the objective of these standards is to minimize loss of life and property at sea. Pollution reduction results primarily as a byproduct of the minimization of property loss objective. Although most of the problematical aspects of oil pollution from vessels in adequately covered by this arrangement of objectives, there still remain serious flaws. It appears necessary to achieve the goals set by the NATO conference that minimization of pollution potential must become a major objective in the Coast Guard Marine Safety program.

2. Intergovernmental Marine Consultative Organization (IMCO)

IMCO was formed in 1959 as a specialized agency of the United Nations for maritime matters. Its creation provided a forum for continuing exchange among the world's maritime safety administrators. The organization soon became the sponsor of international conventions regarding the sea. One of its first activities was the 1962 amending Convention for Prevention of Pollution of the Sea by Oil. Because of the importance of international body in eliminating oil pollution of the sea by vessels it is appropriate to gain some insight as to the mandate, structure, and procedures of the body.

The Assembly is the central body of the Organization and is composed of 72 member nations. The Council is the administrative arm and generally meets twice a year. Beneath the Council are the three working committees. They are Legal, Facilitation, and Marine Safety (MSC). MSC is the primary action group and of basic interest for purposes of this study. The committee is made up of 16 member nations and at present there are 11 working subcommittees handling technical matters as related to the specific disciplines. A complete organization chart is shown in Figure III-C1.

FIGURE III-CI

INTERGOVERNMENTAL MARITIME CONSULTATION ORGANIZATION (IMCO)

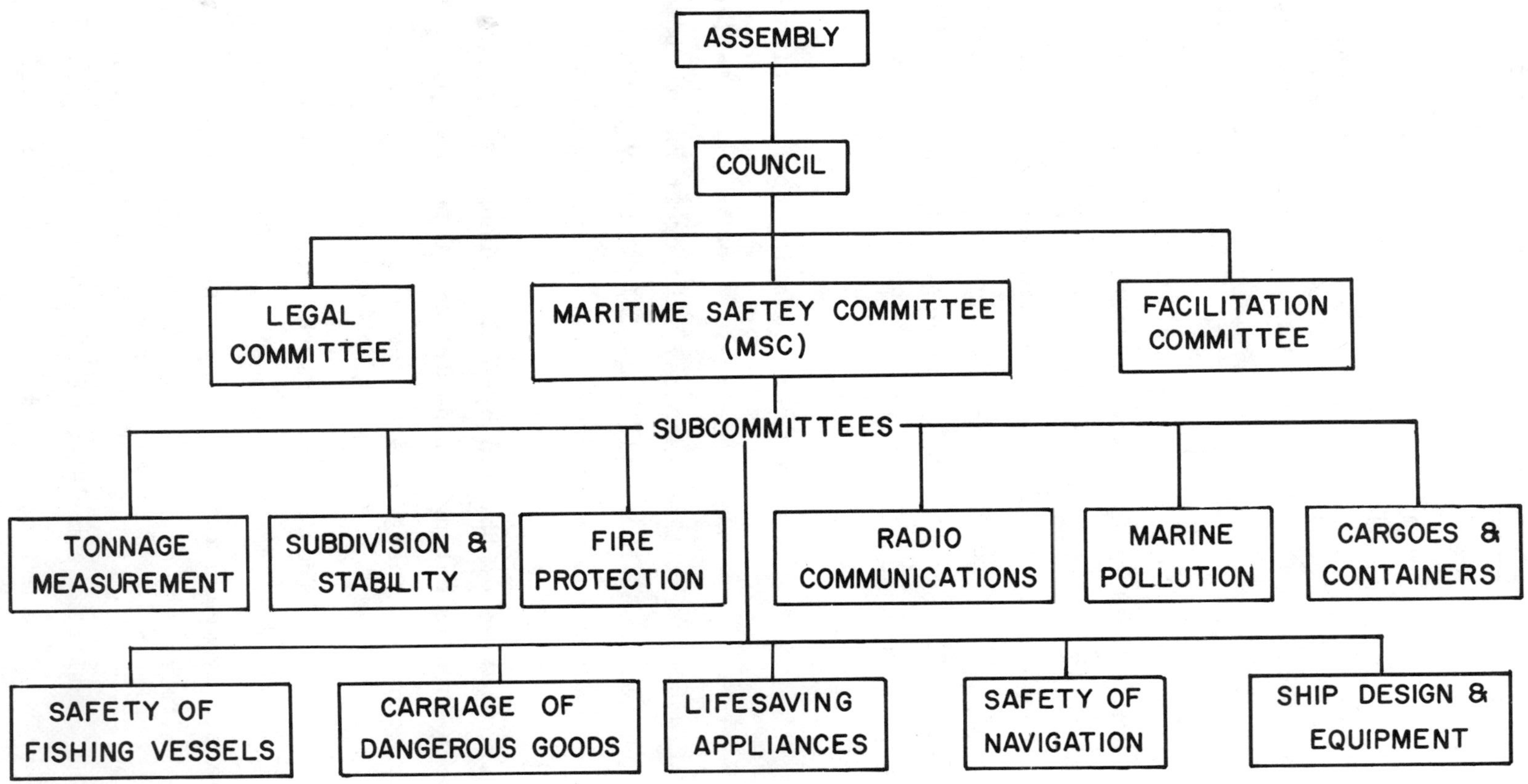

The representatives of the member nations working primarily at the subcommittee level conduct the bulk of the effort. Each member nation is responsible to undertake basic research which is used in support of proposals placed before the various subcommittees.

A proposal by a member nation becomes an IMCO resolution after it has been carefully considered for its technical viability at the subcommittee level and approved by a vote of the Assembly. All member nations have an equal representation in the Assembly.

It should be noted that all IMCO resolutions are not automatically binding on the member nations. The resolution actually takes the form of a recommendation upon which the various member governments may or may not adopt. In the United States an IMCO resolution relating to an International Convention or an amendment to an International Convention would not become binding on private industry until the following steps were completed.

a. IMCO resolution ratified by U. S.
b. Implementing any required legislation by Congress.
c. Writing of regulations as required (as promulgated in the normal manner this requires a public hearing before the regulation becomes binding).

Where an IMCO recommendation does not relate to an International Convention, ratification is not necessary and the matter is generally referred by the State Department directly to the appropriate agency. In cases of recommendations originating out of MSC that agency would be the Coast Guard.

The present efforts of IMCO, expended basically through MSC, are specifically directed at reduction and eventually complete elimination of oil pollution of the sea. Following the *Torrey Canyon* incident the pace was quickened considerably with measures, to be discussed, broken into three areas:

a. Preventative

   i. Sealanes
   ii. Shore Guidance
   iii. Speed Restrictions
   iv. Navigational Equipment
   v. Officers and Crew Training
   vi. Automatic Pilots
   vii. Construction and Design of Tankers
   viii. Identification and Charting of Hazards

b. Remedial

i. Procedures in the Event of Accidents
ii. Research in Oil Clearance

c. Legal

i. Legal Rights of a Coastal State
ii. Official Inquiries
iii. Liability in Event of Accidents
iv. Compulsory Insurance for Tankers
v. Movement of Salvage Equipment

Official action with regard to these measures came in the form of resolutions presented by the Assembly in 1968 and 1969.[6] By far the most significant action, resolution 175, is directed at promoting the load on top procedure in tankers. Although the LOT procedure has its shortcomings it was recognized that it offered a means of significantly reducing the amount of oil discharged to the sea.

The majority of these proposals do not directly affect Puget Sound. Their primary impact will be quite subtle, only to be seen over time as a better safety record and lower incident of pollution resulting casualties. The only resolution that would have immediate effect on the Puget Sound area is #161, calling for the establishment of traffic separation schemes. At the international level, this would be adoption of a traffic separation scheme in the Strait of Juan de Fuca. At present this particular measure has not been considered by IMCO.

b. <u>Ship Design Standards</u>

1. Hull Damage and Oil Spill Statistics

Oil has been moved in large quantities for years on the sea. Almost every vessel, whether a tanker or not, carries enough oil either as cargo or fuel to be a potentially large polluter. Oil has been, and will continue for some time to be, the major energy source for man. Until an adequate substitute is fully developed it will continue to be moved in these large quantities over the water.

Prior to the oil spill at West Falmouth, Massachusetts,[7] it was commonly thought that release of oil into the sea had, at the worst, only short term effects on the environment, and that these were confined primarily to the aesthetic values of that environment. However, a thorough scientific analysis of the effects of the West Falmouth spill by Dr. Max Blumer[8] and his team at Woods Hole not only demonstrated the long term effects of oil spills, but more dramatically the interruption and destruction of marine life and the ecological environment which is the basis of the food chain. As a result of Dr. Blumer's work the potentially harmful effects of oil were better understood and it was realized that potential future spills were more significant ecologically than previously thought.

Much effort began to be directed toward the evaluation of existing methods of transporting oil; the design of vessels began to receive close scrutiny with a perspective of reducing possible pollution sources as well as designing for the safety of personnel and preventing the loss of equipment. Other efforts were directed at learning how collisions or groundings occur, and what were the potential causes of spills.

The Dillingham Report,[9] completed in 1970 for API, on oil spill cleanup procedures attempted to analyze some of the reasons for oil spills and how and where they are most likely to occur. Its analysis is based primarily on historical frequency data of groundings and collisions and is presented in graphical form.

Their findings regarding the patterns of past spills is summarized as follows:

Source	75 percent were associated with vessels, principally tankers.
Composition	90 percent involved crude or residual oils.
Volume	10 percent of the spills were greater than 5000 bbls.
Distance Offshore	80 percent occurred within 10 miles of the shoreline.
Duration	75 percent of the spills incidents lasted more than five days.
Extent	80 percent contaminated less than 20 miles of coastline.
Distance from Port	75 percent were within 25 miles of the nearest port.

On the basis of these findings, the typical situation that might be expected in an offshore spill is that (a) the source is likely to be a tanker, (b) crude or residual fuel oils are likely to be involved, and (c) the size of the spill most likely will be greater than 5,000 bbls. The spill will probably occur a few miles offshore and the duration of the spill incident will be more than five days. Most likely, contamination of the beach will take place. Further it is likely that the spill will occur within 25 miles of a port and most likely not more than 50 miles from such a port.

2. Design Factors

The two principal factors of interest to ship designers are (1) the most likely volume of spill, and (2) the type of vessel most likely to be involved in a spill situation. Dillingham's findings bear out that definite problems exist as to tanker size (the median volume was 25,000 bbls.) and structural integrity of tankers.

In 1967, the Maritime Safety Committee (MSC) of IMCO established a subcommittee on Ship Design and Equipment. Their initial efforts were directed toward the hazards of bulk shipping of chemical cargos, on the strength of a proposal from the United States. However, shortly thereafter the Torrey Canyon incident brought into focus the full potential for pollution from a common cargo - that being the transport of oil. The need for a complete study and re-evaluation of existing vessel design was full realized. It was agreed in the fourth session of the Subcommittee to collect the necessary data base of grounding casualties of major ocean going ships with respect to the frequency of casualties and the position and extent of bottom damage and rupture in order to assess the effectiveness of present double bottoms in large tankers. At their meeting in October, 1970, an interim agreement was made to indicate to tanker interests that no further increase in tank size should be considered, and a limitation was applied to present volumes of 30,000 M^3 for wing spaces and 50,000 M^3 for center spaces. Further, the Subcommittee has declared its intention to lower such values subsequent to evaluation on the impact of additional subdivision on the cost of ship construction and operations.

Data has been collected and analyzed to determine the effectiveness of double bottoms. The concensus of the committee was that the effectiveness of double bottoms would, under present requirements (the current height average 6% of the breadth of the vessel), be limited to low energy groundings and that further study was required to make further advances.

The proposed improvements in ship design standards are looked at from two points of view:

a. those proposals aimed at preventing or avoiding a loss of cargo containment due to a collision or stranding, and
b. those proposals put forth to limit the loss of cargo from a damaged tank due to collision or stranding.

It should be noted that these proposals are not to be considered as independent alternatives. To be effective in improving the existing design standards the proposals from each point of view must be considered as complementing each other (i.e., one of either set is not as effective an improvement as some combination of the two).

a. Improvements to Prevent Loss of Cargo

The suggested improvements for preventing the initial loss of cargo from a tank are as follows:

i. Maneuverability - general.
ii. Stopping and backing power.
iii. Prime movers.
iv. Controllable pitch propellers.
v. Contrarotating propellers.
vi. Vertical axis propellers.
vii. Multiple screws.
viii. Stern anchors.
ix. Remote release for anchors.
x. Rudders.
xi. Twin rudders.
xii. Side thrusters.
xiii. Speed reduction equipment.
xiv. Protection of hull.

Each of the following proposals is discussed further with brief comment as to their present feasibility and effectiveness.

(i) If maneuverability in low energy impact[10] situations is determined to be a major factor in these impacts then the establishment of criteria for maneuverability should be developed. For example, this criteria could include the ability to stop within a given distance when operating at a given speed and under a standard set of conditions; the ability to change a vessel's heading at a specified rate under a standard set of conditions. Some indication of the differences that exist tactical diameters can be viewed in Table III-C1. It is not known at this time what efforts are being directed toward the establishment of such standards, although it does appear some work is needed in this area. An expanded list of other proposals that meet the general maneuverability proposal is discussed below.

(ii) As an interim measure the operators of each vessel could be provided with information for operational limits of stopping and backing capabilities of the vessel they control. The basic constraint on stopping ability is the limitation of the propeller or propellers to absorb energy over time. Thus, it is feasible, knowing the characteristics of the propeller along with empirical data, to determine this stopping and backing capability.[11]

(iii) With regard to prime movers, most of the various types are capable of being designed to provide the necessary power to shorten stopping distances (and thus improve stopping time). Each type presents some difference with respect to response time for maneuvering demands and the choice is generally based strictly on economic considerations.

(iv) Stopping ability can be enhanced through the use of controllable pitch propellers. A greater range of effectiveness is obtainable at the expense of increased cost, more complexity and less reliability (most cargo and tank vessels operate on one screw). Recent technology continues to make this idea more feasible as more of the engineering problems involved are solved.

(v) Contrarotating propellers offer some of the same advantages as the controllable pitch screws but introduce the same problems. These types of screws appear less practical than the adjustable pitch propellers.

(vi) Vertical axis propellers provide for improved maneuverability and backing characteristics, however, the complexity of these devices reduces their reliability. Other disadvantages include a limitation as to practical size, and their susceptibility to damage.

TABLE III-C1: TYPICAL TACTICAL STATISTICS FOR TANKERS[12]

Vessel Parameter	Cedros	N. News Tnk "C"	N. News Tnk "A"	Oriental Giant	Esso Suez	Nissho Maru
LBP	940'	685'	535'	804'	600'	906'
Breadth	142'	93'	75'	108'	82' 6"	141'
Depth	81'	48' 7"	40' 5"	-	42' 6"	
H	54'	36' 8"	31' 9"	-	31' 10"	
dwt	175,940	50,200	25,500	92,980	34,840	162,300
shp	27,500	26,500	15,000	-	12,500	28,000
W_k	18.3	19.5	18.8	17.6	16	16.5
Technical Diameter, ft.	3,105	2,380	3,200	2,500	2,280	-
Technical Diameter, ft/LBP	3.31	3.47	5.97	3.22	3.8	-
Turning Circle, miles	-	-	-	2.34	-	2.06

(vii) Multiple screw arrangements increase the reliability of the propulsion system. If one screw is damaged, fifty percent power remains in the other screw. Twin screws result in improved turning ability particularly at low speed, but may somewhat affect backing ability because of their smaller diameter. Multiple screw arrangements are also more vulnerable to damage than a single screw on the centerline.

(viii) The proposal for stern anchors would allow a vessel, if stopped in a narrow channel, to maintain its heading when used in conjunction with the normal bow anchor. Similarly, if a vessel is broken in half, the stern anchor would allow for securing the aft section of the vessel. The main disadvantage of using stern anchors is the possible damage that may be inflicted to the propeller and rudder, rendering the ship uncontrollable.

(ix) The primary advantage in a remote release capability of the anchors is a reduction of the response time in initiating the action. This advantage does not appear significant, however, when the hazard of accidental release is considered along with the incapability of an anchor effectively reducing the speed of a vessel with any appreciable way on.

(x) Rudders work most effectively when employed in conjunction with a propeller. For multiple screw arrangements, multiple rudders should be used. Improvement of the existing rudder designs could result in improved maneuverability but most likely at a cost such that the marginal benefits would be uneconomical.

(xi) Currently, bow thrusters are used as aids in vessel control at very low speeds. Their effectiveness decreases appreciably at speeds above four knots. At higher speeds the effective employment of bow and/or stern thrusters is limited by their practical size and the sheer mass of large vessels.

(xii) The most significant aspect of speed reducing equipment is the reduction of the number of high energy impact situations to low energy impact situations. The equipment currently takes the form of bow flaps, split rudders, drogues, and reverse thrust devices. Their effectiveness decreases as velocity of the vessel increases and much research and development remains to make many of these proposals suitable for practical application.

(xiii) To reduce the chance of grounding it is proposed to equip vessels with a detection capability of submerged obstructions that lie in its path.

Systems of this type are presently employed in low speed situations such as docking, or in close quarter conditions. At higher speeds such a system would be useful provided it gave warning in sufficient time for the vessel to take corrective action.

b. Improvements to Limit Loss of Cargo

Proposals to avoid or limit the outflow of oil from a container in the event of a collision or a stranding include the following:

i. Limiting the size of cargo tanks.
ii. Move longitudinal oil tight bulkheads near sides of vessel.
iii. Interrupt wing tank spaces with ballast spaces.
iv. Requirement for more longitudinal bulkheads.
v. Place small tanks forward (where there is more possibility of hull damage) in conjunction with larger tanks aft.
vi. Isolate tanks from ship's side and bottom.
vii. Isolate internal cargo piping from side or bottom damage.
viii. Increase strength of hull through thicker hull plating, use of high strength plates, etc.
ix. Rubber linings for tanks.
x. Provision for the rapid discharge or transfer of cargo in the event of an emergency.
xi. Reduction in the overall size of tankers.

A discussion of the relative merits, feasibility, and limitations of each of the above proposals follows.

(i) The resultant effect from limiting of cargo tank size is that less oil will be spilled in the breaching of a tank. However, as bulkhead spacing is reduced, the probability that a given length of damage will rupture more than one tank increases. This increase in probability is somewhat negated when it is considered the additional bulkheads provide extra energy absorbing structure, which could reduce the extent of damage. With a high energy impact, tank size differences would have little significance on the amount of oil spilled. The explosion hazard resulting from tank cleaning may be changed. While the time each tank is in the explosive range during cleaning is decreased, the total time of cleaning would be increased.[13] For the same deadweight (cargo) carried the smaller tanks mean additional bulkheads and weight. This results in increased construction and operational costs. Because loading rates are based more on overall vessel size than on individual tank size, there is more of a chance of overflow using present day loading procedures and piping arrangements. Maintenance expenses would increase as there are more internal surfaces on which tank cleaning residues could collect. Likewise the time required for vessel inspection would increase.

(ii) The strategy of moving the longitudinal bulkheads nearer the vessel sides decreases the size of the side tanks, while increasing the size of the center tanks. A low energy impact on the side or bilge would result in less oil being spilled. However, in a high energy impact more oil would be spilled as the chance of the inner bulkhead being pierced is increased. There would also be increased weight to the vessel resulting from the increased size of structural transverse members between the longitudinal bulkheads. More oil would most likely be discharged as a result of bottom damage due to the increased size of the center tanks. Certain operational restrictions most likely would result because of the requirement to keep the center tanks pressed full to reduce the free surface athwartship effects on ship stability.

(iii) Interruption of the wing tanks spaces with ballast tanks would reduce the probability of multiple cargo tank damage resulting from side collisions. The piping arrangements would become more complex and tend to increase the weight of the vessel.

(iv) The requirement for additional longitudinal bulkheads would result in a more complex structure, with more piping and fittings required. Tank sizes would be reduced, the structural integrity of the vessel would be increased but at the expense of increased vessel weight and the corresponding construction, operating and maintenance costs.

(v) The reduction of tank size forward on the vessel appears to have little merit. Present design standards allow tankers to absorb a considerable amount of damage resulting from head on collisions. However, in the high energy impact situation, such an arrangement may have additional value. The idea also has practical limitations as the probability distribution of collisions over the length of the vessel decreases only for the aft two-thirds of the vessel.

(vi) Isolating cargo tanks from the sides and bottom of the vessel would most likely result in less oil discharge in low energy impact situations.

There are other advantages in such an arrangement. Tank cleaning would be simplified as the internal surfaces of the tanks for the most part would be smooth. The innerbottom and additional longitudinal bulkheads would contribute to the hull glider strength, with the possibility of using thinner less notch sensitive plating. Additionally, the isolation spaces could serve as clean ballast tanks. The disadvantages include an increase in the explosion hazard as gas vapors could accumulate in the void spaces through leaks or vents. These spaces would require an additional system to maintain them in a gas-free state. To carry the same volume of cargo the ship would have to be larger, thus it would weigh more, adding to operational and construction costs. The increased structural complexity and hull size would result in larger maintenance and inspection costs. In a high energy impact situation the cargo tank isolation would make little if any difference.

(vii) In a low energy impact the isolation of pipes from the sides and bottom would most likely prevent their disturbance. There are practical limitations as to piping placement to avoid damage from high energy impacts. Cargo volatility and pumping limitations under low cargo levels in tanks place constraints on the height piping may be placed above the compartment bottom. Cargo stripping from tanks would become more difficult resulting in more involved gas freeing problems. One means to combat these problems is by employing individual deepwell pumps with all piping run vertically to the deck.

(viii) A stronger hull under low energy impact conditions would result in less probability of hull penetration. Under abnormal conditions there would be less chance of overstressing the structure. It is doubtful under high energy impacts that additional structural strength would be at all effective. The thicker plating would result in a heavier vessel with the corresponding increases in construction and operating costs.

(ix) In an emergency situation the rapid transfer of oil from a damaged tank could limit the total outflow. To make this idea effective standardized hull fittings must be provided on all cargo tanks. The cargo could possibly be pumped into a reserve tank aboard the vessel or to portable reserve tanks deployed from the vessel or by some other means.[14]

(x) The use of rubber linings in tanks is not yet at a practical application stage of development to prevent the ouflow of oil in low energy impact situations. Liners would not be effective in high energy impacts. This idea, however, has much potential and requires much more research and development.

(xi) By reducing tanker size there would theoretically be less chance of bottom damage due to the reduced operating draft. Practically speaking, however, there would remain the same operational judgement errors in attempting to traverse channels thought to be a foot or two over the vessel draft. Any one total casualty would result in less oil discharge, larger tankers are less likely to visit congested traffic areas because of their larger drafts, thus reducing their collision potential.[3] For a given amount of cargo to be transferred, fewer vessels (of supertanker size) would be required. This means less drain on the available manpower pool with the result that more experienced crews are likely to be serving on the larger vessels. Because the probability of a collision increases in part as the square of the number of vessels in a given channel, the overall probability of a collision may be reduced through the use of larger tankers. However, this does not take into consideration that these larger vessels are more difficult to control thus raising the probability of their grounding.

c. Operational Procedures

A second area closely related to improvements in vessel design is the improvement of operational procedures and expertise of personnel. The MARAD Report[16] on a pollution abatement plan for oil and the testimony of Dr. Max Blumer[17] state another major source of oil pollution is the result of inadequate operating procedures and human error. Factual data as to the extent of spillage resulting from operations and personnel is nonexistent. However, estimates have been made that show approximately .0002% of the oil handled in any one port can be expected to be lost. This figure was arrived at on the basis of the experience in Milford Haven, a large British oil port.[18] This particular port had taken extensive measures to prevent the spillage of oil because of its close proximity to a national park. This experience strongly suggests that much effort needs to be directed to procedures and personnel, and unloading equipment.

Proposals to deal with the problem of vessel-caused oil pollution are primarily being directed at two facets. The first are sources of pollution resulting from deliberate or accidental discharges of oil in the loading, unloading and transporting operations of oil. The second aspect are sources resulting from unintentional release of oil into the water as a result of collision or grounding. It is the general consensus of opinion that the most significant source of oil pollution is not from collision or groundings, but from the less dramatic continual everyday operational aspects of oil handling. This is not to say that sources of pollution from collisions and groundings can be disregarded as being insignificant. The Liverpool Underwriters Association statistics on ship collisions and groundings do show that losses result all too often to be neglected,[19] however, the loss of oil at .0002% of the total volume handled bears mute testimony to the size of the problem at hand.

Many of the present discharges of oil into the water can be eliminated through improved operational procedures. Presently, both MARAD and IMCO have proposals in this area. For the most part the following suggested improvements are general in nature; however, they can be considered in varying degrees as potential solutions to the specific problems of Puget Sound. These proposals include:

1. Load on top procedure of tank cleaning and ballasting.
2. Clean ballast operation.
3. Navigating equipment standards.
4. Modification of the Maritime Academy curriculum.
5. Training of personnel.
6. Simplified uniform operating standards.
7. Reliable instrumentation.
8. Fail-safe engineering design criterion.
9. Uniform tank and piping construction and arrangement standards.

(1) The development of the Load on Top (LOT) procedure is the direct result of the requirement that cargo be replaced with ballast after a cargo has been discharged to insure stability of the vessel as it proceeds to another port. As a result, there is some mixing of cargo oil residue and sea water. The usual practice has been to wash these tanks down at sea and dispose the oily

ballast and wash water to the sea. Free dumping of this waste ballast is now prohibited except under specified conditions.[20] In order to meet these requirements the LOT procedure was developed. Figure III-C2 depicts the mechanics. The residue that remains after tank washing is pumped into one of the ship's normal cargo tanks designated as a slop tank and allowed to separate into oil and water. The separated water is decanted until only oil remains. This is then combined with the next cargo, hence the LOT designation.

In theory, the LOT idea appears simple and an adequate solution to the problem. However, there are many problems existing that prevent it from being a totally effective solution. It is not a requirement but only an operational procedure practiced but by 25 percent of the tankers in service today. It requires special piping arrangements thus imposing an economic penalty on the owners of those vessels not equipped for it. There are certain technical problems to be overcome. Detection of the oil water interface is necessary to prevent pumping some oil overboard from the slop tank. Time constraints do not always allow for complete separators need to be employed. Product tankers present problems in that they carry various products and must present "clean" tanks at loading. They cannot load on top of tank cleaning slop. This requires completely separate tanks for the slop or extensive facilities at each port.

FIGURE III - C2

LOAD ON TOP PROCEDURE [21]

As the ship eases out to sea, tanks 2,4,6 are filled with sea water ballast to maintain stability. These tanks then contain a mixture of clean water with residual oil.

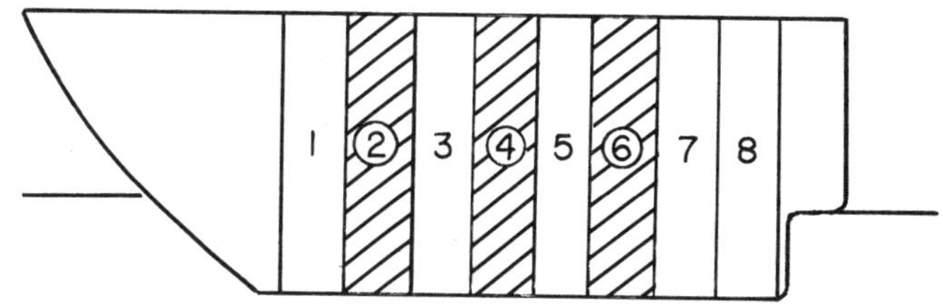

Tanks 1, 3, 5, and 7 are cleaned with high pressure hoses. The oil water mixture in 2, 4, 6 begins to separate with the oil floating to the top of the tanks. A mixture of wash down water and oil residue exist in tanks 1, 3, 5, and 7.

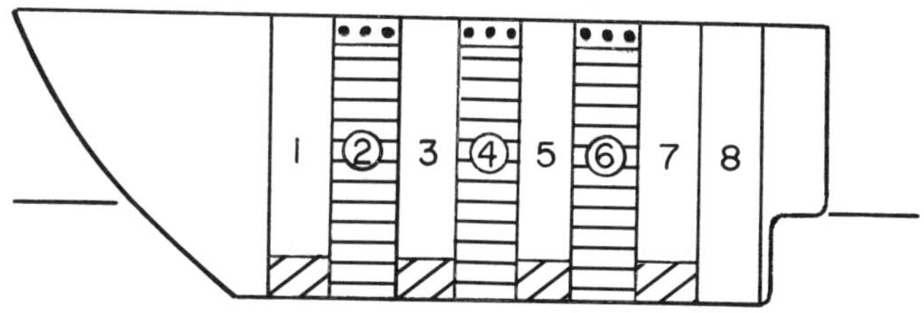

The wash down mixture is pumped to tank 8. Also the sperated oil water mixture in 2, 4, 6, is pumped down to the oil level. This oil is then pumped into tank 8 with the other slop and allowed to seperate once again. Tanks 1, 3, 5, and 7 are filled with clean ballast to maintain stability.

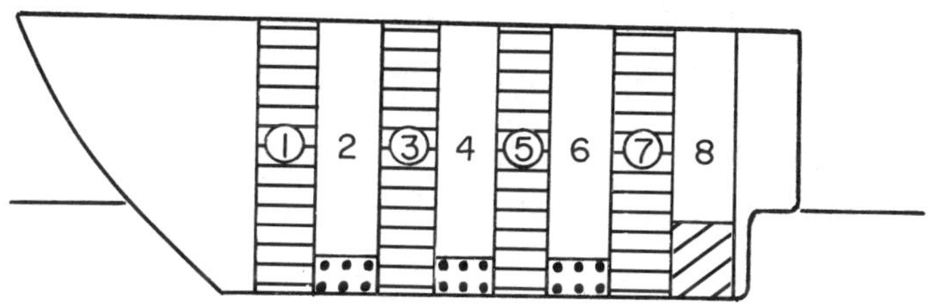

The oil water mixture in tank 8 becomes seperated and the water can be pumped to sea leaving only the oil in tank 8.

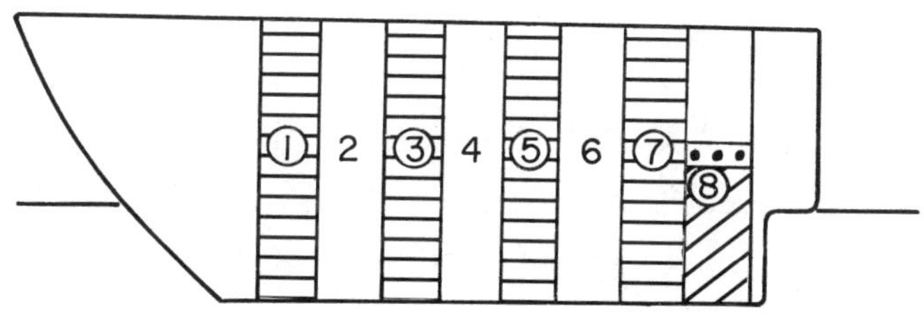

As the vessel enters a port to receive a new load all tanks are clean and ready to receive a new cargo.

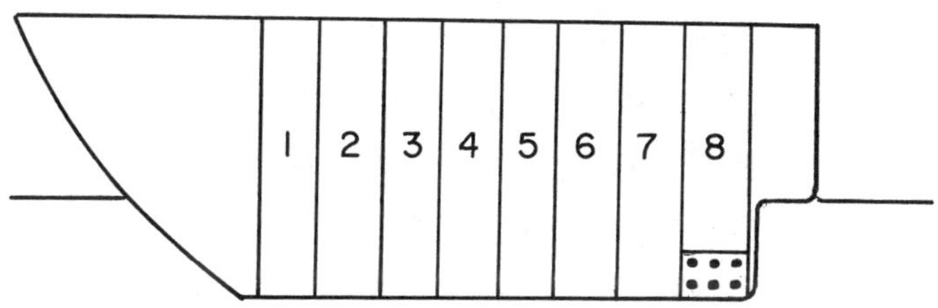

(2) Most vessels do carry some clean ballast because of their excess volume relative to the allowable deadweight for the cargo carried. Clean ballast is not a problem unique only to tankers. Ships of all types encounter the problem of oily ballast water disposal, primarily because of the insufficient importance given to the problem during the design phases. Thus, to sufficiently solve the problem a departure from the current design practices is necessary. The primary argument against changes in design are economic in nature; however, it has been shown that the percentage increases in costs may be small when compared to the potential environmental threat posed by existing ballast operations.[22]

A second argument put forth is that the elimination of ballast water from compensating spent fuel will require a larger beam to maintain intact stability. This results in a ship that is overly stiff and more difficult to control.

The second frequently proposed solution to the problem of oily ballast is the increased availability of shoreside facilities for the processing of oily ballast. IMCO has recommended that all ships be fitted with standard international shore connections to allow for the discharge of oily ballast to the shore. This is most likely to be the interim solution. Most tanker ports are already equipped with some type of oily ballast water disposal capability, however, the situation with respect to other types of vessels is not clear. Presently, the operators find themselves in an economic squeeze in properly disposing of their oily ballast and all too often it finds it is way over the side.

(3) Clearly the present system of navigation equipment standards is far from universal. True, most vessels are equipped with the standard aids to navigation such as radar, LORAN, RDF among others; yet, there are no standards on an international basis that control the degree of effectiveness of these instruments. Many vessels operate today with radar that does not function properly and with LORAN receivers that are not calibrated properly. To this end some system of control over a universal installation of navigating equipment aboard ships, as well as on the shore, would in all probability reduce the number of collisions and groundings. Some efforts now are being made in this direction. For Puget Sound, Honeywell recently proposed a

traffic control system that would encompass all the major shipping lanes. IMCO has adopted such measures as constraints on the use of the automatic pilot, carriage of radar, gyrocompass, and echo sounding equipment (SONAR). They have also recommended more efficient navigation light. The Coast Guard, on the national scene, is currently developing a precision location system for harbors and narrow channels. It is anticipated this system will be implemented in two steps by December 1972. Regardless of these effects and others not mentioned, much more can be done to develop on an international plane the practical standards necessary for a new updated system of vessel navigation.

(4) A more long run solution lies in the proposal to modify the various Maritime Academy curricula to stress pollution abatement in the operation of ships. MARAD is presently working in the forefront of the effort at both the national academy and various state academies.

(5) In addition to the pollution abatement oriented training proposed for the future generation of ship operators, effort needs to be directed at the retraining of existing operating personnel along these same lines. Presently, many operators have been brought up in the "old school" and because of various pressures do not fully appreciate the importance of a complete pollution abatement program. Their retraining can best be done through the unions, the companies that employ them and through the control the Coast Guard exercises in their licensing function. However, regardless of these efforts, no significant gains can be made until the economic advantages to polluting are completely removed.

(6) Presently, a disparity exists between cargo loading and discharge procedures at various ports. The operational procedures followed will vary depending on many factors, most notably the owners of the receiving dock. The recent Anacortes oil spill in Puget Sound[23] points out the necessity of more standardized operations as well as a clearer delineation of the responsibility for the overall loading or discharge operation. Currently, the responsibility is divided between those on the vessel and onshore personnel. Perhaps a model loading and discharge procedure can be developed for all those receiving and discharge points in Puget Sound through the joint efforts of all those concerned.

(7) The proposals for more reliable instrumentation fall into two areas. Those for improving the navigational aspects of vessels, which include the ideas discussed part (b), as well as the idea of requiring more sophisticated instruments aboard vessels. Some of these include inertial guidance systems, satellite guidance, and collision avoidance devices. The technology exists today to make these types of devices feasible; what is needed is some effective measure to require their use on vessels. The second area has to do with better control of the loading and discharge of cargos as well as the detection of possible spilled oil during these operations. Again the spill at Anacortes pointed out the need for more reliable instrumentation.[24]

(8) Fail-safe engineering design criteria can be incorporated into the overall design of a system that handles oil. Fail-safe simply means that through proper incorporation of design standards into a system any fault that develops in the system and results in a failure, that failure will not jeopardize the system or the environment in which it operates any further. Clearly, this idea can be employed to prevent those accidents that result from human error of one sort or another.

(9) The idea of uniform oil handling systems in the form of tank and piping arrangements is aimed at reducing the confusion of personnel in operating and controlling the system. Although many problems exist as to large scale application of this idea, certainly it could prove beneficial within smaller locales. For example, within a company which owns a fleet of vessels, a standardized piping system on all those vessels will reduce the probability of company personnel controlling the flow of oil through the piping system making an error which results in an oil spill.

d. Applicability to Puget Sound

In evaluating which of the above proposals is most applicable to the particular conditions in the Puget Sound region, consideration of the implementation time is necessary. For many of the proposals, the time from inception of the idea to its actual implementation is often many years. The oil-on-water problem exists today, and the initiation of any action will not be felt for some time to come. Thus, it is imperative to consider interim or short range measures that can be applied with only a short time lag.

Both long range and short range proposals are briefly discussed below.

Long Run Proposals

Proposals which are currently under development but, because of their nature, will not become effective in reducing oil pollution for several years are listed below. The primary reason for the delay is that all of the proposals involve changes in the basic design of the vessel. Considering that the economic life of a ship is approximately 20 years, it will take that many years before the present fleet is replaced.

1. Clean Ballast

Present overboard discharge of tank cleaning residues and oily ballast water is one of the most significant sources of oil pollution from vessels, primarily because of the lack of attention to pollution reduction objectives during basic design stages. Design changes are necessary to solve the problems created by the existing overboard discharge systems on present vessels. A study of what the economic consequences are as a result of designing for clean ballast was done by E. Scott Dillon of MARAD.[22] Three types of vessels are considered:

a. a conventional cargo ship,

b. a container ship, and

c. a tank ship.

For the cargo ship two clean ballast alternatives were proposed and the incremental increase in the required freight rate (RFR) was determined over the RFR of the original vessel, to achieve the some economic return. The increase in RFR required for alternative (a) was 7.9%, for alternative (b) 3.4%. Thus, the penalties imposed by a clean ballast operation are not that severe.

Because of the configuration of container ships, clean ballast can be obtained without encroaching on the vessels earning capacity. Considering this flexibility there appears to be no excuse for overboard discharge of oily water from new ships of this class.

Until recently clean ballast operation in tankers was not considered a practical possibility. There is simply not sufficient space for tankage exclusively devoted to clean ballast if the vessel is to take full advantage of minimum freeboard decrees under the 1966 International Load Line Convention. Generally, clean ballast in a tanker is achieved by adding wing tanks or double bottoms. Four schemes are shown of possible alternatives in Figure III-C3 through III-C6, with Figure III-C7 showing the percent increase in RFR vs. percent clean ballast operation.

2. Double Bottoms

Extra protection is afforded oil tanks by double bottoms on tankers. At the same time clean ballast can be achieved. The Subcommittee on Ship Design and Equipment uncertook a study to determine the effectiveness of double bottoms as a result of grounding or stranding in preventing an oil outflow.[30] The basic findings indicate that for an average double bottom height of .06 B, the probabilities indicate oil tanks would not be damaged in 73% of the groundings. In approximately one fifth of this 73% the outer shell was not penetrated. A study of this type has yet to be considered regarding collisions.

3. Limitation of Tank Size

This proposal is currently the most feasible. IMCO has generated a resolution to the effect of limiting tank size. The economic penalties associated with a limitation in tank size are shown in Figures III-C8. The fixed limit of oil outflow decided upon by IMCO is 30,000. Thus, the percent increase in RFR for a 200,000 ton tanker at this outflow is approximately 1/2 of 1%.

4. Maneuverability

The ability of existing large tankers to avoid dangerous situations in a minimal amount of time is quite limited. Presently, much work is being focused on this aspect in an effort to determine practical means to increase ship maneuverability. This aspect of preventative proposals is perhaps more in the developmental stage and as such will involve a longer delay to implementation.

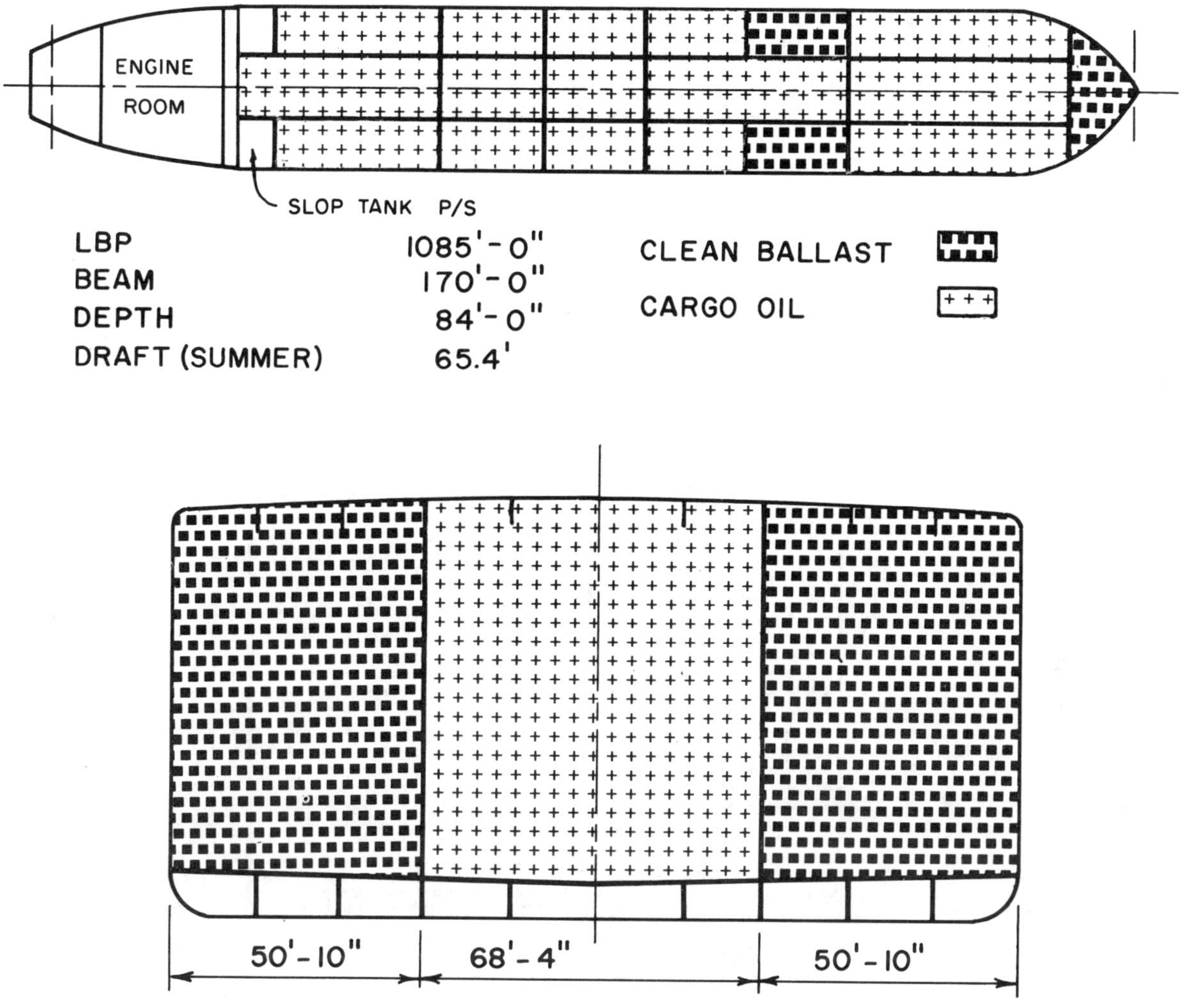

CONVENTIONAL 250,000 TON DWT TANKER, 13.7 % FULL LOAD DISPLACEMENT IN CLEAN BALLAST.

FIGURE III-C3

PRINCIPAL DIMENSIONS OF CONVENTIONAL 250,000 DWT TANKER, PLAN & MIDSHIP SECTION VIEWS [25]

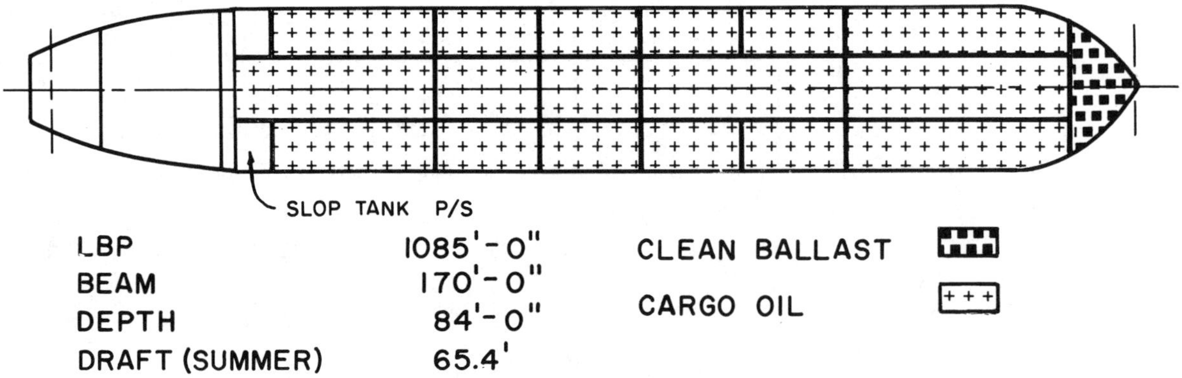

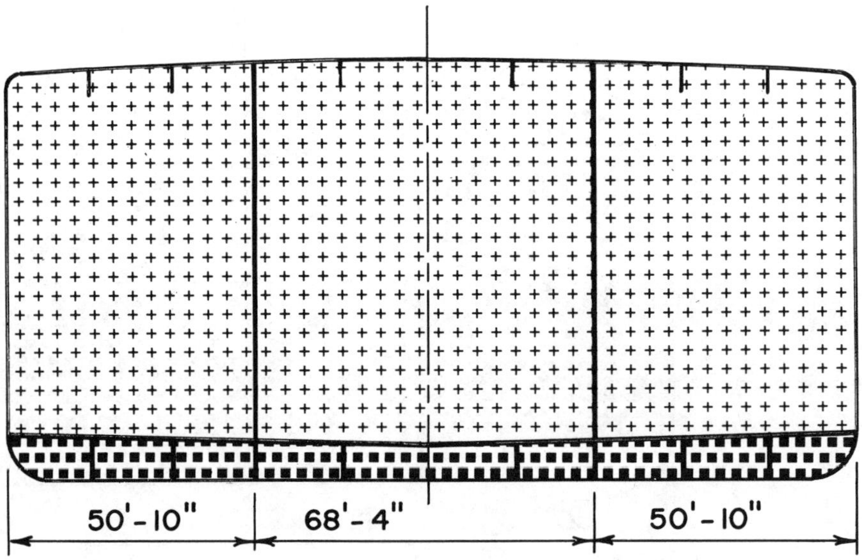

250,000 TON DWT TANKER WITH DOUBLE BOTTOM 13.7 % FULL LOAD DISPLACEMENT IN CLEAN BALLAST.

FIGURE III-C4

PRINCIPAL DIMENSIONS OF BASIC 250,000 DWT TANKER FITTED WITH DOUBLE BOTTOM TANKS, PLAN & MIDSHIP SECTION VIEWS[26]

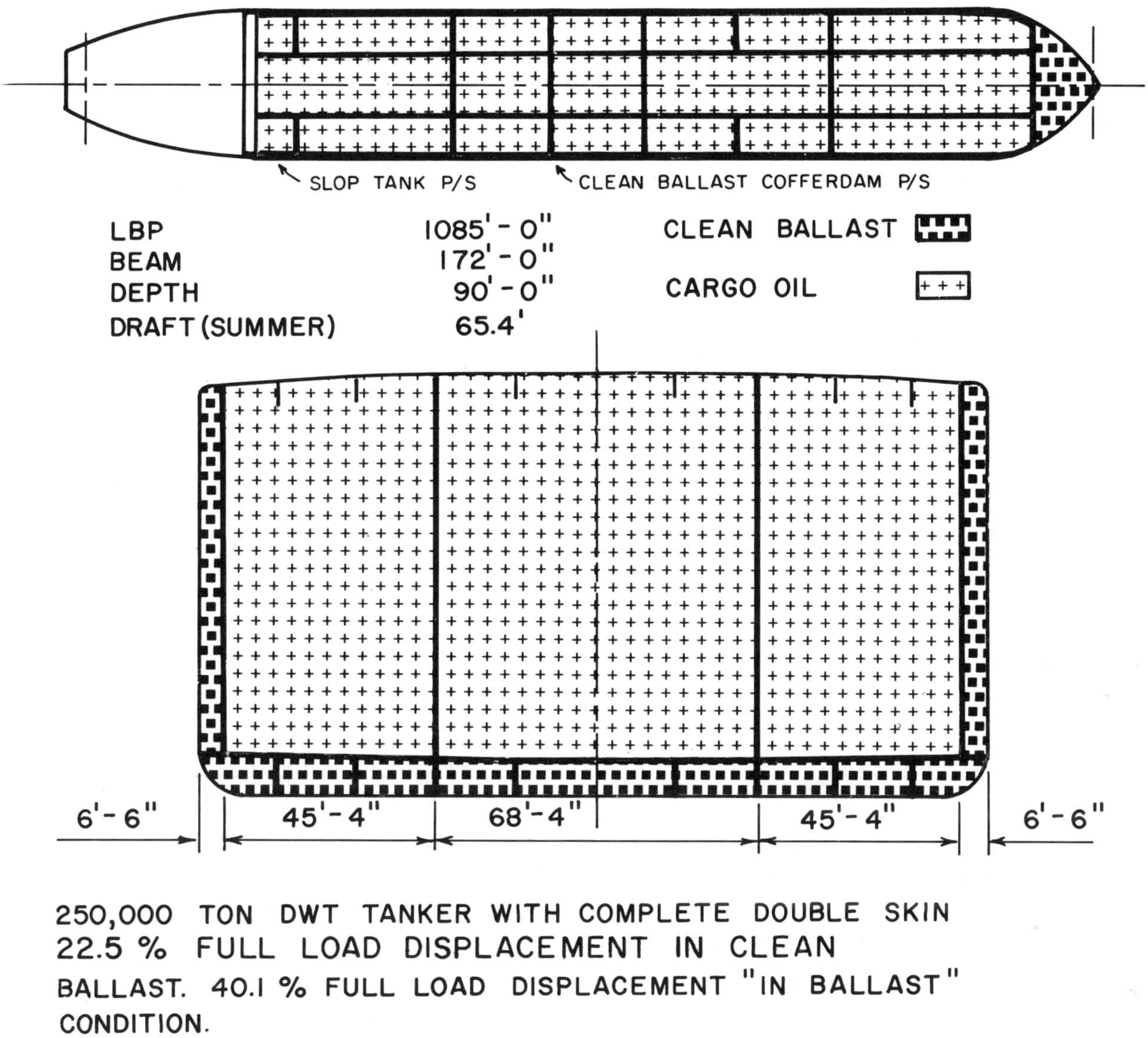

FIGURE III - C5

PRINCIPAL DIMENSIONS OF DOUBLE BOTTOMED TANKER WITH SIDE BALLAST TANKAGE PLAN & MIDSHIP SECTION VIEWS [27]

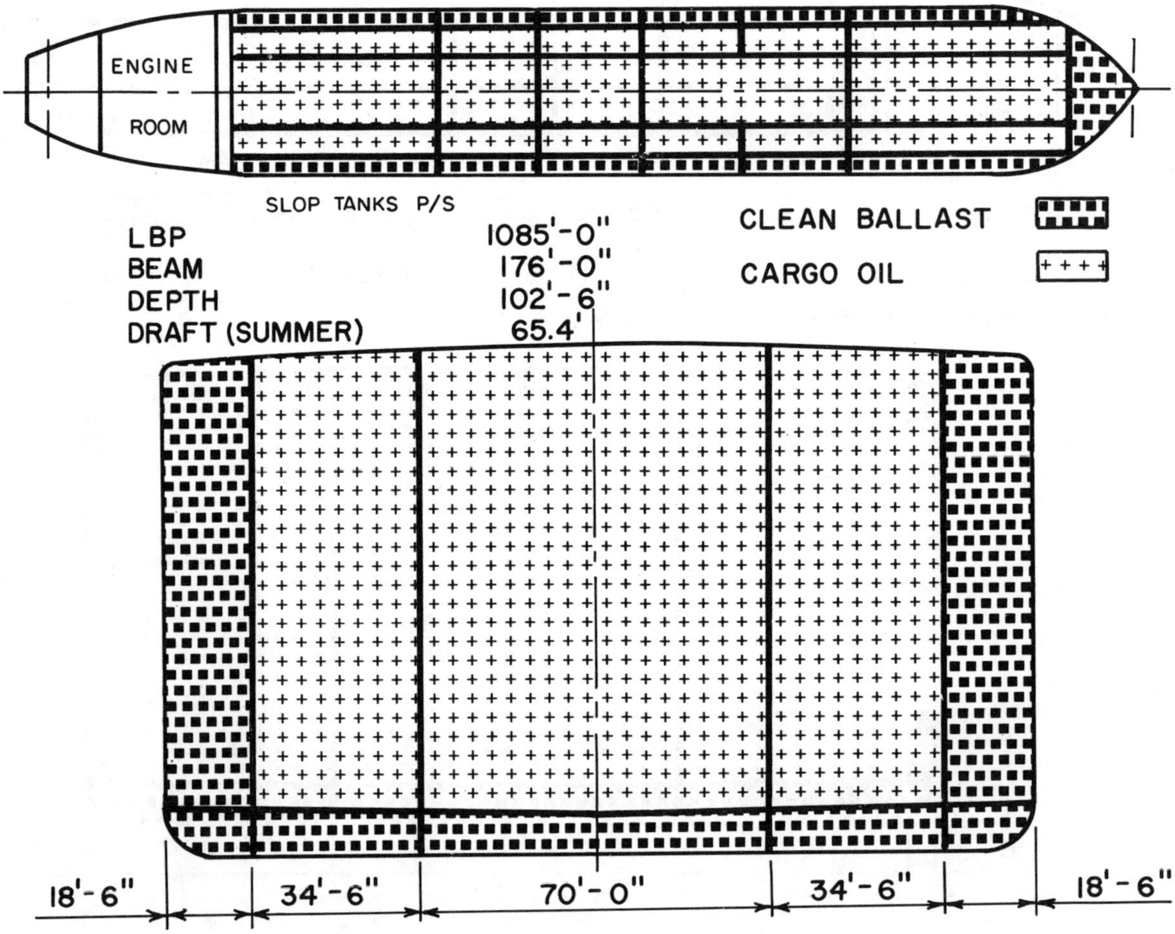

250,000 TON DWT TANKER WITH COMPLETE DOUBLE SKIN (ALL CLEAN BALLAST) 41.1 % FULL LOAD DISPLACEMENT IN CLEAN BALLAST. 60.1 % FULL LOAD DISPLACEMENT "IN BALLAST" CONDITION.

FIGURE III - C6

PRINCIPAL DIMENSIONS OF DOUBLE BOTTOMED TANKER WITH LARGE SIDE BALLAST TANKAGE PLAN & MIDSHIP SECTION VIEWS [28]

FIGURE III-C7

ECONOMIC EFFECT OF CLEAN BALLAST CAPACITY

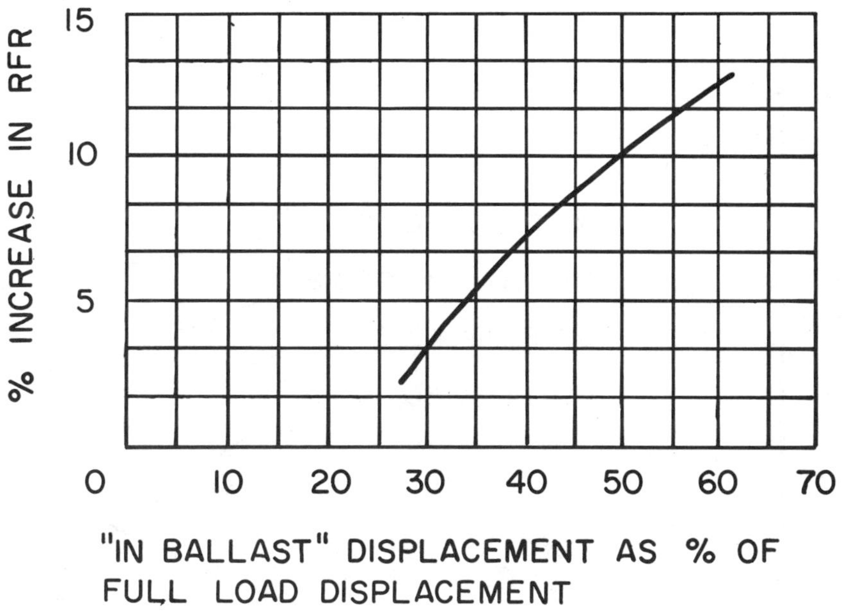

FIGURE III-C8

PERCENTAGE INCREASE IN REQUIRED FREIGHT RATE TO LIMIT OIL OUTFLOW

(REFERENCE HAS 4 WING TANKS ON A SIDE)

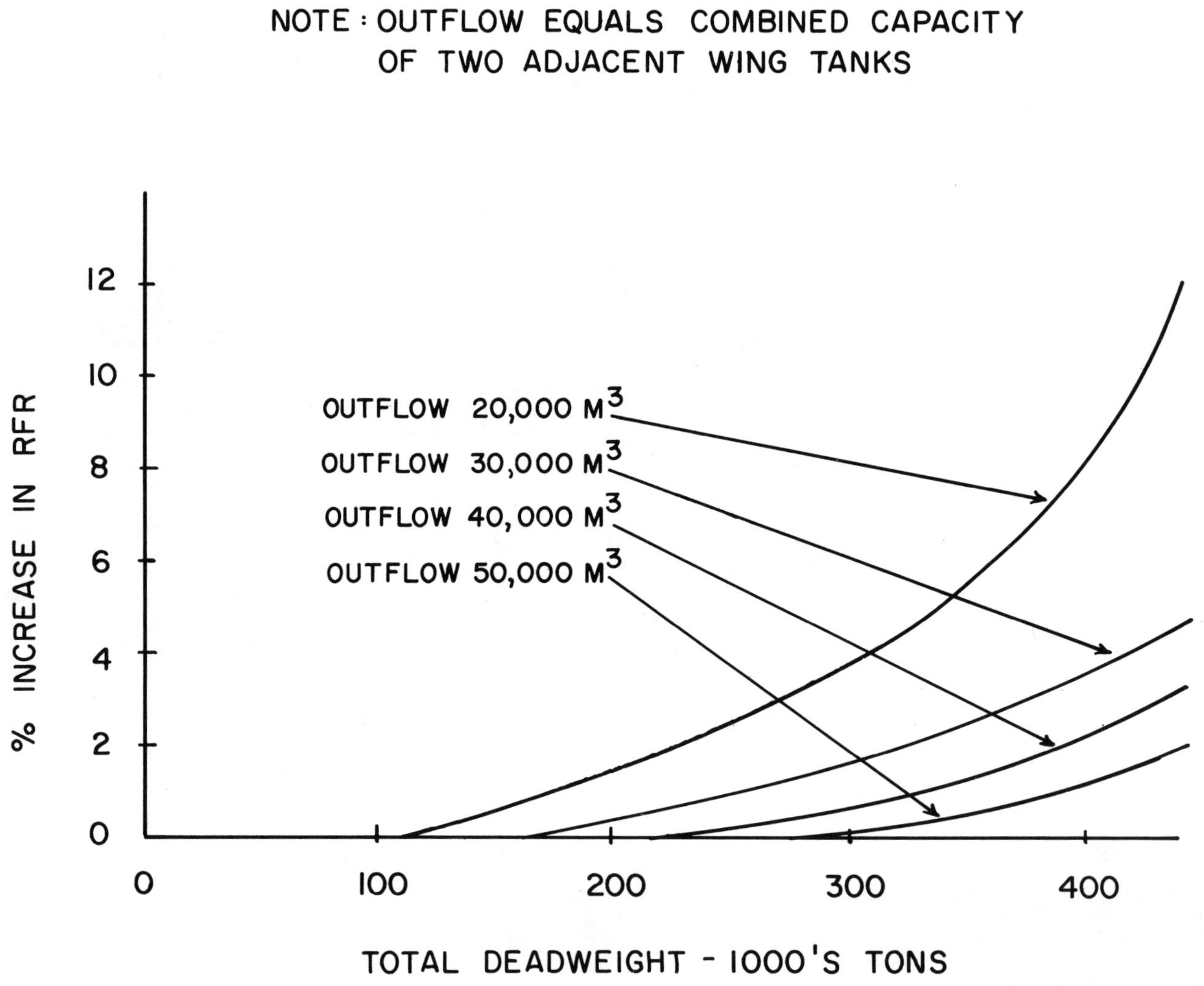

5. Education

This proposal is accomplished through the various state and the U. S. Marine Academies. Emphasis on principles of pollution abatement in vessel operation. A proposal such as this can be achieved at virtually no incremental costs and the benefits derived therefrom could be enormous.

Short run proposals are of more interest as they involve measures over which more control can be exercised from the local level. Of necessity the measures involve the operational aspects of prevention and are as follows:

1. Load on Top Procedures

Although the LOT procedure does not result in 100 percent reduction in oil pollution, it does provide some immediate relief to the problem. Much of the existing local traffic can institute LOT operations through minimum changes in existing vessel makeup. Although it is not known if any specific costs studies regarding the RFR penalty imposed by LOT procedures have been done, it is the opinion of this study that they are not that severe; and thus, all efforts should be undertaken to see that LOT is implemented for all concerned vessels on Puget Sound.

2. Port Facilities

Generally, it is the economic penalty of increased operating time associated with most preventative measures that lead to their perceived ineffectiveness. Provision of oily waste reception facilities at strategic ports in Puget Sound would help alleviate this penalty motive. Presently, a survey is underway of all available reception facilities in U. S. ports. The most recent, taken in 1960, indicated that Puget Sound had available 64 reception facilities both mobile and permanent with a capacity for 30,223 long tons.[32] MARAD is promoting the construction of additional facilities for oily waste treatment. Once the total deficit is determined efforts should be directed toward the filling of the void.

3. Traffic Control

Various traffic control systems are now in operation in ports around the world. In most instances, some type of external control of vessel movement in congested areas has been extremely beneficial in terms of the reduction of casulaties. Although the idea of traffic control goes against the aged traditions of the sea, it appears that the needs of society are now best served through some imposition on this tradition. A complete analysis of traffic control is contained in Section III-C, Part 2 of this report.

e. Overall Assessment as Applied to Puget Sound

The question in everyone's mind at this point is what can be done with the problem of vessels polluting Puget Sound as the result of an oil spill. As has been pointed out the problem is real and a definite threat, yet one that appears well on its way to solution. The magnitude of the problem requires an attack from an international level which by necessity introduces long time lags from inception to implementation of pollution reduction proposals. The full effect of these efforts will not be felt in the Puget Sound area for a number of years.

In the meantime, the transportation of oil on Puget Sound is increasing. The three tankers being built by ARCO to transport oil from Valdez to West Coast markets are virtually being built in accordance with design standards that have in their stated objectives no provisions for minimization of pollution potential. This is not the fault of any one individual or group but simply the result of setting priorities as to which objectives to achieve. Not all is lost however; for much can be done to reduce the pollution threat on the local level without the usual time delay. These proposed measures can for the most part be instituted tomorrow to achieve an interim margin of safety.

2. TRAFFIC CONTROL SYSTEMS

a. Introduction

Puget Sound is one of the world's safest waterways.[1] Commercial vessels have been calling on Puget Sound for over 100 years. For more than half of this period, tankships have been active on the Sound. In fifty years of tanker activity, there has never been a serious collision on Puget Sound involving a tankship nor has the Sound ever experienced a major oil spill as a result of collision or grounding.

This is a truly remarkable record. If, indeed, Puget Sound has never experienced a serious collision involving a tank ship, why should money be spent for elaborate control systems; or even, change the present traffic patterns and operational procedures?

No one can take issue with the fact that Puget Sound pilots are among the best in the world and that the Sound is a relatively safe waterway. But, there have been collisions involving freighters, and a statistical analysis of the Sound[2] indicates that we may be living on borrowed time with regard to a tankship incident.

To underscore the problem that exists today in many European ports, and the problem that may exist tomorrow on Puget Sound, the development of the mammoth tanker must be fully examined. These ships represent not just a new class of ship (C-2, C-5, C-5A, etc.); but, the evolvement of a whole new technology.

It has been 110 years since the first overseas shipment of oil in quantity. By the 1880s the tanker movement was underway. In 1886, the SS Gluckauf, the prototype of the modern tanker, was completed. This ship was 310 feet long with a deadweight of 3,000 tons. True to form, the Gluckauf was one of the largest ships of her time.

World War I saw the building of 18,000 dwt tankers.[3] World War II created a similar need for tankers and the United States responded with the building of over 500 T-2 tankers. These ships survived to form the bulk of the world's postwar tanker fleet, and they still exist today. Many have been "jumboized" and others have been converted to dry cargo service by building new hulls around the ship's machinery space.

The T-2 tanker and others like her effectively served the trade for decades. The period from World War I to the late 1940s did not see many innovations in tanker design. The late 1940's saw a change in the characteristics of the trade.

Increases in consumption of petroleum products made it profitable to refine the product near the consumer instead of importing the product from distant refineries. This innovation naturally created a need for more long haul bulk carriers. It became apparent to tankship operators that the tanker design must be changed to satisfy the need for economies of scale. The first of the so-called supertankers appeared in 1950 and had a dwt of 30,000 tons. In 1953, the Tina Onassis figuratively broke the dam of conservatism by the sheer weight of her 50,000 dwt.

By the late 1950s ships were being built which were eight times as large as the T-2. The mid 1960s saw the emergence of the first 200,000 dwt tanker. Today, ships fifty times the size of the T-2 are being designed. Table III-C2 presents examples of various size tankers and their operational characteristics.

The development of these tankers is not without problems. It is because of these problems that one begins to see the need for safety improvements in ship movements in confined waters.

New construction techniques had to be developed before each great leap in size could be accomplished. Of course, each increase in the size of a tanker required a proportionate increase in the size, capacity, output or strength of its component parts, such as propulsion plant, generators, pumps and propellors. Merely building everything larger, however, was the answer.

A glance at the figure comparing tank ships reveals that the crash stopping distance of a fully loaded T-2 is seven ship lengths (3500'). The stopping distance of the Universe Ireland is reasonable when compared on this basis with the T-2 (i.e. 9-1/2 ship lengths). This distance is about 10,000 feet compared with 3,500 feet for the smaller vessel.

Tank size plays an important part when one considers the need for safety requirements. The center tanks of the Universe Ireland have a

TABLE III-C2: EXAMPLES OF TANKER OPERATIONAL CHARACTERISTICS[4]

EXAMPLES OF TANKSHIPS OF VARIOUS TONNAGES

Class or Name	T-2	T-5	Pennsylvania	World Glory
Dwt	16,800	26,500	28,170	45,500
Length	523'6"	627'6"	624'7"	736'6"
Beam	68'	83'6"	84'	102'
Draft (loaded)	30'2"	32'10"	33'	37'6"
Displacement (loaded)	21,800	34,630	36,280	58,265
Capacity (bbls)	141,000	204,000	241,500	396,000
Shaft horsepower	6,600	20,460	13,750	15,000
Speed (kts.)	14.5	18	16.6	16.0
Turning Radius	3/10 mile	N/A	N/A	N/A
Crash Stop (min)	12 min.	N/A	N/A	N/A
Crash Stop (ship length)	7	8	N/A	N/A

Ship Name or Owner	Barracuda	Niarchos	ARCO	Universe Ireland
Dwt	60,000	106,000	120,000	326,000
Length	810'	949'	N/A	1,135
Beam	104'	134'	N/A	174'10" (moulded)
Draft (loaded)	41'9"	49'	N/A	80'
Displacement (loaded)	76,000	N/A	N/A	--
Capacity (bbls)	430,000	821,000	N/A	2,400,000
Shaft horsepower	25,000	43,000	N/A	37,400
Speed (kts.)	17.2	18.0	N/A	14.8
Turning Radius	4/10	6/10 mile	N/A	3/4 mile
Crash stop	14 min.	N/A	N/A	17 min.
Crash stop (ship length)	N/A	N/A	N/A	9.5

138,000 bbl capacity - almost equal to the total capacity of the T-2. Even a "minor" collision with these larger vessels could be catastrophic in the environmental sense.

An evaluation of the need for improvements in ship traffic control can best be analyzed by examining the ability of the supertanker to avoid grounding or collision should it find itself in such a condition. We have already alluded to the handling characteristics of the mammoth tankship. Let us now bring this problem into proper perspective.

The crash stopping distance of a 200,000 dwt tanker is about 10,000 feet and requires about 20 minutes. The controllability of a ship is, however, a function of two items: dynamic stability and maneuverability.[5] A ship is said to be dynamically stable if, after a slight disturbance, it returns to its original motion without the use of the rudder. A ship that has good maneuverability, responds quickly to the change of the helm. The maneuverability is dependent upon rudder effectiveness.

The controllability of a ship is markedly impaired while stopping; both the dynamic stability and maneuverability are altered, as characterized by the following:

a. The velocity of flow over the ship decreases due to the reduction of the ship speed.

b. The velocity of the flow over the rudder decreases.

c. Moreover, the flow around the ship and the rudder will be disturbed due to the deceleration of the vessel.

d. Rudder effectiveness decreases with decreasing RPM and has virtually no effect after the rotation has been reversed.

e. Dynamic stability is reduced due to lower prop RPM and thrust.

More work in this area was performed by Dr. Masataka Fujino of the University of Tokyo.[6] The following may be concluded from his work:

a. Instability of a ship in a narrow channel is mainly due to the asymmetric hydrodynamic forces and moment caused by the presence of channel walls.

b. The shallower the water, the greater the effect of wind on the unstable condition.

The supertanker, however, is affected by wind regardless of the water depth (more severely in shallow water).

It would seem that, while certain steps can be taken to improve the handling characteristics of large tankers; these ships are extremely limited in their ability to avoid a collision or grounding when maneuvering in close quarters.

Furthermore, the recent spree of large tanker strandings, burnings, collisions and explosions has shaken the previously successful argument that the fewer large ships, the lower the odds on disaster.

b. The Implementation of Traffic Control Schemes

One of the problems involved with the implementation of traffic control systems will be their acceptance. Historically, mariners have been independent, having sole responsibility for the safe conduct of their vessels. There are only two places in the world where the master relinquishes responsibility for the vessel, the Panama and Suez Canals. Everywhere but here, the master rules supreme!

Lights, shore structures, radio beacons and LORAN are all of a nature to provide navigational assistance if a mariner so desires to use them. Pilots aboard vessels in Puget Sound are there in an advisory capacity only.[7] Pilots, therefore, are employed as another aid to navigation.

As vessels become larger and their maneuverability diminishes, the process of navigation becoming too critical and complex to be left to the vessel alone. Development of shore based harbor advisory systems provides an additional source of information for the master. However, acceptance of shore developed information will depend on education and retraining of vessel personnel and agreement between information from the shore and information developed aboard ship.

It has been found that the most important prerequisites to successful introduction of sophisticated techniques and automation are detailed job analysis and comprehensive retraining of staff to operate the systems to full advantage. If one or both of these factors are neglected, then the chances of the system being both economically and operationally viable are drastically reduced. It appears that in field of navigational aids, both of these areas have been neglected to a certain degree and have resulted in over-sophisticated equipment being produced by technologists and its use being forced on under-trained operators.

An analysis recently conducted by the Liverpool Regional College of Technology, showed that a sample of 550 masters and officers who returned the questionnaire "... at most only 25% have sufficient knowledge to use their radar fully in conditions of reduced visibility and the majority of masters at sea take only partial advantage of the full potential of their equipment."[8]

There is obviously a great need for regular, formalized training and refresher courses for officers and masters, possibly of the type undertaken by airline pilots; but an equally large problem with merchant ships is that individual owners and designers tend to express their personal preferences in the selection and siting of equipment. Masters and officers also change ships regularly and take their disciplines and preferences with them. Thus, in many cases, when the officer of the watch is faced with a sudden unfamiliar situation, he has to take "ad hoc" action to deal with it and is not able to put into action a preconceived, tested and rehearsed plan in which all participants have a fully understood part to play.

A training facility under the aegis of the Master, Mates and Pilots Association has been established in Baltimore, Maryland.[9] The Maritime Institute of Technology and Graduate Studies has specially designed equipment simulating onboard experiences in automated shipping.[10]

The simulation system developed by the Link Division of the Singer Company is so realistic that trainees will actually hear fog horns and will receive engine room reports in response to their orders.[11] The equipment consists of a collision avoidance radar simulator, an automated propulsion control center, and a cargo loading trainer. The collision avoidance simulator is comprised of eight cubicles, each containing two radar stations and a control console. The radar sets will be computer programmed to trace the course of the trainees ship, plus as many as 15 other vessels. Radar will record not only passage of ships, but also, buoys and land points of key harbors and waterways. In a typical program, the trainee will be told that he is the pilot of a ship at Ambrose Light (off New York). The weather conditions are: extensive fog; visibility, one-half mile; "Bring your ship into New York Harbor."

The purpose of this training program is to raise the capability of ship's officers to a level consistent with the requirements of new technology. A regular scheduled refresher course utilizing the multimillion dollar Singer facility should be employed by all pilot associations handling the mammoth tankers on a regular basis. It would also be essential for masters and officers of new vessels to attend the simulator program prior to their sea trials.

c. New Shipboard Navigational Aids

Acceptance of traffic control schemes also relies on agreement between shorebased and shipboard information.

Marine radar has been a disappointment to those who thought that its general installation in merchant ships would drastically reduce or eliminate collisions because ships, like autos, simply go faster in poor visibility until the risk is approximately what it was before the advent of advanced radar systems. The collisions involving large tankers have focused public attention on this problem. Radar engineers, meanwhile, have been studying ways in which the new radar information can be processed and displayed to assist decision making for collision avoidance. This amounts to indicating, more or less automatically, the time to and distance from the closest point of approach (CPA) of threatening targets. All the major radar manufacturers have come up with solution of widely varying complexity and cost.

The following is a brief description of some of these navigational aids.

1. Marconi Predictor

The Predictor radar stores the raw radar picture on magnetic tape. The picture from every fourth rotation of the scanner is stored. The tape loop is arranged so that the pictures or frames are played back onto the cathode ray tube (crt) display in a different order to that in which they were recorded. Thus, each group of four frames, covering a 10 second period in real time consists of the radar pictures recorded 6, 4, 2 and 0 minutes ago, in that order. When switched to show relative (or true) tracks every target thus appears in four positions, each of which is sequentially updated every 10 seconds. These rows of dots give course and speed information and indicate changes that have occurred, or are occurring during the previous 6 minutes. In the "true tracks" mode, one's own ship remains at the center, previous positions showing astern. A "predicted relative tracks" mode is also provided to show what these would have been had one's own ship reached her present position from a different direction. This enables an avoidance maneuver to be tested before being carried out.

2. Decca Transar

The 66AC radars of this series are popular and simple. Five interscan lines or markers, of fixed length, are provided. The head of each line can be set to mark any ship that appears to present a potential risk. Once set, the markers point toward own ship and continue to do so, moving bodily across the display (true motion mode) as necessary. If a target echo moves down its own line, it is approaching on a constant bearing and so, after a few minutes the dangerous targets can be identified. The echo tails show true course, or aspect, while the relationship between the head of the marker and the echo itself gives the relative track and a visual indication of the CPA. If an avoidance maneuver is to be made the markers must be re-set.

3. Kelvin Hughes Photoplot

The "Photoplot" display stores the plan position indicator (ppi) radar pictures on 16mm film which is rapidly processed and transported for back projection onto a large screen at intervals controlled by the operator. The longer these intervals are the larger the number of scanner sweeps that will be superimposed on one display, so for moving targets, there will be longer tracks showing past history. Unlike the echo tails caused by the afterglow of a long persistence tube, these tracks give a sufficiently accurate estimate of speed as well as direction to enable the collision risk of any ship within the scope to be assessed. A transport "predictor overlay" is fitted over the projected display to facilitate the drawing of relative (or true) tracks when the set is in the true motion (or relative) mode. The effect of an alteration of course can be assessed in the same manner.

4. GEC - AEI Compact

(Compact stands for computer and automatic course tracking.) This display computes the velocity vectors of selected targets from present information. Up to 12 targets can be manually selected by laying a green circle round them which locks the computer onto them. They are then followed automatically, and their velocity vectors displayed in green. A CPA circle can also be shown round one's own ship, or round the termination of one's own ship vector in true motion. Alarm signals operate for all closing targets when they come within 11 and 9 miles, and also for targets under surveillance which will pass, on present form, within CPA distance selected by the operator. A separate green phosphor display tube is used for the indicators and this is set at an angle to the standard ppi and the image superimposed by a semi-reflecting mirror. In this set also an avoidance maneuver can be tested in advance.

5. Norcontrol Data Radar

This anti-collision function is part of a system known as Data Bridge. The full system provides Data Radar; a dead reckoning, position fixing, and great circle sailing interface; and a third function designed to improve autopilot performance. The collision avoidance function is based on the coordination of radar and computer. The system, controlled by a simple joystick, gives automatic tracking and collision thread assessment of up to 12 targets, presenting relative and true motion vector displays in an easily assimilated form, superimposed on a conventional ppi. A system of time oriented dots, superimposed on the vectors of both own ship and threatening targets enables closest approach and CPA time to be readily assessed. The system automatically gives both visual and audible alarms of collision courses, as well as calculating drift by means of reference to known positions. Data Radar also enables the officer of the watch to simulate maneuver data in the case of dangerous situations.

6. Sperry Collision Threat Assessment

The device is used as an accessory to Mark 12 solid state radars and may be adapted to other marine radars. Capable of operation in both true and relative motion displays either centered or offset. The device automatically supplies the navigator with immediate visual indication of degree of risk for every target intruding into the radar screen and obviates the need to manually track or lock onto radar echoes.

Each echo appearing on the radar display will cause a threat assessment marker (TAM) to appear. This is a 5° wide segment of similar thickness and brightness to a range marker line. The TAM "visual fence" will appear at present positions determined by the officer of the watch (OOW) in light of known speed and maneuverability of their own vessel.

When the echo moves behind the TAM, the target is moving at a relative speed indicating arrival at one's own ship in a time greater than that initially selected as immediately dangerous. When the echo moves coincident with the TAM arrival at one's own ship will be close to that selected. When the echo penetrates and crosses the TAM, it represents a potential collision threat which will require action.

For any one echo, the time control may be adjusted to position the TAM over the echo and thus obtain an approximate CPA.

7. Iotron Digiplot

The "Digiplot," another separate display, can analyze and track target echoes from any standard radar. It differs from the others discussed in that the computer automatically acquires all radar targets. Having separated out the land mass targets, it selects the 40 most threatening small targets for display. The selection is made on the basis of the calculated CPA for each one, operator selected time and distance to CPA being used as a criterion. The 40 selected targets are shown with their velocity vectors and the range, bearing, course, speed and CPA time and distance of any one of them can be shown digitally when picked out by the operator by means of a light-sensitive pointer. The display is scanned 50 times a second, making it completely flicker-free. Automatic alarms are also provided and the effect of changes of course and speed can be completely simulated, the projected maneuver being speeded up 30 times.

8. James Scott Mark II Mini Radar

This device is not truly comparable to the seven other systems in terms of its application. But it is an important part of the new hardware and should be utilized if at all possible.

Several recent accidents have indicated that the inability to estimate sideways movements of ships toward a berth has led to entire jetties being offset and severe hull damage. This light weight portable radar system has been tested at British Petroleum's Finnart Ocean Terminal. The device is able to keep the pilot informed of the tanker's approach speed from an initial 50 ft/min down to 5 ft/min. The mini radar is based on the doppler system and operates from a normal 12 volt battery source. It is provided with connections for computers and data displays.

d. Evaluation of Shipboard Aids

Are these devices really worth the money and do they aid shipboard personnel in their decision making process?

Prices vary depending on which system and what accessories are chosen. The Norcontrol Data Bridge System should run about $96,000. Digiplot would appear to be the most automatic of the systems and probably is the most expensive. The price, whatever it is, is negligible in relation to the risk it is designed to reduce or eliminate. But is the risk really reduced?

Most collisions result from end-on encounters and these are sometimes caused by late course alterations. But every CPA calculation, whether the ships are on alternate courses or not, makes the implicit assumption that no one is going to change course or speed. Several of the displays referred to make provision for trying out avoidance maneuvers on the plot before making them and the explanatory literature that goes with them often includes a series of diagrams which start with ships converging on steady bearings. Disaster seems imminent. The next diagram he sees reveals the always successful trial maneuver. Nobody ever alters course or speed. How many collisions have ever occurred in which only one ship has altered course?

Another problem associated with these devices is the alarm system. Ships maneuvering in heavily congested waters would have alarms going off constantly. A better alarm would indicate course or speed alterations by another vessel. A computer could spot this before it became apparent on the display and this is the moment when maximum alertness may be required.

Of course, it would be even more valuable to know course and speed changes before they occur and this would require bridge to bridge communications.

These new shipboard devices can best be utilized, if a controller was in contact with and could make suggestions or instructions to all ships in the area. This interface between shipboard equipment and personnel, and shore based equipment and personnel is really the key to effectiveness of collision avoidance.

e. <u>Traffic Control Systems</u>

The following evaluates two operational traffic control systems and their effectiveness. One is located in Rotterdam, Netherlands, while the other is in San Francisco, California.

1. Rotterdam, Netherlands[12]

Pilots equipped with a 24 channel VHF set communicate with the appropriate radar stations, use the assigned channel for the radar coverage area concerned and switch to the next channel after "hand over" has taken place.[12] The information supplied to pilots consists of the position of the ship, relative to familiar landmarks and "leading lines," amplified with additional information, such as surface pictures, shipping movements in the area, possible navigational hazards, etc.

Most of this information is extracted directly from the (shore) radar-screens. Rivercraft, which are not under obligation to be piloted, usually monitor these channels to acquire information on the traffic on the river, needed for a safer passage.

The Netherlands Radar Research Establishment at Noordwijk (The Netherlands) was entrusted with the assignment in 1963 to plan and design a new automated radar system with sufficient capacity to serve the Port of Rotterdam and Europort, including present and future harbor complexes and the approaches.

It became obvious that the use of computers to automate a number of functions could also serve to incorporate an information giving capability. This was particularly desirable for coordination purposes in Rotterdam itself, but also at the Hook of Holland, where the entrance to the river and to Europort necessitates a strict traffic control.

A centralized "Plotting Center" in Rotterdam and a "Sub-Center" at the Hook must be an integral part of the system and enable interested parties to obtain filtered information, tailored to their specific activities.

The presence of port and shipping authorities, port services and other agencies at a centralized location would stimulate coordination, necessary for a port of this size.

The new system is therefore not exclusively a radar assistance service, but also a shipping information or data system, intended to facilitate the handling of ships. As an aid to shipping it has to operate at all times, irrespective of climatic conditions.

The new system consists of 10 radar stations:

1. One unmanned station, remotely controlled by the Sub-Center at Hook of Holland, on an artificial island, to cover the dredged channel, the approaches to the river entrance and the emergency anchorage of the Hook.

2. "Pilot Mass" radar station on the same island, remotely controlled by the Sub-Center at the Hook, with the main task of rendering assistance to ships, which have not yet embarked a pilot, or have disembarked their pilot at the pilot's vessel cruising station.

3. One radar station at the Hook of Holland with a triple function:
   a. to control the "island" radar stations;
   b. Sub-Center Hook of Holland and traffic control;
   c. to cover the entrance of the river and Europort, part of the river and parts of Europort.

4. Seven radar stations as part of the chain.

All radar stations are linked and supply position data to the "Plotting Center" at Rotterdam and the "Sub-Center" at the Hook.

Ships are entered and processed into the computer system at these two centers, incoming ships at the Hook and outgoing ships at Rotterdam. Pilots will supply the following ship's data by VHF radiotelephony:

a. the pilot's reference number;
b. the name of the ship;
c. destination;
d. draught;
e. nature of cargo.

All ships' data are passed to and stored in the local computer of the chain. They are also available for presentation to interested parties at the two Pilot Centers.

The Principle of Position Determination

Position determination is based on the use of:

1. a primary radar;
2. a secondary radar system, including a responder, a data-extractor and a local computer;
3. peripheral equipment such as consoles, communication buffers and other communication units.

The secondary radar is linked and synchronized to the primary radar in its identification and determination functions.

Pilots are equipped with a portable X-band responder and a VHF communication unit. This unit contains the data receiver and an "intercom" set. The responder is hoisted or attached in such a way that a free 360° horizon is presented.

The primary radar operates on a frequency in the 9025-9225 MHz band with a beam width of 0.5 degree, if a linear antenna system is used; or a beamwidth of 0.25 degree for a product radar system, depending on the required accuracy in a given area.

The secondary radar transmits an address selective interrogating code-signal to a corresponding responder, which will answer on 8880 MHz after the primary radar's measuring pulse has triggered the responder trasnmitter and a time-gate is opened to pass the responder pulse.

A relatively short response time of 0.1 microsecond helps to eliminate unwanted reflection responses.

Normally, the interrogation sequence of the various targets is programmed by the local computer. New targets are entered by interruption the programmed interrogation during one antenna revolution, until a response has been obtained and position coordinates are processed.

In case of computer failure responder signals can be made directly visible on the screen.

Computer derived positions are presented as a circle round the "paint" of a target. Pilot reference numbers are also indicated on the radar screens outside the fairway, to the south of it for outgoing ships, and to the north of it for incoming ships.

The responder has a capacity for 500 address selective codes.

Position data are passed every 10 seconds to ships via the VHF data receiver.

Positions are supplied every 5 minutes to the two Plotting Centers at Rotterdam and The Hook.

For "hand-over" purpose[13] they are passed to adjacent radar stations. On the tote-display a number of functional indications are available related to the "handover" procedures, in addition to the relevant ship's data.

Communications

The complex automated radar assistance and data exchange system requires an extensive telecommunications network. It can be divided in:

1. VHF communications between shore based radar stations and ships for data exchange and intercommunication purposes;
2. microwave communications between the Hook of Holland Sub-Center and the unmanned artificial island;
3. line communications between the component shore based radar stations and from each of these stations to the two Plotting Centers at Rotterdam and the Hook of Holland.

2. San Francisco Harbor Advisory Radar (HAR)[14]

Placed into operation by the Coast Guard, 21 January 1970, this system does not have stand-by equipment.

This system is really on an experimental basis in an effort to provide information regarding the need for such systems.

The system uses a Raytheon Model 1640 radar located near the center of the harbor on the peak of Yerba Buena Island (elevation 400 ft). A 12 foot slotted, waveguide antenna provides 0.6 degree horizontal beamwidth, and the 11 degree vertical pattern provides coverage to the base of the island. Located at the site, the radar provides coverage of the Oakland terminal complex, and surveys the harbor from the San Rafael-Richmond Bridge, to the Golden Gate Bridge, south to Hunter's Point.

A second radar site is located at Point Bonita and provides coverage from the Golden Gate Bridge to the San Francisco Light ship.

Radar video, trigger, and antenna rotation data is microwave relayed from Yerba Buena Island to the operations center at pier 45.

The ppi at Point Bonita is scanned by a TV camera and the TV signal ouput is microwaved to YBI. From YBI to the operations center at pier 45, the TV signal will be multiplexed to an existing microwave channel and demultiplexed at the Pier 45 receiving site.

The displays at the operation center are the conventional center sweep ppi, 16 inches in diameter, and ppi repeaters with 10 inch slopes. The latter have off-centering capability and the ability to electronically determine the range and bearing of two targets which are not related to the center of the display. This allows the operators to provide the range and bearing of a vessel to any selected shore aid, or to another vessel without manual plotting. The operations center shares its location with the Marine Exchange's Vessel Movement Reporting Center, providing a rapid exchange of information as vessels enter and depart the radar coverage area.

Communications with commercial vessels are via VHF-FM on channel 18A. This simplex channel is used for bridge-to-bridge communications for the entire mid-California area, from Sacramento to San Mateo, and out to the Lightship. Because of the wide coverage area, this single channel was reaching saturation prior to the introduction of HAR, and with additional vessels reporting to the HAR center, increased problems are envisioned. The primary objection to the existing voice communication network concern the use of the channel for non-navigational information, which should be passed on other available channels. Through the services of the Marine Exchange, placards are being developed which provide ready information to the vessels concerning the proper channel usage. Located on or adjacent to the transceiver, the vessel operator will be provided with education material essential to the eventual resolution of the discipline problem.

In addition to channel 18A, UHF-AM on 277.' MHz is used for U.S. Navy vessels which are not equipped with the commercial channel. Installation of both equipments at one operations center provides the first opportunity for indirect radio contact between the two classes of vessels. The transceiving equipment is located on the roof of the 12th Coast Guard District Office building in downtown San Francisco, with remote control facilities via telephone line from the operations center. Ship equipment consists either of fixed stations, or pilot transported hand talkies. A problem with the fixed equipment is that the power output is too high for harbor use. The level of radiation is sufficient to block communications at the far extremes of the coverage area, often unknowingly interfering with other bridge-to-bridge conversations.

Continuous tape recordings are made of all voice communications, with a simultaneous recording of local time. A ppi located at Yerba Buena Island is photographically recorded every three minutes on 35 mm film to provide a record of all harbor activities. This data, and data recorded on ADP cards by the watchstanders, will provide the information deemed necessary for system analysis.

The recent collision that occurred in San Francisco[15] should not be taken as proof that the Harbor Advisory Radar is ineffective. The Coast Guard radar recorded the Arizona Standard inbound for the refinery and the Oregon Standard outbound for Vancouver. The Coast Guard radioed and alerted the Arizona, but could not raise the Oregon. Pictures of the radar images of the vessels appear in the June 1971 issue of the National Geographic.[16] These pictures seem to verify that the problem was mainly with communications.

Traffic Control Advisory Systems are also in operation on the Thames River, Great Britain and the entrances to the Panama Canal. One of the reasons these systems are not in more widespread use is the very high cost. Such a system would be advisable for Puget Sound, but with cost estimates ranging around 6 million dollars, alternative solutions must be researched.

f. Traffic Separation Schemes

The Inter-Governmental Maritime Consultative Organization (IMCO) has made various recommendations. The aim of these recommendations is to produce an orderly flow of traffic for the purpose of reducing the risk of collisions and/or strandings, mainly in areas of converging routes or high traffic density.

Member governments of IMCO have been invited to advise ships under their flags to conform with the principles of traffic separation, to adhere to recommended routes, and to take account of the fact that areas to be avoided by certain classes of ships have been designated in some regions.

The use of traffic separation schemes is not compulsory; it is for ships' masters to decide, after assessing the situation and circumstances, whether or not to follow the recommended routes. It should, however, be understood that ships navigating against the recommended direction of traffic, within designated traffic lanes, are exposed to unreasonable risks because of the possibility of meeting a large number of ships on opposite or nearly opposite courses. Consequently, ships' masters not wishing to follow the recommended routes are advised to keep well outside the boundaries of established schemes and not hamper the organized traffic.

There were 58 such schemes as of 1 October 1970.[17] Traffic schemes are established on the Pacific and Atlantic Coasts of the United States, Arabian Sea, Red Sea, Persian Gulf, South Atlantic and Indian Oceans, Baltic Sea, Western European waters and the Mediterranean Sea.

It is recommended that the following basic schemes developed by IMCO be established wherever applicable on Puget Sound. Figure III-C9 depicts scheme traffic symbols while Figure III-C10 presents five different schemes.

1. Separation of Traffic by Separation Zones or Lines (Frame 1)

In such cases, the separation of traffic is achieved by introducing a separation zone or separation line; ships navigating in opposite or nearly opposite directions are advised to keep to the right-hand side of the zone or line. The outside limits in such a scheme are the outer boundaries of lanes intended for organized traffic. Beyond such limits ships may navigate in any direction.

FIGURE III-C9
TRAFFIC SCHEMES SYMBOLS

DETAIL	PRESENTATION	DESCRIPTION
Outside Limit of Traffic Lanes Inshore Zone Limit		Broken Line - The Symbol Used for Maritime Limits in General
Outside Limit of Traffic Circle		
Separation Zone (any type)		Zone will be indicated by solid colors.
Separation Line		Single Solid colored Line
Center of Traffic Circle With no separation zone inside.		Solid Colored Circle
Arrows indicating dirrection of traffic in Traffic separation schemes	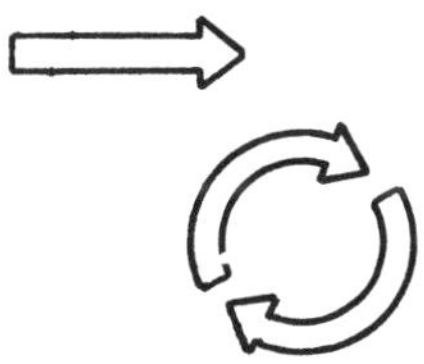	Outlined arrows to Indicate general direction of traffic flow.
Boundary of areas to be avoided by ships of certain classes Limit of Sea Exploration and/or Exploitation regions which may be dangerous to free navigators	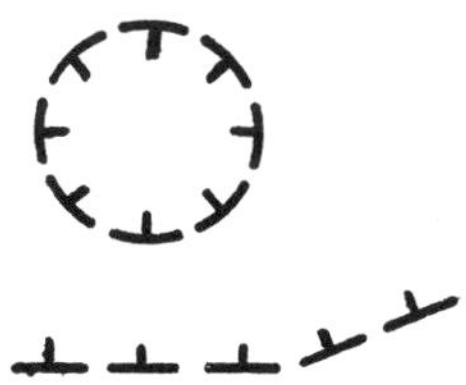	A line composed of a series of T-shaped signs down stroke points to the area in question.
Deep Draft Route	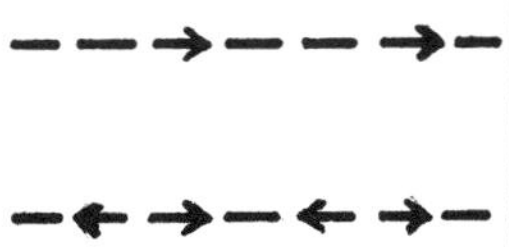	A single-dashed line in which arrow heads are inserted at regular intervals either way track, or in opposing pairs, to represent a two-way track.

FIGURE III-C10
TRAFFIC CONTROL SCHEMES

FRAME #1

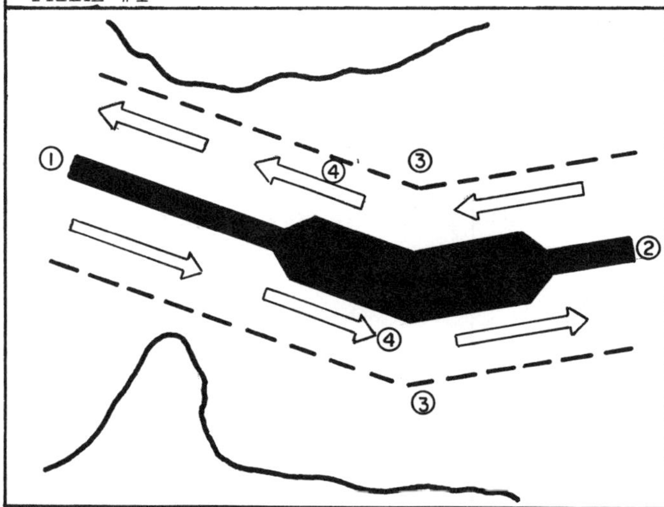

Separation by traffic Zone & Line
1) separation line
2) " " Zone
3) Outside Limits of Lane
4) Arrows indicating main traffic div.

Frame #2

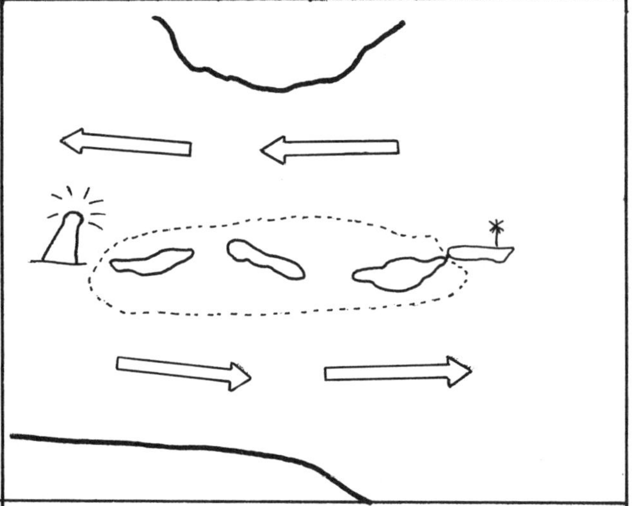

Separation by Natural Obstacles

FRAME 3

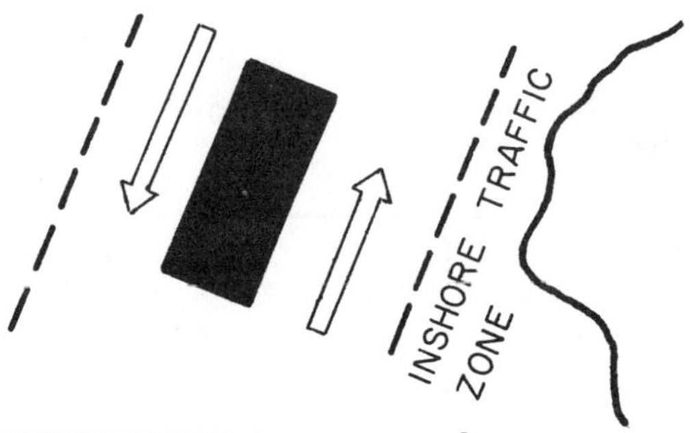

Separation of traffic by Inshore Traffic Zones

FRAME 4

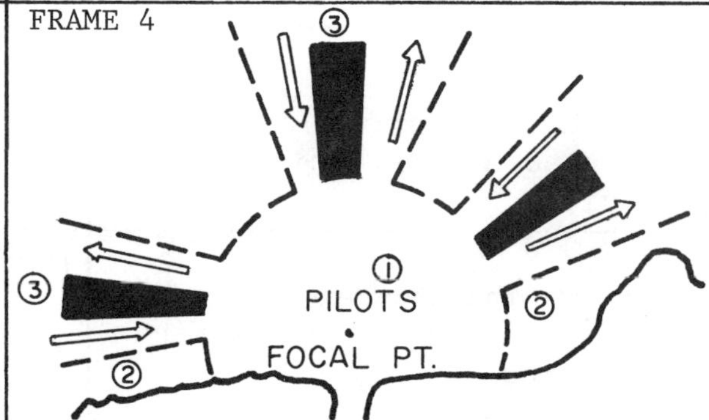

Separation by use of sectors at Approaches to Focal Point

FRAME 5

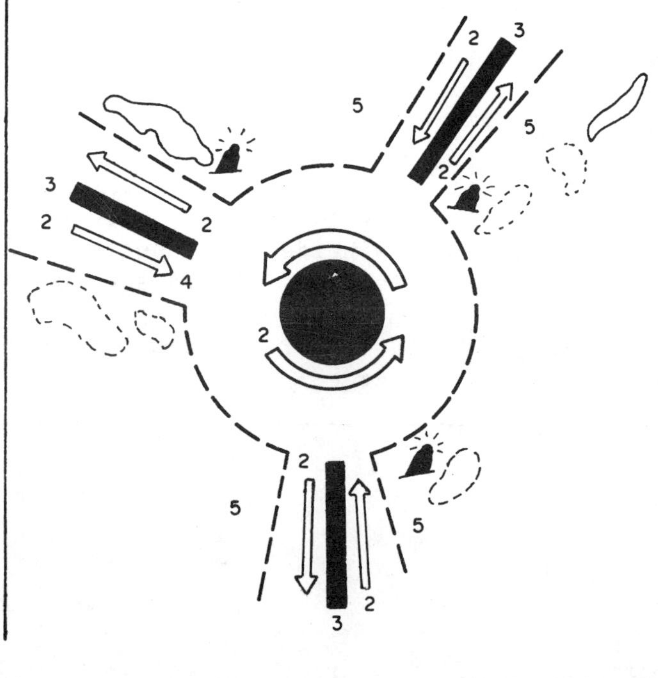

(FIGURE 5)
Separation of Traffic By introduction of Traffic Circle intended to facilitate navigation at Focal Point, where recommended routes meet
1) circular separation scheme
2) Arrows indicating traffic direction
3) Separation schemes
4) Separation LIne
5) Outside limits of Lanes

The width and length of separation zones and traffic lanes are determined after careful examination of local conditions, traffic density, prevailing hydrographic and meteorological conditions, space available for maneuvering, etc., and are kept to a minimum, allowing ships which do not use the lanes to keep clear easily. In narrow passages and restricted waters a separation line may be adopted instead of a zone, for the separation of traffic, to allow for more navigable space.

2. Separation of Traffic by Using Natural Obstacles and Geographically Defined Objects (Frame 2)

This method is used in places where there is in existence a defined area with obstacles such as islands, shoals, rocks or wrecks, restricting free movement, and providing a natural division for opposing traffic streams.

3. Separation of Traffic by Establishing Inshore Traffic Zones Intended for Keeping Coastal Traffic away from Traffic Separation Schemes (Frame 3)

The term 'inshore traffic zone' is understood to cover an area between the landward boundary of the scheme and the adjacent coast, wherein ships navigating in any direction may be encountered.

Generally speaking, all traffic separation schemes are designed for through navigation. However, in areas of considerable coastal traffic, the establishment of inshore traffic zones, intended for keeping coastal shipping away from traffic separation schemes, is adopted. The inshore traffic zone may also be used by ships which prefer, for one reason or another, to keep close to the coast and which at the same time would like to avoid meeting ships using the scheme and moving in the opposite direction.

4. Separation of Traffic by the Use of Sectors at Approaches to Focal Points (Frame 4)

Such a method is used where ships converge on a point or a small area from various directions. Port approaches, sea pilot stations, positions where landfall buoys or light vessels are fixed, entrances to channels, canals, estuaries, etc., may be considered as such focal points. The number of shipping lanes, their dimensions, and directions depend mainly on the type of the local traffic.

5. Separation of Traffic by Introduction of Traffic Circles or Roundabouts Intended to Facilitate Navigation at Focal Points, Where Recommended Routes Meet (Frame 5)

At focal points where several recommended routes meet, traffic is so organized as to correspond to the rotation schemes known to motorists as traffic circles or roundabouts. To facilitate navigation in such places and to comply with steering instructions in Rule 19 of the International Regulations for Preventing Collisions at Sea (commonly called International Rules of the Road),[18] ships are advised to move in a counterclockwise direction around a specified point or zone until they reach turning points where they join the appropriate lane. The radius of the traffic circle or roundabout depends on local conditions; it is generally limited in area to avoid unduly long rotations. If the space available permits, a circular traffic separation zone is recommended at the center of the traffic circle or roundabout to keep ships at a distance from each other. It is considered desirable, where possible, to establish traffic circles or roundabouts around buoys, light vessels, or geographically defined objects.

Although these traffic separation schemes are much less expensive (less effective too) than complete traffic control systems, they do involve significant capital outlays. The Swedish National Administration of Shipping is currently involved with the establishment of a "Big Ship Corridor." They report that the surveying and production of charts are expensive and time consuming operations. The report indicates that surveys and wreck removal for deep draft vessels can easily cost $300,000 to $600,000 per wreck and modern light buoys cost at least $14,000 each plus annual upkeep.[19]

g. Recommendations for Puget Sound

The applicability of traffic control techniques to the Puget Sound waters is best summarized in a lecture presented before the study team by Rear Admiral Kenneth Ayers USN (retired), Chairman of Northwest Marine Safety Council.[20] It is recommended that these ten points be implemented immediately.

1. Use of safety lanes as proposed by IMCO. This has begun with the establishment of a channel designation system through the San Juan Islands to separate northbound and southbound traffic. A system of buoys is being established to designate the routes shipping is to take. Charts have been

prepared and some buoys are already in place. Islands are used to physically separate the routes much of the way. Such traffic separation schemes should be employed for the entire route of the large tankers.

2. Bridge-to-bridge communication system now being used by most vessels on a voluntary basis, should become compulsory for all commercial vessels in the near future. Agreements would have to be worked out with Canadians and others who share these radio circuits with the United States.

3. Speed regulations should be enforced on Puget Sound. Many vessels do not slow down even during periods of low visibility in order to arrive at their berth "on time" and make full use of the contracted-for longshore "gangs."

4. Install the very latest type of navigational aids, e.g., the latest height of fixed lights.

5. Long and short weather forecasts with the latest information should be made available to all ships entering and leaving Puget Sound.

6. Adequate tugs should be made available for the transit and docking of large tankships.

7. Puget Sound pilots should attend training similar to that offered at the Maritime Institute of Technology and Graduate Studies.[21]

8. Tank sizes should be regulated. IMCO has just made some studies in this direction by limiting the upper limits on center and wing tanks of tankships.[22]

9. Improved seamanship would certainly do much to prevent dangerous situations from becoming disasters. One recommended procedure would be to "walk-out" the anchor prior to taking arrival. This practice would insure that the anchor was free and could be dropped in the event that the ship lost steerage or the powerplant "blacked out." Another practice would be to encourage motor ships to switch from burning bunker type fuels to burning diesel fuel prior to arrival at the pilot station.

10. A mid-channel buoyage system should be established. Where both inbound and outbound ships use the same channel, a buoyage system that uses mid-channel buoys rather than buoys down either side of the channel should be

installed. It is felt that this would improve safety by separating vessel traffic traveling in opposite directions through the same channel. Some ship masters voice the opinion that the present buoyage system, where the edges of the channel are marked by buoys, tends to "crowd" ships toward the middle of the channel and in close proximity with each other.

These ten points should be employed immediately. Some of the points would require changes in current practice. Others, like the mid-channel buoyage system, would require funding. It is here suggested that the oil companies and carriers provide this capital (approximately $250,000 for safety lanes, etc.), in turn, the state should place a reasonable limit on the liability to which these carriers may be subject in the event of an oil spill.

For the long run, Puget Sound is to become a waterway regularly traversed by mammoth tankers, then a traffic control system utilizing shore based radar should be established. The first step would be to fund a study to determine what would be required for such a system on Puget Sound.

Honeywell Marine Systems has prepared a proposal for an automated traffic advisory system for Puget Sound.[23] The questions that need to be answered now are: What type of computer would be utilized? Where would be the most cost effective places for radar stations? What portion of existing hardware could be utilized? With these answers, an analysis of the cost effectiveness of the system can be achieved, and its value to Puget Sound will be determined.

3. OIL DETECTION SYSTEM

Techniques and equipment for preventing oil spills have been discussed and evaluated in Parts 1 and 2 of Section III. None of these, however, incorporate the detection of oil on water other than visual. In order to prevent a leak from becoming a large spill, or to track the behavior of a spill, some method must be available to detect oil on water that is effective under adverse conditions.

The several objectives of an oil detection system are distinct from each other, and each utilizes different techniques. To detect oil on any given body of water, local sensing methods can be used, while the tracking of an oil spill requires remote sensors to trace its time dependent behavior. Both objectives are necessary for the protection of the environment.

Oil detection systems are conspicuous by their almost total absence in the Puget Sound region. The same statement holds true for many of the oil handling regions in the United States. In fact, the state-of-the-art of oil detection techniques is still in the experimental stages. A cursory examination of displayed hardware at the 1971 API Oil Spill Conference showed almost a complete void of oil detection equipment. Those that were available were designed for monitoring specific oil transfer operations, rather than monitoring the water.

The discussion presented here is not meant to be an evaluation of the research and technology of oil detection, but to survey the available hardware and systems. It will be obvious from the text that there are few systems on the market. Thus, an initial recommendation is to investigate the adaptability of detection systems, in general, to the oil pollution problem. Aerial reconnaissance, infrared, ultraviolet, and microwave techniques are some examples.

a. Local Sensing

The development of a system of detecting petroleum products should give first priority to areas which are subject to heavy oil traffic. These include harbors, refinery docks, oil handling facilities and marinas. Such a system can utilize devices which are physically located in the vicinity of oil traffic, and can be tied into alarm devices to provide integrated warning systems.[1] Since traffic is usually heaviest on or near waterways, location of such a detection system should be on the water, and thus is dependent on the local current, wind, and other meteorological conditions. Once the optimal location is determined, floats or buoys could be anchored. These floats would house the detection device. When in place, the sensor can provide continuous monitoring of the local water conditions.

There are several local sensing detection instruments on the market which work on the principle of reflectivity. Light on water possesses certain reflective characteristics. These characteristics are significantly increased when there is oil on the water. A photometer is used to measure the reflected light intensity. The change in reflectivity can thus be utilized to signify oil spills in the water. One such device consists of the photometer and light source mounted on floating pontoons, with a remote control section that displays the monitored results plus providing the alarm function.[2] Approximate price for such a device is $3,000, and it is currently available commercially. The unit is very effective in calm waters, but may experience difficulty in rough waters.

Shell has developed and installed a prototype "oil-on-water monitor' at its Dominquez, California refinery to keep a constant surveillance of the quality of water leaving that plant.[3] This unit also works on the principle of reflectivity. The oil firm plans to integrate the monitor with a recorder to determine water quality trends.

Other local methods of sensing include vapor, conductivity, and surface tension. These devices, when combined with an alarm, can continuously monitor the water quality of regions such as Puget Sound. Their primary disadvantage is their inability to track an oil spill after its occurrence. This is due to the fixed location of the devices.

b. Remote Sensing

Another method of detecting oil spills, and a method of continually monitoring its behavior, is remote sensing. Remote sensing can be done in a variety of ways, all involving electromagnetic waves and usually the use of an aircraft.[4,5,6] During an overflight, photographs can be taken by a cluster of different cameras with different types of film. Often used film types include: Infrared Ektachrome, Panchromatic Black and White, Infrared, and Infrared Color. To produce a sharp image of an oil slick any combination might succeed, depending upon the petroleum product spilled. False color images may produce desired effects when dealing with oil on water. By taking a series of photos, or utilizing periodic overflights, an oil spill can be tracked. A camera for such picture-taking will cost between $15,000-$35,000, but one may be improvised for around $1,500. A review of aerial reconnaissance techniques to investigate their adaptability to oil detection is recommended.

Other airborne remote sensing techniques include infrared scanners,[7,8] multispectral scanners,[9] and microwave radiometers.[10,11] All work on the principle of reflected light. The reflected light is broken down into certain wavelength ranges. The right combination of wavelengths will enable oil slick images to appear clearly against a background of water. It would also be possible to predict the thickness of an oil slick by studying its image. These instruments can produce a continuous picture on a screen and record the information on tape for time behavior analysis. This is a definite advantage of these devices over the camera in that continuous sensing is accomplished. An infrared scanner will cost around $160,000 and the other two somewhat more.

The remote sensing methods described up to now have been so-called passive methods, with certain requirements of light and cloud cover. The Side-Looking Airborne Radar (SLAR)[12] is of the active type and is the only reliable method of surveillance in heavy fog and under other extreme weather conditions. SLAR can be used effectively even at night. Here too, the image can be projected on a screen or recorded. There is a possibility of adding a selector that can activate a warning signal if oil is detected on water. The radar that is used is a Four-Frequency Radar, which is in common use by the United States Navy. The smoothing effect of the oil on waves enables the device to discriminate oil from water. Oil contaminated areas appear as dark regions with water as lighter in contrast. At present, research efforts are being expended to determine the thickness of oil slicks by SLAR, and to discriminate between the different petroleum products. The cost of a SLAR is about $250,000.

When evaluating the different remote sensing systems, the SLAR stands out as the most promising, in terms of ease of obtaining information and of all weather feasibility. But it has high costs associated with it.

The availability of integrated remote sensing systems is almost exclusively in government hands due to the prohibitive costs of airplanes. However, utilization of the Coast Guard[13] or Navy to fly daily surveillance missions for a given coastal region appears feasible. At present, no costs for such services have been obtained. A future alternative for remote sensing is to monitor from satellites. This would be the ultimate mode of surveillance, but again the costs would be staggering. In the present socio-economic system, only government action and subsidies could bring about such a system.

c. Evaluation

In the final analysis, interim measures should be geared toward the utilization of local detection devices for the Puget Sound waters. When integrated with recording and warning devices, an effective system of continuous monitoring of water quality will result. This type of system is more than adequate to detect if and when an oil spill occurs. But the problem of tracking spill oil in water still remains. Remote sensing is a step in alleviating the situation. But its cost effectiveness has not yet been fully evaluated. Until it is integrated into the total monitoring system, optimal capability to protect the Puget Sound waters will not be fully realized. Given the current state-of-the-art and the ever increasing urgency of the oil issue, adaptation of existing remote sensing hardware to oil detection appears to be the most feasible approach.[14] Thus, funding should be provided to knowledgeable personnel to initiate this effort.

TABLE III-C3: SURVEY RESULTS FOR OIL HANDLING FACILITIES

Inadequacy	Number of Occurrences (11 maximum)
Unsheltered harbor	6
No lighting on water surface adjacent to dock	6
Inadequate dock construction	9
Inadequate retaining walls	2
Inadequate security system	5
Shortage of valves along pipeline	4
Insufficient personnel for fueling	2
Inadequate monitoring of quantity of transferred fuel	2
Inadequate emergency shutoff system	6
No anti-pollution checkoff list	5
Inadequate inplant communication system	3
Insufficient oil spill cleanup system	5

4. SHORESIDE OIL HANDLING FACILITIES

a. Introduction

The oil handling facilities in Puget Sound constitute a potential source of oil pollution into Puget Sound. To analyze the probability of an oil spill from an oil handling dock facility, a study was made of eleven of the major facilities on Puget Sound. Table III-C3 summarizes the findings, which are detailed in Appendix 1. The objective was to formulate some realistic recommendations that, when implemented, would be expected to reduce the probability of an oil spill.

There are some steps already taken by these eleven facilities to prevent an oil spill. These steps are listed first. They cover both the mechanical devices used and the techniques employed on the dock. Next are listed the additional devices and techniques that would even further reduce the possibility of an oil spill. The use of an oil spill prevention checkoff sheet is a good method of preventing an oil spill; the study team's recommended checkoff sheet is included at the end of this section.

b. A Summary of the State-of-the-Art in Oil Spill Prevention on Puget Sound

The present state-of-the-art in prevention will be sub-divided into two areas: prevention devices and prevention techniques. Prevention devices are mechanisms and hardware that have been installed at the various oil handling facilities to lessen the chance of an oil spill. Prevention techniques are human operations and procedures that are being used at the various oil handling facilities to lessen the chance of an oil spill. The items presented below are those witnessed during the on-site visits. No one facility possessed all the listed items. Appendix 1 delineates the particular shortcomings of each facility inspected.

1. Summary of Existing Oil Spill Prevention Devices
    a. Prevention Devices and Systems on the Vessel
        i. Have oil containment booms on barges.
           Enough boom to encircle barge.
           In ready deployment position.
        ii. Plugs for drains and scuppers.
        iii. Always man the vessel.
        iv. Lash overboard and sea-suction valves.
        v. Have drip pans available.
        vi. Have absorbent material available.

b. Connections from Ship or Barge to Shore

i. Eliminate hoses by having solid "chiksan" connection between ship and dock. These connections are solid pipes with movable joints. The "chiksan" is motorized and contains motorized valves to control fuel flow.

ii. X-ray hoses every six months to determine any cracks or fatigue of hose material.

iii. Hydrostatic check of hoses every three months to 1-1/2 normal working pressure.

iv. Hoses capped after fueling.

v. Hoses tagged with numbered seals on vessel side during product loading to identify product.

c. Prevention Devices on Dock

i. Valve system (header, riser and other valves).
   - Check valve (one-way flow valve).
   - Valves color coded and tagged with the name of the product.
   - Clear visual markings to indicate whether valve is on or off.
   - The control of all motorized valves and pumps is possible from the dock area.
   - Valves locked with a locking device whenever not used for a specific fueling operation.

ii. Piping system
   - More valves installed on the dock piping at critical points where a ship or barge can damage the pipes.
   - Pressure relief valves along piping system.

iii. Other equipment on docks.
   - Troughs under risers.
   - Dock surface sealed and a drainage system installed that connects to a slop tank.
   - Drip pans available for valves on dock.
   - If feasible, install an oil containment boom around the whole dock permanently.
   - Have oil containment boom available close to dock area in sufficient quantity to contain any spill from barge or dock.
   - Have work boats on ready position during fueling operations to deploy booms and sorbents.
   - Have enough sorbent available to start clean up operations until more sorbent can be delivered.
   - Have direct lighting on the water surface around ship or barge. Also lighting under dock to inspect for any spilled oil.
   - Have an integrated communication system between dock and other areas of the plant.
   - Dock area fenced and locked at all times.
   - Install a retaining wall along the shoreline whereever an oil leak can occur.

d. Piping System (from waterline to fuel tanks)

i. Block valves (preferably motorized) on the land side of the dock, with sumps and retaining wall.
ii. Include valves into the piping system that can be automatically shutoff (or have personnel available).
iii. Include relief valves into the piping system.
iv. Have sumps wherever there is a possibility of an oil spill (these sumps should be connected to the oily water sewage system).
v. Install check valves at appropriate places.

e. Tank Farm

i. Have flow meters installed and checked for accuracy.
ii. 100% retaining wall around tanks.
iii. Automatic gauging system used in conjunction with flow meter during fueling.
iv. Valve at tank always closed when not fueling.

2. Summary of Existing Oil Spill Prevention Techniques and Systems

Most of the oil handling facilities visited submitted a copy of their oil handling procedures or instructions and discussed the techniques involved in transfering oil between the dock facility and the vessel. Listed below is a summary of all the various techniques that are used at oil handling facilities here in Puget Sound.

a. Before the Fueling Operation

i. Have adequate line handlers available.
ii. Inspect the receiving tanks.
iii. Be in agreement with the vessel personnel on the quantity of each product transferred and the quantity of any flush required.
iv. Write a plan of oil transfer. Be sure both the dock and vessel personnel understand the plan. Include in this the quantities of oil transferred, their flow rates, system pressures and any expected interruptions.
v. Determine the temperature and quantity of any product already in the vessel's compartments. Inspect all compartments that will be loaded and all unassigned compartments. Record all the temperatures and gauge readings of these compartments.
vi. Set up a workable vessel-to-dock communication system.
vii. Regularly inspect the dock area around the vessel for any trace of an oil spill.
viii. When transfering fuel at night, rig lights around the pier to shine on the water surface; and make lights available on the far side of the vessel.
ix. Plug all deck scuppers on the vessel and have absorbent material available to clean up any oil spilled on the dock or the vessel.

x. Place drip pans or buckets under all hose connections and valves unless there is installed a permanent drip trough.

xi. Make sure all hoses are in good condition and have available their record of inspection.

xii. Close all sea suction and overboard discharge valves on the vessel. Have the officer in charge supervise any opening of these valves. Make sure all seals are leakproof.

xiii. Be in agreement with the vessel personnel on the lineup of valves, product lines and headers or riser valves.

xiv. Have a dock supervisor on the pier at all times. He is responsible to see that his men understand their duties and that there are sufficient personnel available to safely transfer the cargo. He is expected to take the initial action in an emergency.

xv. The dock supervisor oversees the transfer of the dock hoses to the ship and their safe connection to the vessel's valves.

xvi. These dock hoses must be checked to see that they are properly connected to the pier header or riser valves and that they are adequately supported.

xvii. Unused oil connections are to be blanked.

xviii. Make allowance for any relative motion between the dock and the vessel.

xix. Check all system pressure indicators. These include pressure recording charts and pressure gauges.

xx. Give the vessel a 5 to 10 minute warning before starting the oil transfer, changing the flow rate or stopping the oil transfer.

b. During the Fueling Operations

i. Maintain a hose watch at the header or riser valve at all times during the transfer. He shall immediately close the header or riser valve when requested or when a problem becomes apparent.

ii. The transfer operations are to begin at a slow rate. Before increasing to a normal flow rate, check all hose connections, valves and the water around the dock area for possible oil spills. The increase in flow rate is done in coordination with the vessel personnel.

iii. Stop the transfer operation if any instructions are not understood or if there is a sudden change in system pressure.

iv. Notify the vessel personnel of any sudden change in oil flow rate and be prepared to stop the transfer operation if requested.

v. Determine that there are no leaks in oil pipelines by following them back to the tank, if possible.

vi. When receiving oil from a vessel, have a man at the tank to make sure that the oil is going into the proper tanks. Gauge these tanks frequently to determine flow rates and compare these rates with the results from gauging the vessel's tank. If any discrepancy exists, stop the oil transfer operation and resolve the problem before resuming transfer operations.

vii. When topping off a tank, station a man at its top to indicate when the tank is full. Top off one tank at a time.

c. After the Fueling Operation

i. Secure the header or riser valves.
ii. Drain and clean the hoses.
iii. Remove hoses from vessel and cap them.
iv. Make a final check of all tanks that received oil.

d. General Information

i. Both the dock and vessel personnel must be made aware of the penalties imposed by Federal and State regulations on those responsible for willful or negligent oil pollution of Puget Sound's waters.
ii. Both the dock and vessel personnel must have a clear understanding of the procedure for containing and cleaning up an oil spill.

The biggest drawback in the oil handling facility evaluation is that not once was an actual oil transfer operation witnessed. The list of items presented are commendable; but, until these devices and procedures are implemented by authorized and competent personnel, the effectiveness of the overall oil spill prevention system is very low. As an initial recommendation, it is strongly suggested that each and every facility in Puget Sound critically examine their current oil handling procedures and equipment to detect any shortcomings in their systems.

c. Oil Spill Prevention Devices and Techniques that Should be Incorporated by the Facilities

The following is a summary of devices and techniques that should be evident at every Puget Sound faiclity which transfers oil.

1. Prevention Devices

a. On Barges

i. During oil transfer operations, barges should have a lighting system on the far side of the barge powered by the shore facility. These lights are to be directed into the water so that any oil surface film can be visually detected.

ii. Oil containment boom on barge.

- Enough boom to encircle barge.
- In ready deployment position during fueling operations.

iii. Seal or chain-lock sea suction and overboard discharge valves in closed position during fueling operations.

b. Connections

Between the hose and the barge valve install a combination flow and pressure meter. This new device will indicate the total volume through it at any instant and the maximum pressure experienced at this connection since the beginning of the operation.

c. On Dock

i. Install a permanent containment boom around the dock. This is to contain oil spilled from dock.

ii. Have deflection booms readily available with their moorings permanently installed to deflect oil that is spilled outside the dock area (i.e., from a barge or tanker).

iii. Know the tide and current conditions near the dock so that deflection booms can be logically deployed.

iv. Install a lighting system that illuminates the surrounding water area.

v. Chain lock all valves on dock.

vi. Do not leave valves partially open. Instead, put relief valves around them.

vii. Install permanent drip troughs under header and riser valves. Drip troughs should lead into a sump which empties into the waste oil separator system.

viii. Portable vacuum pumps (i.e., truck powered) and hoses with a suction head should be available near fueling area to skim up spilled oil. There should be a pipeline to take this oil to a separator facility.

ix. Have an oil detection device permanently installed under the dock area that can appropriately monitor any oil spilled from the dock.

x. Pave or cement all oil handling areas. These areas should be graded to drain oil into sumps, which can then be pumped to the separator system.

xi. Have an integrated communication system between dock and other areas of the plant.

xii. There should be a retaining wall along the shoreline wherever oil spillage can occur.

xiii. Block valves for the piping system on dock should be installed on the land side of the dock with an appropriate retaining wall and sump system which will take spilled oil to the separator as shown in Figure III-C11.

xiv. Notification to personnel of inoperative equipment on dock.

- Tag equipment
  List reasons why equipment is inoperative; and state when the equipment will be operative again.
- List all defective equipment on spill prevention checkoff list.

2. General Oil Spill Prevention Systems

a. There should be a minimum of two men handling oil transfer operations. One man should be located on the dock area at all times. The other man should inspect the piping connections, tank farm. A specific and comprehensive work schedule should be made for these men by the company that will most appropriately cover the specific problem area of the particular oil handling facility.

b. Have the Coast Guard set up a series of oil spill prevention lectures and oil cleanup methods that are required for all oil handling personnel.

c. An effective inplant communications system should be set up that connects the vital areas of the oil handling facility.

d. The oil handling facilities should have a good fence system and adequate surveillance at all times.

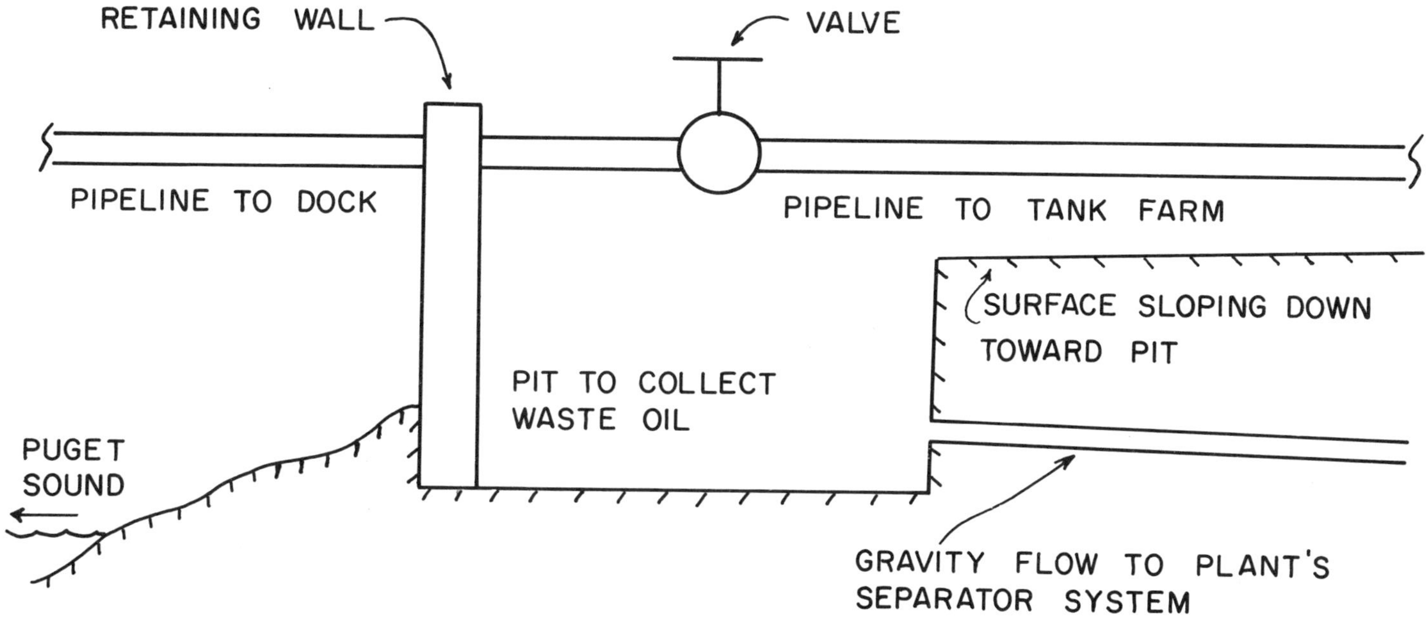

FIGURE III-C11
PROPOSED DOCKSIDE OIL SPILL PREVENTION SYSTEM

e. The oil handling facilities should develop a comprehensive oil pollution prevention and cleanup system plan that could be incorporated into the oil handling procedures plan. Such a plan has been implemented at Shell Oil Company's Harbor Island facility.

f. The Phillips 66 facility in Seattle should be used as a central receiving depot for all oil skimmed off the water by either vacuum trucks or barges during the course of cleaning up a major oil spill in Puget Sound. This additional task could be performed without interrupting their normal operations. The facility has the largest and most extensive gravity separator system of any oil handling facility visited during the course of this study. They have a rather long dock, 650 feet, with a number of unused pipelines. They also have an oil barrel filling building that they are no longer using which could be turned into a facility to incinerate the unrecoverable waste products. Also, the facility is rather large considering the amount of oil they handle.

d. Proposed Oil Spill Prevention Checkoff Sheet

Many shore installations have their own set of standard operating procedures or general instructions, but few have step-by-step checkoff sheets designed to minimize the possibility of an oil spill. In addition to any such written or verbal operating instructions, an oil spill prevention checkoff sheet should be used when transfering oil to or from a vessel.

The following checkoff list is not meant to be blindly followed or thought of as the minimum steps needed to perform the operation. It is not to supersede one's judgement or technical knowledge; rather, it is to work in harmony with one's ability and experience. The two are to be used together making a unified approach in the oil transfer process. The emphasis should be on the man undertaking the fueling operation; he should always think prevention. This is stressed since most oil spills are due to human error and not equipment breakdown.

The first page of this checkoff sheet is an emergency guide data sheet and should be kept with the dock personnel at all times. In the event of an emergency, this sheet is to be given to the supervisor of the crew fighting the emergency. Any special handling precautions are to be checked and discussed in the "REMARKS" section.

OIL SPILL PREVENTION CHECKOFF SHEET

EMERGENCY GUIDE

Name of vessel: ______________________

Berth: ______________________

Date: ______________________

Terminal receiving cargo ______

or (check one)

Vessel receiving cargo ______

CARGO	FLASH PT.	COLOR	ODOR	QUANTITY
1.				
2.				
3.				
4.				
5.				

SPECIAL PRECAUTIONS

FIRE_____ VAPOR_____ ACIDIC_____ CAUSTIC_____ EXPOSURE_____ WEATHER_____

Remarks: ______________________

____

____

____

IN CASE OF EMERGENCY

FIRE: Call 911 (Say FIRE and give location).

OIL SPILL: Activate your oil spill response plan. Call M.O.P.S.

OIL TRANSFER IS NOT TO BEGIN UNTIL ALL QUESTIONS HAVE BEEN ANSWERED AND ALL STATEMENTS UNDERSTOOD WITH THE APPROPRIATE SIGNATURES.

1. When receiving cargo, have the receiving tanks been inspected to ensure adequate storage of received cargo?

____
Tank Farm Personnel Signature

2. Is there an agreement as to the quantity of cargo to be transferred?

Dock Personnel Signature

3. Has a plan of oil transfer been written and clearly understood? Include in this plan the flow rates, system pressures, and any expected interruptions.

Dock Personnel Signature

4. When fueling a vessel, is there agreement as to where on the vessel the cargo is to be loaded and have these spaces been inspected?

Dock Personnel Signature

5. Set up a ship-to-shore communication system.

6. Regularly inspect the dock area around the ship for possible oil spills.

7. During night transfer operations, have the lights been rigged around the pier to shine on the water and are lights rigged on the far side of the vessel?

Dock Personnel Signature

Vessel Personnel Signature

8. Are the deck scuppers plugged (leakproof) and are drip pans or buckets placed under valves?

Dock Personnel Signature

Vessel Personnel Signature

9. Use cargo hoses that are in good condition and make available their record of inspection.

10. Have all overboard discharge valves and sea suction valves been closed and sealed?

Vessel Personnel Signature

11. Have the dock hoses been checked for proper connection, are they properly supported, and do they have adequate drip pans beneath them?

Dock Personnel Signature

12. It is the dock supervisor's responsibility to see that there are adequate personnel available during the entire cargo transfer operation, that they thoroughly understand their duties, and that they remain on duty.

DURING THE FUELING OPERATION, INSURE THAT THE FOLLOWING PROCEDURES ARE UNDERSTOOD:

1. Give the vessel 5 to 10 minutes of warning before changing the flow rate. Warn the vessel immediately of any sudden change in flow rate.

2. Stand by the pier valve at all times in order to immediately close it in an emergency.

3. Oil transfer operations are to be stopped if any instructions are not understood or if any pressure valves show a dangerous system pressure increase.

Dock Personnel Signature

AT THE COMPLETION OF THE TRANSFER OPERATION:

1. Secure valves; then drain, clear, and cap the hoses.

2. Properly dispose of waste oil collected in the drip pans.

Dock Personnel Signature

OIL SPILL PREVENTION IS YOUR RESPONSIBILITY

1. Do both the dock and vessel personnel know the penalties imposed by Federal and State regulations on those responsible for oil pollution of the waters of Puget Sound?

2. Are both the dock and vessel personnel aware of any oil response plan and the steps taken to contain and cleanup an oil spill?

Dock Personnel Signature

Vessel Personnel Signature

e. Conclusion

As with all projects of this type, time works against you; and this was found to be true here. So, it is appropriate to state here those areas not covered and the work that needs further attention. The obvious first remark is that not enough oil handling facilities were visited. Attempts were made to tour the major shore installations (either refineries or distribution centers) but there was not enough time to visit the small boat fueling docks. While at the facilities, there was no oil transfer in progress. It would have been desirable to visit the facilities while a vessel was actually receiving or discharging oil. Understanding the terminology used in the oil handling business would have made some interviews more productive.

A survey of both the biology and the current and tide conditions of the dock area should be carried out. The biological survey would be used to determine the effect an oil spill would have on the environment near the facility. The current and tide information is necessary to predict the movement of any oil spill. This would involve the installation of current and tide meters at the various dock facilities.

One area not covered was the transfer of oil from a barge to a ship anchored in the Sound or moored at a pier. There is a significant amount of oil bunkered by this technique, especially in the central part of Puget Sound. Section II-C of this report showed that thirteen facilities, having a total capacity of 121,050 barrels bunkered vessels from barges.

ENDNOTES SECTION III-C, Part 1

[1]Defined as having more than 100 parts of oil per million parts of water.

[2]1962 Conference for the Prevention of Oil Pollution at Sen, London April 4-11, 1962.

[3]Resolution of Extraordinary Session, 1968, & 1969 IMCO Assembly, 6th Regular Session.

[4]NATO Committee Challenges of Modern Society -- Colloquim on Pollution of the Seas by Oil Spills, Brussels, November 2, 1970, opening session.

[5]Data for IMCO Conventions, originally drawn up in 1948, but not actuated until 1959.

[6]1968 & 1969 sessions, Annex III.

[7]Oil spill in Buzzard Bay, Off West Falmouth, Massachusetts on September 16, 1969.

[8]Blumer M., Sass, J. & Souza, G. "Hydrocarbon Pollution of Edible Shellfish by an Oil Spill", *Marine Biology*, Vol. 5, 1970, pp. 195-202.

[9]*Systems Study of Oil Spill Cleanup Procedure. Vol I.*, *Analysis of Oil Spills and Control Material*, Dilligham Corporation, February 1970, pp. 7-28.

[10]The term "low energy impact" refers to an impact of such energy that only a local puncture or rupture results, whereas a "high energy impact" is one that causes considerably more damage than a break in the container.

[11]For further discussion, see Marine Technology. July 1970.

[12]MARAD Pollution Abatement Plan For Oil, p. 5-14a.

[13]Recent explosions due to tank cleaning operations have occurred to a number of vessels. In December 1969 the supertankers Marpessa, Mactra, and Kong Haakin VII suffered explosions. As recently as the tankers suffered explosions while cleaning tanks. Presently the problem is under intensive investigation.

[14]Presently the Coast Guard has developed an air deliverable anti-pollution transfer system employing rubber bags into which oil can be pumped from damaged tanks. It is anticipated this system will be implemented by December, 1972.

[15]This may not be true in Puget Sound waters because their depth does not constrain the passage of these large vessels.

[16]*MARAD Pollution Abatement Plan For Oil*.

[17]Testimony by Dr. Max Blumer before the Antitrust and Monopoly Subcommittee, August 4-6, 1970; the Conservation and Natural Resources Subcommittee on Air & Water Pollution, June 30, 1970, in Machias, Maine.

[18]See Table III-A5 of this report.

[19]The Liverpool Underwriters Association statistics show that for the period June 1964 to April 1967 there were 196 collisions involving 238 tankers with 22 cases of oil spillage and 91 groundings with 17 cases of spillage or leakage.

[20]The 1954 International Convention for the Prevention of Pollution of the Sea by Oil prohibits the dumping of oil within 50 miles of any coastline of the world excepting Eastern Canada, Australia, and Europe where the prohibited zone extends further offshore. In 1969, IMCO established more rigid amendments with the prohibition of all oil discharge subject only to certain specified exceptions. These exceptions restrict the rate of discharge of oil not to exceed 60 litres per mile.

[21]MARAD, report, p. 5-5a.

[22]Dillon, E. S., Ship Design Aspects of Oil Pollution Abatement, U.S. Department of Commerce, Maritime Administration.

[23]April 26, 1971 oil spill off Texaco Refinery, Anacortes.

[24]There existed nothing more at the dock than a pressure gauge to measure the flow of oil into the barge. The nearest accurate knowledge of the flow was at the blending plant 3 miles away. The only connection to the plant from the dock was via telephone. Since the loading was taking place at night some time elapsed before oil was discovered on the water and noticing a film on the water.

[25]Dillon, p. 1.

[26]Ibid., p. 19.

[27]Ibid, p. 10.

[28]Ibid., p. 21.

[29]Ibid., p. 24.

[30]Ground Casualty Statistics, IMCO Subcommittee of Ship Design & Equipment, 5th Session.

[31]Dillon, p. 42.

[32]R. I. Price, USCG, "International Activity Regarding Ship Control", March 22, 1971, Annex II.

ENDNOTES SECTION III-C, Part 2

[1]Conversation with Captain F. Smith of the Puget Sound Pilot's Association, May 1, 1971 at the University of Washington. This statement is also documented in _Ship Movements and Petroleum Transportation on Puget Sound Area, State of Washington_, a white paper by the Seattle Chamber of Commerce, p. 2.

[2]See Section II-B5 for a discussion of collision probabilities for Puget Sound.

[3]This is about the size of the T-2 tankers, such as the Arizona Standard and Oregon Standard, which collided in San Francisco Bay on January 18, 1971.

[4]H. S. Bell, _Petroleum Transportation Handbook_, McGraw Hill New York.

[5]Dr. J. P. Hooft, "The Steering of a Ship During the Stopping Maneuver," _International Shipbuilding Progress_, Vol. 17, June 1970, #190, p. 191.

[6]Dr. Masataka Fujino, "Experimental Studies on Ship Maneuverability in Restricted Waters, Part II," _International Shipbuilding Progress_, Vol. 17, February 1970, #186, p. 45.

[7]Conversation with Rear Admiral Ken Ayers, Chairman of Northwest Marine Safety Council, at the University of Washington on February 11, 1971.

[8]_Shipbuilding and Shipping Record_, November 1970, Special Supplement, p. 15.

[9]Master, Mate and Pilot, October 1970.

[10]Ibid.

[11]Ibid.

[12]Untitled paper discussing Rotterdam traffic control system.

[13]Hand over - as a ship passes from one radar control area to the next, details of his position and course are transferred to the next areas control station.

[14]T. H. Baetsen, Cdr., USCG, "San Francisco Advisory Radar."

[15]Collision of Arizona Standard and Oregon Standard tankers in San Francisco Bay, January 18, 1971.

[16]_National Geographic Magazine_, June 1971, Vol. 139, no. 6, p. 868.

[17]U.S. Hydrographic Office Chart #50.

[18]_International Rules of the Road._

[19]*Shipbuilding and Shipping Record*, January 1, 1971, p. 35.

[20]Conversation with Rear Admiral Ken Ayers, Chairman of Marine Safety Council, at University of Washington, on February 11, 1971.

[21]*Master, Mate, and Pilot*, October 1970.

[22]*IMCO Proceedings Maritime Safety Committee*, 23rd Session, March 18, 1971.

[23]*A Proposed Automated Marine Traffic Advisory System for Puget Sound*, Honeywell Marine Systems Center, Doc K03/70, November 6, 1970.

ENDNOTES SECTION III-C, Part 3

[1.] D. K. Phelps and G. Fain, "Instrumenting a Buoy for Estuarine Monitoring," Oceanology International, December, 1970.

[2] Model 1479 On-on-Water Detector, manufactured by Hallikainen, 750 National Court, Richmond, California.

[3] "Oil-on-Water Monitor," an unpublished paper provided by Mr. P. H. Walton, of the Shell Oil Company, March 1, 1971.

[4] L. G. Swaby and A. F. Forgiati, "Remote Sensing in Oil Slicks," API Publication No. 4040, pp. 297-307.

[5] G. A. Helgeson and D. S. Ross, "Remote Sensor Imaging for Oceanography," Oceanology International, September 1970.

[6] "Remote Sensors Map Barrier to Oil Slicks," Oceanology International, April 1970.

[7] Kaye, D.N., "Designing for the Pollution Free Industrial Era," Electronic Designs, May 13, 1971.

[8] N.W. Guinard and C. G. Purres, "The Remote Sensing of Oil Slicks by Radar," U.S. Coast Guard, April 1970, project no. 714104/A/004.

[9] Carns, M., "Multispectral Sensing-New Environmental Tool," The MBA, January 1971.

[10] A. L. Edgerton and D. T. Trexler, "Radiometric Detection of Oil Slicks," January 1970, U.S. Coast Guard, Report No. AGC-SD-1335-1.

[11] J. C. Aukland and D. T. Trexler, "Oil Pollution Detection and Discrimination by Remote Sensing Techniques," U.S. Coast Guard Office of R & D, October 1970, report no. 714104/A/006-1.

[12] A. S. Lewis and H. C. MacDonald, "Interpretive and Mosaicking Problems of SLAR Imagery," Remote Sensing of Environment, December 1970.

[13] The Seattle Coastal Region contingency plan of USCG lists various techniques of remotely sensing available, though no specific details are given. See Annex XII, 2202-2203 of that report.

[14] Two papers presented at the Marine Technology Society's 7th Annual Conference and Exposition, Washington, D.C., August 16-18, 1971, should be of value, though they were not available at publication time. These include:

a) "Oil Spill Reconnaissance Using Remote Sensing Techniques," by Watson, Terry, and Buckmeier of Texas Instruments, Inc.

b) "Passive Remote Sensing System for Estuarine Pollution," by D. T. Hodder of North American Rockwell.

SECTION IV: CONCLUSIONS AND RECOMMENDATIONS

IV-A. CONCLUSIONS

The results of the six-month interdisciplinary study on oil spill prevention and control on Puget Sound have been documented in the preceding sections and the appendices. The list of conclusions below present a capsule summary of the subject matter in this report; where applicable, reference to specific portions of the text is noted.

1. The oil industry is but one of several economically vital industries which rely on the Puget Sound waters for their survival in the Northwest. Therefore, all sectors must strive to achieve a unified effort in preserving this asset.

2. The presently accepted method of determining the value of resources is based on economics. But the value assigned to a given resource is that which is perceived by interested groups and not society in general. (Section II-D, III-B, part 2.)

3. The potential sources of oil spills in the Puget Sound area are well known. These include both large and small volume discharges. (Section III-A.)

4. The probabilities of a spill from these sources and/or the expected quantity spilled are not accurately known. (Section III-A.)

5. Estimates of the quantity of oil discharged in an oil industry related accident (i.e., during transport, transfer, or production) is fairly accurate under ideal visibility conditions. Estimates under adverse weather conditions can attain high accuracy by deductive techniques, but these results are achieved too late to initiate any effective cleanup response capability. (Section III-A, Appendix 12.)

6. The quantity of oil discharged by pleasure boats, municipal sewer systems, industrial effluence, or consumer disposal would be difficult to assess by any means, because of the high frequency of occurrence and small amounts involved per incident. (Section III-A.)

7. There presently is no detailed, comprehensive inventory on the biological and other natural resources that inhabit the waters of Puget Sound. This report has attempted to initiate action to fill this void. (Section II-2.)

8. The effects of petroleum products, either refined or crude oil, on the biological resources are not fully understood. There is evidence of both long-term and short-term damage. (Appendix 14.)

9. There presently exists no scientific technique for evaluation of resources inherent to assigning protective priorities during an oil spill. This report proposes a mechanism for achieving this end. (Section III-B, Part 2.)

10. The effects of an oil spill on the profitability of the commercial fishing industry are known. The economic loss of a ban on sale of a given seafood product line can be accurately estimated. (Section II-D.)

11. The effects of an oil spill on tourism, pleasure boating and sport fishing are predicted to be economically detrimental but the extent is not known. (Section II-D.)

12. The response capability required to clean up an oil spill on Puget Sound is at best minimal, and most likely unknown. This applied to contingency plans as well as manpower and equipment. (Section III-B, Part 1, Appendix 12.)

13. Commercially available containment and cleanup devices have demonstrated high levels of effectiveness in calm water, ideal visibility conditions. However, their performances under adverse conditions is questionable. Integrated containment cleanup systems are needed to cope with the variety of environmental conditions on Puget Sound. (Section III-B, Part 3, Appendix 5, 6, and 12.)

14. Data collection techniques on oil spill statistics is at best, primitive and fragmentary. (Sections III-A; III-C, Part 1; and Appendix 13.)

15. Prevention techniques for waterborne transportation of oil are mostly in the planning stages. Their effect on Puget Sound waters will not be felt for several years. (Section III-C, Part 1.)

16. Prevention techniques utilizing low cost, well planned traffic controls are effective short-term solutions to the ship collision problem. (Section III-C, Part 2.)

17. Prevention techniques for vessel/shore oil transfer operations satisfy current Coast Guard regulations. But their effectiveness in preventing an actual oil spill is questionable. Table III-C3 summarizes the findings of the study team surveying eleven facilities. (Section III-C, Part 4, Appendix 1.)

18. Historical data on oil industry efforts in water conservation have indicated that, at best, a minimum level has been expended. (Section II-E.)

19. The value of visually tracking an oil spill or visually assessing water quality subsequent to a cleanup operation is questionable. "Out of sight, out of mind," is not an adequate criteria for environmental protection. (Appendix 12.)

20. A computer-based model for tracking the behavior of an oil spill would be feasible for Puget Sound, given an accurate assessment of the environment. (Appendix 8.)

21. The Washington State government has not backed up its laws on oil spills with the necessary resources to effectively prevent and control these spills. The most apparent shortcomings are manpower and funds. (Appendix 2.)

22. With regard to oil spill prevention and control, the regional offices of federal agencies are merely extensions of the central headquarters. While it is important to have uniform management control functions to cope with oil spills on the national level, operational controls of the regional office should be adapted to the local conditions, and not stated in general terms. (Section III-B, Appendix 2 and 7.)

23. Because of the proximity of Puget Sound to international waters, the technique for oil spill cleanup and prevention should be compatible with international regulations. This applied especially to Canadian regulations governing oil spills. (Appendix 4.)

24. Detailed data needs to be collected and refined before any probability results on ship collision and oil spills in Puget Sound can achieve acceptable levels of accuracy. (Appendix 13.)

IV-B. RECOMMENDATIONS

Recent events are providing evidence that increasing concern on the oil-on-water problem is surfacing in the Puget Sound region. Historically, the time delay between awareness and action has been the most significant roadblock to effectiveness. Hopefully, this barrier is slowly being lifted.

A bill was introduced June 16 of this year, by Senator Magnuson and Hart[1] regarding the environmental quality and protection of ports, harbors, and navigable waters in the U.S. It will increase penalties imposed on transport of hazardous cargo, including oil, and give the USCG powers to establish regulations to prevent damage to water resources. Other recent developments in governmental regulations with respect to oil spills include a set of proposed USCG rules. Though these will not be implemented for some time,[2] they include provisions for double bottomed tank barges, minimum lighting requirements for oil handling terminals, and approved operational instructions for oil transfer.[3] In addition, the 13th Coast Guard District is gearing up its manpower level capable of monitoring 90% of oil transfer operations in the Puget Sound region for the next six months.

EPA has established guidelines to:

1. establish minimum criteria to plan and implement state, local and regional response capability;
2. identify critical water use areas;
3. incorporate interconnecting communication systems between the various agencies;
4. inventory containment and cleanup equipment on the local level;
5. develop agreements with third party contractors;and
6. specify priorities for resource protection.[4]

It must be emphasized to EPA that these goals are indeed honorable, but that until the regional officers are given the authority and necessary resources to initiate action, the guidelines will remain only dreams.

The Washington State Oil Cooperative has set forth plans to examine their response capability, with the goal of integrating manpower and equipment of individual facilities into an interfirm response plan. Implementation date of this plan is not available at the present time.

The Boeing Company has surfaced on two fronts of the oil-on-water problem. There are reports that the firm has developed a surveillance and cleanup system utilizing a 727 airplane and sorbents.[5] No details are presently available. In addition, it is engaged in the design of an oil waste process and plant in Boardman, Oregon, a city located on the Columbia River in the northeast part of that state. Current plans call for transporting oil wastes from as far as Portland to that site.

Based on the findings summarized in Section IV-A, recommendations have been made in this report on improving the total effort of oil spill control and prevention on Puget Sound, along with the predicted benefits. The following outline summarizes these suggestions, assigning recommendations to the party <u>responsible</u> and <u>capable</u> of their implementation.

1. <u>Oil Industry</u>

    a. Improve inter- and intra-communication and eliminate ambiguity when disseminating information.

    b. Critically examine oil spill response capability of individual facilities. This includes the assessment of manpower, equipment, and environmental conditions. Update capability where necessary.

    c. Provide funding and guidelines to evaluate state-of-the-art oil spill containment, cleanup, and prevention equipment. Assistance and advice to such equipment manufacturers will increase the overall effectiveness of abatement techniques.

    d. Examine the value of, and compliance with the present prevention regulations set forth by state and federal agencies. Suggest improvement where deemed necessary.

    e. Point up the necessity for improved surveillance techniques, both for oil-on-water sensing, and the tracking of an oil spill. This entails taking a wider viewpoint, than which exists presently, on the facility location studies.

2. Other Industries, Special Interest Groups, and Academia

   a. Critically examine all possible sources of potential oil spills, and initiate corrective action where necessary. This includes prompt and accurate reporting of spill incidents.

   b. Assume a wider perspective than that of self-interest in viewing the oil-on-water problem in Puget Sound.

   c. Establish criteria for environmental quality that are both realistic and relevant. Include the valuation of resources in the criteria, where available.

   d. Provide manpower and equipment to manage and operate a continuous monitoring of the inventory of Puget Sound resources.

   e. Provide manpower and equipment to conduct studies on the consequences of oil on Puget Sound resources. This includes prediction of short-term and long-term effects.

   f. Develop techniques to monitor waterways to enhance the capability for detection of oil-on-water.

   g. Develop positive incentives for the proper disposal of waste oils.

   h. Critique the resource protection techniques presented in this report. Suggest improvements or modifications where necessary.

   i. Evaluate the cost-effectiveness of a traffic control system for Puget Sound, and the possibility of expanding the utilization of such a system to other fields.

   j. Carefully research and accurately document technical data when providing support evidence in reports and testimonies. Nothing discredits an argument quicker than to misquote quantitative information.

3. Washington State Government Agencies

   a. Provide the necessary funding and guidelines for a detailed study to inventory Puget Sound resources.

   b. Provide funding and manpower to adequately enforce State statutes related to oil pollution.

c. Make public the state oil spill response plan.

d. Integrate its role in oil spill response capability with that of the federal government.

e. Critically examine oil transfer and discharge operations of state controlled vessels (i.e., ferries, patrol boats). Insure that proper techniques and devices are onboard to prevent the possibility of a spill.

f. Examine the State of Maine's Coastal Conveyance of Petroleum Act[7] for applicability to the waters of Puget Sound. In particular, consider their plans for funding cleanup capabilities.

g. Resolve conflicts with federal agencies on the definition of water quality and environmental standards for Puget Sound.

h. Update waterborne commerce log procedures. Require vessels to provide timetable of travel on Sound, including all destinations, cargo transfers, and fueling operations.

4. <u>Federal Agencies</u>

a. Adapt oil spill related operational controls to the local environment of Puget Sound.

b. Develop and train personnel to become locally oriented in their responsibility and decision making logic.

c. Insure compatability between state and federal statutes.

d. Insure prompt implementation of international oil spill prevention resolutions.

e. Provide liaison to international organizations on the efforts and progress of oil spill abatement in the United States. Obtain committments from foreign governments to reciprocate accordingly.

f. Provide funds and guidelines to investigate short- and long-term effects of oil on biological and other natural resources. This includes the selection of technically capable parties to participate in such studies.

g. EPA should evaluate the cost effectiveness of their Northwest regional post-spill assessment capabilities. This includes both the management and operation involved.

h. Provide stiffer regulations on the shipping industry to prevent purposeful discharge of oil onto the waters of Puget Sound. This may require the implementation of an integrated surveillance network for the area.

i. Evaluate the cost-effectiveness of computer based oil spill surveillance as compared to the state-of-the-art. Job security should not be a criterion in the retention of outdated techniques.

j. Initiate a mechanism to provide prompt retreival of information for interested parties. This includes, historical oil spill data, incidences for violation of regulations and ship collisions.

The following general recommendations are made for regional studies and data required for evaluating the oil spill contingency planning in any given region. The information necessary is identified only according to its impact in a potential spill situation and hence only certain aspects of a particular body of data are indicated. No ordering of importance or priority is implied by the order of items presented in this note.

1. Oceanographic Data. The extent and accuracy to which current patterns in a given region are known must be defined. Available information must be presented in a form easily used in field work by personnel not necessarily trained in the reading of tide and current tables; thus graphical or chart representation is desirable. Gross current patterns in mid-channel of large passages usually suffice. Considerable detail in terms of surface current, turbulence, sub-surface reverse currents must be available in the vicinity of ecologically sensitive areas which must be protected from oil influx. The location, extent and strength of tide rips should be known throughout the potentially threatened area. Fresh water inflow and amount of silt carried (silt acts as a sinking agent, causing possible sub-surface suspended oil concentrations; c.f., San Francisco spill) must be known by time of year. Detailed current data should be available for all terminal locations. Potential wave conditions

under varying extremes of wind known to occur in the region must be identified by area so that effectiveness of available mechanical devices may be assessed prior to deployment.

2. Meteorological Data. Wind and visibility conditions in the threatened area must be monitored so that this information can be integrated with current information to assess potential effectiveness of available equipment. Data should be sufficiently dense to formulate local forecasting. The accuracy of existing weather forecasting for the given area should be evaluated on a continuing basis. Peculiarities in wind patterns due to local topography should be readily available.

3. Topography, Bathymetry. The bathymetry of a given area should be available in a form such that identification of shallow areas of less than one meter or 1/2 fathom depth by tidal conditions is readily available. (No effective clean-up equipment exists which operates in shallow water areas.) All areas of less than 20-fathom depth should be identified - the bathymetry information should be available in chart form for rapid retrieval. General bottom characteristics - rocky, muddy, etc., should be identified. The surrounding shoreline topography at high and low tide should be identified by type and constituent material - rocky, sandy, mud flats, reeds and grass, etc. Vehicle and personnel access to shoreline from the land must be identified and where nonexistent, feasibility established.

4. Biological Resource Inventory. The biologically sensitive areas in region must be identified by type, seasonal and geographical sensitivity. For each sensitive area, detailed topography, bathymetry and current information must be available. Since usually there are only few such areas in a given region, detailed protection plans for each should be available and appropriate equipment identified. Biological baseline data - such as would be determined in evaluation of post-spill effects - should be available for these areas.

5. Protection Priority Evaluation. Predetermined priorities must be established as to resource protection in the event of a spill. Areas that may be intentionally contaminated to protect a resource should be identified and the social and legal consequences of such action studied.

6. Oily Waste Disposal. Areas for disposal of oily waste, contaminated sorbents and debris must be identified. Studies of ground water flow, drainage and soil stability at potential disposal sites should be made to insure that waste disposal poses no further hazards.

Specific contingency plans should be checked out via "fire drills" in order to identify communications, administrative and logistical problem areas. Theoretically estimated reaction times under ideal to adverse conditions should be verified or modified by use of fire drills - these also serve actual personnel training functions. The plans should then be modified based on the drill experience. Problems of tracking, monitoring and predicting spill movement under adverse conditions - fog, darkness, sea state - should be evaluated. The listings of available equipment and services should be periodically checked, updated and verified. Checks on at least one contingency plan revealed 20 - 30% of the telephone numbers listed were invalid. In addition, availability of equipment and mobilization times from neighboring regions should be determined and updated (this implies close liaison between regions in terms of industry, state and federal agencies responsible in oil pollution problems). Provisions, both financial and logistical, should be made for immediate activation of damage assessment studies to provide knowledge and understanding of the effects of spills. Such knowledge can then be used to modify accepted contingency plans and clean-up procedures.

ENDNOTES SECTION IV

[1]Senate Bill: S. 2074 -- To Promote the Safety and Protect the Environmental Quality of Ports, Waterfront Areas, and the Navigable Waters of the United States, June 16, 1971.

[2]Initial public hearing scheduled for October 4, 1971.

[3]*The American Waterways Operators Inc. Weekly Letter*, Volume XXVIII, No. 24, June 19, 1971, pp. 3-4.

[4]*Federal Registrar*. Vol. 36, No. 145, Wednesday, July 28, 1971, pp. 13918-13919.

[5]*Seattle Times*, August 19, 1971, p. 1

[6]*Boeing News*, August 5, 1971, p. 1.

[7]H. P. 1459 - L. D. 1835, *An Act Relating to Coastal Conveyance of Petroleum*, State of Maine, 1970.

SECTION V: BIBLIOGRAPHY

A. Books and Reports

A. D. Little, Inc., Oil Spill Treating Agents, Selection Based on Environmental Factors, 1970, API Publication #4068.

American Petroleum Institute. Oil Spill Control Survey for Onshore and Offshore Facilities, March 1970. Prepared by Crossley Surveys, Inc. API Report #4023.

American Institute of Mining, Metallurgical, and Petroleum Engineers, Preprints, 1967 Offshore Technology Conference, Vol. I.

American Institute of Mining, Metallurgical, and Petroleum Engineers, Preprints, 1969 Offshore Technology Conference, Vol. II.

American Petroleum Institute. Proceedings of Joint Conference on Prevention and Control of Oil Spills, 1971. Library of Congress Catalog Card #74-124324.

American Petroleum Institute. Proceedings, Joint Conference on Prevention and Control of Oil Spills, 1967. Library of Congress Catalog Card #74-124324.

American Petroleum Institute. Report on Air and Water Conservation Expenditures of the Petroleum Industry in the U.S., 1966-70, prepared by Crossley Surveys, Inc.

American Petroleum Institute. The Petroleum Industry Oil Spill Control Survey for Onshore and Offshore Facilities, 1969.

Baetsen, R. H. (Commander, USN). San Francisco Advisory Radar.

Battelle Memorial Institute. Oil Spillage Study Literature Search and Critical Evaluation for Selections of Promising Techniques to Control and Prevent Damage, 1967, AD 666289, US Department of Commerce.

Battelle Memorial Institute. Study of Equipment and Methods for Removing Oil from Harbor Waters. Report #CR.70.001, 1967.

Battelle Memorial Institute. Impact of Non-Resident Travel on the Washington State Economy, 1969.

Battelle Memorial Institute. Oil Spill Treating Agents - A Compendium, May 1970.

Battelle Memorial Institute. Oil Spill Treating Agents: Test Procedures: Status and Recommendations, 1970.

Battelle Memorial Institute. Review of Santa Barbara Channel Oil Pollution Incident, 1969, DAST 20.

Beers, W. J. Analysis of World Merchant Ship Losses, 1968.

Bell, H. S. Petroleum Transportation Handbook, McGraw-Hill, New York.

Blumer, Max. Critique of Santa Barbara Oil Spill Study by Allan Hancock Foundation, 1971.

Blumer, et al. Hydrocarbon Pollution of Edible Shellfish by an Oil Spill, 1970.

Blumer, Max. Oil Contamination and the Living Resources of the Sea.

Bulletin of Fish Reserve Board, Canada.

Carthy, J. D., and Arthur, D. R., editors. Biological Effects of Oil Pollution on Littoral Communities, June 1968.

Coast Guard. Load Line Regulations, Sub-Chapter E, 1966, CG-176.

Coast Guard. Rules and Regulations for Cargo and Miscellaneous Vessels, Sub-Chapter 1, 1962, CG-257.

Coast Guard. Rules and Regulations for Passenger Vessels, 1962, CG-256.

Coast Guard. Rules and Regulations for Tank Vessels, 1962, CG-123.

Coast Guard. Seattle Coastal Region Oil and Hazardous Materials Pollution Contingency Plan, December 1, 1970.

Code of Federal Regulations.

Committee of Scientists on the Scientific and Technological Aspects of the Torrey Canyon Disaster, The Torrey Canyon. London: Her Majesty's Stationery Office, 1967.

Corino, Dr. Edward R. "Chemical Gelling Agents and Dispersants." Esso Research and Engineering Company, May 20, 1970.

Cornell, Howland, Hayes, and Merryfield. Seattle Area Oil Waste Disposal Study, 1967, Record #S 5064.0.

Cowan, Edward. Oil on Water. The Torrey Canyon Disaster. Philadelphia/New York: Lippincott, 1968.

Cross and Hoult. Collection of Oil Slicks, 1970. Preprint 1936.

Cross and Hoult. Oil Booms in Tidal Currents, M. I. T.

Defense Documentation Center. In Tank Gelations of Petrochemicals, a Report Bibliography. Search Control # 056709.

Defense Documentation Center. Oil Spills, A Report Bibliography. Search Control # 055043.

Defense Documentation Center. Report on In Tank Gelation of Petrochemicals, 1971. Report # CT 4865.

Defense Documentation Center. Report on Oil Spills, 1971. Report #CT 4538.

Department of Commerce. Tidal Current Tables, 1971, Pacific Coast of North America and Asia, 1970.

Department of Commerce. U. S. Nautical Chart Catalog #2, Pacific Coast, 1970.

Department of Ecology. Laws and Oil Spill Emergency Procedures.

Department of Ecology. Water Pollution Control Laws.

Department of the Navy. ARROW, Navships 90094-008-1010.

Dillingham Corporation. Volume I. Systems Study of Oil Spill Cleanup Procedures - Analysis of Oil Spills and Control Materials. API Publication #4024.

Dillingham Corporation. Volume II. Systems Study of Oil Spill Cleanup Procedures - Proposed Industry Response Plan. API Publication #4025.

Dillon, E. S. Ship Design Aspects of Oil Pollution Abatement, 1971, US Department of Commerce.

Dolan, C. W., I. F. Savage, & R. C. Clark, Jr. Environmental Statement on the Proposed Leasing of Puget Sound Shorelands and Beds of Navigable Waters for Oil and Gas Exploration, 1970.

Dominion Bureau of Statistics. Shipping Report, 1967.

Harris and Ilfrey. Drilling in 1300 Feet of Water - Santa Barbara Channel, California, from 1/70 Journal of Petroleum Technology.

Harris, R. G. Surface Winds over Puget Sound and the Straits of Juan de Fuca and their Oceanographic Effects. University of Washington, 1954 (unpublished MS thesis).

Henrickson, D. L., and R. E. Hughs, Jr. Travel Activity in Washington State. Department of Commerce and Economic Development, 1969.

Honeywell Marine Systems Center. A Proposed Automated Marine Traffic Advisory System for Puget Sound, 6 November 1970. Doc. # K03/70.

Hoult, D. P. Containment and Collection Devices for Oil Slicks, M. I. T.

Hoult, David P. Containment of Oil Spills by Physical and Air Barriers, 1970. PB 192629, US Department of Commerce.

Hoult, David. *Oil on the Sea*, 1969.

Hoult, *et al*. *The Spreading and Containment of Oil Slicks*, M. I. T.

Hoult and Suchon. *The Spread of Oil in a Channel*, 1970.

Kolpack, R. L., Editor. *Biographical and Oceanographical Survey of the Santa Barbara Channel Oil Spill, 1969-1970, Volume II, Physical, Chemical and Geographical Studies*. Alan Hancock Foundations, U. S. C., 1971.

Legislative Digest and History of Bills of Senate and House of Representatives for Washington State.

Lindenmouth, *et al*. *Analysis and Model Tests to Determine Forces and Motions of an Oil Retention Boom*, Volume I. 1970, AD 702512, US Department of Commerce.

Lindenmuth, *et al*. *Analysis and Model Tests to Determine Forms and Motions of an Oil Retention Boom*, Volume II. 1970, AD 702 513, US Department of Commerce.

Macklin, J. "Effects of Crude Oil and Bleedwater on Oysters and Aquatic Plants," *Progress Report*, Texas A&M Research Foundation, 1950.

Master, Mate and Pilot. October, 1970.

Milz, E. A. *An Evaluation - Oil Spill Control Equipment and Techniques*. April 14, 1970.

Moody's Investor Service. *Moody's Industrial Manual, 1970*.

Mungall, J. C. H., and Matthews, J. B. "A Variable-Boundary Numerical Tidal Model," Institute of Marine Science, March 6, 1970.

Nelson-Smith, A. *A Classified Bibliography of Oil Pollution*, 1968.

Newsfont. *25,000 Leading US Corporations*. 1970.

Oberle, F. N. *The Role of the Marine Salvor in the Control of Water Pollution*, 1969.

Puget Sound and Approaches, A Literature Survey. University of Washington, Department of Oceanography, 1953.

Puget Sound Task Force, Pacific Northwest River Basin Committee. *Pleasure Boating Study, Puget Sound and Adjacent Waters*, 1968.

Puget Sound Task Force, Pacific Northwest River Basins Commission. *Puget Sound and Adjacent Waters, State of Washington*, Summary and Appendices I through XV. 1970.

Puget Sound Tug and Barge and Pac-Mar Services. Puget Sound Oil Spill Response Plan. April 1, 1971.

Roe, R. W. "Report on Barge 17 Oil Spill," United Transportation Co.

Scherer, J. "Design Requirements for Booms," 1969.

Smithsonian Institution. Notices of Research Projects on Oil Spillage on Water.

Sperry Piedmont Company. Final Report, Lookout Assist Device Feasibility Studies, August 1965.

Straughan, D. Biological and Oceanographical Survey of the Santa Barbara Channel Oil Spill, 1969-1970, Volume I, Biology and Bacteriology. Alan Hancock Foundation, U. S. C., 1971.

Texas Instrument, Inc. "Biological Assessment of Diesel Spill, Anacortes, Washington, May 1971 Preliminary Report," June 14, 1971.

University of California, Santa Barbara. Oil Pollution of the Sea: Selected References, 1970.

US Army Corps of Engineers. Port Series, US Government Printing Offices.

US Army Corps of Engineers. Waterborne Commerce of the United States.

US Department of Commerce. Tidal Currents Table, 1971.

US Department of Commerce. Statistical Abstract of the United States, 1970.

US Department of Interior. Pacific Northwest Economic Base Study for Power Markets, 1966.

US Maritime Commission. A Statistical Analysis of the World's Merchant Fleets.

US Naval Oceanographic Office. Glossary of Oceanographic Terms, Special Publication 357, 2nd Edition, 1966. US Government Printing Office.

Washington State Census Board. Population Forecast - State of Washington 1965 - 1985. Department of Commerce and Economic Development, 1966.

Washington State Department of Fisheries. Groundfish Data Report #9, 1970.

Washington State Department of Fisheries. 79th Annual Report.

Washington Department of Fisheries. Plan for Washington State Food Fisheries, 1970.

B. Periodicals (Please see specific endnotes for detailed reference)

Advances in Water Pollution Research

American Waterway Operator, Inc. Weekly Letter.

Boeing News.

Business Week Magazine.

Electronics Design.

Environmental Science & Technology.

Federal Register.

Fordham Law Review.

International Legal Materials.

International Shipbuilding Progress.

Journal of Maritime Law & Commerce.

Limnology & Oceanography.

Marine Law Review.

Marine Biology.

Marine Technology.

Michigan Law Review.

National Geographic Magazine.

Ocean Industry.

Oceanology International.

Remote Sensing of Environment.

Science.

Sea Frontiers.

Sea Secrets.

Seattle Post-Intelligencer.

Seattle Times.

Shipbuilding and Shipping Records.

Standard Oil Company of California.

Time Magazine.

The Journal of the Institute of Navigation.

The MBA Magazine.

Undersea Technology.

US News & World Report.

Wall Street Journal.

Water Pollution Control Federation.

C. Sources of Unpublished Information

American Petroleum Institute.

Atlantic Richfield

Battelle Northwest

Bennett Pollution Controls, Ltd., Vancouver, B. C.

British Petroleum, Technical Development Division.

Chemical Processors, Seattle.

Environmental Protection Agency, Northwest Regional Office.

Esso Research and Engineering Co.

Hallikainen Co., Richmond, California.

Honeywell Marine Systems Center, Seattle.

Intergovernmental Marine Consultative Organization.

Liverpool Underwriters' Association.

Lloyd's Registry.

Manchester Navy Installation.

Marine Technology Society.

Maritime Administration.

METRO of Seattle.

Mobil Oil Co.

Northwest Marine Safety Council.

Oceanographic Institute of Washington.

Office of U.S. Senator Warren G. Magnuson.

Oil Spill Services, Inc., Bellevue, Washington.

Oregon State University.

Phillips Petroleum Co.

Puget Sound Pilot's Association.

Seattle Chamber of Commerce.

Shell Oil Company

Superior Refineries, Woodinville, Washington.

Texas Instruments, Inc.

Union Oil Co.

U.S. Attorney, Western District of Washington.

U.S. Coast Guard, Captain of the Port, Seattle, Washington

U.S. Coast Guard, 13th District.

U.S. Marine Fisheries Service, Seattle Office.

United Transportation Company.

University of Washington, Department of Oceanography.

Washington Oil Spill Cooperative

Washington State Department of Ecology.

Washington State Department of Fisheries

Washington State Legislature.

Western Oil and Gas Association.

Western Washington State College..

APPENDIX 1

OIL HANDLING FACILITIES OF PUGET SOUND

A. Introduction

The recommendations presented in Section III-C4 of this report are the culmination of an extensive study on oil handling facilities in the Puget Sound area. This appendix reports the details of that study. A preliminary set of criteria was formulated upon review of Coast Guard regulations, which are summarized in the next section. From these, a set of questions were drawn up to integrate into an onsite inspection tour. The results of this are shown in part C. The final portion of this appendix presents a comparative rating of selected oil handling facilities, defining both strong and weak points.

B. Oil Handling Facility Regulations

Rules and regulations which pertain to the handling of dangerous liquid cargoes by the onshore oil handling facility are defined in Title 33 of the Code of Federal Regulations.[1] These apply both when the facility is receiving oil for storage in their tanks or when distributing oil to a vessel. Enforcement of these rules falls under the jurisdiction of the U.S. Coast Guard. The significant points are summarized here. The onshore oil handling facility shall:

1. Continuously control and supervise the oil transfer.
2. Train the men who control and supervise the transfer operation.
3. Post warning signs.
4. Insure that no repair work is done to the transfer system or tanks during transfer.
5. Place adequate drip pans under hose connections.
6. Insure that couplings are leak free.
7. Insure that bolted couplings have sufficient bolts to prevent leaking.
8. Insure that the person on the vessel reports when he is ready to transfer oil.
9. Obtain a cargo information card.
10. Obtain a Declaration of Inspection.
11. Set up communications with the person on the vessel.
12. Stop transfer operations if:

    a. severe electrical storms occur,
    b. a fire breaks out,
    c. a break occurs in the oil transfer system,
    d. requested by the person on the vessel.

When the onshore facility delivers cil to a vessel the supervisor handling the onshore transfer operation is responsible for:

1. The operation of all oil valves.
2. Notifying the receiver of his intent to pump oil.
3. Slowly commencing the pumping operation.
4. Maintaining oil line or hose connections to prevent leaks.
5. Observing the operating pressure in the system.
6. Stopping the pumping quickly when requested by the receiver.

When the shoreside facility receives oil from a vessel the supervisor handling the onshore transfer operation is responsible for:

1. The operation of all oil valves.
2. Maintaining oil line or hose connections to prevent leaks.
3. Observing the oil flow rate to prevent overfill or overload of the transfer system, and
4. Not securing the system until the person on the pumping vessel has been notified.

When the transfer operation has been completed the onshore facility shall drain all hoses and secure all piping.

Oil handling equipment shall be maintained in good operating condition at all times:

1. The cargo hoses must not leak through the body of the hose due to pressure.
2. Test the oil pumping system once a year.
3. Calibrate the oil pump pressure gauges once a year.
4. Hydrostatically test the hoses and pipelines once a year to 1-1/2 times the maximum working pressure.
5. The cargo hoses shall not be used with pipelines whose maximum working pressure is greater than that of the hose.
6. The relief valve operation shall be checked at the time of each system hydrostatic test.
7. Record and make available, to the Captain of the Port, the date and result of test.
8. The escape pipelines from relief valves shall return the product to the supply or other receiver.
9. Mark all hoses and pipelines to tell their product at multiproduct facilities.

The person in charge should use his judgement in handling an emergency.

C. Oil Handling Facility Procedures and Equipment Status

In order to evaluate industry compliance to the Coast Guard regulations, on-site visitation to various oil handling facilities was conducted. Cursory examination indicated over sixty facilities in the Puget Sound area which had a storing capacity of more than 10,000 barrels. Because of time constraint, the following eleven sites were chosen, their selection being based on randomness and convenience:

I. ARCO, Pier 11, Harbor Island
II. ARCO Refinery, Cherry Point
III. Manchester Navy Installation
IV. Mobil Oil Refinery, Ferndale
V. Phillips Petroleum, Pier 33, Seattle
VI. Shell Oil Company, Pier 19, Harbor Island
VII. Shell Oil Refinery, Anacortes
VIII. Standard Oil of California, Pier 32, Seattle
IX. Standard Oil of California, Point Wells, Richmond Beach
X. Union Oil Dock, Pier 71, Elliott Bay
XI. Union Oil Company Pier, Edmonds

To initiate the visits, a letter of introduction, shown in Figure 1-1, was forwarded to the facilities. Figure 1-2 presents a summary of the questionnaire used on the visits. It delineates the various aspects of the facility operations on which information was sought.

The remainder of this section presents responses of the eleven visits to the questionnaire. The numbered responses for each visit correspond to the topic numbers on the questionnaire.

UNIVERSITY OF WASHINGTON
SEATTLE, WASHINGTON 98105

College of Engineering
Department of Aeronautics and Astronautics

May 21, 1971

To Whom It May Concern:

The Ocean Engineering Systems Design class at the University of Washington is conducting a study of the methods and procedures to prevent oil spills on Puget Sound. As part of this study, it is necessary to visit various oil handling docks in the Puget Sound area and interview supervisory personnel to obtain information as to methods and procedures currently used to prevent oil spills and any recommendations or suggestions. The students involved in this series of interviews are:

Jacques Duringer
Charles Gundersen

Sincerely,

Juris Vagners
Assistant Professor
Department of Aeronautics
and Astronautics

JV:nh

Figure 1-1
Letter of Introduction, On Site Visitation of Oil Handling Facilities

QUESTIONNAIRE

NAME OF DOCK:

1. CONSTRUCTION OF DOCK:

2. LIGHTING OF DOCK AND WATER:

3. PIPING SYSTEM AND STORAGE:

4. HOW MANY VALVES FROM DOCK TO TANK:

5. HOW MANY VALVES OF THE ABOVE ARE MANUAL, MOTORIZED, REMOTE CONTROL OPERATED AND HOW ARE THEY CONTROLLED:

6. HOW IS WASTE OIL TREATED:
   a. On Grounds:
   b. On Dock:
   c. Ballast:
   d. What Kind of Processing:

7. WHAT IS THE FUELING SYSTEM:
   b. Check System:

8. FUEL MONITORING:

9. PIER SECURITY:

10. CLEANUP EQUIPMENT AND SYSTEMS:

11. COMMENTS:

Figure 1-2

FACILITY VISITATION QUESTIONAIRE

QUESTIONNAIRE RESPONSES

I. <u>ARCO, Pier 11 -- Harbor Island</u>

1. Timber pile with two approaches, 460' of berthing space.
2. Floodlights for the dock, floodlights under dock at the ends of the pier.
3. 2-10", 1-6", 4-8" and 3-4" going to 23 steel tanks with a total capacity of 617,800 barrels, have 100% retaining walls, but no retaining wall at the truck loading area, which is about 20' from water.
4. Five sets of valves: 1) header, 2) pier, 3) waterline fence, 4) pumping pit, and 5) tank. There are no check valves.
5. Dock valve is manually controlled. No motorized or remote control valves. There are some relief valves in the pumping pit. Pumps are controlled from dock, pumping pit and switchboard.
6. a. storm drains in the oil handling areas go to separator, others drain into sound;
   b. drip pans and sawdust;
   c. no provisions for oily water ballast;
   d. oil from drains and tank pit goes to separator.
   Oil is then stored in tanks and shipped out.
7. One man on dock controls the valve and pumps. There is no communications system.
   There is no check system.
8. Fuel monitoring is done from barge and when full, the message is "yelled" to the man on shore.
9. Waterline fence, but is left open most of the time.
10. None at the moment (no need to have detailed current analysis).
11. Plan to buy a 14' Alumacraft work boat for cleanup purposes, plus 400-500' of Clearwater Inc. Boom and plan to install a drip trough on the dock. Also a log boom will be installed around the whole dock area.

II. ARCO Refinery -- Cherry Point (under construction)

1. Timber pile with concrete decking, built-in trough on dock for spilled oil leads into collecting tank. Berth for supertankers.

2. Information not available.

3. Information not available.

4. Approximately three to four sets of valves.

5. Valves on dock are solid "chiksans" (no hoses) and are completely motorized. All valves are motorized and can be remotely operated at a central computer.

6. a. All water on the grounds is treated before discharge goes into separating ponds.
   b. Sump-trough system.
   c. Goes through their separator facilities.
   d. Have a multiple treating system of ponds, gravity separators, biodegradation, etc.

7. One man on dock handling the "chiksans". The valves can be remotely controlled at dock or a computer room, more men on refinery grounds or dock as needed.

   b. Information not available.

8. Fuel monitoring is done by flow meters and is computer operated.

9. Information not available.

10. Two purse seine workboats, 520' of Uniroyal boom, straw or sorbents will be available. A study of current conditions should be made to determine where the oil will go.

11. This installation is not finished. Some information is vague and other was not available at the present time.
    Adverse weather conditions represent a hazard.

III. Manchester Navy Installation

1. Timber pile, 1000 ft. of berthing.
2. Lighting of dock and lighting of water area around the ship.
3. There are 2-18", 1-10", 1-6", 1-14", 3-8" pipelines. All tanks are underground and there is complete wall retainment along shoreline.
4. Large number of valves along dock to tank.
5. All valves are manually operated with men standing by at three valves during fueling. The oil is gravity fed.
6. a. Goes to separator system and stored in tanks.
   b. Drip pans.
   c. Retaining tank.
   d. Gravity separator.
7. Man on dock controls valve and other men are in charge of other valves on dock and shoreside.

   b. There is a checkoff list for dock and instructions.
8. Fuel is monitored at the ship and at the pump house.
9. Complete Navy security.
10. Oily sorbent.
11. Comprehensive new cleanup system which includes boom installation under dock, 1000 ft. of Uniroyal boom, a navy skimmer, two work boats, currents study and new retaining walls. Also a waste oil incinerator system is being built.

IV. Mobil Oil Refinery -- Ferndale

1. Timber pile, asphalt surfaced. Capability for 800,000 lbs. tankers. Reconstructed due to weather damage.

2. Dock is lighted and some lights shine on water.

3. There are 3-20", 4-16", 1-10" and 1-6" pipeline, to 40 tanks (2,370,000 lbs. capability). Gravity fed. 100% retaining walls on tanks.

4. There are three valves: 1) header, 2) at the shoreline, and 3) at tank (left open all the time).

5. Two sets of manual valves and one remote operated. The valves at the shoreline can be shut off from the dock by a switch.

6. a. No provisions.
   b. Drip pans.
   c. To holding tank, then processed.
   d. Gravity separators, API separator, bio-treatment.

7. One man on dock and another on the shoreline. Communication by intercompany phone.

   b. There is no check-off sheet or system.

8. From the tank, then communicated by telephone to dock. The oil is gravity fed, no pumps are used.

9. Security is adequate. Fences around installation.

10. Some hay. Some boom (not adequate for use on oil spills).

11. Currents are very strong. A study should be made of the currents and tides in the area.

V. Phillips Petroleum, Pier 33 -- Seattle

1. Dock is steel, timber decked. There is 650' of berthing space.
2. Lighted, but no lights shining on water.
3. There are 6-10" and 5-6" pipes. The tanks have 100% retaining wall.
4. There are three sets of valves between tank and dock, check and relief valves are at the pump.
5. All valves are manual. Oil is either gravity fed or pumped. One man shuts off the dock header valves.
6. a. Storm drains in loading rack go to separator.
   b. Drip pans.
   c. No facilities.
   d. Six gravity separators, but some of them not in use.
7. One person on dock and the other person where needed. There is a public address system around the plant.
   b. There is no checkoff list.
8. Barge gauging and tank gauging.
9. No waterline fence. Outside fence locked at 12 p.m., opened at 6 a.m.
10. None - call M.O.P.S.[2]
11. Have a large unused separator system that can be used.

VI. Shell Oil Company, Pier 19 -- Harbor Island

1. Timber pile dock. 290' pier used for fueling Washington State Ferries. 340' berth for barges and tankers.

2. Well lighted, some lights can illuminate the water.

3. There are 2-2", 7-10", 2-8", and 5-6" pipes. Tanks are away from water line, pipelines run under roads and railroad tracks. Dike system holds 100%.

4. There are five sets of valves - 1) header, 2) warehouse (beginning of pier), 3) at manifold, 4) at dike wall, and 5) at tank there are relief valves, but no check valves.

5. Header valves are manually controlled. At manifold - motor operated valves, these are controlled from the fillstand. All others are manually controlled.

6. a. Rain water on grounds flows into the bay. Oil spilled near fillstands goes to separator.
   b. Drip pans for oil spilled from valves.
   c. No facilities.
   d. Gravity separators and the waste oil is burned in boilers.

7. One man on dock, one man on a line watch, and one man in the lab. Men are trained one year before working alone.

   No check list. There are instruction sheets for men.

8. Fuel is monitored three ways: 1) flow meter on dock, 2) tank gauge, and 3) barge gauge.

9. There is no fence around pier area, facility fenced but not locked.

10. 300 ft. of boom 18" skirt, 17' Boston Whaler. Have the most comprehensive oil spill contingency plan of all places visited.

11. Have detailed instruction sheets for the three men in charge of fueling operation.

VII. Shell Oil Refinery -- Anacortes

1. Timber pile, 65' wide. Have adequate berth space for any size vessel.

2. Lighted, no direct lights shine on water but there are lights under the dock.

3. There are 1-16" crude, 1-16" and 5-12" pipelines to 54 tanks (3,390,000-barrel capacity). There is 100% dike system, tank area is removed from the water.

4. There are three important sets plus many others.

5. There are two motorized valves remotely controlled from dock or control house, with a fast shutoff system. The tank is manually closed except for fueling.

6. a. Storm drainage goes to a settling pond.
   b. Drip pans.
   c. Goes into separator.
   d. Gravity separators, and reprocessing of waste oils.

7. There is one man per ship on dock and another at the tank. The man on dock can shutoff the pumping and valves by push-button.

   b. There is no checkoff list.

8. Fuel is monitored at control house continually while fueling.

9. Grounds are adequately fenced, only way in is by boat.

10. Two boats 14' and 16'. Hay is available. They do have a knowledge of current conditions in the area.

11. Hoses are X-rayed for inspection purposes.

VIII Standard Oil of California, Pier 32 -- Seattle

1. Timber pile, timber decked, 433 ft. of berthing space.
2. There is lighting on dock but no lighting of the water.
3. There are 3-10", 5-6", 5-8", and 3-4" pipelines. The lines are kept full. Tank dike wall contains 100% of the oil.
4. There are three sets of valves: 1) header valves, 2) block valves at dike, and 3) tank valves. There are no dock valves.
5. All valves are manual and are operated by one man.
6. a. Storm drains in oil handling facilities go to the separator.
   b. Drip pans on dock.
   c. No facilities.
   d. Gravity separators.
7. a. Usually one man handles the fueling operation. He also does the fuel gauging. Fuel transfer is done by pumping.

   No checkoff list or personnel instruction list.
8. Fuel is monitored by person on the dock, through barge gauge or tank gauge.
9. There are no gates in the vicinity of the dock area.
10. No cleanup equipment although relatively close to M.O.P.S.
11. No planned improvements.

IX. Standard Oil of California, Point Wells -- Richmond Beach

1. Timber pile, concrete paved. There 1,054' of usable berthing space.

2. Dock is lighted but there are no direct lights onto the water.

3. 14-12", 10-6" and 6-8" pipes going to tanks with storage of 1,926,000 bbls. Between tank and shoreline is a big manifold system and pumphouse. Some tanks do not have 100% retaining dikes.

4. There are many sets of valves, generally: 1) header, 2) shoreline, 3) manifold, 4) pump station and 5) tank.

5. Manual dock valves, one set of motor operated valves along pipes, the motor valves are air driven. These valves can be shut off at a control point. There are no check valves.

6. a. Rain water goes to Sound.
   b. Dock has drip trough and pans.
   c. Ballast from ships goes to separator.
   d. There are six small gravity separators and then to a holding tank.

7. One man per vessel at dock, one man at pumphouse, manifold and tank area. There is an interplant phone system.

   There is no check system.

8. Tank gauging and barge gauging.

9. Outside fence locked at 4 p.m., no gate at approaches to dock.

10. 100 sacks of sorbent C on deck. No other cleanup equipment at the moment.

11. Planning to get 3000' of Bennett Boom[3] and a 15'-20' work boat to deploy the boom. Tugs which the company owns can also be used for cleanup. Also completely modernizing their oil-water separator system, and storm drains will be hooked up to this.

X. Union Oil Dock, Pier 71 -- Elliott Bay

1. Timber pile L shaped pier. 450' berthing space for barges.
2. Dock is lighted. No lights shine on water.
3. There are 4-4", 4-6" and 4-8" pipes. Pipes are underground from tanks to dock. There is a 100% dike wall around tanks.
4. There are six sets of valves, two sets of valves on the pier, and four sets on the pipeline.
5. All valves are manual. Two valves on dock controlled by the man on dock and tank valves controlled by man at the tank farm.
6. a. All oil spilled on grounds goes through separator system.
   b. Drip pans.
   c. No facilities.
   d. There are three oil-water separator units, waste oil is burned at Edmonds facility.
7. There is one man in dock area and one man in tank farm. Coomunication via walkie-talkies. Dock valves can be shut off at two places; and tank can be shut off at the tank valve.

   Yes, a checkoff list is used for dock and ship.
8. Fuel is monitored at the tank and at the barge.
9. Valves are locked. Dock is locked. Security check every hour.
10. None.
11. Men working with oil have safety meetings and training periods.

XI. Union Oil Company Pier, Edmonds

1. Timber pile, very old pier (1920). 276' of berthing space, 170' for barges.

2. Dock is lighted, but there are no lights shining on the water.

3. There are 3-12", 3-10", 1-8", 4-6" pipes from docks to 20 tanks (750,000 barrel capacity). The dike has 100% capacity. There are check valves on tanker risers.

4. There are four sets of valves: 1) at riser, 2) near pier end, 3) at bridge (bottom of the hill), and 4) at the tank there are check and relief valves installed.

5. All valves are manually operated. They are all closed when not in use and dock valves are locked with chain and lock.

6. a. Rain water goes to Sound. Oil from oil handling areas into separator.
   b. Drip pans.
   c. Goes to separator.
   d. Gravity separator. The waste oil is burned as fuel for their refinery.

7. One person on the dock is in charge of two valves and another is in charge of two more valves on shore. There is an interphone communications system, and horn signal.

   Yes, there is an anti-pollution check list.

8. Fuel monitoring is done from tank gauge and barge gauge.

9. Dock is well locked and so are grounds.

10. No cleanup equipment.

11. Should have a detail study of the current conditions in the area.

D. Evaluation of Oil Handling Facilities

To perform a comparative analysis on effectiveness of the various facilities, with regard to prevention and control of oil spills, some of the more important factors in dock oil handling facilities were selected to make a rating of these facilities. The sole purpose of this rating is to indicate some basic weaknesses of an existing oil handling dock and how it compares to others in the Puget Sound area. For analytical purposes, the following characteristics were selected. When possible, quantification is shown; otherwise, the evaluation was subjective:

1. Having a sheltered harbor facility without strong current or wave action. This is important because inclement weather on currents can make fuel transfers more hazardous and therefore increases the probability of an oil spill.
2. Lights shining on the water surface during the night operations to detect possible oil spills.
3. Construction of dock -
   a. ability to withstand damage without oil leakage;
   b. troughs, or other permanent installations that minimize the amount of oil lost during fueling operations.
4. Adequate retaining walls around the facility to prevent oil spilled on the grounds from reaching the water.
5. Adequate security system -
   a. locked fence system;
   b. locked valves on dock areas;
   c. guard system.
6. Adequate number of valves along pipelines.
7. Adequate number of personnel involved during fueling.
8. Adequate monitoring of quantities being transferred.
9. A fast, shutoff system for the valves in case of emergency.
10. An anti-pollution checkoff list.
11. Type of communication system in the plant.
12. Oil spill cleanup equipment and system.

The rating system was subjective, but based on an adequate - not adequate evaluation of the above twelve characteristics. Table 1-1 is a summary of the findings. The system covers only some of the more important aspects of prevention and control, and is not meant to be all-inclusive. The facilities are listed in order of fewest weaknesses, beginning with the best. Also shown are the number of inadequacies per facility.

TABLE 1-1. SUMMARY OF SURVEY (BY FACILITY)

Facility	Inadequacies
1. Arco Refinery	1
2. Manchester	3
3. Standard - Pt. Wells	4
4. Shell Refinery	4
5. Shell, Harbor Island	5
6. Union, Pier 71	4
7. Mobil Oil Refinery	5
8. Union, Edmonds	5
9. Arco, Harbor Island	8
10. Phillips, Pier 33	7
11. Standard, Pier 32	9

One other interesting comparison is the frequency of occurrence for inadequacies of each of the parameters. These are summarized below in Table 1-2. This listing is in the order of the characteristics described above.

TABLE 1-2: SUMMARY OF SURVEY (BY CHARACTERISTIC)

Inadequacy	No. of Occurrences (11 maximum)
Unsheltered harbor	6
No lighting on water surface adjacent to dock	6
Inadequate dock construction	9
Inadequate retaining walls	2
Inadequate security system	5
Shortage of valves along pipeline	4
Insufficient personnel for fueling	2
Inadequate monitoring of quantity of transferred fuel	2
Inadequate emergency shutoff system	6
No anti-pollution checkoff list	5
Inadequate inplant communication system	3
Insufficient oil spill cleanup system	5

The following is a detailed evaluation of each facility, noting pertinent comments to each of the designated characteristics, and the adequacy rating of each of these.

ARCO REFINERY, CHERRY POINT (under construction)

Characteristic	Comment	Adequacy
1	Unsheltered harbor	No
2	Not yet installed	**
3	Concrete pier with built-in troughs	Yes
4	Retaining walls adequate, no spill into water	Yes
5	Fence system, motorized "chiksans" are more tamper-proof than conventional valves. There is a guard system.	
6	At least 3 valves, computer operated	Good
7	Adequate number of men	Fair
8	Computer operated fuel flow	Good
9	Computer operated valve shutoff system also at dock	Yes
10	Not available yet	**
11	Intercom system	Yes
12	Cleanup system for current conditions	Fair*
	TOTAL NUMBER INADEQUATE	One

* Should have larger and better suited booms and equipment due to current and weather conditions. Also a study of the current conditions should be made.

** Will be incorporated at a later date, since dock is not in operation.

MANCHESTER NAVY INSTALLATION

Characteristic	Comment	Adequacy
1	Partially sheltered	No
2	Light system on dock and into water	Yes
3	Construction is barely adequate and no trough system installed	No
4	Complete wall system and additional walls under construction. Tanks are underground.	Yes
5	Navy Security System	Yes
6	Six or more valves in pipeline	Good
7	Three or 4 men during fueling	Good
8	Fuel gauged at barge and tank	Fair
9	Enough men to shut off valves adequately in emergency situations	Yes
10	Checkoff list plus instruction sheets	Yes
11	No communication system on dock	No
12	Cleanup system (will be installed)	Good
	TOTAL NUMBER INADEQUATE	THREE

STANDARD OIL OF CALIFORNIA-PT. WELLS, RICHMOND BEACH

Characteristic	Comment	Adequacy
1	Unsheltered harbor	No
2	Lighted but no direct lights to water	No
3	Newly rebuilt dock, concrete, paved with troughs	Yes
4	Retaining walls adequate, some new walls being built for tanks	Yes
5	No locked gate at pier, valves not locked	No
6	At least five in pipeline	Good
7	Two men during fueling	Fair
8	Tank gauge and barge gauge	Fair
9	Fast shutoff system operated from dock or station	Yes
10	No checkoff list	No
11	Interplant communication system	Yes
12	New budget will include new booms, boats, etc. and good prevention program	Good
	TOTAL NUMBER INADEQUATE	FOUR

SHELL COMPANY REFINERY, ANACORTES

Characteristic	Comment	Adequacy
1	Partially sheltered facility	No
2	Lights on dock and some under dock barely adequate	No
3	Dock construction adequate but no provisions for spills on dock, nor troughs	No
4	Walls are adequate, facility is moved from shoreline	Yes
5	Security system is good	Yes
6	At least four sets of valves	Good
7	At least two men during fueling operations	Fair
8	Constant fuel monitoring and barge gauging	Good
9	Good, fast, shutoff system; operable from three different stations	Yes
10	No checklist, there is an instruction sheet	
11	Intercom system	Yes
12	Cleanup equipment available, no adequate boom	Fair
	TOTAL NUMBER INADEQUATE	FOUR

SHELL OIL COMPANY, PIER 19, HARBOR ISLAND

Characteristic	Comment	Adequacy
1	Sheltered harbor facility	Yes
2	New light system, but marginally lights up water surface	Yes
3	Dock construction adequate, but no provisions for spills on dock, no troughs	No
4	Adequate retaining walls on tanks, but spill near waterline might go into water	No
5	Valves are not locked, no fence around pier area	No
6	There are five sets of valves on the pipelines	Good
7	Three men involved during fueling	Good
8	Three ways to monitor fuel	Good
9	No fast shutoff system	No
10	No checkoff list, there is a good instruction sheet	No
11	Communication system between dock and tanks	Yes
12	One of the best cleanup systems and equipment of all facilities visited	Very Good*
	TOTAL NUMBER INADEQUATE	FIVE

* Because of the good cleanup system this facility was rated higher than others with a smaller number of deficiencies.

MOBIL OIL CO. REFINERY, FERNDALE

Characteristic	Comment	Adequacy
1	Extremely unsheltered harbor	No
2	There are adequate lights shining on dock and into water	Yes
3	Dock is not strong enough to withstand weather conditions without damage, there are no troughs	No
4	There are adequate retaining walls on the refinery tanks	Yes
5	No locked valves, no fence around dock area	No
6	Three sets of valves	Fair-Bad
7	Two men during fueling operations	Fair
8	Barge and tank fuel monitoring	Fair
9	There is a fast shutoff system	Yes
10	No checkoff list	No
11	There is a telephone communication system	Yes
12	Some cleanup equipment	Bad*
	TOTAL NUMBER INADEQUATE	FIVE

* Not enough to meet the current and weather conditions in the area.

UNION OIL COMPANY PIER, EDMONDS

Characteristic	Comment	Adequacy
1	Unsheltered facility	No
2	No lights shining on water surface	No
3	Dock is old and is not adequate, no troughs on dock	No
4	Retaining walls are adequate	Yes
5	Valves are not locked, but fence system is good	Yes
6	Four sets of valves	Fair
7	Two men during fueling	Fair
8	Tank and barge gauging	Fair
9	No fast shutoff system	No
10	Checkoff list available	Yes
11	Interphone system	Yes
12	No cleanup system	No
	TOTAL NUMBER INADEQUATE	FIVE

ARCO, PIER 11, HARBOR ISLAND

Characteristic	Comment	Adequacy
1	Sheltered harbor	Yes
2	There are adequate lights under dock that shine on water	Yes
3	Troughs and a permanent boom are going to be installed	Yes
4	Adequate retaining walls around tanks, but oil spilled on grounds will spill into the harbor	No
5	Valves are not locked, fence is left open	No
6	Five sets of valves*	Bad
7	One man during fueling operations	Bad
8	Fuel is monitored at barge only	Bad
9	No fast shutoff system	No
10	No checkoff list	No
11	No communication system	No
12	Cleanup system adequate	**Fair-Good
	TOTAL NUMBER INADEQUATE	EIGHT

* No relief valves, therefore valves are left partially open.

** Because of good cleanup system is rated ahead of Phillips Company dock.

PHILLIPS PETROLEUM COMPANY, PIER 33, SEATTLE

Characteristic	Comment	Adequacy
1	Sheltered harbor	Yes
2	No lights shining on water	No
3	Construction of dock adequate. There is no trough system	No
4	There are adequate retaining walls	Yes
5	The valves are locked, but no waterline fence	No
6	There are three sets of valves	Bad
7	Two men during fueling operations	Fair
8	Barge and tank gauging	Fair
9	No fast shutoff system	No
10	No checkoff list	No
11	P.A. system in plant	Yes
12	No cleanup system	No
	TOTAL NUMBER INADEQUATE	SEVEN

STANDARD OIL OF CALIFORNIA, Pier 32, SEATTLE

Characteristic	Comment	Adequacy
1	Sheltered harbor	Yes
2	No lights shining into the water	No
3	Dock construction adequate, but no troughs	No
4	There are adequate retaining walls	Yes
5	Valves not locked, gate not locked	No
6	Three sets of valves	Bad
7	One man during fueling operations	Bad
8	Barge and tank gauging	Fair
9	No fast shutoff system	No
10	No checkoff list	No
11	No communication system	No
12	No cleanup system	No
	TOTAL NUMBER INADEQUATE	NINE

E. Endnotes for Appendix 1

1. Code of Federal Regulations, Title 33. Navigation and Navigable Waters, 1 to 199, revised September, 1970. Part 126.

2. Marine Oil Pickup Service, a joint venture by Puget Sound Tug and Barge Company and Pacific Marine Services.

3. Manufactured by Bennett Pollution Controls, Ltd., Vancouver, B.C. See Appendix 5 for discussion of booms.

APPENDIX 2

STATE LAWS FOR OIL SPILLS PREVENTION AND CONTROL

A. Introduction

The Washington statute is of considerable importance in the regulation of oil spills. The basic statutory pattern is one of strict liability without regard to fault for damages arising from oil spills plus penalties for wrongful discharges of oil. The following is a summary of certain provisions of the statute followed by a brief commentary. This paper is limited to Washington jurisdiction. No consideration of conflicts of law, either state, federal, or foreign will be undertaken. Also considered herein are certain suggestions for reform of the Washington statute, especially by comparison to Maine statutes.

B. Summary of Washington Statute[1]

1. Discharges of Oil Unlawful

   RCW 90.48.320 made it unlawful to allow entry of oil into state waters regardless of whether the persons having control over the oil were at fault except when the entry of oil:

   a. was authorized by the pollution control commission, or
   b. was after a failure of the commission to act on a permit request, or
   c. was caused by -
      i) an act of war or sabotage, or
      ii) the negligence of the state or federal government.

2. Obligation to Collect-Dispersants Control

   RCW 90.48.325 makes it the duty of any person owning or having control over oil entering state waters in violation of 90.48.320 to immediately collect and remove same. When collection and removal are not feasible such persons must contain, treat, and disperse the oil. This section makes the decision to use chemicals or dispersants strictly a matter that is within the discretion of the director (Director, Department of Ecology). While the statute is unclear as to whether a polluter may utilize dispersants or chemicals before being instructed by the director, it is clear that he does so at his own risk. Any increased damages caused by the use of chemicals only expands civil liability.

3. State May Clean Up Oil Spills.
Under RCW 90.48.330, the state may take affirmative action to investigate and clean up a spill subject to the limitation that such action will be undertaken only to protect public interest or property.

4. Civil Liability for Cleanup Expenses Incurred by the State.
RCW 90.48.335 makes any person who fails to immediately collect the oil in accordance with RCW 90.48.325 civilly liable to the state for all the expenses incurred by the state in the cleanup of such oil spills.

5. Strict Liability for Damages from Oil Spills.
RCW 90.48.336 makes any person owning or having control over oil which enters state water, "strictly liable, without regard to fault, for damages to persons or property, public or private caused by..." oil entering the waters of the state, in violation of 90.48.320. Section 336 goes on to relieve such persons of liability if such persons can prove that oil entered the waters of the state by causes set forth in RCW 90.48.320 (3).

6. Civil Liability of Third Parties.
RCW 90.48.338 gives the state a right of action for the necessary expenses in cleanup against third parties who caused an oil discharge. Also, persons liable for cleanup under 90.48.325 and .350 are extended assurance that they retain a right to recover expenses from persons actually causing discharges.

7. Expenses Due the State.
RCW 90.48.340 provides a procedure for investigation, recovery of expenses, and review of findings made by the director. The director has five years within which he must notify responsible persons with an appropriate order to pay the state after completion of the complained of activity. The order specifying the necessary expenses owed to the state must be contested within thirty days. The commission

may reduce or modify orders after further review. The attorney general, upon request of the director, will after thirty days of the first order specifying the necessary expenses owed the state or will after fifteen days following the final order (if modified), bring suit in behalf of the state to recover the amount of the final order. The section also includes investigatory expenses as a necessary expense due the state.

8. Permits.

   RCW 90.48.343 requires that permits are to be given by the director prior to any discharge of oil into the water. Commercial discharges are in accordance with RCW 90.48.180. RCW 90.48.345 gives the commission rule-making powers to effectuate the policies of this act.

9. Penalties.

   RCW 90.48.350 provides penalties for persons negligently or intentionally discharging oil, in addition to any other penalties provided by law. Penalties may be set up to $20,000 for each violation. Every act of omission or commission which causes the violation is considered a separate violation under the act. The director of the commission has wide power to set the amount of penalties up to the $20,000 limit for each violation. Persons have 15 days to contest penalty orders and the director has wide power to review such penalty orders; however, if payment is not made within thirty days of the final order, the attorney general, upon request of the director, will bring suit in the name of the state to recover such penalties.

10. Right of Entry.

    RCW 90.48.355 provides the commission with the right of entry to search for and/or to investigate for possible violation subject to the limitation that trade secrets will not be divulged.

    This right of entry will probably run afoul of the Supreme Court Decision in See vs. Seattle (387 U.S. 541(1967)), if search warrants are not issued prior to searches wherein consent is not given by the party being inspected. Such warrants, however, would be issued by any court of competent jurisdiction on a proper showing of the reasons and purposes of such investigation. When there exists an apparent, visible, or demonstrable emergency, the necessity of warrant would, however, probably not inhere.

11. Notice of Spill Required

RCW 90.48.360 imposes upon the one owning or controlling oil a duty to give notice of any discharge of oil to the commission.

C. Reforms to Washington Statute as Suggested by Maine Oil Pollution Control Act

Maine has recently enacted laws which provide a significant development in the regulation of oil pollution. Maine's new laws will:[2]

"1. Prohibit oil discharges into coastal waters or on their edges and into any waters which drain into them;

2. enable tankers, ships, refineries, terminals, and even oil spills and disaster-fighting to be regulated to a distance of 12 miles from shore;

3. empower the Environmental Improvement Commission to make emergency rules, good for 30 days, and extend the Governor's civil defense powers to imminent or fresh oil disasters;

4. Provide for damage claims agains a new Maine Coastal Protection Fund of $4 million and have arbitral hearings on these claims;

5. fund damage claims, or insurance against them by the state, commission and spill abatement expenses, and research projects up to $100,000 each year approved by the legislature, out of a license fee of one-half cent on each barrel "transferred" at an oil terminal, and out of a $4,000,000 bond issue;

6. punish oil polluters and scoff-laws by fines of $100-$5,000 per day of violation, and put the fines collected into the Maine Coastal Protection Fund;

7. require reimbursement for costs of abating pollution, give power to the Attorney General to be their ultimate collector, and limit discretionary waivers to cases of unforeseeable acts 'exclusively occasioned by the violence of nature without the interference of any human agency' and acts of government or of war;

8. empower the Environmental Improvement Commission to recommend control of air, water, and coastal flat pollution, as well

as 'other aspects of the natural environment,' to aim at preservation of their 'highest and best use;'

9. empower the Environmental Improvement Commission to forbid or condition any developer's commercial plan for any area over 20 acres or any parcel whose structures are to cover more than 60,000 square feet, or for any scheme whose waste disposal already requires a license, if necessary to prevent irreparable damage to the environment;

10. shift the burden of proof to the developer on the adequacy of his plans to fit in 'harmoniously,' so as to not adversely affect existing uses, scenery, resources, or property values, and to provide the finances and technical ability to control pollution commensurate with state standards;

11. force developers to notify the commission before construction and give the commission 14 days to approve the project or schedule a public hearing with at least ten days' newspaper notice;

12. prohibit dumping of waste from outside Maine 'in any structure or on any land within the state;'

13. add to the commission's powers the duty to take into account financial ability to control pollution according to state standards before allowing effluent discharges in general;

14. enable enforceable conservation restrictions to be placed on certain lands;

15. permit state water quality standards to preserve designated spawning beds in inland waters; and

16. specify criteria for loans to begin community plans for pollution abatement.

The anti-pollution plan also recommended that responsibilities now spread over 52 Maine agencies be combined into one agency called the Department of Natural Resources."

As noted in the above summary, Maine has taken a far more enlightened view of the need to assure that funding will be available for the control of oil pollution disaster.

Washington has presently defaulted to the discretion of oil companies the responsibility to assure that funding will be available to provide for oil pollution disasters. Thus, the citizens of Washington have been subjected to a scheme that cannot be extended beyond the oil companies' narrow self-interests. The notion that the oil companies will regulate their affairs in the public's best interest ignores the enormous problems already created by those very same companies.

The State of Washington must create, as the State of Maine has, an environmental defense fund. Any agreement between the oil companies to create a pool of resources for use after oil spills is inadequate. An oil industry fund could only be resorted to when the oil industry consented or when a fund member was liable. This gives the public no assurance that coverage will be extended to all oil spills. Based on past practice, it would seem foolish to rely on the enlightened view of the public interest not yet displayed by the industry. Indeed, the oil companies' self-interest is probably quite at odds with the public interest in the regulation of oil. It should be the policy of government to increase the cost of oil to encourage more efficient utilization of energy resources. Increasing the cost of oil would both reduce the present hazardous levels of oil consumption and encourage the development of efficient alternative sources of power. Any increased cost will create, however, what amounts to a monopoly profit. This profit should not create for the oil industry a fund which would expand its monopoly power. Any regulatory scheme providing for the public's welfare should not be controlled by the narrow self-interest of the oil industry, but should begin to provide the source of funds necessary to undo some of the great harm that has been created by society's addiction to an expanding dependency on oil resources.

The Maine statute for providing a fund should be a pattern for the State of Washington:

> "§551. Maine Coastal Protection Fund
> The Maine Coastal Protection Fund is established to be used by the commission as a nonlapsing, revolving fund for carrying out the purposes of this subchapter. The fund shall be limited to the sum of $4,000,000. To this sum shall be credited all license fees, penalties, and other fees and charges related to this subchapter, and to this shall be charged any and all expenses of the commission related to this subchapter, including administrative expenses, costs of removal of discharges of pollutants, and third party damages covered by this subchapter.
>
> Monies in the fund, not needed currently to meet the obligations of the commission in the exercise of its responsibilities under this subchapter, shall be deposited with the Treasurer of State to the credit of the fund, and may be invested in such manner as is provided for by statute. Interest received on such investment shall be credited to the Maine Coastal Protection Fund.

1. Research and development. The Legislature may allocate not more than $100,000 per annum of the amount then currently in the fund to be devoted to research and development in the causes, effects and removal of pollution caused by oil, petroleum products and their by-products on the marine environment. Such allocations shall be made in accordance with the provisions of section 555.

2. Third party damages. Any person claiming to have suffered damages to real estate or personal property or loss of income directly or indirectly as a result of a discharge of oil, petroleum products or their by-products prohibited by section 543 may apply within 6 months after the occurrence of such discharge to the commission stating the amount of damage he claims to have suffered as a result of such discharge. The commission shall prescribe appropriate forms and details for such applications. The commission may, upon petition, and for good cause shown, waive the months limitation for filing damage claims.

A. If the claimant, the commission and the person causing the discharge can agree to the damage claim, the commission shall certify the amount of the claim and the name of the claimant to the Treasurer of State and the Treasurer of State shall pay the same from the Maine Coastal Petroleum Fund.

B. If the claimant, the commission and the person causing the discharge cannot agree as to the amount of the damage claim, the claim shall forthwith be transmitted for action to the Board of Arbitration as provided in this subchapter.

C. Third party damage claims shall be stated in their entirety in one application. Damages omitted from any claim at the time the award is made shall be deemed waived.

D. Damage claims arising under the provisions of this subchapter shall be recoverable only in the manner provided under this subchapter, it being the intent of the Legislature that the remedies provided in this subchapter are exclusive.

3. Board of Arbitration. The Board of Arbitration shall consist of 3 persons, one to be chosen by the person determined in the first instance by the commission to have caused the discharge, one to be chosen by the commission to represent the public interest and one person chosen by the first 2 appointed members to serve as a neutral arbitrator. The neutral arbitrator shall serve as chairman. If the 2 arbitrators fail to agree upon, select and name the neutral arbitrator within 10 days after their appointment then the commission shall request the American Arbitration Association to utilize its procedures for the selection of the neutral arbitrator.

A. No member of the commission shall serve as an arbitrator.

B. Arbitrators shall be named by their principals within 10 days after the commission receives notice of claims arising from a discharge prohibited by section 543. If either party shall fail to select its arbitrator within the said 10 days the other party shall request the American Arbitration Association to utilize its procedures for the selection of such arbitrator and the 2 arbitrators shall proceed to select the neutral arbitrator as provided in this section.

C. One Board of Arbitrators shall be established for and hear and determine all claims arising from or related to a common single discharge.

D. Hearings before Boards of Arbitrators shall be informal, and the rules of evidence prevailing in judicial proceedings shall not be binding. The board shall have the power to administer oaths and to require by subpoena the attendance and testimony of witnesses, the production of books, records and other evidence relative or pertinent to the issues represented to them for determination.

E. Determinations made by a majority of the board shall be final, and such determinations may be subject to review by a Justice of the Superior Court but only as to matters relating to abuse of discretion by the board.

F. Representation on the Board of Arbitration shall not be deemed an admission of liability for the discharge.

4. Funding.
A. Annual license fees shall be determined on the basis of 1/2 cent per barrel of oil, petroleum products or their by-products transferred by the applicant during the licensing period and shall be paid monthly on the basis of records certified to the commission. License fees shall be paid to the commission and upon receipt by it credited to the Maine Coastal Protection Fund.

B. Whenever the balance in the fund has reached the limit provided under this subchapter license fees shall be proportionately reduced to cover only administrative expenses and sums allocated to research and development.

5. Disbursements from fund. Moneys in the Maine Coastal Protection Fund shall be disbursed for the following purposes and no others:

A. Administrative expenses, personnel expenses and equipment costs of the commission related to the enforcement of this subchapter.

B. All costs involved in the abatement of pollution related to the discharge of oil, petroleum products and their by-products covered by this subchapter.

C. Sums allocated to research and development in accordance with this section.

D. Payment of 3rd party claims awarded in accordance with this section.

E. Payment of costs of arbitration and arbitrators.

F. Payment of costs of insurance by the State to extend or implement the benefits of the fund.

6. Reimbursements to Maine Coastal Protection Fund. The commission shall recover to the use of the fund all sums expended therefrom, including overdrafts, for the following purposes: provided that

recoveries resulting from damage due to an oil pollution disaster declared by the Governor pursuant to section 547 shall be apportioned between the Maine Coastal Protection Fund and the General Fund so as to repay the full costs to the General Fund of any bonds issued as a result of such disaster.

A. Costs incurred by the fund in the abatement of a prohibited discharge including 3rd party claims when the person permitting the same shall have failed to promptly report the discharge as required by rules and regulations of the commission, and such costs where the person permitting the prohibited discharge is not a licensee.

B. In the case of a licensee promptly reporting a discharge as required by this article, costs involved in the abatement of any single prohibited discharge including 3rd party claims in excess of $15,000, over and above payments received under any federal program.

C. Requests for reimbursement to the fund for the above costs if not paid within 30 days of demand shall be turned over to the Attorney General for collection.

7. Waiver of reimbursement. Upon petition of the person determined to be liable for reimbursement to the fund for abatement costs under subsection 6, the commission may, after hearing, waive the right to reimbursement to the fund if the commission finds that the occurrence was the result of any of the following:
A. An act of war.
B. An act of government, either State, Federal or municipal.
C. An act of God, which shall mean an unforeseeable act exclusively occasioned by the violence of nature without the interference of any human agency.

Upon such finding by the commission immediate credit therefore shall be entered for the party involved. The findings of the commission shall be conclusive as it is the legislative intent that waiver provided in this subsection is a privilege conferred not a right granted."

Also to be noted is that the Maine legislature has given the governor emergency power to deal with eminent or existent oil disasters. Such power would be too late if granted after the fact.

The Washington legislature recently gave special emergency powers to the governor in riot situations (RCW 43.06.210 and .220). It would seem appropriate for the same concern for public welfare to manifest itself when the potential harm in an oil disaster could mean the destruction of Puget Sound or perhaps some part of the Columbia River.

D. Further Comments on the Washington Statute

1. Uncertain Extent of Strict Liability

In addition to the problems created by the omission of funding, discussed in section II above, it should be noted that RCW 90.48.336 creates strict liability for damages if oil enters state waters in violation of section 320. But section 336 goes on to specifically relieve civil liability only when the immunities of subparagraph (3) of section 320 can be proved. One might conclude that omission of the immunities of subparagraphs (1) and (2) was intended and that the legislature intended to retain strict civil liability of polluters even when such polluters had obtained prior authorization under subparagraphs (1) and (2). The rationale for such a construction would be threefold:

a. Polluters would not be encouraged to exert wrongful influence on state officials to immunize themselves from the expected consequence of their enterprise.

b. When an emergency dictates authorization to discharge oil the public should not be robbed of its right to indemnification caused by an industrial emergency.

c. When authorizations are given, following proof of their environmental safety, the burden of monitoring their continued safety should be retained with the polluter.

Whether or not the above construction is adopted by the courts, however, the immunities created by the statute are inconsistent with and unnecessary to the statutory scheme of strict liability. The statute has created what amounts to an insurers' duty which imposes on the oil industry the obligation to indemnify the public for the expected hazards of their enterprise. The narrow immunities created by section 320 and 336 do not, however, preclude liability for the numerous other causes of oil spills which also do not involve negligence or fault. Some fund must be created to indemnify all the other faultless discharges, and the special immunities in all likelihood do not significantly diminish the total insurance cost to the industry.

If the benefits gained by the exceptions do not balance the problems created by the confusion, uncertainty, and litigation that results from the immunities, then the immunities should be removed.

The problem could easily be resolved by:

a. omitting the immunities created by section 336,and

b. establishing a publicly regulated fund that insures uniform distribution of economic risk, protects the small businessman, and assures continuing protection of the public interest (see discussion in section II above).

An alternate construction would absolve polluters from strict liability if any governmental approval was obtained and would make no distinction between subparagraph (3) immunities and subparagraph (2) immunities. This construction, however, seems to be at odds with the Environmental Policy Act of 1971.

Further examples of the uncertainty bred by section 336 are:

a. The definition of sabotage: RCW 9.05.060 describes criminal sabotage as a felony:

   9.05.060 Sabotage defined-Penalty. Whoever, with intent that his act shall, or with reason to believe that it may, injure, interfere with, or obstruct any agricultural, stockraising, lumbering, mining, quarrying, fishing, manufacturing, transportation, mercantile or building enterprise wherein persons are employed for wage, shall wilfully injure or destroy, or attempt or threaten to injure or destroy, any property whatsoever, or shall wilfully derange, or attempt or threaten to derange, any mechanism or appliance, shall be guilty of a felony.

   In State vs. McLennon 116 Wash. 612 (1921) the Washington court indicated that sabotage was not expressly defined by 9.05.060. The meaning therefore is uncertain and will in all likelihood remain so.

b. What constitutes governmental negligence is perhaps even more uncertain and will in the end be defined by the court as a factual question based on the merits of each case, or at the very least will be determined only after considerable litigation.

The courts also will be faced with question of whether the negligence immunized includes official action or discretionary decisions.

2. Collection and Containment after an Oil Spill.
Additionally it should be noted that the liability of members of the oil industry to take affirmative action under RCW 90.48.325 in order to contain existing oil spills is limited to persons owning or having control of the oil. However, there may be doubt as to the liability or statutory duty to collect oil in all cases because of the immunities created in RCW 98.48.320. The common law imposes no liability for inaction when there is no duty of care. Therefore, when there has been an oil spill an affirmative duty should be created in persons engaged in the oil business to take all reasonable action with the facilities at their disposal to collect and contain oil spills no matter who is at fault or liable. The risk of inaction by failure to collect or prevent the discharge of oil is clearly on the owner or persons having control over the oil; however, when liable persons are unable to control the oil spill the risks to the environment do not justify the retention of the archaic immunity created by the common law. This view would be consistent with the balance of the act which makes the owner/polluter strictly liable.

The problem created by the present statute is the lack of financial assurance that expenses incurred during collection will be indemnified. This problem must be met by further legislation. (See discussion in section B.)

E. Summary

The scope of RCW 90.48.320 through 90.48.360 is pervasive. The entry of oil into state waters is made unlawful per se. The scope of the legislative purpose under the act is amplified by the Environmental Policy Act of 1971. Inherent in the statutes is the purpose of preserving the environment. It must be concluded that the Director of the Department of Ecology will make rules under 90.48.343 and 350 to assure that all discharges of oil, from whatever source, will be subjected to scrutiny and proper regulation.

The oil spill problem, however, is merely peripheral compared to the problem of inherently wasteful levels of oil consumption which now seems near to creating an environmental disaster. Prudent conservation requires that the levels of oil consumption be reduced. The classic method of reducing consumption is through price increases. It is unlawful under the nation's anti-trust laws to allow an ologopoly to reap a monopoly profit. Therefore, it is essential that any fund created to indemnify oil spills be created by a tax. The alternative to a tax is the accumulation of a cash reserve by the industry from monopoly profits which will be used to insure itself. Such an industrial fund, however, will only extend the ologopolistic power of the oil industry and will give the same industry that has encouraged the dangerously wasteful levels of oil consumption in the past, the power of the purse strings over the most viable source of financing the repair of the damage done.

The single clear question remains for the legislature to face. The bold policies of the Environmental Policy Act of 1971 will not be implemented without funds. The oil industry will be expected to resist bitterly any attempt to restrain its present source of profit, but the Maine Statute could guide the adoption of a more forward looking approach to funding the regulation of industry.

Endnotes for Appendix 2

1. Washington Water Pollution Control Laws, (Chapter 90.48 RCW), Washington Department of Ecology.

2. Marine Law Review, #23, 1971.

APPENDIX 3

FEDERAL LAWS FOR OIL SPILL PREVENTION AND CONTROL

A. Introduction

Historically, federal oil pollution programs have been designed to support and stimulate the existing state and regional efforts, rather than to pre-empt the field. There were two principal reasons for this approach. First, local control of water pollution was thought to be more efficient than national control. Second, Congress had been strongly influenced by the arguments of those who contended that water pollution is a uniquely local problem and that the federal government should not intrude upon state prerogatives with regard to it.

This approach has been largely abandoned, however, and has been replaced by more active involvement by federal agencies, most notably the Coast Guard.[1] Greater federal leadership is even now being discussed in Congress,[2] due mainly to glaring gaps in both state and federal programs revealed by recent oil spill incidents.[3] These incidents, and their consequences, are beginning to be viewed in a more national perspective.[4]

State authority to control and regulate certain aspects of oil spills still remains - shared concurrently with the federal government. United States Attorneys have the ability to defer action or to refrain from acting, under applicable federal laws, if they deem state action adequate to protect the federal interest. The federal government has indicated, however, that it will be prepared for any situation in which the state involved is either slow to act, or fails to act.[5]

B. Federal Laws

1. Water Quality Improvement Act of 1970

   The primary federal law governing oil spills is the Water Quality Improvement Act of 1970.[6] Through this Act, Congress has declared it to be national policy that "...there should be no discharges of oil into or upon the navigable waters of the United States, adjoining shorelines, or into or upon the waters of the contiguous zone."[7] Discharges in "harmful quantities" are prohibited in these areas.[8]

To the President is given the task of determining "those quantities of oil the discharge of which, at such times, locations, circumstances and conditions, will be harmful to the public health or welfare of the United States..."[9] In accordance with this delegation, the President has determined that discharges which "(a) Violate applicable water quality standards, or (b) Cause a film or sheen upon or discoloration of the surface of the water or adjoining shorelines or cause a sludge or emulsion to be deposited beneath the surface of the water or upon adjoining shorelines" are "harmful" within the meaning of the Act.[10]

Certain discharges are expressly exempted from coverage. These include "discharges into the waters of the contiguous zone...permitted under Article IV of the International Convention for the Prevention of Pollution of the Sea by Oil, 1954 as amended" and discharges "when permitted in quantities and at times and location or under circumstances or conditions as the President may, by regulation, determine not to be harmful."[11] Article IV of the Convention permits, among other things, discharges from ships for the purpose of securing the safety of a ship, preventing damage to a ship or cargo, or saving life at sea, and discharges resulting from damage to a ship or unavoidable leakage, providing all reasonable precautions have been taken to minimize escape.[12] The President has determined that discharges from a "properly functioning vessel engine" are not harmful.[13]

Various penalties are provided in the Act. Owners or operators of vessels from which oil is discharged in violation of the Act are liable for the costs incurred by the government in cleaning it up, to a limit of $100 per gross ton of the vessel or $14,000,000, whichever is less.[14] Where the United States can show that the discharge was the result of willful negligence or willful misconduct, liability is unlimited.[15] Liability may be avoided only by proving that the discharge was the result of "(A) an act of God, (B) an act of war, (C) negligence on the part of the United States Government, or (D) an act or omission of a third party without regard to whether any such act or omission was or was not negligent, or any combination of the foregoing causes."[16] The Act also provides for civil penalties for knowingly discharging, and for criminal penalties for failure to report a spill to the appropriate federal agency.[17] Finally, for certain vessels over 300 gross tons, the act requires evidence of financial responsibility sufficient to meet the liability which might be imposed under the Act.

2. Rivers and Harbors Act of 1899

The federal oil pollution abatement program is primarily based on the provisions of the Water Quality Improvement Act of 1970. As an adjunct to this law, the federal government has relied on the Rivers and Harbors Act of 1899,[18] and, in particular, Section 407 of this Act, commonly referred to as the Refuse Act. It is used as a supplement for accidental or continuous discharge of oil. The Act was originally designed to ensure navigability of the nation's waterways, and was not formulated to cope with modern forms of pollution. The Act provides for the allowable deposit of refuse in navigable waters. Penalties and enforcement are provided for in Sections 411 and 413, respectively.

The only permissible deposits of refuse are those from streets and sewers, flowing therefrom in a liquid state, those connected with improvement of navigable waters or construction of public works, and those permitted by the Secretary of the Army.[19] Oil has been interpreted to be within the meaning of the term "refuse."[20] The Act provides criminal penalties for violations.[21]

3. Oil Pollution Act, 1961

This Act was passed by Congress to implement the provisions of the International Convention for the Prevention of Pollution of the Sea by Oil, 1954, as amended. The significant portions of that Treaty are analyzed in Appendix 4.

4. Federal Water Pollution Control Act

The procedures to be followed in the federal enforcement process are specifically delineated in the Federal Water Pollution Control Act,[22] as amended: (1) a conference; (2) a public hearing; (3) court action.

Where pollution emanating from sources in one state endangers the health or welfare of persons in another state, initiation of the enforcement process is mandatory upon the request of a state governor and/or an official state water pollution control agency (in Washington, the Department of Ecology), or a municipality in whose request the state agency and governor concur.[23] It is similarly mandatory in intrastate pollution situations upon the request of the governor of the state concerned, when the effect of the pollution on the legitimate uses of the waters is judged sufficient by the Administrator of the EPA to warrant federal action.[24] The exercise of federal jurisdiction to abate interstate pollution is required when reports, surveys, or studies indicate that such pollution is occurring.[25] The Administrator may also call an enforcement conference where pollution of interstate or navigable waters has damaged shellfish so that substantial economic injury has resulted from the inability to market shellfish products in interstate commerce.[26] This includes the inability to market shellfish in interstate commerce where pollution occurs in interstate waters.

The Administrator of the EPA is also empowered to abate pollution where water quality in interstate waters or portions thereof is not in compliance with standards established under the Water Quality Act of 1965.[27] That is, once the states have adopted water quality standards and the Administrator has found them acceptable as federal standards, a violation of the standards is subject to federal enforcement action. States have the initial responsibility for taking action if standards are violated. If the stated do not act, the U.S. Attorney General, at the request of the Administrator of the EPA, may initiate an enforcement action.[28]

5. National Contingency Plan for Oil and Hazardous Materials Pollution

Exemplifying the increased involvement of the federal government in treatment of oil spills is the formulation of the National Contingency Plan for Oil and Hazardous Materials Pollution.[29] The purpose of this plan - required by the Water Quality Improvement Act of 1970 - is to provide for efficient, coordinated, and effective action to minimize damage from oil discharges, including containment, dispersal, and removal of oil.[30] The plan includes "(a) assignment of duties and responsibilities, (b) identification, procurement, maintenance, and storage of equipment and supplies, (c) establishment of a strike force and emergency task forces, (d) a system of surveillance and notice, (e) establishment of a national center to coordinate identifying, containing, dispersing, and removing oil, and (f) a schedule identifying dispersants and other chemicals that may be used in carrying out the plan, the waters in which they may be used, and quantities which may be safely used."[31] The plan will be revised from time to time.

The National Contingency Plan concentrates on providing a pattern of coordinated and integrated responses to spills by agencies of the federal government, and on promoting coordination of federal, state, and local response systems. More detailed planning for actual spills is carried out in regional contingency plans. The National Plan contains guidelines for the establishment of regional contingency plans and regional response teams. The regional plan for the Puget Sound area is analyzed in Appendix 7.

C. Breakdown of Federal Agency Responsibilities and Powers under Federal Laws Relating to Oil Spills

The primary federal agencies in this area are the Department of Transportation/Coast Guard and the Environmental Protection Agency. Other federal agencies assigned duties and powers with respect to oil spills are the Department of the Interior, the Federal Maritime Commission, and the Department of Defense.

1. Coast Guard

Under the Water Quality Improvement Act of 1970, the Secretary of Transportation is charged with assessing and compromising civil penalties for knowing discharges in violation of the Act;[32] issuing rules and regulations concerning cleanup, development and implementation of local and regional contingency plans, prevention of discharges, and inspection of vessels and cargoes;[33] removing and eliminating pollution hazards, and removing and destroying vessels;[34] removing or arranging for the removal of oil discharged into or upon navigable waters, adjoining shorelines, or the contiguous zone;[35] assessing civil penalties for violations of DOT regulations made pursuant to the Act;[36] and boarding and inspecting vessels, as well as arresting for violations of Section 11 of the Federal Water Pollution Control Act, as amended.[37] The Coast Guard has been designated the "appropriate agency" to receive notification of discharges of oil as required by the Act.[38]

Under the Oil Pollution Act of 1961, as amended,[39] which implements the International Convention for the Prevention of the Pollution of the Sea by Oil, as amended, the Secretary of DOT is charged with issuing regulations governing the reporting of reasons for certain discharges;[40] issuing regulations governing ship fittings and equipment and operating requirements under Article VII of the Convention;[41] providing Oil Record Books for entry of information required to be kept by the Convention, and issuing regulations governing entry of such information, keeping of the books, and their surrender;[42] issuing regulations concerning, and publishing notice of, extensions or reductions in the zones within which discharges are prohibited under the Convention;[43] issuing regulations for the administration of Sections 1002, 1003, 1004, 1007 (a), 1008, and 1011 of the Oil Pollution Act of 1961, as amended;[44] issuing regulations empowering persons to board vessels and require production of certain records;[45] boarding, inspecting, and arresting for the enforcement of Sections 1001-1011 of the Act;[46] and, subject to the provisions of 46 U.S.C.A. 239, suspending or revoking masters' and officers' licenses for violations of the Act.[47]

Other powers and duties of the Secretary of Transportation include establishing anchorage grounds for vessels, and adopting and enforcing rules and regulations relating thereto;[48] establishing, maintaining, and operating aids to navigation;[49] marking obstructions;[50] saving life and property;[51] generally enforcing all laws of the United States in waters over which the United States has jurisdiction, and on the high seas;[52] and making and enforcing regulations concerning artificial islands and fixed structures erected on the outer continental shelf.[53]

2. Environmental Protection Agency
Under the Water Quality Improvement Act of 1970, the Administrator of the EPA has the responsibility of determining, in consultation with the Secretary of the Interior, the quantities of oil, the discharge of which will be harmful;[54] of determining which dispersants and other chemicals, if any, may be used in carrying out the National Contingency Plan;[55] of determining when there is a substantial and imminent threat due to actual or threatened oil discharges, and securing such relief as necessary to abate the threat;[56] and, in consultation with the Secretary of Transportation, issuing regulations establishing procedures, methods, and requirements for preventing discharges from vessels and facilities.[57]

   The Administrator also has the enforcement powers outlined above under the Water Pollution Control Act and the Water Quality Act of 1965.

3. Department of Interior
The Office of Water Resources Research, Department of the Interior, is engaged in research, experiments, and demonstrations relating to prevention and cleanup of oil spills.[58]

4. Federal Maritime Commission
The Federal Maritime Commission has the responsibility, under the Water Quality Improvement Act of 1970, of issuing regulations under and administering the financial responsibility provisions of the Act.[59]

5. Council on Environmental Quality
The Council on Environmental Quality has the responsibility to prepare, publish, revise, or amend the National Contingency Plan for Oil and Hazardous Materials Pollution.[60]

6. Department of Defense

The Army Corps of Engineers enforces regulations regarding anchorage grounds, under the direction of the Secretary of Transportation, in ports and places where there are no Coast Guard vessels available;[61] advises the Secretary of the Army regarding permits to deposit refuse in navigable waters;[62] and,arrests certain violators of Section 407 of the Refuse Act of 1899.[63]

D. Endnotes for Appendix 3

1. 33 C.F.R. 1-199. Pages 587-599 list the yearly revisions to this Title since 1964.

2. 116 Congressional Record H. 4571 (H. Doc. 91-340). Presidential Special Message to Congress of May 20, 1970.

3. Anacortes Oil Spill Incident, April 26, 1971.

4. National Contingency Plan for Oil and Hazardous Materials Pollution, Federal Register, Vol. 35, No. 106 (1970).

5. Public Law 91--224, Sec. 11 (c) (1); 14 U.S.C.A. 141; 33 U.S.C.A. 1161 (e); National Contingency Plan, 201.2

6. Public Law 91-224, 33 U.S.C.A. 1161 (1970).

7. Public Law 91-224, sec. 11 (b) (1).

8. Public Law 91-224, sec 11 (b) (2).

9. Public Law 91-224, sec. 11 (b) (3).

10. 18 C.F.R. 610.3 (a) & (b).

11. Public Law 91-224, sec 11 (b) (2).

12. Amendment to the International Convention for the Prevention of Pollution of the Sea by Oil, 1954, 17 UST 1523; TIAS 6109 (May 18, 1967).

13. 18 C.F.R. 610.6.

14. Public Law 91-224, sec. 11 (f) (1).

15. Id.

16. Id.

17. Public Law 91-224, secs. 11 (b) (4) & (b) (5).

18. 33 U.S.C. 401-413.

19. 33 U.S.C. 407.

20. United States v. Standard Oil Co., 384 U.S. 224 (1966).

21. 33 U.S.C.A. 1160.

22. 33 U.S.C.A. 1160

23. 33 U.S.C.A. 1160 (d) (1).

24. Id.

25. Id.

26. Id.

27. 33 U.S.C.A. 1160 (c) (5).

28. 33 U.S.C.A. 1160 (g) (1) & (2)/

29. FederalRegister, Vol. 35, No. 106, Tuesday, June 2, 1970.

30, National Contingency Plan, sec. 101.1.

31. Id.

32. Public Law 91-224, sec 11 (b) (5).

33. Public Law 91-224, sec 11 (j) (1) (A)-(D).

34. Public Law 91-224, sec. 11 (d).

35. Public Law 91-224, sec. 11 (c) (1).

36. Public Law 91-224, sec 11 (j) (2).

37. Public Law 91-224, sec 11 (m).

38. Public Law 91-224, sec. 11 (b) (4); Executive Order No. 11548, sec. 6 (1970).

39. 33 U.S.C.A. 1001-1015.

40. 33 U.S.C.A. 1002 (c).

41. 33 U.S.C.A. 1007 (b).

42. 33 U.S.C.A. 1008.

43. 33 U.S.C.A. 1011 (a) & (b).

44. 33 U.S.C.A. 1009.

45. 33 U.S.C.A. 1010 (a).

46. 33 U.S.C.A. 1007 (a).

47. 33 U.S.C.A. 1006.

48. 33 U.S.C.A. 471.

49. 14 U.S.C.A. 81.

50. 14 U.S.C.A. 86

51. 14 U.S.C.A. 88.

52. 14 U.S.C.A. 89

53. 43 U.S.C.A. 1333 (e) (1) & (2).

54. Public Law 91-224, sec 11 (b) (2) & (b) (3).

55. Public Law 91-224, sec 11 (c) (2) (G).

56. Public Law 91-224, sec. 11 (e).

57. Public Law 91-224, sec. 11 (j) (1) (C).

58. 33 U.S.C.A. 1155 (i).

59. Public Law 91-224, sec 11(p) (1) & (p) (2).

60. Public Law 91-224, sec. 11 (c) (2).

61. 33 U.S.C.A. 471.

62. 33 U.S.C.A. 407.

63. 33 U.S.C.A. 413.

APPENDIX 4

INTERNATIONAL LAWS RELATED TO PREVENTION AND CONTROL OF OIL SPILLS

A. Introduction

The international community has only recently begun to regulate the marine movement of oil on a scale appropriate to the risks posed by worldwide transport of enormous quantities of oil. The regulatory problem can be divided into three topical areas: Prevention, damage minimization, and liability assessment. The last area has received the most attention to date, but none of the areas are yet satisfactorily regulated. Regulatory approaches can also be divided into three segments: Multinational, involving treaties, declarations, or conventions supported by a significant number of the world's nations; regional, wherein concerned countries attempt to solve problems associated with a particular body of water, or unilateral, where one country proclaims regulations it intends to enforce near its coasts. Most recent oceanic pollution controls have been multinational. However, the most comprehensive set of controls so far promulgated is unilateral. Unilateral regulation is the method least approved by the international community. Multinational and unilateral controls are the topics of this Appendix.

The following excerpt, which delineates the problem faced by the world community was written in 1968, and the statistics cited therein are outmoded. As of January 1, 1970, 63 tankers of over 200,000 tons D.W.T. capacity were operating on the world oceans; 198 such tankers were on order.[1]

> For the present and foreseeable future there is no adequate substitute for the world-wide demands for crude oil to fuel industrial society, and due to formidable natural or marine barriers, there are many important industrial states which cannot be served by pipelines. Ocean transport of petroleum products is largely a one-way trade from the principal supply areas in the Middle East, Southeast Asia and the Caribbean, to Western Europe, Japan and North America. This vital resource is carried in the busy coastal shipping lanes along the United States' east coast north of Hatteras, the English Channel, the Caribbean, and the Sea of Japan - all major tourist areas. In many instances the shipping lanes are outside the territorial limits of any state, thereby creating international law problems to add to the domestic conflict. To transport these cargoes the industry is presently using many vessels of Second World War vintage with carrying capacities of about 20,000 Dead Weight Tons (D.W.T.). However, vessels of 100,000 D.W.T. are

> common; vessels of 200,000 D.W.T. and 300,000 D.W.T. are being constructed, and it may be possible to build vessels of 450,000 to 500,000 D.W.T. in the near future. Thus, the possibility grows that a spill from one of these giant tankers might impair the maritime resources of several nations.
>
> The dangers from oil pollution are: the destruction of fish, shellfish, sea birds, fishing gear or beach installations; the creation of fire hazards in ports; the fouling of small boats; and, the loss of natural beauty with resulting financial losses to resort owners and the dependent tourist industry. [2]

The problem of actual and potential oil pollution of the Puget Sound region is a microcosm of the general situation described above. No substitute power source to replace crude oil is presently available, and it is presently thought impractical to supply all Puget Sound oil through pipelines. Many relatively small tankers presently supply the region with oil. Future plans call for large tankers to bring Alaskan crude oil from Valdez through the sometimes treacherous Gulf of Alaska and the waters of the Northeast Pacific off the coast of Canada into the Strait of Juan de Fuca and Puget Sound. The possibility of spills will increase with the volume of traffic; the potential damage from spills will increase with the capacity of tankers.

B. Multinational Regulations

That even the most recent international conventions dealing with aspects of oceanic oil pollution do not provide entirely adequate remedies is due at least in part to the continuing division between coastal states and maritime states as to proper goals in regulating sea uses. Coastal states, interested in protecting fisheries and near-shore environments, generally favor strict controls on the transport of oil and strict liability for the spiller. Maritime states seek to insure the broadest possible right of passage for commercial and military vessels of all types. It is important to note that maritime states have on their side the traditional rule of freedom of navigation; to impose preventive measures on oil shipments, as coastal interests advocate, is to change one of the most basic tenets of the law of the sea. However, the "worldwide" tradition of freedom of navigation was a rule devised exclusively by maritime states in earlier days when maritime interests were the only interests concerned with oceanic events. Leading maritime states are now becoming aware that they too have coastal interests to protect;[3] consequently, there is now movement, gradual but visible, toward responsible governance of oil transport. The two special interest groups now most opposed to regulation are the shippers themselves, who are interested in incurring as few expenses as possible, and the great powers' navies (especially the United States and Russian), which want to preserve the freedom of vessels to move on the seas.

1. Geneva Conventions

Article 24 of the 1958 Geneva Convention on the High Seas[4] is an elemental statement of national responsibility as viewed by the world community:

> Every State shall draw up regulations to prevent pollution of the seas by the discharge of oil from ships or pipelines or resulting from the exploitation and exploration of the seabed and its subsoil, taking account of existing treaty provisions on the subject. [5]

Another relevant statement is contained in Article 1, paragraph 2 of the companion 1958 Geneva Convention on Fishing and Conservation of the Living Resources of the High Seas:[6]

> 2. All States have the duty to adopt or to co-operate with other States in adopting, such measures for their respective nationals as may be necessary for the conservation of the living resources of the high seas. [7]

Neither article sets down specific international community standards for pollution prevention. Neither article discusses the possibility that to achieve the goals set out, States might have to abridge other international rights, such as the afore-mentioned freedom of navigation. The articles amount to a declaration by the international community that it was not yet able to assume primary responsibility for pollution prevention, that a potential source of pollution could only be regulated by the nation otherwise governing it. The Geneva Conventions are normally the basic source of international law of the sea, but in controlling oil pollution they do not offer much help.

2. International Convention for Prevention 1954

The first multinational convention intended specifically to deal with the problem of oceanic oil pollution was the International Convention for the Prevention of Pollution of the Sea by Oil, 1954,[8] which was significantly amended in 1962 and again in 1969.[9] The 1954 Convention was adopted to cope with hazards caused by tank cleanings and other intentional discharges of oil. Massive spills caused by such events as tanker founderings and collisions were not thought to be a major problem in 1954.

The 1954 Convention required that some ships install separators for oily water, and encouraged all ships to install such devices. The Convention established zones, generally of 50 miles' width, off the coast of the world; discharges were prohibited in these zones. Discharge in the zones was not considered a violation if it was necessitated by an emergency or if reasonable precautions were taken to minimize the effects of a discharge which resulted from damage to the ship or from an unavoidable leak.[10]

The effect of the 1962 amendments was to extend the width of prohibited zones in certain especially vulnerable or congested coastal regions from 50 to 100 or 150 miles, and to extend the Convention's application to some ships previously exempted.[11]

The following paragraphs examine a few of the key provisions of the convention as it is presently constituted.

> Article III - Subject to the provisions of Articles IV and V:
>
> (a) the discharge from a ship to which the present Convention applies, other than a tanker, of oil or oily mixture shall be prohibited except when the following conditions are all satisfied:
>
> (i) the ship is proceeding en route;
> (ii) the instantaneous rate of discharge of oil content does not exceed 60 litres per mile;
> (iii) the oil content of the discharge is less than 100 parts per 1,000,000 parts of the mixture;
> (iv) the discharge is made as far as practicable from land;
>
> (b) the discharge from a tanker to which the present Convention applies of oil or oily mixture shall be prohibited except when the following conditions are all satisfied:
>
> (i) the tanker is proceeding en route;
> (ii) the instantaneous rate of discharge of oil content does not exceed 60 litres per mile;

(iii) the total quantity of oil discharged on a ballast voyage does not exceed 1/15,000 of the total cargo-carrying capacity;

(iv) the tanker is more than 50 miles from the nearest land;

(c) the provisions of sub-paragraph (b) of this Article shall not apply to:

(i) the discharge of ballast from a cargo tank which, since the cargo was last carried therein, has been so cleaned that any effluent therefrom, if it were discharged from a stationary tanker into clean calm water on a clear day, would produce no visible traces of oil on the surface of the water; or

(ii) the discharge of oil or oily mixture from machinery space bilges, which shall be governed by the provisions of sub-paragraph (a) of this Article.

Article III is largely self-explanatory. It does not prohibit the discharge of oil on the seas. It does set specific standards by which a violation may be judged. The standard set forth in sub-paragraph (b)(iii) seems unnecessarily loose in the light of the carrying capacities of supertankers. It would permit a 300,000 ton D.W.T. tanker to discharge 20 tons of oil at a time (less continuous discharges) in the course of ballast flushing.

Article IV exempts discharges in the course of saving the ship, preventing damage to ship or cargo, or saving life at sea. Article V stays article III's application to bilge discharges for one year from the time a party ratifies the convention. Article VI, detailing penalties, is unchanged from the form described supra, p. 5.

Article VII requires ships to have devices inhibiting the escape of oil into bilges, or to provide means which prevent illegal discharges; article VII also suggests that "carrying water ballast in oil fuel tanks shall be avoided if possible."[12] Contracting parties are required to furnish appropriate facilites for accepting residues at ports, terminals, and repair yards.[13]

The International Convention for the Prevention of Pollution of the Sea by Oil deals generally with the day-to-day operating discharges which actually account for most oceanic oil pollution rather than with large, dramatic, accidental spills such as those of the Torrey Canyon or the Ocean Eagle. Its provisions apply only to certain ships of ratifying states. It does not provide for community punishment of violators. Yet it does set specific standards for violations and does require the installation of pollution-inhibiting equipment. It is also a convention whose coverage has twice been expanded since its inception. Though the Convention does not yet prevent discharge, it is evidence of increasing community responsibility in coping with the most serious cumulative source of oceanic oil pollution.[14]

3. Intergovernmental Maritime Consultative Organization (IMCO). The impetus to the 1969 Brussels Conference[15] which resulted in IMCO conventions' recognizing coastal states' rights to intervene in certain instances involving oil spills and also declaring certain measures of liability for damage from oil pollution was the Torrey Canyon disaster.[16] Shortly after the Torrey Canyon wreck, the British government sent a note to the IMCO Council suggesting revisions in the law of the sea aimed at lessening the likelihood of future catastrophic spills and also defining clear lines of legal liability for damages. Specific proposals included designating mandatory sea lanes, requiring additional navigation aids to be placed aboard ships, establishing shore radio control of tankers moving in coastal zones, setting tanker speed limits in areas near land, restricting the use of automatic pilots, and providing for design controls, equipment testing, route markings and crew training. Additional proposals included compulsory insurance, recovery of expenses incurred by parties fighting pollution, and strict liability. IMCO promoted a series of studies and conferences which culminated in the adoption of the 1969 Brussesl Conventions.

4. Convention Relating Intervention to Casualties . The Preamble to the International Convention Relating to Intervention on the High Seas in Cases of Oil Pollution Casualties[17] claimed that

the contracting nations agreed to the Convention because they were aware of the need to protect coastal interests from oil pollution damage; the Preamble also indicated that though the dangers presented by marine oil pollution demanded exceptional preventive measures, the rules the Convention set forth to meet the demand would not offend traditional freedom of the seas.[18]

Article I of the Convention reads:

> 1. Parties to the present Convention may take such measures on the high seas as may be necessary to prevent, mitigate or eliminate grave and imminent danger to their coastline or related interests from pollution or threat of pollution of the sea by oil, following upon a maritime casualty or acts related to such a casualty, which may reasonably be expected to result in major harmful consequences.
>
> 2. However, no measures shall be taken under the present Convention against any warship or other ship owned or operated by a State and used, for the time being, only on government non-commercial service.

Note that the article makes Convention provisions applicable only "following upon" or _after_ an accident has occurred, and even then only when "grave and imminent danger" to coastal interests becomes apparent. The Convention does not empower coastal states to take actions intended to _prevent_ maritime casualties from occurring in waters off their shores; under the Convention, action can be taken only in response to a threat posed after a ship has been incapacitated.

No oil has to be leaking from a distressed ship before a coastal state can justify taking preventive actions (which presumably might include burning the oil slick or sinking the tanker) but there must be evidence of danger to the coasts. Evidence must consist not merely of the existence of a large amount of oil aboard a ship in trouble, but, at a minimum, of indication that the oil is imminently likely to escape the ship and spread toward the coast in a harmful manner. When such circumstances can be shown the Convention permits the instigation of protective actions.

It can be argued that the right to take protective actions as permitted in Convention article I has been long recognized in international law as the general concept of a national right to self-preservation[19] and that the Convention is not necessary to the right's exercise. One of the most frequently mentioned reasons for Britain's initial reluctance to destroy oil flowing from the Torrey Canyon was that, as a major maritime nation, it did not want to appear to be exercising or supporting broad coastal state rights against ship movement.[20] That the responses which the Convention sanctions are quite restricted is attributable to maritime nations' unwillingness to recognize a greater protective right in coastal states.

The Convention is similar to the International Convention for the Prevention of Pollution of the Sea by Oil in that its provisions are not extended to naval vessels. (Provisions of the instant Convention are not extended to any non-commercial government vessels.)

Further inhibitions are placed on the coastal state's ability to respond to spills by the clauses of article III. In coping with what might be described as an "ordinary" oil pollution threat, the coastal state must comply with subsections (a) and (b) before it can take substantive action: It must notify other interested states; it must inform such people as vessel owners and cargo owners and must take their opinions into account in determining the course of its response. Only where "extreme urgency" exists may the coastal state act under the Convention to prevent damage without having held prior consultations with other nations and individuals. It is in the coastal state's interest to move rapidly to minimize damage; however, the coastal state who acts without prior notification, but cannot thereafter demonstrate the situation's extreme urgency may be held liable for damage caused by its preventive acts to both ship and cargo.

Article VI contains the direct statement that a state which takes measures contrary to those stated in the Convention must pay compensation for damages caused by its excesses. Article VII provides that Convention provisions are to be strictly and narrowly construed whenever they come into conflict with recognized rights of other states or individuals. This rule is another case where deference was paid to "freedom of the sea." Intrusions by Convention provisions into traditional maritime law are countenanced only so long as they are minimized.

As noted above, a coastal state cannot act under the Convention to prevent spills from occurring. The Convention does not permit a coastal state to act in anticipation of danger from an existing casualty. When conditions sufficiently threaten for the coastal act, it must not act independent of advice from those states and persons who control the offending ship. There is argument that the coastal state's right to act in self-preservation exists independent of the Convention. The Convention may accurately describe the interests of the parties to a spill, but it deals incompletely with intervention and not at all with prevention. The Convention leaves the problem of coping with the substantive problems of massive spills largely unanswered.

5. Liability for Oil Pollution Damage

The International Convention on Civil Liability for Oil Pollution Damage[21] was created to establish uniform international rules for compensating persons damaged by marine oil pollution.[22] The Convention, which applies only to damage occurring on the territory or territorial sea of a contracting party,[23] holds the owner of a polluting ship liable for damage caused by escaping or discharged oil from any vessel[24] which carries oil as a cargo.

A shipowner can be held liable under the Convention up to the lesser figure of $134.40 per adjusted net ton or $14,000,000,[25] unless the owner is either at fault or at privity with the person at fault, in which case liability is unlimited.[26] The owner may establish a fund equal to the limit of his liability absent fault.[27] If he does so, no claims for damage can be made against his other assets as long as he is entitled to the liability limitation.[28] Claims must be brought against the shipowner within three years after damage occurs or within six years after the damage-causing incident.[29] The latter provision sets an outer limit on claims which can be brought against owners of long-sunken, leaking ships. The owner of any vessel which carries more than 2,000 tons of bulk oil must provide insurance or a substitute to the limit of his non-fault liability.[30]

The Convention provides considerable compensation for damage from almost all spills as well as unlimited compensation for some spills. The circumstances which characterize liability are fully described, and the procedures for adjudication and claim enforcement are satisfactory. There has been much informed speculation that the exceptions to strict liability and unlimited compensation included in the Convention were written in at the behest of insurance companies who were unwilling to guarantee against all possible spill causes and/or to issue policies much in excess of $14,000,000.[31]

6. Tanker Owners Voluntary Agreement

Certain of the world's tanker owners have banded together in a voluntary agreement which they have dubbed with the acronym TOVALOP (Tanker Owners Voluntary Agreement Concerning Liability for Oil Pollution).[32] TOVALOP would by itself leave tanker owners liable for fewer spills and for a lower total dollar amount than would the 1969 Liability Convention. Principal features of the TOVALOP agreement include: Reimbursement to national governments for reasonable cleanup expenses; a rebuttable presumption that any discharge is negligent, and agreement of the owners to pay governmental cleanup costs to the lesser figure of $100 per gross registered ton or $10,000,000; compensation is only for cleanup of land adjoining tanker-traversed waters, and no compensation is offered for actual damage to marine or other life, to the environment, or to property.

C. Unilateral Controls

Perhaps the diplomatically most explosive and pragmatically most comprehensive set of oceanic oil pollution controls was announced by the Canadian government on April 8, 1970. The Arctic Waters Pollution Prevention

Act[33] claims jurisdiction for Canada in all waters within 100 miles of any Canadian land north of 60° North latitude; jurisdiction is not claimed generally, but rather for the particular purpose of guarding the coastal and marine resources of the Arctic against pollution.

The Canadian legislation was prompted by the voyages of the tanker Manhattan along the proposed route from the Alaskan oil fields at Prudhoe Bay through Arctic waters north of the Canadian mainland to the East coast of the United States. More basic Canadian motivations for asserting pollution control jurisdiction in Arctic waters are described by Prof. Richard Bilder:

> For one thing, the hazards of Arctic navigation substantially increase the risk of maritime accidents. Moreover, the peculiar ecology of the Arctic region - an environment in which life exists only precariously - coupled with the slow rate of hydrocarbon decomposition in a frigid climate and the difficulty of dispelling oil in Arctic areas, might cause any major oil spill to have disastrous and irreversible ecological consequences. Indeed, under these conditions, even small amounts of oil pollution could be extremely damaging. Thus it is not surprising that Canada takes the view that it has both a special interest in its Arctic environment and a particular responsibility to protect it from the threat of oil pollution.[34]

The Act prohibits pollution in the affected region from any source, except as specifically approved through other of the Act's provisions.[35] Ships in distress which are depositing waste or which are likely to deposit waste may be removed or destroyed.[36] The definition of waste stated in §2(h) is strict and comprehensive:

> (h) "waste" means
>
> (i) any substance that, if added to any waters, would degrade or alter or form part of a process of degradation or alteration of the quality of those waters to an extent that is detrimental to their use by man or by any animal, fish or plant that is useful to man, and
>
> (ii) any water that contains a substance in such a quantity or concentration, or that has been so treated, processed or changed, by heat or other means, from a natural state that it would, if added to any waters, degrade or alter or form part of a process of degradation or alteration of the quality of those waters to an extent that is detrimental to their use by man or by any animal, fish or plant that is useful to man,

and without limiting the generality of the foregoing, includes anything that, for the purposes of the Canada Water Act, is deemed to be waste."

"Shipping safety control zones," in which navigational regulations are made mandatory for any entering ship may be declared in any area of the Arctic.[37] Regulations may pertain to navigational aids, safety equipment, personnel qualifications, hull and fuel tank construction standards, passage time and routes, cargo limitations, mandatory pilots, and other specific items.[38] Zones may be closed to ships when conditions so warrant.[39]

"Pollution control officers" who can board and inspect ships as well as stay the movement of ships suspected of regulation non-compliance, are given their authority by the Act.[40] Fines to be levied against depositors of illegal waste are set at maximums of $5,000 for individuals and $100,000 for ships.[41] Failure to obey regulations in general can bring a $25,000 fine.[42] Ships and cargo suspected of violating regulations may be seized.[43]

Illegal depositors of waste are held absolutely liable for all costs and expenses incurred in prosecuting violators and cleaning up waste, and are likewise absolutely liable for all actual losses and damages.[44] The firmness of this regulation may be softened by taking into account a ship's size and the nature of its cargo.[45] A ship may be required to prove financial responsibility before it is permitted to enter affected waters.[46]

If the goal of a regulatory system is to preven marine oil pollution in a given region, the Canadian legislation can be regarded as an ideal prototype. The legislation contains a wealth of specific provisions designed to facilitate that movement of ships through the particular region which would most likely minimize the threat of damage from oil spills. Navigational aids, ship construction, route designations and many other factors equally important in promoting shipboard safety were included. If a similar regulatory system were to be constructed for a part of the world where the environment was less harsh and ecology was less fragile, the severity which characterizes the Canadian regulations could be eased considerably without sacrificing a significant margin of safety.

The Canadian legislation is a model of proper methodology in pollution control, but it is also anathema to leading maritime nations. Two reasons account for its unpopularity in the policy centers of such nations as the United States, which has vigorously protested the legislation as without foundation in international law. First, the Canadians have redefined and restricted traditional freedom of the seas. Even if it is assumed that Canada has some territorial claim to the waters, the international community has in the past held that such waters were subject to the "right of innocent passage" by vessels of foreign countries. Innocent passage was thought to include any vessel which did not pose an active threat to the coastal state. What Canada has said is that there are some circumstances which it has defined under which no passage would be treated as innocent because of the potential threat contained in prospective passage. The Canadian position is not unjustified, given the region's circumstances.

The second feature of the Canadian legislation objected to by maritime countries is that the action was taken unilaterally, and the problem should have been attacked and solved by the world community. This is another contention actively pressed by the United States, which fears that other nations may use the Canadian legislation as a pretext for more general jurisdictional extensions.[47] Canada took the action after the second test voyage of the *Manhattan* and after completion of the 1969 Brussels Conference, which produced conventions woefully inadequate for the problems Canada faced in the Arctic.

D. Importance to Puget Sound

The international conventions, none of which deals adequately with the problem it attacks, pertain to Puget Sound insofar as the Sound may be affected by spills occurring outside zones subject to U.S. national regulation.[48]

The Canadian legislation is important to Puget Sound in at least two ways: first, it could serve as a generalized model for solutions to the problems associated with controlling tanker movements in Puget Sound; second, its passage and future enforcement mean that a larger percentage of Alaska crude will find its way into the Puget Sound market.

E. Endnotes for Appendix 4

1. "Large Tankers - Jan. 1970 and the Year 1969" Fearnley and Egers Chartering Co. Ltd. Radhusgt, 23 - Oslo 1, Norway p4, 11.

2. Sweeney, J.C. "Oil Pollution of the Oceans" 37 Fordham L. Rev. 156 (1968) (footnotes omitted)

3. Neuman, R.H., "Oil on Troubled Waters: The International Control of Marine Pollution," 2 J. of Maritime Law and Commerce 350-351, (1970)

4. "Convention on the High Seas" done at Geneva, April 29, 1958, (1962) 2 U.S.T. 2312, T.I.A.S. No 5200, 450 U.N.T.S. 82.

5. Art. 24

6. "Convention on Fishing and Conservation of the Living Resources of the High Seas" done at Geneva, April 29, 1958, (1966) 17 U.S.T. 138, T.I.A.S. No 5969, 559 U.N.T.S. 285.

7. Art. 1, par 2

8. International Convention for the Prevention of Pollution of the Sea by Oil, 1954, opened for signature May 12, 1954, (1961) 3 U.S.T. 2989, T.I.A.S. No. 4900, 327 U.N.T.S. 3.

9. 17 U.S.T. 1523, I.M.C.O. Document A VI/Res. 175 of Jan. 16, 1970.

10. Sweeney, supra n. 2 at 190-192.

11. ibid at 193.

12. Art. 7, par. 2.

13. Art. 8.

14. This is not to be taken as an indication of overall approval of the treaty as presently constituted. Among the shortcomings: Flag state enforcement is not effective, designating violators is almost impossible, and comparatively few nations have consented to adhere to convention provisions.

15. ...during the course of which the 1969 amendments to the International Convention for the prevention of Pollution of the Sea by Oil, discussed supra, were adopted.

16. Healy, N.J. "The C.M.I. and I.M.C.O. Draft Conventions on Civil Liability for Oil Pollution, 1 Journal of Mar. Law and Commerce, 93, (1969).

17. I.M.C.O. doc.C/ES III/3, "Lessons arising from the incident of the Torrey Canyon, April 18, 1967.

18. International Convention Relating to Intervention on the High Seas in Cases of Oil Pollution Casualty, _9 International Legal Materials 25, 1970_.

19. Now enshrined in article 51 of the United Nations charter. See _Sweeney_, supra not 2. at 202.

20. Cowan, Edward, _Oil and Water: The Torrey Canyon Disaster_, Philadelphia/New York, Lippincott, 1968. It should be mentioned that the "initial reluctance" was eventually overcome. British planes bombed the _Torrey Canyon_ on the 28th, 29th, and 30th of March, 1967. (Spillage began March 18.) the slick was set afire though it did not burn for long.

21. _9 International Legal Materials 45_. (Done at Brussels, 11/29/1969)

22. Preamble

23. Art. II

24. Art III

25. Art. V (1) Art. V(10) describes the method of calculating adjusted net tonnage.

26. Art. V(2)

27. Art V(3)

28. Art VI(9)

29. Art. VIII

30. Art. VII (1)

31. See _Cowan_, supra n. 20.

32. _8 International Legal Materials 497_

33. 18-19 Eliz. 2, c47 (Can., 1970)

34. Bilder, R., "The Canadian Arctic Waters Pollution Prevention Act," 69 Mich. L. Rev. 1, (1970)

35. Sec. 4

36. Sec. 13

37. Sec. 13

38. Sec. 12

39. Sec. 12(1) (c)

40. Sec. 14, 15.

41. Sec. 18

42. Sec. 19

43. Sec. 23(1)

44. Sec. 6, 7.

45. Sec. 9.

46. Sec. 8

47. The United States is perhaps the world's leading promulgator of jurisdictional extensions intended to achieve specific purposes. Anti-smuggling zones were declared out to 62 miles from the coasts in 1935 and are still in effect. In 1945, The Truman Proclamation claimed U.S. control of its continental shelf resources. In 1962 the U.S. established a quarantine around the island of Cuba. The U.S. in times past often closed portions of the Pacific for the purpose of conducting atomic weapons tests.

48. See Appendix 3

APPENDIX 5

EVALUATION OF DEVICES FOR CONTAINMENT OF OIL SPILLS AND THEIR APPLICABILITY TO PUGET SOUND

If an oil spill incident does occur, the cleanup response is initiated by preventing the oil from spreading. This task includes both diverting the oil from areas vulnerable to damage, and containing the oil so that recovery of the product is effective.

This appendix evaluates devices specifically designed to prevent spreading of oil. These primarily include gelling agents and physical barriers. The latter group is comprised of floating booms, pneumatic barriers, and chemical barriers. Each group is discussed separately, and where information is available, effectiveness of the equipment in the Puget Sound waters is analyzed.

A list of commercially available containment devices, their manufacturers, and prices appears in this appendix.

A. Gelling Agents

1. Introduction

Gelling agents have been considered for containment of oil spills by application of these agents to oil in water to prevent spreading. Basic difficulties with post facto application of gelling agents are in obtaining adequate mixing to obtain complete or sufficient gellation of the oil and the delay in application. Some form of agitation energy must be applied to obtain the required mix; therefore, the logistics of application are difficult and satisfactory means have not been proved. Also, this mechanism requires time delays in application which are unacceptable from a damage prevention standpoint.

Accordingly, it is proposed that the gellation mechanism most likely to have some measure of success requires shipboard deployment of gellants to the oil either:

a. before the accident (i.e. during or prior to loading of the cargo or ships fuel oil);

b. after the accident but before significant spillage has occurred; or
c. during the spill apprehending gellation on water surface later.

At the very minimum, it is apparent that dispersal of gelling agents to provide gellation of petrochemicals on the water surface versus the preferred gellation in the tank has significant value as the second alternative. The advantages gained by gellation of oil within a short time of loss of containment are: (1) spread prevention, (2) easier pick up of lost oil, (3) prevention of loss of aromatic solvents to the water body which are highly toxic to plants and animal life.

The only immediately apparent danger in gellation appears to be the creation of a denser mass of gelled oil than that of the water it is immersed in, with the consequent loss to the sea floor or to some intermediately dense water layer. In San Francisco, it appears significant "mud" gellation of cargo oil occurred and remained suspended below the water surface. Information on specific gravity of the various gelled materials is critical.

The economics of gellation mechanisms herein suggested will be cursorily considered. The study team was astounded by the apparent lack of research on the subject, which apparently has only developed quite recently. Such delay can only suggest an industry bias for reasons of self-interest. Therefore, economics will be considered lightly, to rebut objections presented and to suggest models which seem feasible in face of such objections. It is clear that the burden of proof or disproof must be directed toward the industry whenever possible to assure the fullest exposition of alternatives.

This discussion is developed primarily from the Esso article,[1] the Defense Documentation Center Bibliography,[2] and from discussions with various persons in the industry and with various chemists around the country. At present, the information is very incomplete and most of what will follow is based on conjecture of what now seems feasible in the way of alternative applications of gelling agents for containment and/or prevention of petrochemical spills.

2. Alternative Modes of Operation

a. Mixing gelling agents prior to shipment

This mode of operation would seem most desirable from the ecologist's viewpoint. During an emergency, the deployment of gelling agents must be assumed to, at best, involve some time delay and therefore some oil losses.

It seems apparent, from the Defense Department Bibliography,[3] that at least ships fuel oil could now have some temperature sensitive gellant. For example, an additive known as N-coco-gama-hydroxybutyramide was found to make a solid gel when mixed in the ratio of 1.5 per cent to weight of fuel. This agent, when mixed with jet fuel, was found to not impede burning nor to create undesirable deposits. It also was found that when heated to 130°F, the gel is easily liquified. While further study of its overall cleanliness in burning was not reported, its desirable properties are immediately apparent. Also, current data seems mostly related to aircraft applications, therefore should be closely considered for adaptation difficulties.

The use of such a gelling agent for cargo oil is just as feasible. While the cost of the gelling agent was not reported, nor was the nature of the chemical bonding and its consequent refining/ reclamation properties, this agent would be effective for the continuous bunkering of cargo with a temperature sensitivity which would gel the oil on loss of tank heating or when lost to the sea at ambient temperatures below 130°F.[4]

As with all the other gellants to be discussed, the following should be considered and reported on later:

1. The availability of the chemical.
2. Its ability to be stored for considerable periods of time, perhaps under unfavorable conditions.
3. Its reclaiming capability at a refinery.
4. Cost, either under an expendable or reclamation cycle.
5. The stability of the gel formed.
6. Temperature effects on viscosity, etc.

7. Density of the gelled oil.

8. Flammability.

Many other gellants are used in the oil and cosmetic industry which are commonly available to gel various oils to prepare greases, etc. It should be noted that it is quite possible to develop agents which would gel oil when put into contact with water. This, of course, would provide economies of scale over the temperature sensitive mode by saving heating costs and preventing inadvertent gellation. No information, however, has been uncovered on the feasibility of this mode other than a discussion with Dr. Wendal L. Slabaugh of Oregon State University[5] and others.[6]

b. Mixing gellants during an emergency to contain oil shipboard. This mode of operation seems a far less desirable situation in that it apprehends some loss of ungelled oil prior to containment and would always be limited by the effectiveness of the mixing mechanism during deployment. It is clear some ungelled residual oil will remain and the unforeseen or cataclysmic casualty could never be effectively controlled.

The Esso paper of May 20, 1971 provides some insight to one study of the feasibility of this mode. The mechanism appears to be a two part mixture creating a matrix that will capture the crude oil although the details are not made specific. Because this study utilized the aircraft application of gellants, the assumption that the previously discussed gel is a one-part mixture should be questioned. More data should be obtained on these reports.

In any case, this mechanism was considered effective in creating stable gel, but was limited by cost and deployment times. The cost indicated was \$3.00 per barrel of oil at a 3% mixture. No comment is made, however, on the reclamation capability with this agent and assuming such gellant could be recovered, that cost might be considerably diminished. Also, it should be noted that the mixture is twice that reported in the prior model which suggests more economical systems might be developed. However, the comparative strength of the prior gel must be determined to check this assumption.

The prime limitation on this emergency deployment mode is the difficult compromise to be made with the alternative gelling agent available. The fast gelling agent (less than one second) is limited by the fact that local gellation interferes with mixing necessary for a solid gel. The slower gelling agents would not provide shipboard containment of the oil or would have limited utility for same.

Suggested weakness in the Esso scheme's unfavorable conclusions was their inflexible approach. It would seem a bimodal approach might be utilized, wherein baffled voids would be used for rapid gelling agents to provide local chemical bulkheads with the slower gelling agent deployed in the large volume areas to provide solid gel of the bulk of the oil.

While the effectiveness of using the rapid gelling agent for localized reduction of fluidity to diminish spill rate is difficult to judge, its use should not be discounted without consideration; inasmuch as it provides the mechanism necessary for the limitation of the unexpected casualty. The slower agent could be deployed alone where adequate reaction time was available.

c. Gel development by flow energy; mixing during loss from hull. This mechanism is suggested by the above mentioned study but seems a simple strategy of containment. Here the mixing energy would be provided by the dispersing nozzles located near ship's hull and the agitation created by flow from the hull and from the local wave action and movement of oil away from the ship.

This mechanism has most of the limitations of effectiveness described above but might be a more feasible approach in terms of the current state of the art. It is clear that it is the least desirable of the above inasmuch as it provides only the minimization of spread through gellation but apprehends large amounts of ungelled oil lost at sea.

3. Conclusion

The bimodal approach seems, to the study team, the most feasible alternative which present technology would seem to support.

The industry objections of cost should be met with the simple observation that they presently take advantage of, and depend on, the efforts of others to contain this problem. It is easy to lose sight of that fact when confronted with the potential costs of containment. The duty of care should be placed on the person most able to provide solutions and where, as here, the preventive measures seem the only satisfactory solution to the extreme hazards of oil transport, that duty must be placed on the industry. Taxes would accordingly be assessed on all vessels not showing an effective containment mechanism that could assure environmental safety. Certainly, if the cost of petrochemicals should rise as a consequence of such measures, those benefiting from the use of oil would be forced to share a more equitable part of the burden of risk that is now borne by those least capable.

Table 5-1 summarizes the description of various commercially available gelling agents. Included are manufacturers and costs.

TABLE 5-1: GELLING AGENTS

(Based on available information;
no actual physical testing was conducted)

NAME	DISTRIBUTOR OR MANUFACTURER	DESCRIPTION
Spill-Away[7]	Amerace-Esna Corporation Chemical Specialties Division 74 Hudson Ave. Tenafly, New Jersey 07670	A viscuous liquid which is sprayed onto oil slick to form a gel in presence of water, available in 53 gallons drums at $2.50 per gallon.
Strickite[8]	Strickman Industries, Inc. P.O. Box 208 Hillsdale, N.J. 07642	Finely divided granular material which is sprayed or blown onto an oil slick, forms a stable gel for an indefinite period. Amount required for oil treatments runs 1/10 to 1/20 the weight of the oil. Cost: $0.42-$0.45/lb FOB, Hillsdale.
	Esso Research and Engineering Company	See text, pp. 5.2-5.4. Cost $3.00 per barrel of oil at a 3% mixture.

B. Physical Barriers

1. Introduction

The presentation of containment methods with physical barriers (booms) has been developed as follows:

a. A discussion of the present state-of-the-art in boom development, and

b. a list and evaluation of present boom systems.

The present boom hardware was developed mainly by a hit or miss method. Booms and containment devices have usually been built to suit only specific spill incidents. Most of the effort has been channeled in the development of booms in sheltered areas such as harbors and ports. An effective boom for open sea states and currents is not yet available commercially.

An effective boom must meet the following criteria:

a. Provide a barrier above the water surface that can keep oil from splashing over it.

b. Provide a barrier under the water surface to prevent the oil from leaking under.

c. Flexibility of the barrier to follow the contours of the waves.

d. Must be strong enough to be towable and withstand wave action and current action.

e. Maneuverability and ease of deployment of the boom; must be towable at speeds of up to 10 knots.

Under open seas or heavy weather conditions and in the presence of current, the following additional criteria are required:

a. Boom must withstand extensive wave action and must retain oil in sea states with high winds (20 knots) and waves (at least 5-10 feet).

b. Should retain oil in currents of up to 2 knots.

c. Must be flexible enough to follow the wave contour of high sea states.

d. Anchoring system must withstand the sea forces acting on the boom.

Special requirements have to be met by booms serving specific purposes.

a. Permanent booms, which remain in the water indefinitely, must be resistant to sun, oil, fire, and abrasion.

b. Towed booms, such as used for sweeping operation, require special properties that will keep the oil from going under. Also, they have to retain their shape and be effective under towing speeds of up to 5 knots. The boom must also be strong enough to withstand the load from the oil, water, and debris.

Physical barriers for oil retention are generally grouped in the following classification: 1) floating booms, 2) pneumatic barriers (underwater air barriers), and 3) chemical barriers.

2. Floating Booms

A floating boom is the most common oil containment device available today. Floating booms generally consist of a buoyant section which floats on top of the water, a curtain or barrier that extends down into the water to prevent the flow of oil under the boom and a mooring system which permits the boom to be towed or anchored at a specific location.

Many operational problems can occur during placement and use of booms. Floating booms on their moorings failed at Santa Barbara Harbor.[9] (Inflatable, log, plastic.) The log booms seemed particularly susceptible to structural failure of attachment cables, generally because of rough seas. Four ships were required to maintain position of a 2,000-foot log boom extending south from the end of Stearn's Wharf.[10] Also, constant removal of oil accumulating behind log booms was necessary to prevent its leakage. Straw, vacuum trucks, and manpower can facilitate the removal of contained oil from the booms.

Another problem existed on beaches. Most booms observed could not accomodate the effect of tidal flow undercutting the sand at the point where the boom entered the water, thus creating an opening through which the oil could pass under the boom. Rigid-skirted booms or those strung tightly could not fill the passage created.

One of the most common failures in oil retention is due to the inability of the boom to keep oil from being swept under the curtain by the prevailing currents or by towing the boom. This is an area where more extensive research could prove fruitful in designing booms that minimize the amount of oil that is lost under the curtain. (See part 5 of this Appendix.) These are but a few of the many problems that will have to be solved before floating booms become a reliable method of oil containment.

3. Pneumatic Barriers

Pneumatic barriers (underwater bubble barriers) can provide a sufficient surface current to contain oil spills in harbor waters if winds and surface currents are not excessive. The operation entails injection of air through a perforated hose or pipe into a water column. The water section, into which the air is injected, is reduced in density with respect to the surrounding water, and a resultant upwelling occurs. The vertical motion of the water produces a "mound" on the surface which results in a surface current flow from the point of air emergence. Water is drawn from the region immediately above the submerged pipe, thus producing the circulation. The advantages of pneumatic barriers are:

a. Unrestricted passage of ships across the barrier.
b. Relative immunity to environmental forces.
c. Invulnerable to fire on the water surface.
d. In certain instances, such as when the barrier is biased across the direction of water flow, the oil can be guided to a single location for pick up.

The disadvantages of pneumatic barriers are:

a. Extensive installation hardware and maintenance.
b. Possible slight penetration of accumulated oil as ships pass across the barrier.
c. Complete ineffectiveness of the barrier in the absence of power, compressor, or due to pipe failure.

d. Limited use, due to ineffectiveness, of the barrier in currents and rough sea states.

Presently, pneumatic barriers have been used only in port facilities; no sea going models have been used but research is going on in this field. The barriers are permanently installed and are custom designed for each particular application. Such factors as type of oil, amount of debris, tidal velocity, depth of water, orientation with the prevailing winds are important consideration in the design of barriers. Bubble barriers become inefficient in currents above 1.3 ft./sec.

4. Chemical Barriers

Chemical barriers can be formed with such agents as fatty acids spread at the periphery of the oil spill. The spreading force of the fatty material will repel the non-polar petroleum oil and push it into a smaller area. Hydrophobic hydrocarbons generally do not spread well on water and can also be used for this purpose.

Chemical barriers are not widely employed because of extreme sensitivity to environmental factors, possible toxicity to marine life, and operational considerations such as deployment and recovery problems. Shell Oil Company is undergoing extensive research in an "oil herder" that might be feasible, although the toxicity of this chemical has not been established and will require more research before it can be used extensively.[11]

5. Evaluation and Discussion of Physical Barriers

Since no physical testing of the actual booms was made, the principal criteria for boom evaluation was derived from:

a. oil spill accidents, where some booms were employed;
b. data obtained from different companies which undertook studies on booms;[12,13]
c. data obtained from the different boom manufacturers (see part 6 for detailed information on booms).

The evaluation was developed in light of criteria specifically chosen for the Puget Sound area, where currents are a large factor.

Other parameters considered were moderate sea states (up to 5 foot waves), wind (up to 20 knots). Parameters of lesser importance were strength, handling, ability, storage, and cost of the boom.

No consideration was given to the evaluation of booms in terms of their ease of deployment, weight, section lengths, weatherability, fireproofing, resistance to sun, oil or rot, or to the visibility of the boom (color of the boom). This was due to lack of technical information plus the lack of actual performance testing of booms under oil spill conditions.

Booms were categorized into five different groups according to their use. These different categories were:

a. light duty booms,
b. medium-light duty booms,
c. medium duty booms,
d. heavy duty booms, and
e. special duty booms.

The first four categories are listed on Tables 5-2 through 5-5, respectively.

The booms were rated on a point system as follows: For current, wind and sea state, 4 points were assigned if the boom surpassed the specified requirement, 2 points if it met the requirement, and 0 points if it did not meet the listed requirements. For strength, handling ability, storage, the point system was 2, 1, and 0 respectively. For cost, the points were, 0 if it was more costly than the average boom for that class, 1 if it was average, and 2 if it was less expensive. Evaluation was given to booms which had specialized use. If the "special use" column in the rating system is marked "yes," then the total number rating shown is less than the true value for that particular boom.

The enclosed rating is not completely thorough for the following reasons:

a. no actual testing of booms under oil spill conditions was conducted.

b. Factors such as ease of deployment, weight, ease of cleaning, durability, rot proofing, were not considered due to lack of information and test data.
c. Some booms listed in part 6 are of specialized use, such as permanent or semi-permanent installation. These booms are usually heavy duty, have extensive mooring and anchoring systems, are fireproof and sun, oil, abrasion resistant.
d. Complete information was not available on some booms listed in part 6. It was therefore not possible to rate these booms.

Table 5-6 provides a detailed list of commercially available booms, including manufacturers and prices, when information was provided.

TABLE 5-2: EVALUATION OF LIGHT DUTY BOOMS - Limited to operations where currents were negligible (less than 1/4 knot), relatively calm water (waves less than 6"), wind less than 5 knots.

BOOM	Current	Wind	Sea-state	Strength	Handling	Storage	Cost	Special use	COMMENTS	TOTAL
Cork Float Boom	0	3	4	2	1	1	0		High cost	11
Firehose Boom	0	2	2	2	2	2	2		Very inexpensive	12
Slickbar Mark IV 4" with 10" skirt	3	4	3	2	1	1	0			14
Log Boom	3	4	4	2	0	0	1		Problems with handling and storage	14
Sea Curtain light duty	4	4	4	2	1	1	1			17
Spillguard 411	4	4	3	2	1	0	0	Yes		14
Scandinavian OS	4	4	4	2	1	1	0			16
Galvaing 1B	3	4	4	2	1	0	*-2	Yes	*Very high cost	12
Uniroyal Miniboom	4	4	4	2	1	1	0		High cost	16

TABLE 5-3: EVALUATION OF MEDIUM-LIGHT DUTY BOOMS - Limited to operations where currents do not exceed 1/2 knot, wave action less than 1 foot, and wind less than 10 knots.

BOOM	Current	Wind	Sea-state	Strength	Handling	Storage	Cost	Special use	COMMENTS	TOTAL
Galvaing PB	1	4	4	2	1	1	-1	Yes	Very high cost	12
Galvaing Long Skirt	3	4	4	2	1	1	-1	Yes	Very high cost	14
Rode Orm light duty	3	4	3	2	1	1	1			15
Slickbar Mark IV 6" float, 12" skirt	3	4	3	2	1	1	0			14
Scandinavian OS	4	4	3	2	1	1	1			16
Spill guard 411	3	3	1	2	1	1	1			12
Sea Curtain light duty	3	4	2	2	1	1	2			15
Sea Curtain regular	4	4	4	2	1	1	2			18
Sea Curtain harbor type	3	4	4	2	1	1	1	Yes		16
Uniroyal miniboom	3	4	3	2	1	1	0			14
MP Boom	3	4	3	2	1	1	1			15
Warne Type E	4	4	4	2	1	1	1			17

TABLE 5-4: EVALUATION OF MEDIUM DUTY BOOMS - Limited to operations where currents do not exceed 1 knot, wave action is less than 2 feet and wind less than 20 knots.

BOOM	Current	Wind	Sea-state	Strength	Handling	Storage	Cost	Special use	COMMENTS	TOTAL
Bennett Harbor Type	*5	4	3	2	1	1	1	*yes	*Extremely good in currents	17
Fir Trees (makeshift)	3	4	3	2	0	0	2		Requires time to assemble	14
Muletex	4	4	3	2	1	1	0			15
Rode Orm (Large)	4	4	3	2	1	1	1			16
Uniroyal midiboom	4	4	3	2	1	1	1			16
Sea Curtain heavy duty	4	4	4	2	1	1	1			17
Slickbar 6" float, 18" skirt	4	4	2	2	1	1	1			15
Slickbar Mark IV	4	4	3	2	1	1	1			16
Warne Type S	4	4	2	2	1	1	-1	Yes	Permanent use boom	13

TABLE 5-5: EVALUATION OF HEAVY DUTY BOOMS - For open water operations, currents of 2 knots, wave action 4 or more feet, 40 knot winds.

BOOM	Current	Wind	Sea-state	Strength	Handling	Storage	Cost	Special use	COMMENTS	TOTAL
Bennett Offshore	*4	4	3	2	1	1	2	*Yes	*Extremely good in currents	17
Fabri Boom	2	4	2	2	1	1	1			13
Gates Boom	2	4	2	2	1	1	0			12
Merrit Boom	4	4	4	2	1	0	0	*Yes	*Very strong,permanent use	15
Uniroyal High Seas	4	4	2	2	1	1	2			16
Sea Curtain heavy duty	2	4	2	2	1	1	*3		Very inexpensive	15

Special Duty Booms

Some booms listed in the appendix could not be adequately classified in the previous sections. Such booms include fire-proof booms, heavy duty harbor booms (some permanent installations).

Galvaing Booms: Extremely good under fire conditions. (Fire control barrier.)

Merrit Booms: Very heavy duty, can withstand strong sea actions and are semi-permanent installation.

Warne Boom Types T & S: Are very good harbor booms. Permanent installation and are very strong and abrasion resistant.

TABLE 5-6: LIST OF COMMERCIALLY AVAILABLE PHYSICAL BARRIERS
(Based on available information, no actual physical testing was conducted.)

NAME	DISTRIBUTOR OR MANUFACTURER	DESCRIPTION
Acorn Boom	Acorn Marine Boom Ltd.	No information.
Arrowhead Boom	Arrowhead Products Divison Federal-Mogul Corporation 4411 Katella Ave. Los Alamitos, Ca.	Constructed with a series of 10 ft. long rubber coated fabric sleeves, is a flexible boom system in 100 foot sections. Still in development - no costs available.
Bennett Filtration Boom	Bennett Pollution Controls Ltd., Suite 980 Guinnes Tower, 1055 W. Hastings, Vancouver 1, B.C.,Canada	Harbor Type - 1 foot freeboard, 2 foot draft made of filtration material which allows water to flow through but stops oil. Comes in 50' sections which are connectable. Cost of $12.00 per foot. Offshore Type - 2'6" freeboard, 3'6" draft made of filtration material. The filtration material is easily cleaned for re-use. Cost of $25.00 per foot.
Burlap Tube filled with straw	Roscoff Marine Laboratory, France	No information.
Containment Device (inflatable)	Water Pollution Control,Inc. 2035 Lemoine Ave. Fort Lee, New Jersey 07024	No information.
Cork Float Boom	Various harbors (Norfolk harbor) Anchor Canvas Co. 247 W.6th Street Post Office Box 728 San Pedro, California	Constructed from a series of cork discs 6" in diameter by 2" thick and 3" in diameter by 1" thick, strung alternately on a 3/8" diameter polyvinyl coated steel or bronze cable. The cork floats are enclosed in a polyvinyl or canvas cover. These booms are available in 50 ft. sections and are unskirted. The cost is @ $6.00 per foot.

TABLE 5-6 (Cont'd)

NAME	DISTRIBUTOR OR MANUFACTURER	DESCRIPTION
Fabri Boom	Firestone Coated Fabrics Division 2525 Firestone Blvd. P. O. Box 1592 South Gate, California	Three parallel PVC tubes, two air filled, one water filled in an inverted pyramid with a reinforcing wire rope at bottom of ballast tube, available in one model. Cost \$35/foot, section length - 250 ft. freeboard of 22" and draft of 22".
Firehose Boom	Hose Booms Inc. 111 John Street New York, N.Y.	Make-shift air inflated firehose is employed in many port and harbors from oil spill containment in calm waters. The metal hose couplings are removed and replaced with plastic or aluminum inserts to join succeeding sections. Air valve fittings must be provided in the ends.
Fir Tree Booms (Make-shift)	Arrow Spill Task Force Report Operation Oil - Report of the task force to the Minister of Transportation, Volume II.	Fir trees were found to be successful in absorbing and containing oil. A make-shift boom could be manufactured using whole fir trees tied end to end.
Flex Booms	East Coast Services, Inc. 343 Washington St. Braintree, Mass.02184	No information.
Galvaing Booms	Gamlen Chemical Co. 321 Victory Avenue So. San Francisco, California	PB Type - 20 foot sections, 1'4" Navy plywood sections with flotation and ballast weights with 8" freeboard and 6" draft. Cost \$13.60/ft. or \$16/ft. with krypton signals. Fire Control Barrier Type - same as the PB type employing fireproof floats and a skirt of asbestos cloth coated with PVC and strengthened with mosquito net mesh cloth. Cost \$16/ft or \$18.50 with krypton signals. Long Skirt Barrier Type - Same as PB type, with the addition of a neoprene treated nylon cloth skirt. Cost \$14.80/ft. or \$17.25 with krypton signals.

TABLE 5-6 (Cont'd)

NAME	DISTRIBUTOR OR MANUFACTURER	DESCRIPTION
Gates Boom	Gates Rubber Company 6285 E. Randolph Street Los Angeles, California 90022	A 25" diameter reinforced rubber hose with pleated skirt, available in four models. Cost - $50/ft. section length - 25 ft., freeboard of 20" and draft of 25".
Guardian Oil Boom	Guardian Chemical Corp. 41-45 Crescent St. Long Island, N.Y.11101	No information.
MP Boom	Metropolitan Petroleum Co., Inc. 25 Caven Point Rd. Jersey City, New Jersey 07034	A plastic foam filled, canvas covered 6" diameter boom with 12" skirt. Available in 100' sections at $7.25 to $7.85 per foot.
Merrit Boom	Murphy Pacific Marine Salvage Company 4300 Eastshore Highway Emeryville, Ca. 94608	Build 4 x 8 sheets of plywood with four oil barrels strapped to the sides. A curtain made of Fusilon or Herculite is attached to the bottom of the plywood. Freeboard is 3 ft. and draft is 4 ft. 2 ft. of canvas hold the plywood sections together, 1/2 steel rope and chain run continuously along the boom. @ $40/ft.
Muehleisur Muletex Boom	Muehleisur Manufacturing Company 1100 No. Johnson Avenue El Cajon, Ca. 92020	Comes in 100 ft. sections, vinylbase fabric 36" width with 1 ft. freeboard. Ethafoam floats are attached to the outside and the skirt is weighted with a continuous link chain, cost $15/ft.
Oil Absorbing Boom	Oil Absorbing Booms, Ltd. 108 London Road Bromley, Kent ENGLAND	No information.

TABLE 5-6 (Cont'd)

NAME	DISTRIBUTOR OR MANUFACTURER	DESCRIPTION
Pellet Containment	Collins & Aikmen Cavel, North Carolina 27512	Conceptual.
Perma Foam Boom	Perma Foam Insulation & Engineering 231 Island Ave. Wilmington, California	No information.
Rode Orm Boom	Manufactured in Sweden	Comes in two sizes - 1) 36" PVC skirt with built-in floats and weighted along the bottom, and 2) 24" PVC skirt with built-in floats.
Scandinavian Oil Service Boom	Surface Separator Systems 103 Mellor Avenue Baltimore, Maryland 21228	PVC plastic boom with a 6" inflatable float and 15" draft. Weighted with a continuous chain along the bottom. Comes in 80 ft. sections, cost $7/ft.
Sea Curtain	Kepner Plastics Fabricators Inc. 4221 Spencer Torrance, California	Heavy Duty Ocean Type - 20" diameter foam filled float with a 30" PVC, chain ballasted skirt. Cost $10-15/ft. Heavy Duty Harbor Type - 12" diameter foam filled float with a 11" PVC, chain ballasted skirt. Cost $6-9/ft. Regular Duty Type - Inflatable 19" float with a 32" PVC, chain ballasted skirt. Cost $4-6/ft. Light Duty Type - Inflatable 12" float with a 11" PVC skirt. Cost $2-4/ft.
Sea Fence	Alcoa Aluminum Company	Rigid vertical barriers of aluminum sheet, held together by steel cable, with foam floats and neoprene joints between panels.

TABLE 5-6 (Cont'd)

NAME	DISTRIBUTOR OR MANUFACTURER	DESCRIPTION
Sealdboom - Uniroyal	Uniroyal, Inc Engineering Systems Dept. 10 Eagle St. Providence, R.I. 02901	Miniboom - paracryl coated nylon, vulcanized with floats built-in. 6" freeboard and 12" draft comes in 40' sections, cost 40-480' $10/ft., for orders of 1000' or more $8.50/ft. Midiboom - paracryl coated nylon, rubberized, vulcanized with built-in unsollate flotation. 24" draft, 12" freeboard, the skirt is lead weighted. 40-480' $14/ft. for orders of 1000' or more $12.50/ft. A high seas boom is being developed with 2 ft. cylindrical floats and a 6 ft. skirt. Cost will be @ $20-30/ft.
Slickbar Mark IV Slickbar Mark V	Fluid Controls Inc. 3016 NE Blakeley Seattle, Washington	Foam filled plastic section in 4" or 6" diameter with assorted skirt depths - 6", 8", 10", 12", 13" and 24". The skirts are plastic and are lead weighted according to operational requirements, cost from $3 to $12. The Mark V is a heavier boom with polyester fabric and polyurethane flotation. Has a 12" freeboard and 24" draft and is lead weighted. Has a 3/8" steel cable under flotation. Cost @ $12/ft.
Spill Guard Boom	Johns-Manville Corp. Box HJW-23 22 East 40th Street New York, N.Y.	A semi-stiff asbestos compound sheet fence with attached foam plastic floats and chain ballast in 10' lengths. Model 411 provides 4" freeboard and 11" draft. Cost - $7.50 per foot.
Warne Booms	Surface Separator Systems 103 Mellor Avenue Baltimore, Maryland	Type T - 8" float and 20" skirt made of rubberized canvas, the float is filled with plastic tubes. The skirt is weighted with a continuous chain along the bottom. Cost of $20/ft. Also comes with a 16" float and 24" skirt. Cost is $32/ft.

TABLE 5-6 (Cont'd)

NAME	DISTRIBUTOR OR MANUFACTURER	DESCRIPTION
Warne Booms	(continued from previous page)	Type S - Inflatable 8" diameter rubberized canvas with a 22" skirt weighted with a continuous chain. Cost $18/ft. Type E - Solid filled tube with a 24" skirt made of lighter material, cost of $8/ft.
Wooden Timber Booms	Makeshift - built on the spot.	Log booms attached to each other by chain or steel cable. Can be implemented by using seine net, 3/4" mesh nylon twine 5 fathoms deep. This can be further implemented by making two rows of booms with straw in-between and seine net to catch the sorbent.

C. Endnotes for Appendix 5

1. Edward R. Corino, "Chemical Gelling Agents and Dispersants," ESSO Research and Engineering Company, May 20, 1970.

2. Defense Documentation Center, Search Control No. 056709. In Tank Gellation of Petrochemicals, A Report Bibliography.

3. Defense Documentation Center, Search Control No. 056709. Report reference number AD-652 918.

4. Ibid.

5. Dr. Wendal L. Slabaugh, Dean of the School of Chemistry, Oregon State University, Telephone: (503) 754-0123.

6. Included are Drs. George Halsey, University of Washington, Chemistry Department, Telephone: (206) 543-1664; Donald Schwarts, a mud specialist in Bocaratean, Florida, Telephone: (305) 395-5100.

7. Dillingham Corporation, Analysis of Oil Spills and Control Material, Vol. I of Systems Study of Oil Spill Cleanup Procedures. February, 1970, p. C-31.

8. Battelle Memorial Institute, Oil Spill Treating Agents - A Compendium. May, 1970.

9. Battelle Memorial Institute, Review of Santa Barbara Channel Oil Pollution Incident, 1969, pp. 6-18.

10. Ibid.

11. Developed by Shell Oil Company Research and Development Laboratory, Houston, Texas. As reported in January 26, 1971, issue of Seattle Post Intelligencer, p. 1.

12. Dillingham report, Vol. I, pp. C-3 through C-4.

13. Battelle Memorial Institute, Study of Equipment and Methods for Removing Oil from Harbor Waters, Report #C12-70-001, August 25, 1969.

APPENDIX 6

EVALUATION OF DEVICES FOR CLEANUP OF OIL SPILLS AND THEIR APPLICABILITY TO PUGET SOUND

Control of an oil spill encompasses two phases, namely, containment and cleanup. This appendix concerns itself with the latter aspect. Devices for oil spill cleanup are broadly classified into four groups - mechanical, chemical, chemo-mechanical and biological.

Under each group, there are various techniques for removing oil from water. Examples of such include sinking, evaporating, dispersing, recovery, and biodegradation. Because of the possible adverse consequences of sinking and dispersing, these two techniques are not fully developed in this appendix.

For each group, evaluation of commercially available devices is presented, including manufacturers and prices, where available.

A. Mechanical Cleanup

1. Introduction

This rather controversial aspect of an extremely controversial issue should be divided into three classifications:

a. Collection of the spilled oil and subsequent mechanical recovery.

b. Sinking of the oil and recovery by suction devices.

c. Mechanical removal of the oil from the water.

Vast amounts of data are available; thus, the problem is not the quantity but the quality of the information. The data supplied could best be described as inconsistent. Specifications supplied by vendors did not correspond with research laboratory test results; which, in turn, differed from information supplied by salvage masters and others directly involved with oil spills.[1]

Before beginning a discussion of how the various devices work, it is admitted that they all work. Each proponent of a particular device can legitimately claim a measure of success. But if these devices do pick up oil, why do most knowledgeable people disregard mechanical recovery as a practical answer to the problem? The answer is probably twofold:

a. Lack of concentrated effort;

b. Time lag.

Slide presentations of the Santa Barbara[2] and San Fransisco[3] spills point out the lack of concentrated effort in this area. Visualize a projection screen which shows miles of oil stringers and slicks. The moderator then points out a device the size of a pin head which is attempting to clean up this oil. This problem does not lend itself to the efforts of individual entrepreneurs. As an illustration of the type of effort and capabilities necessary to solve this problem, it is pointed out that private industry took 19 years and hundreds of thousands of dollars to develop and market Minute Rice. The Government took approximately three and one half years from idea to delivery for the atom bomb. This is not to say that another Manhattan Project is required to develop a mechanical cleanup device, but that the effort must be on the order of magnitude of the problem involved.

When certain experts were questioned regarding the effectiveness of a device, the usual answer was that the device was not used or the results were inconclusive. As the situation exists today, most skimming devices have to be transported many miles and, in some cases, reassembled at the spill site. It is hours, sometimes days, before these devices begin to attempt the cleanup. As a result of this time lag, much of the oil is on the beach. The oil that remains on the water bears little resemblance to the substance these devices were designed to pick up.

2. State of the Art

Pacific Northwest laboratories of the Battelle Memorial Institute performed a series of tests on mechanical cleanup devices. Most of the data herein presented is a result of their research.[4] The Battelle report is important because it represents the efforts of an independent research group - an unbiased third party. Of even greater importance is their consideration of sensitivity factors (wind, waves, debris, etc.) in rating various devices. But, like all information concerning oil cleanup devices, the Battelle report is at times vague and inconsistent.

An operational description of the devices currently available follows. Special attention will be drawn to those devices available in the Puget Sound area.[5]

a. The Weir Skimmer

Several of these skimmers are in use. All of these skimmers have an adjustable lip which must be located at just below the water level. The weir can be advanced or the polluted water can be brought to the lip by flushing or drawing with a boom. Many of the devices employ a centrifugal pump with suction at the lip to draw the slick.

Norfolk Skimmer - 600 GPH oil: The Norfolk Naval Shipyard skimmer separates the oil from the water by gravity. The oil and water are drawn into a small diffusion box with numerous holes. After separation, the oil is stored in a tank, (10,000 gal. cap.), and the water passes out the bottom.

Such a device is available at the Puget Sound Naval Shipyard. Recent information indicates that the device is no longer operable. This converted LCM is self-propelled and is currently used as an all purpose barge to carry absorbents which can be mechanically retrieved.

Battelle rated this type of skimmer extremely high. The device, however, was given a "0" rating for sensitivity to environmental factors, i.e., rendered inoperable by five foot waves and 20 knot winds.

French Technocean Skimmer: This device is not yet operational. Similar in concept to the Norfolk skimmer, this device, however, is self-propelled. It is designed to operate in reverse when picking up oil. The bow of the vessel is standard and the stern section is "v" shaped. The point of the "v" is an intake valve which automatically adjusts to the water level. The vessel will treat 13,000 cu.yds./hr. of oil and water and store the same amount of oil. (Inflatable) Draconi barges are also carried aboard for oil storage. Technocean, 45 Boulevard Haussman, Paris, France.

Slick Sled: Inverted column of water open to sea at the bottom, supported by pontoons. The entrance to the raised column is an inclined, horizontal funnel which permits the craft to skim the undistrubed surface of the oil. The oil rises to the top of the inverted column, displacing water at the bottom - thence overboard. Water Pollution Control, Inc., 2035 Lemoine Avenue, Fort Lee, New Jersey.

Waterwisser - Shell Oil: Forty-seven foot self-propelled barge has attached vee booms in front making an effective 65 foot sweep. Storage capacity 20 tons.

Union Oil Skimmer: Portable device suitable for use on a YP or other such auxiliary vessel. The device utilizes two self-priming centrifugal pumps. The pumps take suction from a trough. The trough receives the oil-water mixture after it passes over the lip.

This device was used in Santa Barbara. It was most effective at cleaning up stringers and removing thick slicks. Maximum rating is 25 bbls/day.

b. Suction Devices

One of the most easily obtained devices consists of one or more vacuum trucks on a barge. This design is not for use as a skimmer, but can be used in conjunction with "v" booms to obtain a similar effect. This appears to be the most readily available device in Puget Sound area. The trucks are available from: Airo Company, 2103 E. 112th, Tacoma, Washington; Christel Septic Service, 5626 Waller Road, Tacoma, Washington; A-1 Sanitary, Puyallup, Washington; or Ryan & Haworth, Lynnwood, Washington.[4]

Rheinwerft Oil Removal Unit: Small devices useful for harbor operation. These units consist of three floating pontoons. A well is centrally located and receives the oil-water mixture.

One pump lowers the water level and a second pump removes the oil. No operating statistics are available. This device requires that the oil be pumpable. Debris will render the unit inoperable. Made in Germany, by Mainz-Mombach.

Esso Recovery: This converted LCM is equipped with four suction devices and a Victor Oily Water Separator.

Buda I: Similar in size to the Esso Revocery. Two suction devices feed a four tank oily-water separator. This self-propelled vehicle is equipped with a debris catcher. Made by Marine Pollution Control Corp., 9010 Roselawn Avenue, Detroit, Michigan.

Slickskim Oil Recovery Device: Similar in design to the other vehicles. Slickskim includes a skimmer head, suction hose, hose bridge and discharge hose. Supposedly operates in rough water by automatically positioning the floating suction head and utilizing a boom to contain the slick. Made by Slickbar, Inc., Sangatuck, Connecticut.

Manufacturers ratings:

i) Model 60 - max. 86 bbls/hour

ii) Model 160 - max. 254 bbls/hour

c. Oleophilic Devices

The workings of such devices are analogous to sponge in water; that is, the fluid is absorbed by the material which is then removed into a container.

Floating Oil Skimmer: Floating discs in series are rotated. Oil is removed from the water surface. Scrapers remove the oil from the discs and the oil is stored in tanks. The device is highly sensitive to environmental factors and is adequate for use on minor spills. Made by Centri Spray Corporation, 39001 Schoolcrast, Livonia, Michigan.

Oil/Water Separator: Plastic cylinders in series. This device is similar in principle to the floating disc. Highly sensitive to environmental factors. Rate of removal is not deemed adequate for spills greater than 2700 gallons. Does not seem to be highly sensitive to oil viscosity; however, the manufacturer recommends different cylinders based on the following parameters:

i) Maximum oil viscosity.

ii) Minimum pumping temperature.

iii) Head pressure.

iv) GPH.

Made by Surface Separator Systems, Inc., 103 Mellor Avenue, Baltimore, Maryland.

MV "Port Service": Owned by Port of Baltimore. Designed by Surface Separator Systems. This vehicle has been highly publicized. It is similar in design to their floating skimmer. Recovery rates are given as 200 to 500. Highly sensitive to environmental factors. These devices require less carrying capacity than other devices because only 5% of water by volume is picked up.

Mop Cat: Catamaran capable of sweeping a 15 foot swath and reclaiming oil at 1.2 to 3 knot. Revolving drum of hydrophobically treated polyurethane foam. The drum is squeezed free of oil by a metal strip, the oil falls into a sump and is pumped to a container. It supposedly operates in two foot waves. Made by Worthington Corporation, Pioneer Products Division, Livingston, New Jersey.

Oilevator: Continuous belt of oil absorbing material is rolled from a free wheeling pivot extended ahead of the bow of a barge to an elevated roller/scraper and back. The device is capable of recovering 480 bbls/day. The device is operable in 5 foot waves and 20 mph winds. Belt life is 80-100 hours. Made by Bennett International Services, Inc., 5645 Topanga Canyon Boulevard, Woodland Hills, California.

Straw and Straw Recovery: Wheat straw is one of the most readily available absorbents. It will absorb five times its weight in oil. One of the best ways to distribute the straw is the straw mulcher stationed on a barge. Each mulcher is capable of distributing forty-eight tons per day. Approximately 3,000 tons will be required for a 50,000 bbl spill.

A clam shell on an open barge (scow) is capable of picking up tremendous amounts of straw and oil fouled debris.

Several commercial vehicles are available for this purpose. One commonly seen on the East Coast is called the Water Witch. This vehicle is small and maneuverable. It has a pitch fork type device powered by power cylinders on the bow.

3. Cleanup Methods Readily Available on the Sound

Puget Sound offers little in the way of mechanical cleanup devices. If a spill should occur before more devices are available, the use of jury rigs (e.g. wheat straw, mulchers, clam shells, etc.) is recommended.

Pac-Mar[6] has a commercially distributed vehicle similar to the oilevator. The device is capable of being air lifted to the site and is equipped with 1000 feet of plastic boom.

The Puget Sound Naval Shipyard has a skimmer barge similar in design to the Norfolk Naval Shipyard skimmer. Recent reports indicate that this device is no longer operable and is now being used as a personnel and absorbent carrier.

There are numerous vacuum trucks in the area. These trucks should be placed on barges and used as stationary skimmers.

It is estimated that to handle a 50,000 barrel spill, the following items would be required:

a. 3,000 tons of wheat straw.
b. 20 mulchers on barges (480 tons/day cap.).
c. 12 clam shell cranes on barges.
d. 20 tugs.
e. 20 oilevators.
f. one suction-weir type skimmer (Chevron).

B. Chemical Cleanup

The purpose of this section is to evaluate the various chemical agents that can be used in the cleanup of oil spills in Puget Sound region. These agents employ the use of chemicals which, by their own action, will remove the oil from water. Unfortunately, there is, at present, a deficiency of objective quantitative data by which to judge the effectiveness and toxicity of the chemicals. Consequently, this discussion can only serve to collate the available information and outline recommendations and conclusions based on the present state of the art of chemical cleanup agents and procedures. The main types of chemical agents are as follows:

1. Dispersants.
2. Combustion promoters.
3. Biological degrading agents.
4. Beach cleaners.
5. Miscellaneous types.

Perhaps the most widely used chemical agents in the past have been the dispersants. Dispersants cause the formation of oil-in-water emulsions or solutions and thus break down the oil slick and speed up the biodegradation process. Also, dispersants can prevent oil from adhering to surfaces.[7] The effects of dispersants on marine life are usually harmful to some degree[8] and can be disastrous.[9] As a result, the use of dispersants is not recommended by the State of Washington.[10] For this reason, this discussion does not dwell at length on chemical dispersants. But, more thorough treatment of dispersants, as well as the other types of chemical agents discussed here, is available.[11]

Combustion promoters also have been moderately successful in certain cases.[12,13,14] These agents act as a wick enabling the oil agent to sustain combustion until all the oil is burned. However, there are several problems inherent in the use of this particular type of agent. The burning causes air pollution and can present a danger to human life. In addition, the burning agents do not work effectively in windy or rainy conditions - weather conditions common to Puget Sound. Consequently, the potential use of burning agents in the Puget Sound area seems to be extremely limited.

The various beach cleaners are usually of the dispersant or sorbent (discussed later in this appendix) type with the associated advantages and

and disadvantages of each type. The effectiveness of these agents has not been established satisfactorily at this time.

A new type of chemical agent is currently under investigation and shows a great deal of promise. Monomolecular surface films have been used in experimental conditions to prevent the spreading of oil and reduce the oil slick in area.[15] These agents consist of fatty alcohols, glycerol esters of fatty acids, and similar substances. Since the concept of reducing an oil slick to manageable size is an attractive one, the developmental progress of this product should be followed with interest.

Because of the wealth of marine life in the Puget Sound region and the possible harm to this marine life from chemical cleanup agents, it is recommended that any chemical agent introduced into the marine environment be removed.

C. Chemo-Mechanical Cleanup

Chemo-mechanical agents include those chemical materials which, after application on oil spill, must be recaptured physically from the spill site. Within this category of cleanup devices are sinking agents and sorbents.

1. Sinking agents

   Even though sinking agents have proved effective in removing some relatively small oil spills from the surface of the water[16] this method is not recommended for the Puget Sound region. There are two primary reasons for this:

   a. the oil would be harmful to any marine life living on the sea floor, and
   b. the oil may surface at some later time.

   Sinking does not eliminate the oil problem from water, it only temporarily removes it from the surface. After sinking, the oil will contaminate the bottom of the Sound, and most probably will damage the biological species which live on the sea floor. Also, the oil will tend to dissolve into the sea water over a period of time. This is especially critical in the Puget Sound area due to the extensive tidal current system which would tend to spread the contamination over a large area.

A final disadvantage of using sinking agents is the virtual absence of development of recovery devices for recapturing the oil soaked sinking agent.

2. Sorbents

The second type of chemo-mechanical agent, the sorbent, is the one that appears to be the most useful for actual oil spills in the near future. For this reason, the bulk of this discussion will include an analysis of the various types of sorbents currently on the market. Because of cost and availability, straw has been the most commonly used sorbent material in the past. The main problem in using straw and other sorbent materials is that there is as yet no effective way to pick up large quantities of the oil-sorbent product. Also, there is a tendency for a portion of the sorbed oil to drain from some sorbents during harvesting operations. This drainage is usually not accounted for in the manufacturer's effectiveness data so the dosage and effectiveness data supplied by the manufacturers should be viewed with skepticism.[11,17] Disposing of the sorbent-oil product can also be a problem in some cases.

However, given these limitations, it is still believed that sorbents are the best type of chemical agent with which to deal with oil spills in Puget Sound. This is due to the fact that sorbents have the capability of removing the bulk of the oil from the marine environment with little damage to marine life. The newer synthetic sorbents such as polyurethane, polystyrene and styrofoam look particularly promising since they have a very large capacity for absorbing oil and there is a possibility of reclaiming the oil and agent after harvesting.[15,18] The oil is reclaimed usually by the mechanical pressing or heating of the sorbent-oil product.

Floating sorbents include a great number of different natural and synthetic materials which have an affinity for petroleum products and do not have an affinity for water. Sorbents are normally employed as part of a recovery system to prevent slick spreading and facilitate recovery of the petroleum product. A list of the different types of sorbents is included in this section.

Sorbents Evaluation

Sorbents can be grouped into three different groups:

1. Commercial bulk materials (non-foam) \$.30/gallon oil.
2. Straw, \$.03/gallon oil.
3. Polymer foams, \$.15/gallon oil.

A. Commercial bulk materials such as perlite, vermiculite talc, shreaded bark, peat moss. The cost of commercial bulk materials is typically \$100 to \$250 ton and will absorb three or more times their own weight of oil so the approximate cost is \$.30/gallon of oil.

B. Straw is almost universally available at about \$30/ton. It is assumed to absorb five times its own weight of oil and therefore the cost is approximately \$.03/gallon of oil.

C. The employment of "soft" polymer foams (polyurethane, polypropylene, polystryene, and polyethylene), from which oil can subsequently be recovered, is of great potential if systems for broadcasting the sorbent and efficiently recovering it from the water surface become available. Polyurethane foam can theoretically absorb in excess of 90% of its volume and over 100 times its weight in oil. The cost of polyurethane is approximately \$.75/lb. and one pound will typically absorb about five gallons under field conditions making the cost about \$.15/gallon. Some foams can be reused if the oil is squeezed out, but the extra operation and handling facilities might make this method cumbersome.

In the Puget Sound area care must be taken in the use of sorbents. Due to the high currents and wind states, equipment should always be available to handle the sorbent-oil mixture; otherwise this mixture will cause a pollution problem in itself, which might be harder to clean up than the original oil spill by itself.

It might be beneficial to work out a system specifically designed for the Puget Sound area (sorbent-boom combination) that can handle the high currents that exist in the area. Such a system might be sorbents in conjunction with nets, fences, flow through curtains, fir booms, etc. (see Part B of Section III of this report).

A chart listing the sorbents that have possible use in the cleanup of oil spills in Puget Sound is shown in Table 6-1.[19] The parameters in the chart were chosen so as to give an indication of each agent's relative cost and effectiveness. Unless otherwise noted in the Comments column, it can be assumed that the sorbent can be applied manually or by mechanical blowers or mulchers and that the sorbent-oil product can be picked up manually, with the aid of nets or screens or with any of the common mechanical devices currently on the market. The data in the chart includes only those sorbents which are believed to be potentially effective in environmental conditions that are typical of Puget Sound. A list of the manufacturers of these products along with the manufacturers' addresses is also included.

TABLE 6-1: COMMERCIALLY AVAILABLE SORBENTS

Sorbent	Manufacturer or Distributor	Chemical Composition	Price per lb.
Agglomoweb	Collins & Aikman Corp. Roxbord, North Carolina 27575	Polypropylene Fiber web	$.54 FOB N.C.
Bibbipol	Bibby Chemicals Ltd. 8 Stanley St. Liverpool, L1 6EZ England	Polyurethane Foam	By Quantity $.84-$.51
Capillardiamin	U.F. Chemical Corp. 37-20, 58th St. Woodside, N.Y.11377	Urea Formaldehyde Foam	By Quantity $.40-$1.20
Ekoperl	Grefco Inc. 630 Shatto Pl. Los Angeles, Calif.	Water Repellent Aluminum Silicate	$.18
Emtal XT-1	Englehard Minerals Chemical Corp. Memo Park, N.J.	Talc Stearate modified	$60/ton in carload lots
3M Brand Oil Absorbing Microfibers	3M Company Industrial Special Products Dept., 3M Center St.Paul, Minnesota	Polymeric fiber	-
Oil Blotter	Rittiner Equipment Co. 2200 Hancock St, P.O.Box 385 Gretna, La.70053	Carbamide urea Foam	$4.00
Oil Gobbler	Polymer Research Corp. 2186 Mill Ave. Brooklyn, N.Y.11234	Proprietary	$.50 FOB Brooklyn
Perlite King SRD-32	Filter Media Co. P.O.Box 19156 Houston, Texas 77024	Silicone treated Perlite	$2.50/ 4 cu.ft.bag
Mistron Vapor	Cyprus Mine Corp. UMTSO Sierra Division Box 1201 Trenton, N.J.	Talc	4¢-8¢ per lb. in carload lots

Dosage Rate lbs. agent: lbs. oil	Applicable to All Oil Types	Recovery Possible for Oil	Agent	Comments
1:143	Yes	Yes	-	
1:50	Not weathered crude. Less effective on thicker oils.	Yes	-	Sold as 2 component liquid system. Mixed 1:1 in foam dispensing machine. Foam shredded before application.
1:84	Not effective on high viscosity oils.	Yes	Yes	Can be produced from 2 aqueous solutions.
1:5	Yes	No	No	Can be harvested by suction hoses.
2:1 gallon oil	Yes	No	No	
1:20	Less effective for heavier oils at low temperature.	Yes	Yes	
1:50	Yes	Yes	-	
1:50	Yes	Yes	Yes	
N/A	Less effective on lighter oils.	No	No	Perlite may be recovered by burning, though its use is then limited to soil conditioning.
2-3 parts agent to each pt. oil	Not effective on Bunker C or heavy fractions	No	No	Must be agitated.

TABLE 6-1: COMMERCIALLY AVAILABLE SORBENTS (Cont'd)

Sorbent	Manufacturer or Distributor	Chemical Composition	Price per lb.
Powdered Pine Bark	American Modoc, Inc. 5532 Beach Blvd. Buena Park, Calif.	Shredded fibers	$5/100 lb.bag
Scott Industrial Foam	Wilshire Foam Products, Inc. 2665 Columbia Torrance, California	Complete line of open and closed celled foams	$.50 per board ft.
Sorbent Type C	Clean Water, Inc. P.O.Box 1002 Toms River, N.J.08753	2 types clay with Binder and Perlite	By Quantity approx.$.22 FOB N.J.
Stepanpol RM 15	Stephan Chemical Co. Edens and Winnetka Ave. Northfield, Ill.60093	Polyurethane foam	$.53 liquid $.58 foam
Straw		Wheat straw	$.015
Strickite	Strickman Industries, Inc. P.O.Box 208 Hillsdale,N.J.07642	Proprietary white granular solid	$.42 FOB N.J.
Typar	E.I. du Pont de Nemours, Inc. Center Rd. Wilmington, Delaware	Polypropylene porous fabric	$1.00 delivered
Tyvek	E.I. du Pont de Nemours, Inc. Center Rd. Wilmington, Delaware	Polyethylene porous fiber	$1.00-$1.10
Vermiculite	California Zonalite Co. 5440 San Fernando Rd.West Los Angeles, California	a porous, granular silicate	$1.00-$1.65 per 4 cu.ft. bag

Dosage Rate lbs. agent: lbs. oil	Applicable to All Oil Types	Recovery Possible for Oil	Agent	Comments
2:7	N/A	N/A	N/A	
1:5	N/A	N/A	N/A	
approx. 1:4	Yes	No	No	Used product suitable for highway road beds.
1:60	Yes	Yes	Yes	Foam made by mixing liquid components. Five minutes cure time before shredding.
1:5	Yes	No	No	
1:10	Yes	Yes	Yes	Can be put in porous fabric tubes for containment of spills.
1:10	Yes	Yes	Yes	Use on belt type recovery machine or tow behind vessel.
1:10	Yes	Yes	No	Also available in loose fiber form.
1:3	N/A	N/A	N/A	

D. <u>Biological Cleanup</u>

In most waters, and particularly in areas where frequent small oil spills occur, there are a population of microorganisms that are able to degrade oil. "Organisms known to attack hydrocarbons number in the thousands."[11] The microorganisms are selective, and may degrade only a certain type of hydrocarbons, but generally they break down several kinds. The hydrocarbons utilized first are the simple type short chains, then the longer chained and the branched hydrocarbons are used. The cyclic and aromatic hydrocarbons are the most resistant to microbial attack and the most toxic to aquatic life.

A substantial amount of the hydrocarbons can be transferred to protein by the microorganisms. The fraction is determined by the type of organism, the oxygen available, sufficient nitrogen in form of NH_3 or NO_3, phosphorus available and the type of hydrocarbon used. An estimate of the protein fraction is 25-35% of the original carbon content. This protein is then incorporated in the natural food chain, as the predators take care of the microorganisms.

The rate at which the degradation of hydrocarbon occurs is temperature dependent. Some microorganisms are active even at a temperature of $0^{o}C.$, but the activity is very slight. As an illustration of the effect of temperature, it has been found that some organisms decompose seven times as much paraffin at $55^{o}C.$ as at $22^{o}C.$[7]

After the recent oil spill in San Francisco Bay, it was roughly estimated that about one-half of the escaped oil was recovered. The remaining portion had sunk or been emulsified, and was impossible to recover. This part was left for natural microbial decomposition.

It is conceivable that the use of microorganisms, as an aid in oil spill cleanup, would be valuable in lessening the strain on marine flora and fauna. Research is already being done by several firms, but the information is proprietary The procedure is to spread a mixture of fertilizer (ammoniumsulphate, etc.) and dried bacterial culture. The culture should be adjusted to the particular oil of the spill. The cleanup will proceed in the same manner as before to recover as large portion as possible. The remaining oil would be subject to a much faster and more complete breakdown than before. The addition of microbes will lower the dissolved oxygen of the water, but this could be corrected by introducing air to the water through perforated hoses.

The public information about this kind of "treatment" is not very complete, and Zobell's words from 1946 ring very true today, 25 years later: "It is an anomalous situation that the petroleum industry, praiseworthy for its many outstanding scientific and technological achievements, has devoted so little attention to the effects of microorganisms on petroleum or its products."[7] But, through the efforts of individual firms without connection with the industry, some progress is being made.

E. Endnotes for Appendix 6

1. Conversation with Captain Bob Thurman, Oil Spill Services, Inc., at University of Washington, February 11, 1971.
2. Santa Barbara spill slide presentation by Ward Swift, Battelle Northwest Laboratories, to OE 551 class on January 28, 1971.
3. San Francisco spill slide presentation by Jules Mayer, Standard Oil of California, at University of Washington, February 23, 1971.
4. Battelle Memorial Institute, Study of Equipment and Materials for Removing Oil from Harbor Waters, Battelle Memorial Institute Report #C12-70-001, August 25, 1969.
5. Puget Sound Tug and Barge Company, and Pac-Mar Services, Puget Sound Oil Spill Response Plan, (a list of cleanup devices is included), April 1, 1971.
6. Pac-Mar Services, Inc., 3406 11th S.W., Seattle, Washington.
7. David Hoult, Oil on the Sea, 1969, p.35.
8. H. B. Tracy, et al.,"Relative Toxicities and Dispersing Evaluations of Eleven Oil Dispersing Products,"State of Washington Water Pollution Control Commission and State of Washington Department of Fisheries, December 1969.
9. Committee of Scientists on the Scientific and Technological Aspects of the Torrey Canyon Disaster, The Torrey Canyon, Report of the Committee, London, Her Majesty's Stationery Office, 1967.
10. Department of Ecology, Oil Spill Laws and Oil Spill Emergency Procedures, October 1970, p. 10.
11. Battelle Memorial Institute, Oil Spill Treating Agents - A Compendium, May 1970.
12. "Controlled Burning of Heavy Fuel Oil Spill at Sea," Ocean Industry, June 1970, p. 61.
13. Navy Ship Systems Command, The Arrow, Navyships Document #0994-008-1010, pp. 87-89.
14. Hoult, pp. 81-89.
15. Ocean Industry, August 1970, pp. 39-47.
16. Ocean Industry, July 1970, p. 37.

17. *Ocean Industry*, July 1970, pp. 43-44.

18. *Ocean Industry*, August 1970, p. 36.

19. Principal sources of data include Battelle's compendium and the Dillingham report on oil spills.

APPENDIX 7

CRITIQUE OF COAST GUARD CONTINGENCY PLAN[1]

In the aftermath of the April 1971 spill at Anacortes, Washington[2] criticisms are being voiced from a multitude of quarters as to what should be done to prevent another occurrence and what was done incorrectly in respect to cleanup of the pollutant. If one is able to take a position comparable to that of detached objectivists it will be readily apparent that the concentrated efforts to correct the error factor that resulted in a spill are only clouding the issue, primarily because of the very low probability of a spill happening in the same fashion in the future. Also, the criticisms of what should have been done to handle the spill serve the same purpose in many respects. The effort here is not to judge but to evaluate a system approach that is presently the primary modus operandi for handling oil spills. However, it must be recognized that our present systems have come from drawing boards and very limited experience, not from the long-term experience and practice that is so beneficial to the development of a societal, regulatory and protective service. Also, with our massive and sophisticated technology, oil handlers still are baffled and forced to treat oil spills with primitive technology. Furthermore, for those experts of the Torrey Canyon, the Arrow, and other spills, let it be duly recognized that a spill in Puget Sound waters is different than a spill in the Gulf of Mexico or off the coast of California and a spill on inland waters is different than a spill on open waters. The efforts of many to create a world-encompassing plan of attack to handle all oil spills leads to generalizations that provide little in the way of concrete assistance on the working level. Therefore, this critique is geared to the evaluation of the United States Coast Guard plan as it relates to Puget Sound, an inland body of water, and the existing resources of the Puget Sound region. The section numbers in the sequel refer to the U.S.C.G. Plan (1).

Sections 105.6, 105.7, 105.8;

MINOR/MODERATE/MAJOR APILL DEFINITION

These sections treat degree classifications in regard to spillage in terms of gallonage or barrels with little consideration of toxicity and solubility of particular pollutants. Possibly a more sophisticated classification procedure should be developed where the classification is a function of two factors or more. A matrix chart could be developed that will classify spills by considering amounts, types, and chemical characteristics. Such an approach would serve to refine the input to those responsible and key a greater response from those who are operationally oriented as to clean up rather than scientifically oriented. The people, most of them, who clean up a spill are not spill experts, biologists, or chemical engineers. Many are bargemen, crewmen, dock workers, and supervisory personnel. All efforts must be made to provide these people with the tools to make adequate evaluations so that all concerned do have full opportunity to meet the demands of the occasion.

Sections 1212, 1213, 1214;

MINOR/MODERATE/MAJOR SPILL RESPONSE

As was seen in the Anacortes spill initial response the spill was classified as minor and not requiring immediate inspection or mobilization. The plan states that when in doubt treat as a spill of consequence. True, the Coast Guard had no reason to doubt the validity of their initial report in the Anacortes incident. However, due to an error mobilization for the handling of a major spill was delayed by precious hours. It is recommended that a review be made of the practicality of inspectors coming only from Seattle on all spills. When considering the human element involved and the propensity not to respond to a minor spill in the darkness of early morning hours and the compounding that may take place due to the reporting of inaccurate or incomplete information a follow-up plan of verification on all but the most miniscule spill would be advantageous. Now, if, with the cooperation of other agencies and other Coast Guard facilities, there was a stand-in On Scene Commander (OSC), who would be called upon to go directly to the scene of the spill upon notification by the Sub-Regional Response Center (SRC) or the Regional Response Center (RRC), and verify the information the less than adequate procedures that now exist might be properly reinforced. As in the case of the Anacortes spill, with the information of #2 Fuel Oil and of 5,000 gallons, the Coast Guard could have contacted their stand-in OSC (from State Fish & Game, Department of Defense, Health Education and Welfare, Department of Transportation, United States Coast Guard, or any other designated representative) who would have gone immediately to the site and talked with the shore and launch facilities to ascertain the true nature of the spill. Upon verification of original data and/or acquisition of additional or different data the stand-in OSC would have reported to the proper SRC or RRC thus instituting adequate response or allaying fears and doubts. This approach could be used for all spills using aircraft, launches, or autos. The only criteria that must be met is cooperation by agencies that all derive support from public tax dollars and that all have mutual interests.

Section 1500;

COMMUNICATIONS

Consideration should be given to the utilization of a private communications facility. At present, the Coast Guard carries the burden of notifying all concerned agencies. In turn, all of the various agencies have their own ways they wish to be notified. Recognizing that there is a certain amount of friction between segments of the government on the national and state levels as well as within these levels, the idea of a 24-hour seven-day-a-week service by a neutral party seems most advantageous. The task then would be for the informer of a spill to notify the Coast Guard who would then notify the central communications service so that they might notify others. By using a private answering service each agency could institute its own procedures for receiving notification as well as making timely adjustments and requests. This would be all possible because such a service would be geared to meeting the varying demands of many customers. Also, cost would be minimal. At present many of the agencies do not have the funds to maintain twenty-four hour service on all days. By using the central communications service approach costs would be shared by all giving full service at minimum expense. Furthermore, there would be a buffer between agencies thus reducing the friction that must increase during times of stress and crisis.

In addition, to make sure that verifications are made by the service to all concerned, a random number table could be used by each agency as a code source. When the agency is notified, the answering service would request a verification code that they would record and that could be double checked when questions arise. One code number could be assigned for each day.

This system also provides the conditions more conducive to double checks and verifications, thus reducing the human error factor. It also provides multiple channels for communications and could greatly facilitate the installation of temporary on-site communications facilities. Without a doubt, the people involved in spill reporting and cleanup are not trained communicators and few of their budgets provide for hiring a full-time communicator. The central communications service idea provides for on-tap communications expertise as well as communicators experienced in dealing with those that often provide communications facilities; private companies.

Section 2503.20;

FUEL OIL GRADE

This section should be expanded to take into account the fact that all refined products are not the same and that they should be treated differently. Mention should be made of the increased solubility and toxicity of the more refined products, especially in light of the fact that the bulk of the petroleum products shipped on the waters of Puget Sound are refined.

Section 2511.1-1;

REMOVAL

Consideration here should be made of specific and unique beach types. The implications are that where a highway exists or some other access heavy equipment can be used for cleanup. The danger involved is that beaches and shorelines vary in richness of life forms and terrains. Some are sandy and sterile,very conducive to dumping a spill and cleaning up with front end loaders. Others are rocky and veritable habitats of life, not condusive to the equipment nor the technique. The OSC should verify the ecology of a shore before he authorizes or institutes dumping or cleanup techniques. This aspect of the problem re-emphasizes the dire need for baseline information and levels of safeguard priority. The most feasible approach to the mammoth task of determining such information would be a joint venture of national, state, and local agencies as well as industry to finance and manage such a study. The thought of all working together may frighten many but the result if all do not work together and share the burdens together will be that all will lose.

Section 3110;

CRITICAL WATER USE AREAS

In this section the Coast Guard should reconsider their priorities and criteria for ordering the sites. Reference might be made to the Puget Sound and Adjacent Waters Task Force Report. The ports of Seattle and Tacoma have great value with respect to capital goods that are replaceable through the use of our society's factors of production. The idea here is that damage to these areas would be more easily repaired and possible to repair. Also, these areas in terms of environment are already adversely affected by heavy shipping, shore facilities, sewage, motor boating, and other pollutants. On the other hand, the Anacortes region, rich in biota, is extremely susceptible to oil pollution damage that cannot be repaired through the use of the factors of production. Furthermore, there is strong evidence that the natural environment of the North Puget Sound Region might very well be, potentially, the most valuable part of Puget Sound from purely an economic point of view. This will come to pass if projections are correct and tourism becomes the number one industry of the Pacific Northwest and the major contributor of jobs and income to the state's economy.

In essence, the priorities of this section might very well be reversed. If they had been in a more meaningful order at the time of the Anacortes spill, even 5,000 gallons of #2 Fuel Oil would have elicited a greater concern and response on the part of all concerned.

Section 3160.1;

SCIENTIFIC ADVISORY GROUP

The plan calls for a team of experts ready to mobilize in the event of a spill. In the plan three faculty members of the University of Washington are listed as the Puget Sound Region team. Yet, a group of students under the auspices of Professor Juris Vagners are doing more, apparently, to monitor the Anacortes situation than the Coast Guard. Furthermore, after interviewing representatives from the Environmental Protection Agency, Washington State Department of Ecology, and the United States Coast Guard there was no mention of utilization of the stated standby team. As was stated by R.E. Hammond in Commandant Notice 3010, the responsibility of the Coast Guard is ongoing and they have the responsibility to monitor all oil activity. The implication here is that the oil is active in the water and that the Coast Guard should utilize its scientific team during and after a spill to organize monitoring of the impact of oil on the environment. Having an ongoing responsibility, the Coast Guard should monitor oil far beyond the immediacy of the spill.

TRAINING MANUAL AND TESTING TOWARDS CERTIFICATION

The National Safety Council now has a program used by most large companies and government agencies; the Defensive Driving Course. It might be worth considering a similar conceptual plan for oil spill prevention, containment, and cleanup. The plan would be very practical, as costs would be spread over all of the users rather than with a central point such as the limited Coast Guard budget. Initially, the Coast Guard could prepare a manual for training purposes or a course, that would cover such areas; Oil Spill Laws and Liabilities, Types of Oil and Their Performance under Varying Weather and Sea State Conditions, Prevention of Oil Spills, and Notification and Implementation of Cleanup Operations. Actually, four levels of instruction could be developed that would cover the requirements from the workers position through that of the OSC. Then, in conjunction with the Coast Guard coordinated educational program, industry, agencies, and private sectors could be encouraged to send a representative for training to the Coast Guard, at a nominal cost to help spread the burden, and then back to his sponsor to administer training programs to other members of his firm or agency. Along with the above efforts, the Coast Guard could authorize certification as to competence in oil handling and spill conditions. Again, a program that is not new but one that is successful in other areas, and might well be so here given the same crucial criteria as the communications service center and Spill Response section: cooperation.

Section 3130;

EQUIPMENT AND SERVICES

Another area where the plan is not fully followed when it should be is in equipment and services. For example, the plan calls for a firm commitment of personnel, equipment, and supplies. However, it appears that the commitment is less than firm. First, personnel are only specified for the Coast Guard by site and how many man the site. In reality, not all can respond to a spill. Of the sixteen men at the site, what is a skeleton crew? Agencies did not submit quantities or types of personnel available. If the report included even a range of personnel such as 3-5 biologists from

Fish & Game or five men from the Lopez Island Coast Guard Station, the tasks of the OSC and the Spill Cleanup Coordinators would be greatly facilitated. In the plan, equipment and supplies are mentioned. The list seems most comprehensive though it is less than firm. A range of prices is not set and a range of quantities available is not stated. Somehow the OSC must be able to gauge his resources and his capabilities if he is going to be able to most effectively meet the challenges of a spill in Puget Sound waters.

GENERAL COMMENTS

The contingency plan seems most adequate at the upper levels of responsibility. It is clear who is to contact who, how they are to do it and when. In fact, if anyone who is confined to RRC or above follows the clear guidelines he cannot get into trouble. However, for those who must actually cope with the problem on an operational level the picture is not so clear. For instance, under what conditions does an OSC take full charge of cleanup and what are the legal liabilities when he does so. The comments have been to the effect when the "spiller's efforts at cleanup are grossly inadequate." What is "grossly?" What is "inadequate?" The OSC is left out on a limb where action will seldom be praised and more often than not detrimental because it belongs in the legislature rather than the position of OSC. However, there should be common guidelines that he can follow. If the OSC has complete authority and responsibility should he take-over control? Then how far will he be backed; i.e. what is the breadth of complete authority and responsibility. If the OSC, as an advisor, recommends the closest and most adequate supply of hay or equipment for cleanup does he have to worry about the petty arguments after the fact about the Coast Guard recognizing or catering to a specific product or supplier? In essence, the OSC is sitting on a hot box of issues that leave him little room for anything but doing as little as possible. Such negative incentive can only serve to hamper rather than encourage resourcefulness, creativity, and positive and immediate action that may be discounted after the fact but appropriate at the time under the circumstances. In summary, there should be a section specifically relating to the position of OSC that specifically states the do's and don'ts of the position and the official Coast Guard position, should difficulties follow. There should also be adequate reference to the specifics of a spill in respect to ordering of equipment and supplies, mobilizing resources, and coordinating efforts in the field.

Endnotes for Appendix 7

1. U. S. Coast Guard, Seattle Coastal Region 0.1 and Hazardous Materials Pollution, Contingency Plan, 1 December, 1970.
2. Oil spill incident at Texaco Refinery, Anacortes, April 26, 1971, involving the discharge of approximately 242,000 gallons of #2 diesel oil into Guemes Channel during the loading of a United Transportation barge.

APPENDIX 8

UTILIZATION OF COMPUTER MODELS IN PUGET SOUND

The possibility for simulating Puget Sound on a computer will more than likely fascinate the layman; but to those more knowledgeable in computer science, it may evoke more criticism than plaudits. The many varied physical characteristics of Puget Sound, together with its multitude of islands and inlets, requires the programming of an enormous quantity of data.

This appendix presents three separate, but related, operations utilizing a computer model for the Sound. Preliminary analysis of one operation is presented in detail. The selection of the three tasks was influenced by the current state-of-the-art technology (whether computer-oriented or not) which can be summed up to be, at best, minimal. The three operations are:

1. tracking the behavior of an oil spill;
2. traffic control/advisory system;
3. response capability to an oil spill incident.

These are discussed individually in the following pages. The response capability model integrates the outputs of the first two operations in addition to utilizing other input data.

A. Oil Spill Movement Prediction

1. Introduction

One of the major difficulties in the present state-of-the-art of oil spill containment and cleanup is the tracking of the behavior of an oil slick. Various techniques of remote sensing have been discussed in this report; but as yet, their effectiveness has not been proven.[1] If the meteorological and hydrographic conditions of a given portion of Puget Sound are known, the movement of oil on water at a given site could be predicted by utilizing a computer model. The computer system can ask the model a pre-arranged set of queries, and oil spill response operation could be based on the model output. For example, the model will be asked where a spill will spread, and equipment could be deployed to this site by the time the oil reaches this location. It must be remembered that answers from the computer are only as good as the input data.

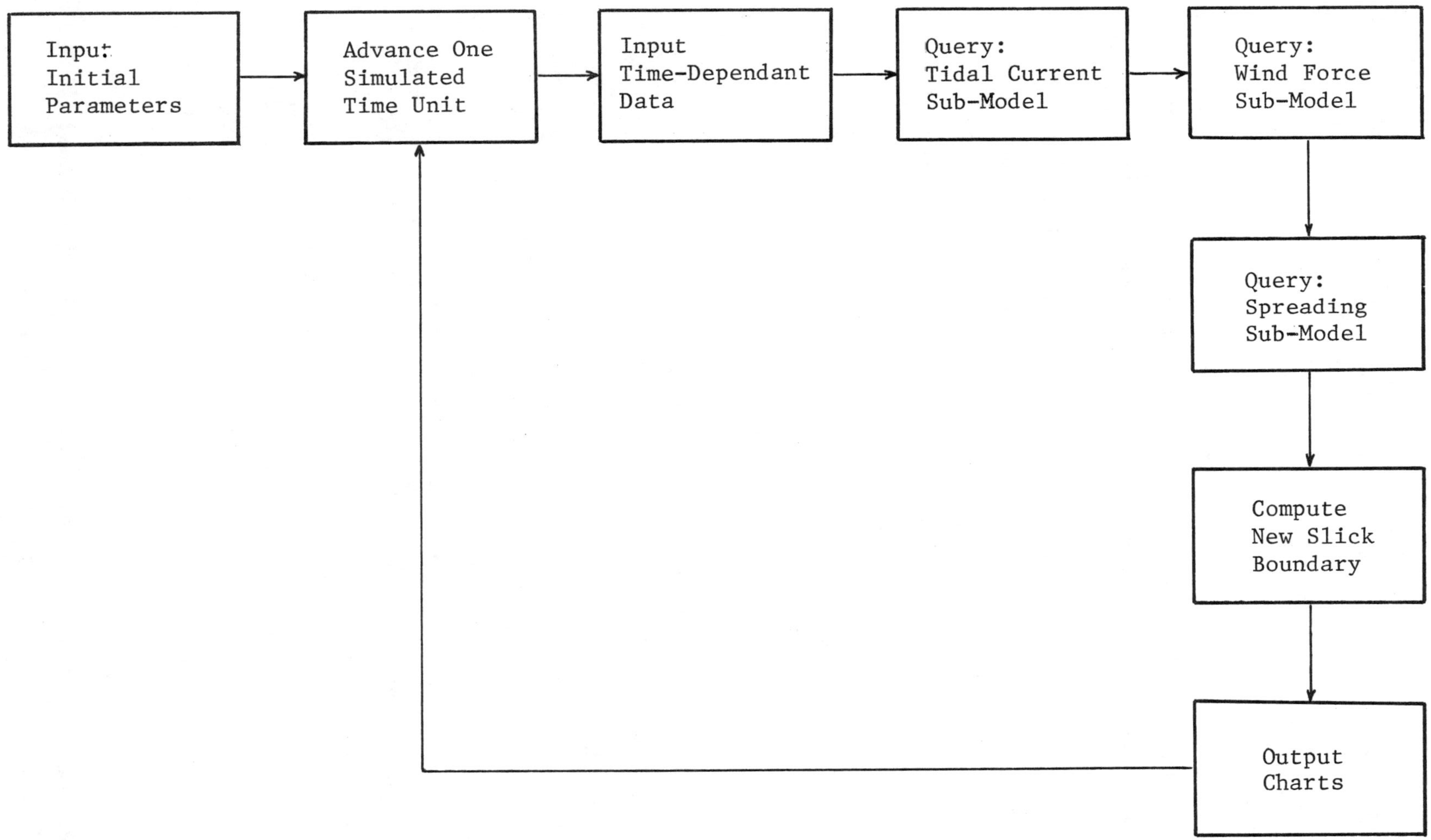

Figure 8-1

BASIC FLOW CHART FOR COMPUTER MODEL

Thus, for the discussion in this appendix, it is assumed that accurate information is available on Puget Sound meteorology and hydrography.

The following is a discussion of a model to simulate the spreading of oil on estuarine waters, with direct application to Puget Sound. It is emphasized that this was a pilot project, and that further refinement will increase the model's accuracy. To the study team's knowledge, it is the first attempt known to use computers for oil spread predictions in complicated estuarine areas.

The model is intended to be used by personnel involved in an oil spill cleanup situation. Among other things, this leads to the selection of computer-drawn charts as the output medium. This in turn leads to the concept of a data base using computerized maps, especially where land-areas are concerned. The emphasis is on a tool for oil spill response capability, rather than a study of the mathematics of oil slick movements.

2. Model Structure

The model was programmed in FORTRAN for the CDC 6400 at the University of Washington Computation Center.

Figure 8-1 is a block diagram showing the basic sequential steps in model operation. Each of these steps will be discussed below to provide an overview of the structure of the model.

Input Initial Parameters -

The model is currently designed to operate without permanent data files in a batched-job environment. This means that, at the beginning of each model run, the data base must be loaded into the system from punched cards. These are called parameter cards and their syntax is laid out in Table 8-1. The sort key is used to order parameters into sequence based on when, in simulated time, that parameter takes effect. All "initial" parameters are identified by a date of year 00, day 00 and a time of 00 hours and 00 minutes. The data base information loaded at this time is as follows:

TABLE 8-1: SYNTAX FOR COMPUTER MODEL

```
<parameter card>                   : = <sort key> <variable data>
<sort key>                         : = <year> <day> <hour> <minute> <code>
<year> <day> <hour> <minute>       : = <number> <number>
<code>                             : = <major code> <minor code> <card count>
<major code>                       : = [A|C|D|I|P|T|W]
<minor code>                       : = [G|S]
<card count>                       : = <number> <number>
<number>                           : = [0|1|2|3|4|5|6|7|8|9]
<variable data>                    : = depends on parameter type
```

a. Geographical bounds of data base expressed as North latitude, South latitude, East longitude, and West longitude.[2]

b. Tidal currents for sea areas, and denomination of land areas. Current information is referred to a standard point and consists of the difference in time and current magnitude between the area specified and the reference point. Thus an area may have a flood tide which occurs 30 minutes later than flood at the reference point and with a current of 1.2 times the current at the reference point. Land areas are noted as those areas having a special, reserved, reference point.

c. The simulated date and time at which the spill begins and ends.

d. The amount of simulated time that is to elapse between calculation of slick boundaries, and the simulated time lapse between output displays. The latter must be greater than or equal to the former. The minimum time expressable is 1 minute.

e. The latitude and longitude of the source of the spill. The model cannot except a slick already on the water.

f. The rate at which oil is being spilled in barrels per minute, and the total amount of oil to be spilled in barrels. The spillage rate may be set to zero, in which case the oil is considered to have appeared instantaneously at a point source.

Currently, the type of oil spilled is assumed by the system to be an oil with a specific gravity of 0.886.

Advance One Simulated Time Unit -

All internal clocks are reset according to the simulated time that is to pass between computations. The model operates on a time-line rather than an event-line basis.

Input Time-Dependent Data -

All parameter cards which specify a date and time less than or equal to the current simulated time are loaded into the system. Parameters which may be entered here are:

a. Tidal currents for standard reference points. Each card specifies the standard reference point, time of day (in addition to being part of the sort key), type of current situation: ebb, flood or slack tide, and the current in knots (0 for slack tide).

b. The elapsed simulated time between computations and the elapsed time between outputs may be altered.

c. The force and direction of the wind over various' areas may be entered.

Query: Tidal Current Sub-Model[3] -

For each point on the edge of the slick, the tidal current sub-model determines the tidal current at that point. The tidal sub-model may determine that a point is on land, in which case, no further calculations are done for that point.

A point is identified by its latitude and longitude. These parameters are sent, along with the fact that we desire a search of the tidal data, to special routines which look in the data base for the entry that contains the point desired. All data associated with this entry is returned to the tidal sub-model.

The data returned in the current version of the model are:

a. The standard reference point, identified as a number.

b. The time at which slack water occurs and the time at which maximum current occurs, relative to the standard reference point. For example, the maximum current entry may be +30 minutes. This means that the maximum flood or ebb tide occurs 30 minutes after flood or ebb at the reference point.

c. Two multipliers, one for ebb and one for flood, which when applied to ebb and flood predictions at the reference point, yield ebb and flood predictions for the points covered by this data base entry. These multipliers apply only to maximum flood and ebb.

d. The true compass heading for the flood direction and the ebb direction.

The tidal sub-model then looks up the data for the standard reference point. This data consists of the time of day for each slack, flood, or ebb tide, along with the current force. The entry closest in time to the current simulated time is used; and this, with the data base information results in a vector in knots and compass degrees representing the tidal current for the point initially specified.

Query: Wind Force Sub-Model –
This is a simpler version of the mechanics of the tidal sub-model. The data base contains wind force and wind direction and this vector is the output of this sub-model.

Query: Spreading Sub-Model –
This sub-model is based on an equation provided by Battelle Memorial Institute[4] which gives slick thickness as a function of the elapsed time, the volume of oil, and the specific gravity of the oil. Assuming the oil to lie in a circular mass, the spreading sub-model obtains the radius of the circle thus:

$$\text{AREA} = \text{Volume/Thickness}$$
$$\text{RADIUS} = [\text{Area}/\pi]^{1/2}$$

The difference between this radius and the radius computed for the previous time segment is returned as the amount of spreading that has occurred over the current time slice. Obviously, this calculation makes the assumption that the amount of spreading on calm seas with no winds approximates whatever the spreading is under real circumstances. For the sake of convenience, this sub-model returns its results in units of knots as a scalar quantity. Other sub-models produce vectors with units of knots.

Compute New Slick Boundary –
The edge of the slick (as intuitively defined) is referred to as the boundary. The boundary set consists of 360 points on the boundary. Each point corresponds to one true compass degree around the edge of a presumably circular slick (see Figure 8-2). Initially, all 360 points are the same, i.e. the source of the spill. For each point P_i $(1 \leq i \leq 360)$ in the boundary set (see Figure 8-3):

FIGURE 8-2

HYPOTHETICAL UNIFORM SLICK & SOME OF THE BOUNDARY POINTS

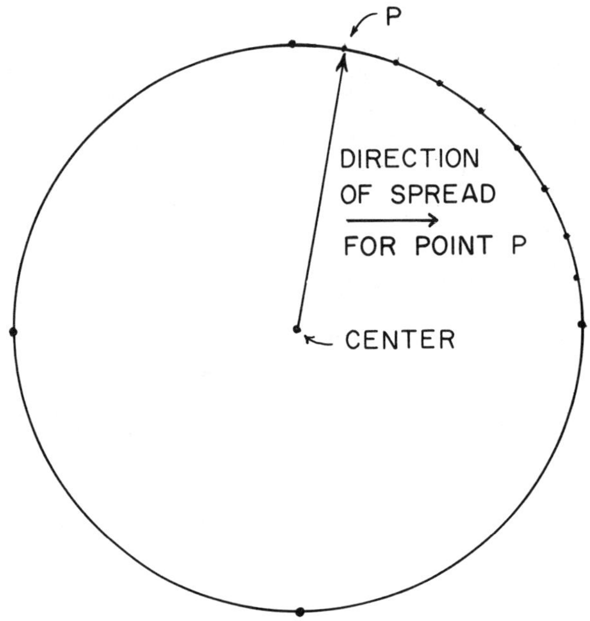

FIGURE 8-3

BOUNDARY CHANGE

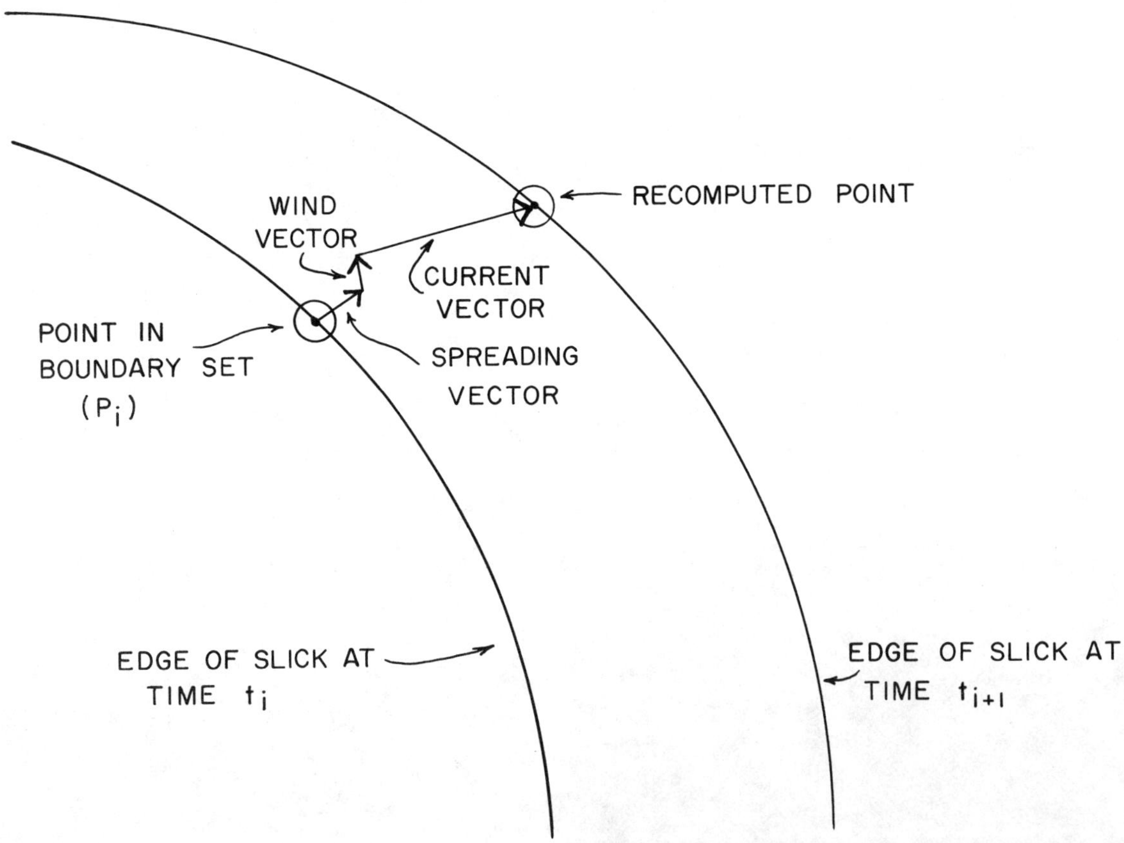

$$P_{i\,new} = P_i + 0.033\,\overline{(wind)} + \overline{current} + \overline{spread}$$

$\overline{WIND}$ is returned by the wind sub-model

$\overline{CURRENT}$ is returned by the tidal current sub-model

$\overline{SPREAD}$ is (s.sin(i), s.cos(i))

where s is returned by the spreading sub-model, and (-) denotes a vector.

The wind and current factors are noted on in the Battelle report,[5] the spreading factor is as reported under the spreading sub-model applied along the line from the center of the slick to P_i. The model does not recognize the change in direction which occurs as the slick goes from circular to elliptical under constant wind and current vectors with spreading being applied in all directions. This error is not considered significant since the spreading velocity itself is small (about .2 knots for 5,000 bbls of crude one hour after spill).

This computation is done to all these points in the boundary set except those which have moved onto land areas as a result of previous calculations. If the current calculation moves a point onto land this is not noted until the next iteration. This makes it possible for the model to "move" oil far onshore, since the "beach" is not recognized.

Output Charts -

This stage is executed at intervals specified by the user. This stage of the model may run less frequently than the computation stage but not more frequently. The most effective way to display the output of the model (the contents of the boundary set) is to mark all the points of the boundary set on a nautical map. Many satisfactory schemes are available. For simplicity, the geographical boundaries were produced on a computer printer. This is equivalent to trying to produce maps on a typewriter.

The model currently produces maps in sections one nautical degree by one nautical degree. For example, a map section will cover

FIGURE 8-4

SAMPLE AREAS FOR DATA BASES

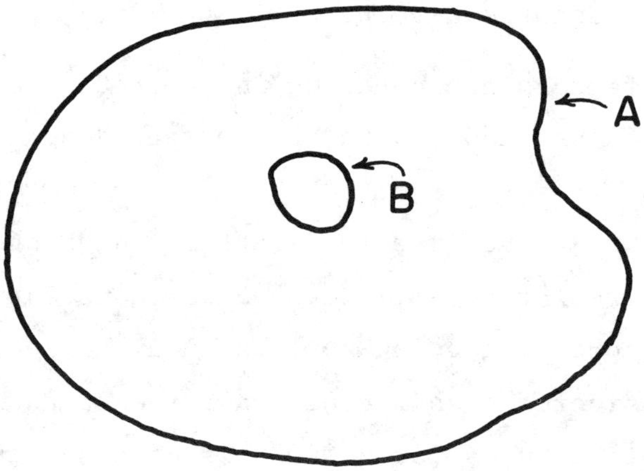

E = ENTIRE AREA

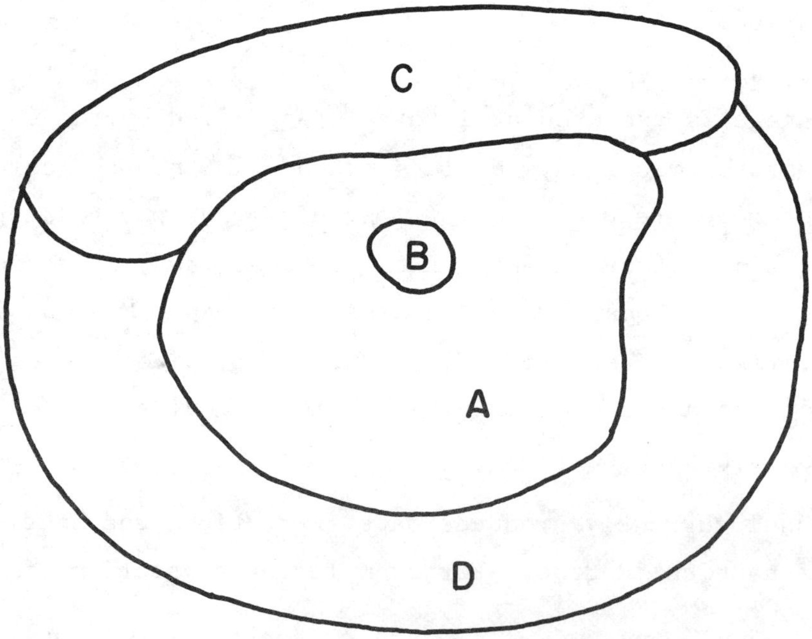

48°-49°N. lat. by 122°-123° W. long. If more than one section is required, they can be produced. Also, land points are marked in to aid recognition. The resolution of a map is one square nautical minute. The resolution is a function of page size. This produces a very crude map, since any land point in a square minute causes the whole square to come out as land. The overall quality of the maps could be improved only by better resolution, producing enough land points to cover a wall.

Some trace values are also produced. The average tidal current and the average net wind effect are displayed in x-y coordinate form. Other values, namely scalar, are expressed with a zero-y coordinate. These include the spreading force from the spreading sub-model, the thickness in centimeters, and the volume of oil spilled.

3. Data Base Structure

Organizing the data base for this model is not unlike organizing any random access data base, except for one index: geographical location. In general, a specific datum, or attribute, applies not just to one point, but to an entire area. An island, for example, represents a collection of points with a common attribute: they are all land points.

As a consequence of this point of view, data base entries are associated with an "area," and there exists a simple mechanism for associating a point with the area that contains it. In addition, two concepts of area organization are introduced: covering, and exceptions.

An exception is an area which is a proper subset of another area. In the top frame of Figure 8-4, area B is an exception in area A. Conceptually, one might think of A as a water area, and B as an island. Thus all points in area A, with the exception of area B, have the same attribute: water.

In point of fact, the tidal current data base assigns to each area a specific current characteristic, with land points represented by special current data. In the bottom frame of Figure 8-4, areas

C and D have been added. C, D and A represent water with three different tidal current characteristics, and B is an island. Areas A, C and D, together (A∧C∧D) provide a "covering" of area E. Note that each of A, C and D, are "exceptions" to area E.

The data base coordinates all this by a series of linked pointers. Let us take area E as an example. Among the data entered for area E is a pointer to a "neighboring" area, and a pointer to an exception area. Thus:

```
E ────────→ neighbor
│
↓
exception
```

There is no neighbor to area E, so the pointer is empty: E──→ ∅. For exception areas we have A, C or D. Thus we can write:

```
E────────→ ∅   or   E────────→ ∅   or   E────────→ ∅
│                   │                   │
↓                   ↓                   ↓
A                   C                   D
```

Let us choose A as the exception. What about C and D? They are neighbors of A; thus:

```
E────────→ ∅
│
↓
A ────────→ C ────────→ D
```

and we can add area B:

```
E ────────→ ∅
│
↓
A ────────→ C ────────→ D
│
↓
B
```

This method of organization provides some notable advantages. First, without the linking of exception points, area A, for example, would have to be described as an area with a "hole" in it; very difficult. Additionally, the number of sequential searches required is much reduced, saving a considerable amount of computer time. The more areas there are in the system, the more the linked list tends to "zero in" on the right area.

The only difficulty is that the computer is not very good at representing arbitrary geometries. As a result, all areas are represented as rectangles. Figure 8-5 shows a land area being approximated by rectangles. Each rectangle is a separate data base entry with two latitudes and two longitudes supplied. Of course, data must also be supplied for each rectangle. Defining and entering rectangles into the system is a very tedious operation.

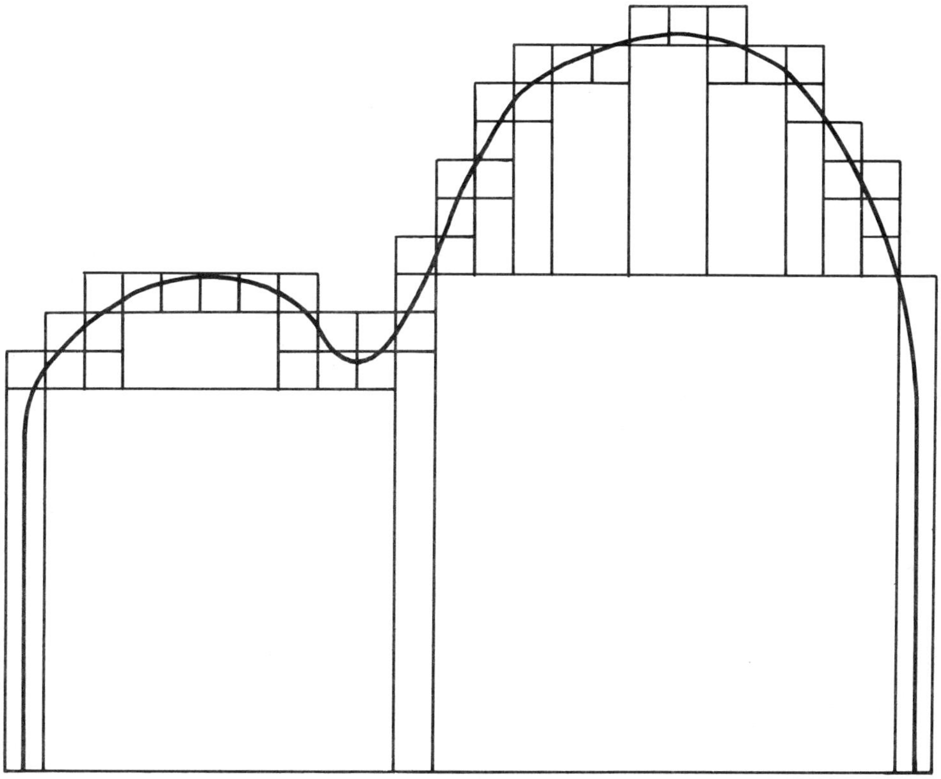

FIGURE 8-5: APPROXIMATION OF LAND AREA BY RECTANGLES

The model's current data base consists of three parts: the wind data-geographically keyed, the tide data-geographically keyed, and the standard reference points-time keyed. The two geographical section are represented by four arrays:

a. AWK - Array Wind Keys,
b. AWD - Array Wind Data,
c. ATK - Array Tide Keys, and
d. ATD - Array Tide Data.

These are FORTRAN arrays defined as:

a. DIMENSION AWK (100,7),

b. AWD (100,2),

c. ATK (100,7),

d. ATD (100,9).

This limits the system to 100 data entries and 100 keys, the latter being the more serious restriction.

Both key arrays are alike and are structured as follows:

(i,1): Westerly longitude·
(i,2): Easterly longitude·
(i,3): Southerly latitude· .defines rectangular area
(i,4): Northerly latitude·
(i,5): j such that ATK(j,-) is exception to entry ATK(i,-)
(i,6): k such that ATD(k,-) is data entry associated with ATK(i,-)
(i,7): m such that ATK(m,-) is a neighbor to ATK(i,-)

Each parameter card specifies enough data to fill both a key and a data entry. The model provides programs which automatically link key entries correctly.

Time is represented in the tidal data base by array DAYCRT. This array contains one day's worth of tide predictions for each of the standard reference points. DAYCRT is a FORTRAN array:
DIMENSION DAYCRT (10,11,4).
DAYCRT(i,j,k) is defined as: the standard reference point, (i), from data entry in ATD; at the j-th time entry, up to 10 different times for flood, ebb, and slack tide; and k is one of four data items. Table 8-2 shows a sample DAYCRT entry. The times shown are monotonically increasing since the input parameter cards are sorted. In this case, however, the sort is on the card count field (see Table 8-1), since the hours and minutes fields are 00. This is an arbitrary setup based on internal programming ease. The net result is that the user must sort this data by hand for all data with the same year and day. DAYCRT is reloaded every day of simulated time at 00 hrs. 00 minutes.

Wind data changes with time by reloading the data array. To date, only the simpliest wind data has been used and new wind data has always gone in with the same keys as previous data. In this case a simple update of AWD occurs. The only problem in updating the data base in this fashion is that old entries are not deleted. Thus, if areas are redefined, old definitions and new definitions are both present in the data base.

TABLE 8-2: DAYCRT Entries for Standard Reference Point 1 - DAYCRT(I,J,K)

	DATA K = 1 Type of Tide	K = 2 Hour Tide Occurs	K = 3 Minute Tide Occurs	K = 4 Force in Knots
J = 1	Ebb	00	00	0.9
J = 2	Slack	03	48	0
J = 3	Flood	05	00	0.3
J = 4	Slack	06	00	0
J = 5	Ebb	10	48	2.5
J = 6	Slack	15	12	0
J = 7	Flood	18	06	1.8
J = 8	Slack	22	06	
J = 9	to J = 10: all zero.			

In this example reference point 1 is Rosario Strait. The data is for April 15, 1971.

Parameter Cards -

A parameter card is physically implemented by at least two cards. The first card contains the sort key in columns 1-13, columns 14-80 are blank. Any number of additional cards are used to carry the parameter data. These cards are interpreted by the FORTRAN NAMELIST facility, and therefore follow the rules set forth in the CDC 6600 FORTRAN manual.

Initially, it was intended that each parameter card would be entered as a variable length logical record, and would then be sorted - hence the sort key. This plan was never carried out, so the user must manually arrange his cards. Another unfulfilled plan involved the use of minor codes G and S (see Table 8-1). Advances in the data base programming obsoleted this arrangement, although some remnants of it still remain. In particular, the minor code is ignored in all cases except major codes I and W, where the minor code is G. Codes are defined as follows in Table 8-3.

FIGURE 8-6
SAMPLE PARAMETER DECK

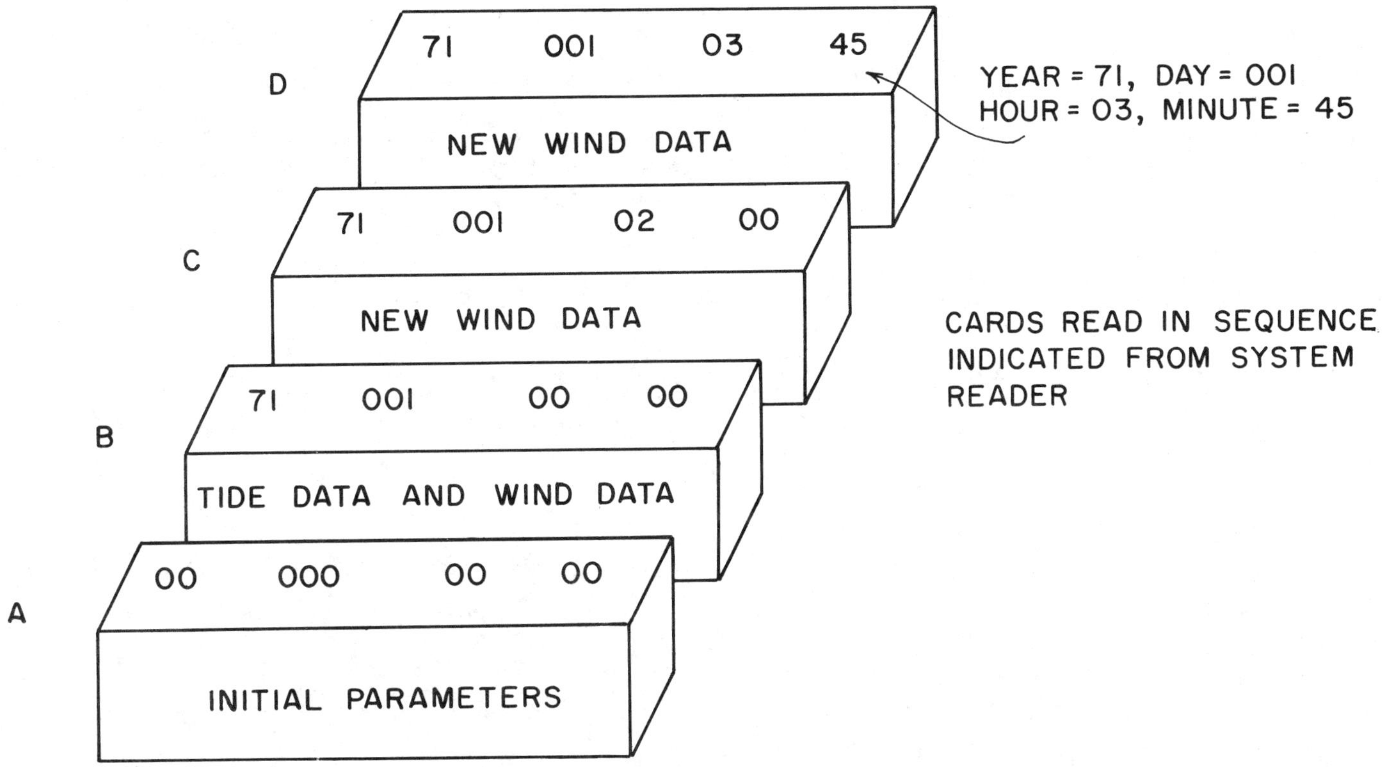

TABLE 8-3: CODES

Code	Namelist Name	Use
A	A CODE	Defines absolute geographical limits of data base in terms of latitude and longitude.
C	C CODE	Data for DAYCRT. One parameter card per standard reference point.
D	D CODE	Obsolete.
IG	IG CODE	Defines key and data entries for tidal data base (ATK,ATD).
IS	Not used	
P	P CODE	Problem parameters such as time spill starts and stop, spill location, spill rate and volume, computation interval, etc.
T	T CODE	Re-defines computation interval and output interval.
WG	WG CODE	Defines key and data entries for wind data base (AWK,AWD).

Figure 8-6 shows a sample parameter deck divided into four sections. Section A provides initial data base information and has a sort key of year = 00, day = 000, hour = 00, and minute = 00. Section B provides data base information of the standard reference points for the tidal data base and the initial wind data. There is usually one such section for each day of simulated time to be run. Hours = 00 and minute = 00 puts this data ahead of all other data for the day 001 of year 1971. Sections C and D are two changes in the wind date. The sort key specifies when this data takes effect. Section A contains cards with codes A, IG and P; Section B has codes C and WG; sections C and D contain code WG. In addition, a code T card may occur in any section. To date, wind data has been represented in the model by only one parameter card (WG), thus the wind has uniform velocity over all geographical points in the system.

Figures 8-7 through 8-12 are a series of reproductions of the output obtained from a typical run. The problem concerns an oil spill off March Point, under the influence of wind only with no tidal currents. The problem parameters are: start of simulation at 01 hours 00 minutes on day 001 of year 71; end of simulation at 01 hours 00 minutes on day 002 of year 71. Location of spill is 122° 34'0" W longitude, 48° 30'30" N latitude; rate of spill is 42 barrels per minute; total volume is 5000 barrels. The model computes the slick boundary once every 30 minutes of simulated time, and produces output once every hour. The trace data along the top includes the spreading force in knots, the volume in barrels, and the thickness in centimeters. Each of these is a scalar quantity and the figure "0" is extraneous. The average tide is, of course, zero, and the average wind is shown in x-y coordinates in units of knots. On the map 'L' stands for land, '0' for oil, '-', 'I', and '+' are grid symbols, and water is represented by blanks. The block of 'L's on the bottom of each chart represents an area not defined in the data base. The maps tend to spill over a few lines onto the following page due to not having the printer exactly lined up.

4. Model Improvements

In the course of operating what is in effect a 'pilot' model several needed improvements have been recognized. These are summarized below. For the most part they fall into two different parts, improvements related directly to the oil slick model, and improvements to the data base structure. This division is a result of considering those problems which apply to environmental modeling in general, and to the oil slick model in particular.

The data base problem is very critical although it has not restricted the work done so far. This is simply due to the fact that the

YEAR = 71 DAY = 1 AT 2 HOURS AND 0 MINUTES. CHART FOR 123 TO 122 DEGREES WEST LONGITUDE, 48 TO 49 DEGREES NORTH LATITUDE. PAGE 1

SPREAD 2.37155E-01 KNOTS AVE TIDE KNOTS AVE WIND -3.96000E-01 6.37638E-05

VOLUME 2.52000E+03 BARRELS THICKNESS 3.63119E-02

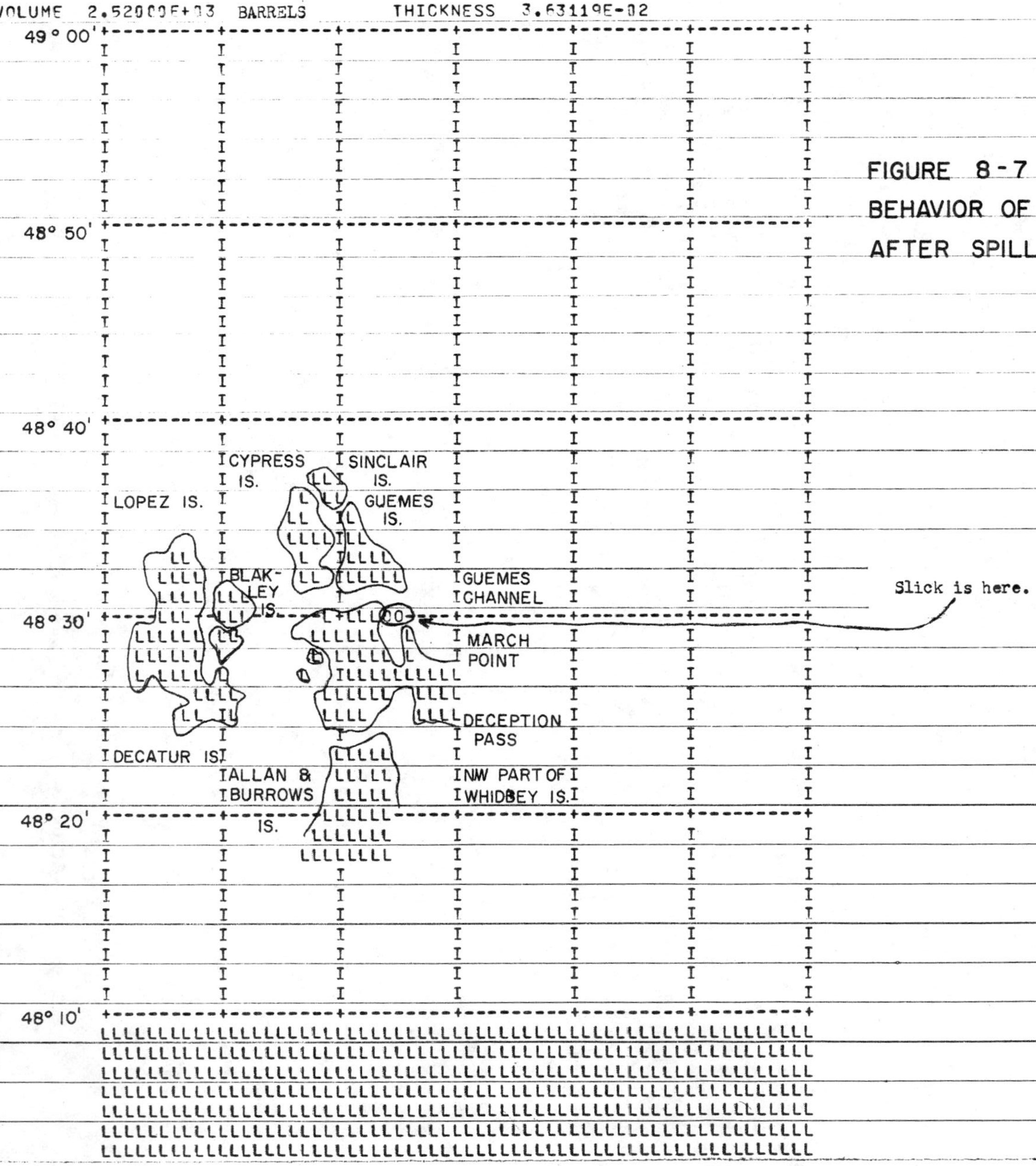

FIGURE 8-7
BEHAVIOR OF OIL SLICK ONE HOUR AFTER SPILL AT MARCH POINT.

YEAR = 71 DAY = 1 AT 4 HOURS AND 0 MINUTES. CHART FOR 123 TO 122 DEGREES WEST LONGITUDE, 48 TO 49 DEGREES NORTH LATITUDE. PAGE 1

SPREAD 6.84860E-02 KNOTS AVE TIDE KNOTS AVE WIND -2.98100E-01 4.79999E-05

VOLUME 5.00000E+03 BARRELS THICKNESS 2.19360E-02

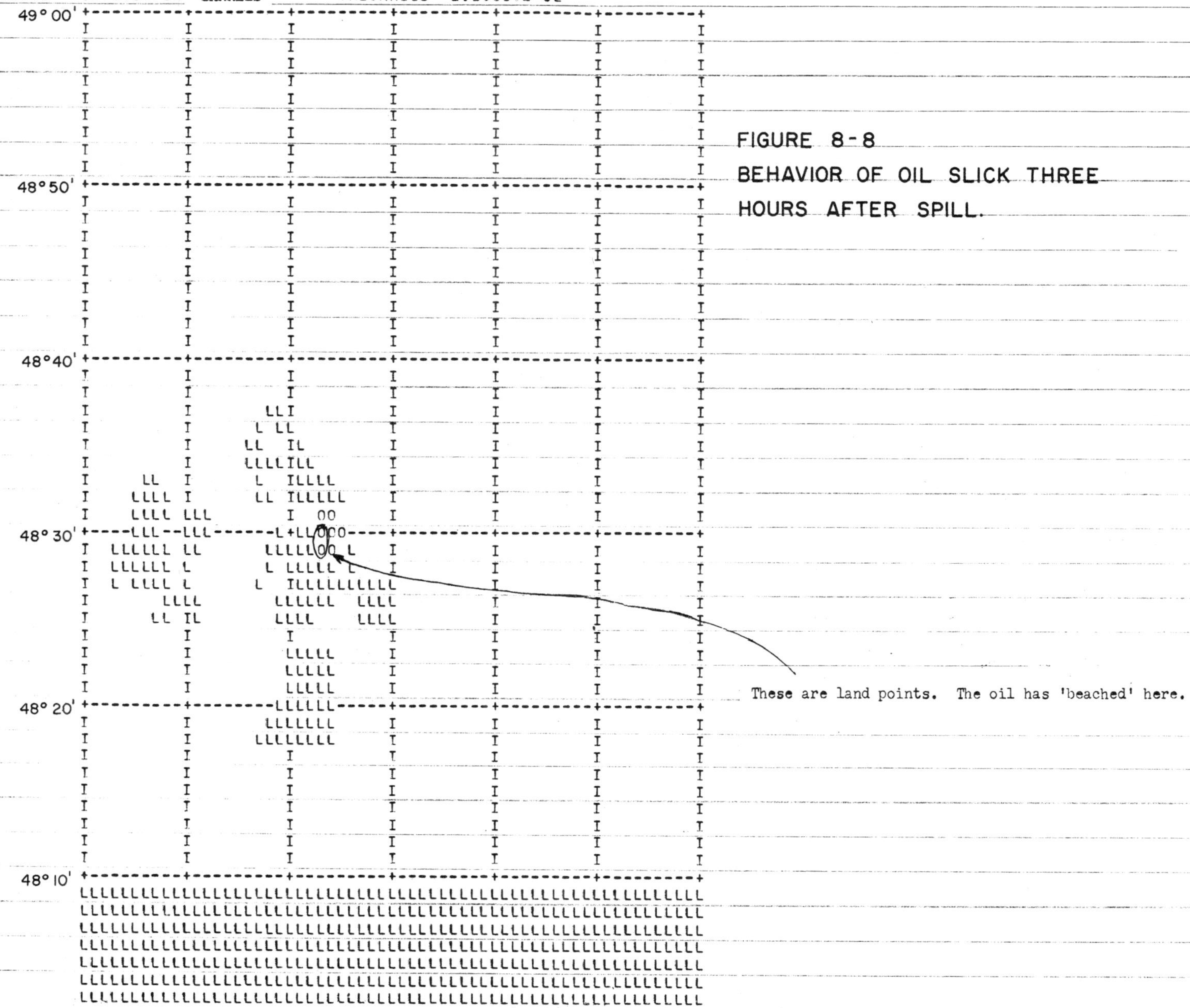

FIGURE 8-8
BEHAVIOR OF OIL SLICK THREE HOURS AFTER SPILL.

YEAR = 71 DAY = 1 AT 6 HOURS AND 0 MINUTES. CHART FOR 123 TO 122 DEGREES WEST LONGITUDE, 48 TO 49 DEGREES NORTH LATITUDE. PAGE 1

SPREAD 4.75245E-02 KNOTS AVE TIDE KNOTS AVE WIND -1.93600E-01 3.11734E-05

VOLUME 5.00000E+03 BARRELS THICKNESS 1.56048E-02

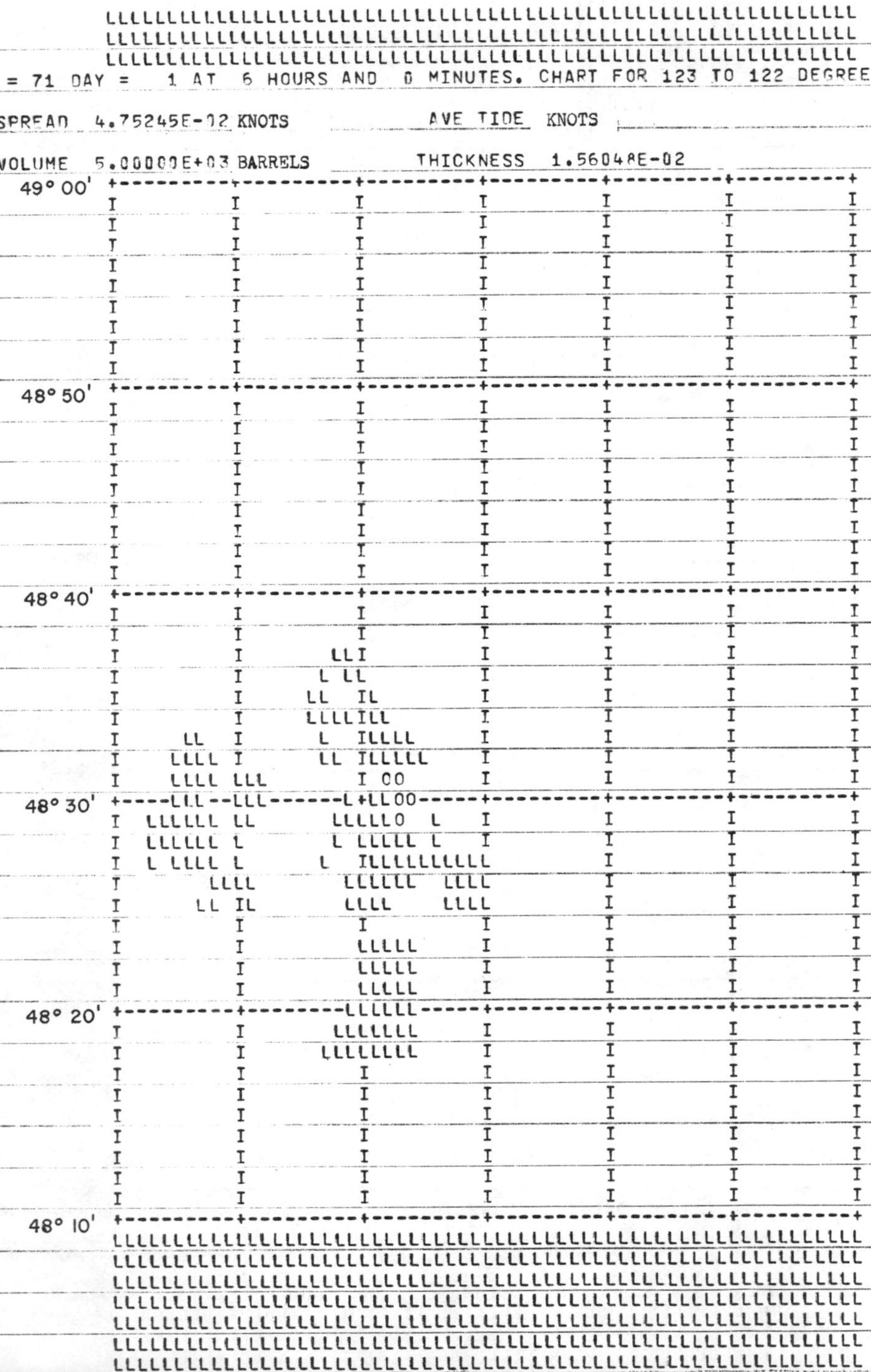

FIGURE 8-9

BEHAVIOR OF OIL SLICK

FIVE HOURS AFTER SPILL.

Wind from the East begins to push oil through Guemes Channel.

YEAR = 71 DAY = 1 AT 8 HOURS AND 0 MINUTES. CHART FOR 123 TO 122 DEGREES WEST LONGITUDE, 48 TO 49 DEGREES NORTH LATITUDE. PAGE 1

SPREAD 3.75894E-02 KNOTS AVE TIDE KNOTS AVE WIND -8.69000E-02 1.39926E-05

VOLUME 5.00000E+03 BARRELS THICKNESS 1.24693E-02

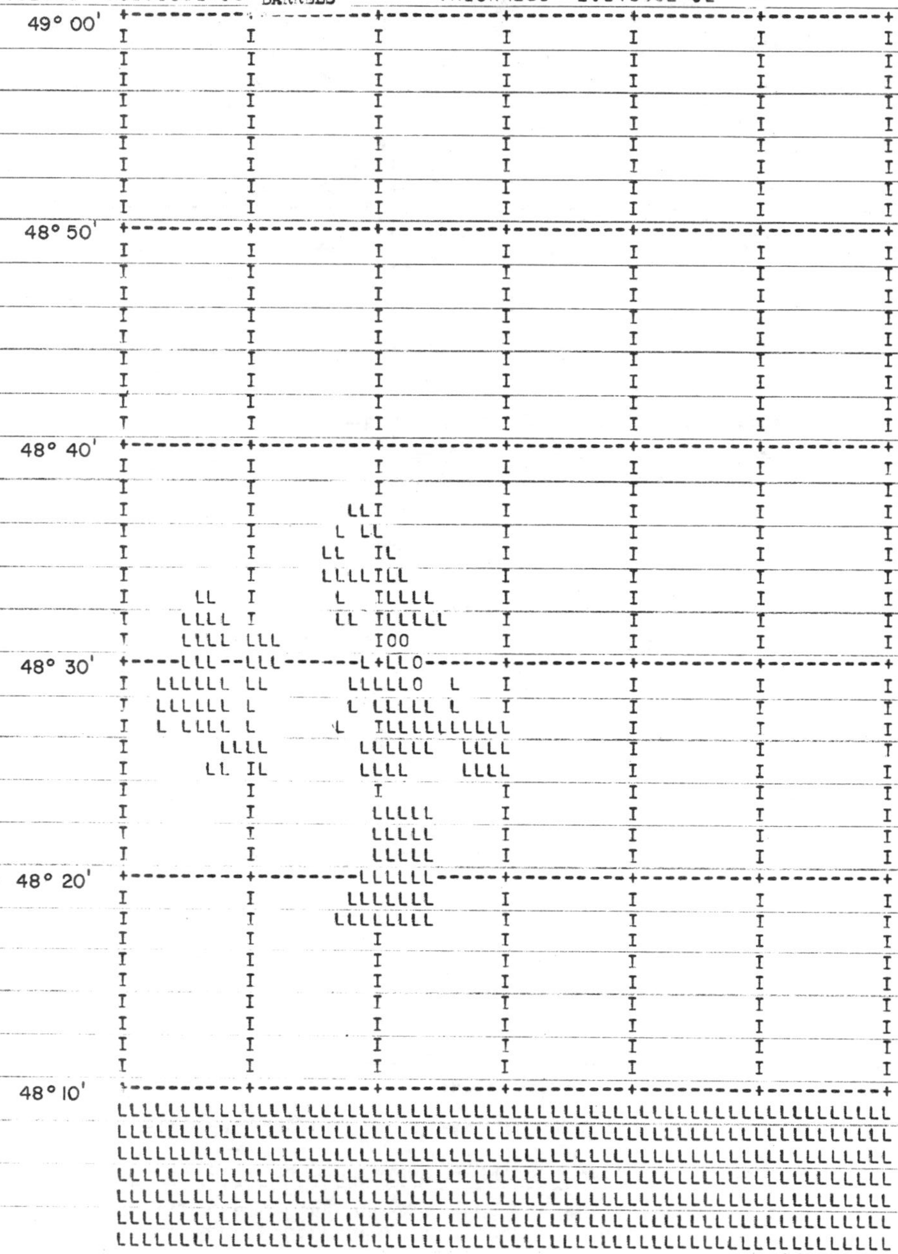

FIGURE 8 - 10

BEHAVIOR OF OIL SLICK

SEVEN HOURS AFTER SPILL.

Wind from the East.

YEAR = 71 DAY = 1 AT 10 HOURS AND 0 MINUTES. CHART FOR 123 TO 122 DEGREES WEST LONGITUDE, 48 TO 49 DEGREES NORTH LATITUDE. PAGE 1

SPREAD 3.16143E-02 KNOTS AVE TIDE KNOTS AVE WIND 1.83026E-05 8.52500E-02

VOLUME 5.00000E+03 BARRELS THICKNESS 1.05457E-02

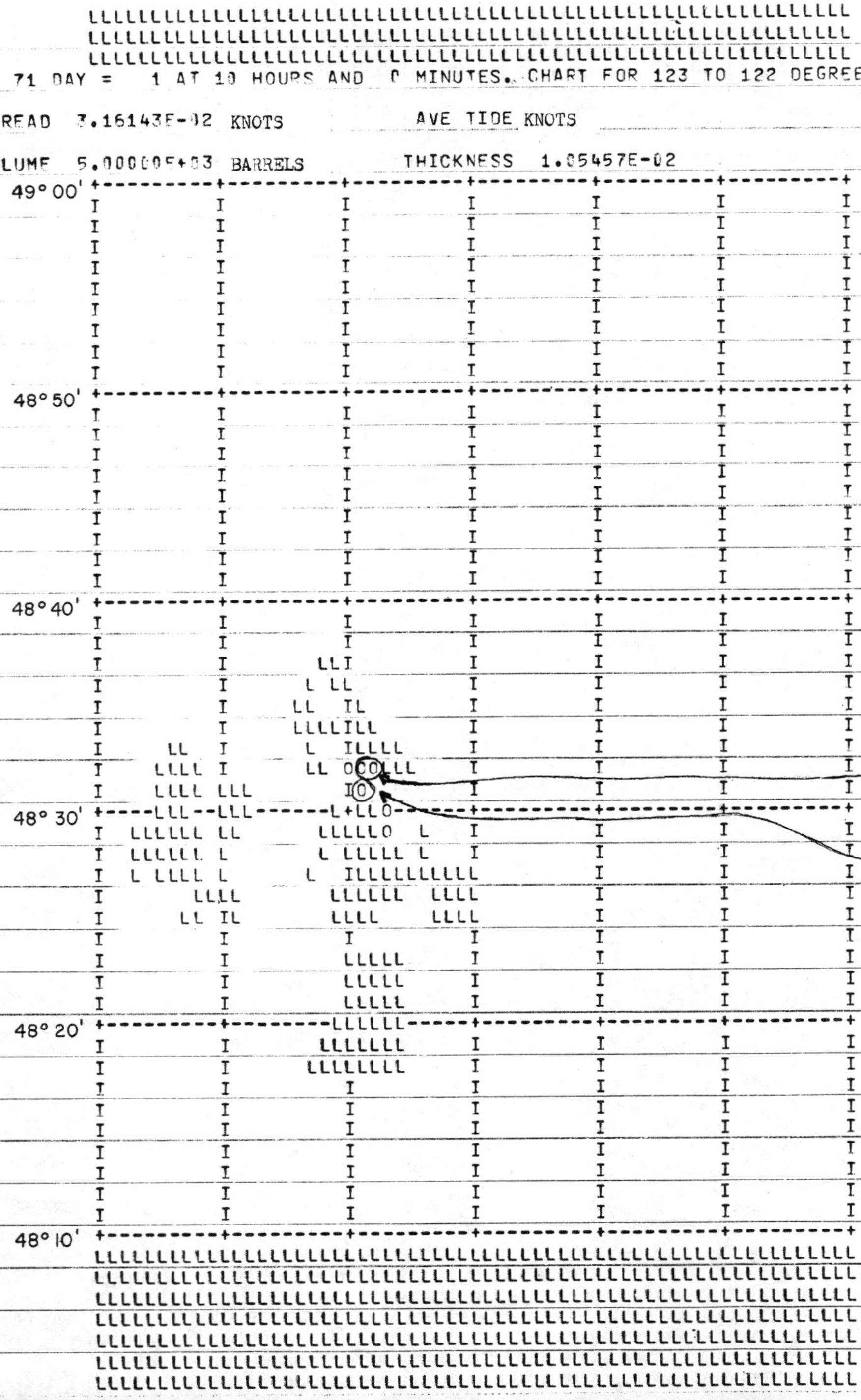

FIGURE 8-11

BEHAVIOR OF OIL SLICK

NINE HOURS AFTER SPILL.

Wind from the South.

(Note all the oil 'beached' on the South side of Guemes Island.)

The map does not show that in the data base the tip of tip of Guemes Island extends south to 48° 30' 30". Thus this "O" represents "beached" oil.

YEAR = 71 DAY = 1 AT 12 HOURS AND 0 MINUTES. CHART FOR 123 TO 122 DEGREES WEST LONGITUDE, 48 TO 49 DEGREES NORTH LATITUDE. PAGE 1

SPREAD 2.75589E-02 KNOTS　　AVE TIDE KNOTS　　AVE WIND

VOLUME 5.00000E+03 BARRELS　　THICKNESS 9.22523E-03

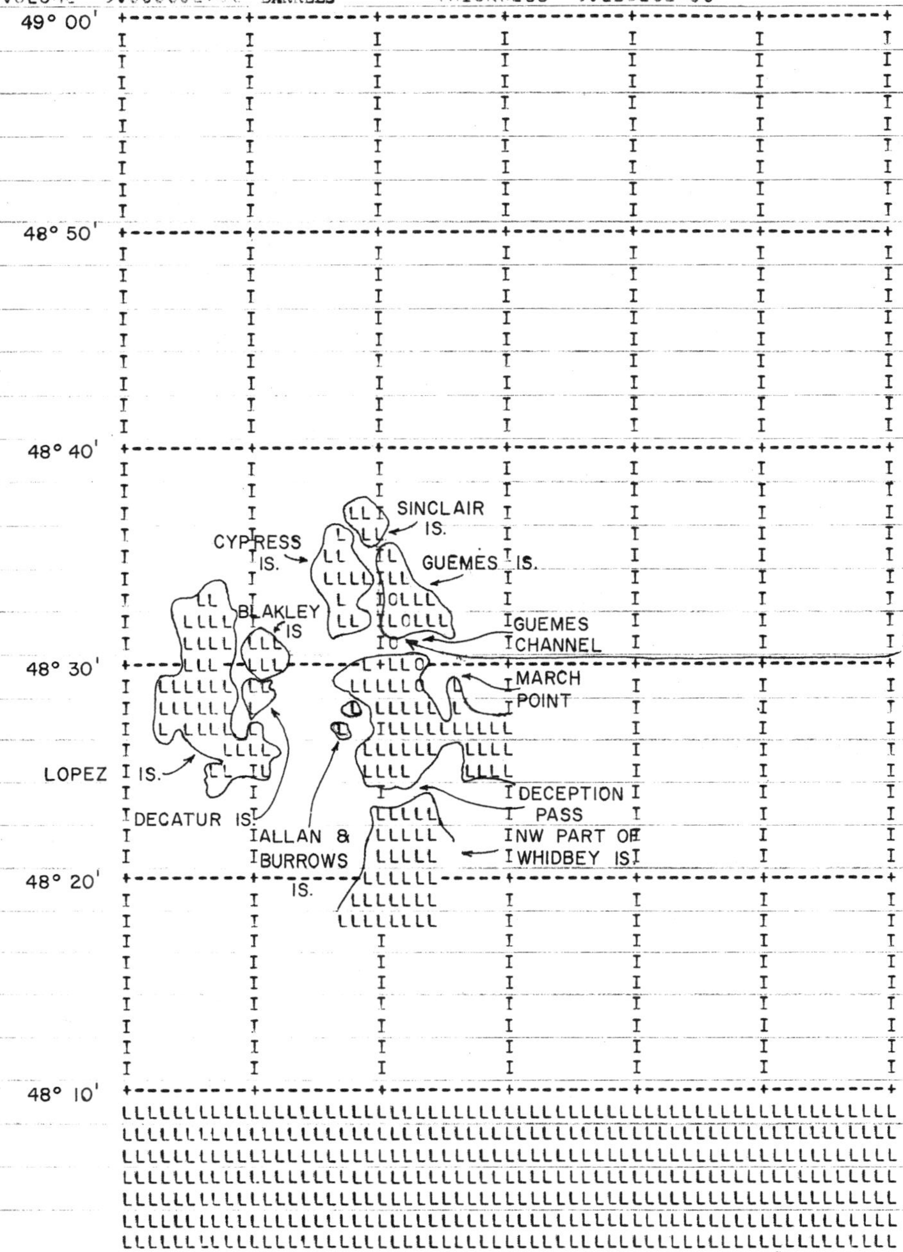

FIGURE 8-12
BEHAVIOR OF OIL SLICK
ELEVEN HOURS AFTER SPILL.

(The westerly wind has 'beached' the remainder of the oil on the West side of Guemes Island. At this point all the points of the boundary set are on land and computations cease. Note that 'Ave. Wind' → 0 as there are fewer points on the water to be effected by the wind.)

present output has a low degree of geographical resolution, plus the very preliminary nature of the testing done to date. The formulation of a detailed data base necessitates the assistance of clerical help.

There are three basic problems to constructing a data base for a general environmental model.

The first is diversity. Consider any given point in a bay, for example. How many different things might you want to know about? How deep is the water? At what time? What is the salinity? How much friction to tidal currents is presented by the underwater geography? What is the long wave action? What is the marine life? At which time of year? The only sure thing is that as an environmental model is constructed, the number of attributes will multiply.

The second problem is resolution. In many cases the data base must model a continuous datum. How does one represent the shoreline? This is a typical problem. As mentioned above, the approach taken here was representing 'areas'. Another approach which is more common is to represent an area as a number of fixed, uniformly distributed points, foregoing the notion of boundary. This technique is used in conjunction with finite difference equations. Mungall and Matthews[6] present a fine exposition of what is involved. Clearly, whatever form the data base takes it will have to support finite difference calculations. And yet another untried approach would be the use of an analog-digital hybrid computer.

The third problem is size. There is a minimum geographical area for which random access is necessary. An oil slick for example may cover a 3 square-mile area. In working with points within this area the model can be assumed to require random access to any of several attributes. If points are allowed to be spaced over a grid at 1000 foot intervals, there will be approximately 30 points to a square yard. Further, in the present model each tide point has 9 data items associated with it yielding 270 words of storage per square-mile. Also, since the slick will not be uniform (perhaps it is a large streamer)the output will have to represent more than

3 square-miles to hold the spill. Experience suggests 15 square miles, yielding about 4000 words of storage. And this is just one attribute. With little thought one can readily imagine the volume of data involved.

Below are listed those improvements to the oil spreading model which are felt necessary and feasible.

The major error in model formulation is the concept of tracking only the edge of the slick. Like a hollow egg, when the edge of the slick touches land the shell bursts and nothing comes out. Thus, although in fact the slick may still be spreading, the model operates as though all the oil is on shore.

The model should keep track of a set of points representing the entire spill. This would allow a much more accurate picture of oil transport to be developed. Some of the improvements would be:

a. ability to note differing thicknesses in slick, thus providing better fluid flow information;
b. a basis for developing the mechanics of slick breakup;
c. an estimation of oil volume in various areas of interest, such as bays, and channels;
d. a more accurate representation of slick shape, especially with regard to streamers.

The model should also be upgraded to include information on the chemistry of oil. Evaporation and solution of oil should be calculated. A composite map, produced at the end of a model run, showing the predictions of the amount of oil in solution at various points would be of great assistance in damage assessment. The model could also greatly assist in cleanup by predicting the nature of the oil on the beach; is it thick enough to be picked up with pitch forks, or is sorbent needed?

It is also apparent that a whole new oceanographic scheme must be developed. For example, slick breakup would seem to be a function of long wave action, amongst other things. Also, the concentration of oil in water channels will require that the model predict their existence. Cross channel transport should also be included to increase model accuracy. Tide heights will have to be added for damage prediction where the amount of beach polluted is a function of the difference in tide heights.

The model must also be rearranged to follow an event-line instead of a simple time-line. Currently if a wind change occurs twice in a time slice, the first wind shift is lost.

A mechanical improvement is that the model should be able to accept aerial reconnaissance pictures in order to correct model errors. In this circumstance, the model will have to be able to restart when cleanup personnel determine that predictions are not holding up.

Some improvements in the data base could be made. First, land points should be separated from tidal data; this to allow improvements in tide modeling. And second, the area definition mechanism should be improved to allow: 1) easier specification, 2) more accurate definition of boundaries, and 3) faster response for points on or near boundaries.

An exploration should be made to see if the tidal sub-model could be improved. The tidal model would be much more accurate if, for any given point, the tide at that point could be determined by interpolating from the nearest reference points. This would probably eliminate the geographically keyed data base and expand the number of standard reference points.

Finally, an improved output technique must be devised. Perhaps by using a mechanical plotter, or a video display, charts with better resolution could be produced.

B. Traffic Control Systems

1. Introduction

The basis of a computerized oil spill prevention system is an automated traffic control system. A later section describes the technical details of such a system. Basically, a series of radar stations are built to provide a small, centralized computer with input. The computer compares vessel courses, looking for a potential collision.

2. Two Types of Prevention Systems

At this point it is important to note the distinction between 'traffic control' and 'traffic advisory'. It is, in essence, the difference between 'you must' and 'you should'. There are two important obstacles. One is the considerable reluctance of ship masters to have their traditionally absolute command superceded. The other difficulty is the question of who is to assume liability for any damage that may occur as a result of commands issued by traffic control. If the agency operating the traffic system is liable, then we have a true 'control' system. In an 'advisory' system the master, and through him the steamship company, retains liability. An advisory system has no affect on the master, it merely is a tool to be used by pilots.

There would appear to be two solutions to the problem of liability insurance. One is to have the agencies which contract for the traffic control service provide the insurance. Currently, the U.S. Government assumes liability for ships going through the Panama Canal, since a Government pilot is mandatory. The other solution is a trifle more speculative. It is based on the supposition that a traffic control system is sufficiently safer than no system so that insurance premiums would be reduced. Perhaps it is possible that the insurance companies would use some of the reduction to insure the traffic control company. It seems that everyone might benefit from such an arrangement.

So far objections from the pilots have not been mentioned. Some people might think that the pilots would be the strongest opponents of a traffic system. This is not necessarily true. In fact, the pilots would seem to benefit from the system. One point to note is that there is nothing to support the idea that a traffic system would eliminate onboard pilots. Systems running now have retained onboard pilots and there is no evidence that anyone is trying to design a system which would replace pilots. It is, in fact, a distinct advantage to the system to have pilots onboard, since they would be experienced in interacting with the system; 'onboard' pilots because the person operating the traffic system on shore should also be a pilot. This is the feature that should do the most to make a traffic control system palatable to the pilots. Like any professional, pilots dislike taking commands from someone not in the profession. Not to mention that having a pilot at the computer console provides a reliable check that the computer is operating correctly. Finally, there are several small things the computer can do to aid pilots. One example is that the computer knows where buoys are supposed to be. On a map, a buoy has an approximate location; in the computer, it is an exact location. Likewise, the computer is not bothered by a missing buoy, while at the same time the loss of a radar response would provide the earliest indication of a missing buoy. Another example, is that the computer could provide better control over places like Rosario and Haro Straits. In general, it is not good practice for ships to go in opposite directions on these waters at the same time. The computer, by means of radar, can detect this situation with greater reliability than is now possible. The computer could also easily schedule ships to use such 'one-way' waters when there is possibility of crowding. In all, it seems that the pros outweigh the cons for the pilots.

3. Marketability

One suggestion for a scheduling application is to use satellites to provide position data. Thus ships which run up and down the coast could be accurately scheduled. It is argued that it would be better to start with such a system and then expand to provide traffic systems in ports such as Seattle. Economically, the satellite system would seem to be more desirable to steamship operators and thus could be sold easily, providing revenues to finance expansion. At this point, there is no way of knowing if this service could be provided at a marketable price.

A better market for the services of the traffic system than the steamship companies would be organizations that operate only on the Sound, such as tugs, barges, and some government operations.

To expand the marketability of scheduling services we shall make a slight technical improvement in the traffic system. The object of this improvement shall call a 'responder'. A responder is a small electronic radio transmitter and receiver. It constantly monitors a set frequency for a code, and when it hears it, responds with an answering code. Up to one thousand responders can operate on one frequency since the entire response cycle takes only one thousandth of a second. At the main station, all the codes are sent out and the responses listened for. Typically, each responder is queried once a second. The responder is small, and may be battery powered if necessary.

The responder is used in two ways. It provides identification to the traffic system, i.e., the radar blips are identifiable. Also, the responders can be located by triangulation and thus can provide position data for things that do not show up on radar.

What is desired is to put a responder on each tug and barge operating under the system. The computer could keep track of the individual characteristic of each tug and barge and by using this information along with knowing the vessels' location, cargo, and status, could provide extremely efficient scheduling. Job orders could be put into the computer and the system would see to it that the orders were executed at the right time, eliminating the possibility of losing orders or incorrectly executing them.

There are some interesting capabilities of a computer-based traffic system; in particular, its ability to deal with emergencies. Consider a vessel in distress. The system can both route rescue vessels to the ship, and can re-route other traffic around the site. A dock fire presents a similar situation which could be readily controlled by a prepared emergency response plan. In each instance, the service would be contracted for by the responsible agency; the Coast Guard, and the Port of Seattle for example.

The responders are of great use in such situations. They could be built into buoys and dropped from airplanes at the sight of a distressed vessel or downed aircraft, thus entering these locations into the traffic system. Such buoys might also be used to track oil spills. Presumably they would float with the slick and thus provide more accurate detection, especially during the night. Since the buoys will not necessarily float the same as oil, buoys which are no longer following the slick will have to be excluded by putting their codes into the computer with instructions to stop polling these buoys. There are many uses to which the responders could be put, all depending upon the idea that the responder allows one to mark any location quickly by means of a dropped buoy.

C. Role of Computer in Oil Spill Response

The final touch to the system, which does most to expand the services to be offered, is a large-scale computer.

Perhaps the most important opportunity given us by a large computer is the ability to run models. The oil spreading model, in particular, could be run on the large computer at the request of the small computer. Thus the model could be used when a pollution incident occurs to get predictions to aid in cleanup. This, and other models, could be used to provide firms with a planning service; i.e., studies on the environmental impact of various projects could be performed.

If one imagines a hypothetical spill incident, he can see some of the capabilities of the system.

First, we must recognize that the operating center for the traffic system will also be the center for cleanup operations. The operating center already has, as a necessary adjunct of its operations, radio facilities for all types of sea-going vessels. It has a medium-duty telephone system, normally used for client contact. It has a dispatcher and a pilot on duty. It has some type of apparatus, perhaps a video unit, for keeping track of ships which can now be used for showing the slick as well. It has the equipment for putting instruction and data into the computer. It also has automatic equipment which allows the computer to place telephone calls. To this one needs to add the ability to expand into a heavy duty telephone net, and perhaps some pre-recorded messages.

The system starts by being notified of the spill. This message is received by the traffic center by radio or telephone. The traffic center enters all known data into the computer and activates the appropriate response plan. The computer automatically notifies all agencies specified on a list stored in memory. It does this by automatically dialing the telephone numbers on the list and sending either teletype messages, or broadcasting one of the pre-recorded messages. The pre-recorded message may be made by the dispatches and thus it will include all pertinent information. The computer will continue to call all the people on its list until it receives a positive response. This can be simply done by having people call a pre-arranged number and then dialing in extra numbers as a response code.

In the meantime, the computer has assessed all available information and provided the traffic center with a list of ship dispatches to be made for rescue vessels and cleanup equipment. The computer can do this because it knows the disposition of all ships that can be used and it can select the vessels that can get there in the least amount of time. In the case of tugs, the computer can determine the availability of vessels by computing the optimum route for disposing of the tug's current cargo and then sending it on cleanup duty. The computer can even initiate communication to get dockside personnel ready to receive the tug's current cargo and to make ready the cleanup equipment the tug needs.

The computer has also activated the oil spreading model and thus has useful information concerning the amount of oil on the water and where it is going. This information goes into the plan held in the computer's memory and controls the scheduling of equipment. Thus, a tug may be instructed to drop off its cargo at one port, go to another port to pick up a specified amount of sorbent and proceed to a certain beach. There it may meet with another tug, similarly dispatched to pick up boom perhaps, and the two of them will go to work to protect the beach, before the oil gets there. The computer can provide such scheduling because it knows the speed of the tugs, the route they must take, the projected route of the oil, and in a minute can compute all possible intercepting paths.

As spill cleanup progresses the computer keeps up its scheduling activity. Since the cleanup vessels have responders on them, the computer knows exactly who is where and what they are supposed to be doing. As additional equipment arrives, perhaps from other cities, responders are assigned, and the equipment's characteristics are entered into the system. This permits the computer to send the right equipment to the right place, such as skimmers and oilevators. Thus the system provides that which experience indicates is most important: control.

It should be emphasized that much of what the system is supposed to do is not at all 'far out'. It is in fact, quite a common accomplishment in process control applications.

Finally, it should be noted that such a system, made operational in the Sound, would be highly marketable in other areas providing the State of Washington with a profitable engineering export.

This would appear to be the kind of public utility that makes money for everyone by filling a need created by advancing technology. To get it started requires only that a thorough study be made of all the possible applications to the maritime industry to provide a solid economic base to the system.

D. Endnotes for Appendix 8

1. See Section III-C, Part 3 for a discussion of oil-on-water detection equipment.
2. All latitudes are represented on cards in degrees, minutes, and seconds, based on North Latitude units. Longitudes are also in degrees, minutes, and seconds, based on West Longitude units.
3. This sub-model is an automated version of the manual hookup technique associated with Table 2 of "Tidal Current Tables, 1971," U. S. Department of Commerce, ESSA, Coast and Geodetic Survey.
4. Battelle Memorial Institute, Study of Equipment and Methods for Removing Oil from Harbor Waters, Battelle Memorial Institute Report #C12-70-001, August 25, 1969, pp. 4-5, 4-6.
5. Ibid.
6. J. C. H. Mungall and J. B. Matthews, A Variable-Bounding Numerical Tidal Model, Institute of Marine Science, March 6, 1970.

APPENDIX 9

CRITIQUE OF 1971 CONFERENCE ON PREVENTION AND CONTROL OF OIL SPILLS

The Joint Conference on Prevention and Control of Oil Spills was convened in Washington, D.C. on June 15-17, 1971. Sponsored by the American Petroleum Institute (API), Environmental Protection Agency (EPA), and the U.S. Coast Guard (USCG), the sessions were attended by over one thousand persons from industry, academia, and government. The setting was not exactly conducive to working meetings though it did provide for a minimal of informal exchanges among attendees. In addition to the formal sessions, oil spill cleanup systems and devices were exhibited.

This appendix presents a critique on the conference including sessions and exhibits. The writeup has been divided along the lines of the official program, which used the following six topic areas:

1. Law and Enforcement
2. Oil Spill Prevention, Control and Monitoring
3. Treating Agents
4. Physical Removal and Containment
5. Physical-Biological Effects
6. Oil Spill Cleanup

The critique will proceed in the same order. All papers presented at the conference were published in the proceedings.[1]

A. Law and Enforcement

The laws governing oil spills can be divided into international, federal, and state. Captain R. I. Price, USCG, discussed the history of international regulations, noting the roles of the League of Nations, IMCO, and the United Nations.[2] It was pointed out the IMCO recommendations are not automatically binding upon the various national governments. Maritime matters related to an International Convention must be ratified by the U.S., passed by legislation, and promulgated into regulations by USCG. Matters not related are similarly handled, except no ratification is needed. Thus, for any IMCO recommendation related to maritime activities to become a federal regulation, USCG is the responsible agency to initiate action.

Example of IMCO resolutions include:

#157) use and testing of shipboard navigational equipment;

#158) port advisory services related to arrival times of ships carrying noxious or hazardous cargoes;

#159) define ships subject to mandatory pilotage services;

#160) data on ship maneuvering capabilities under various conditions made available on the bridge;

#161) establish traffic separation schemes;

#162) lighting of channels; and

#188) updated training manual for masters, officers, and crew.

All these recommendations have also been suggested in this report for the waters of Puget Sound. (See Section III-C.)

The conference reviewed the significant breakthroughs in federal legislation. Included were the Water Quality Improvement Act of 1970 and the National Contingency Plan. One point brought up by members of EPA[3] is of interest to Puget Sound and concerns itself with an illegal discharge of oil. The 1965 Water Quality Act states that "no oil..... should be discharged... that (1) can be detected as a visible film or sheen...." This statement was part of a set of adapted recommendations of the subcommittee for Fish, Other Aquatic Life and Wildlife, which consisted of "29 scientists and leaders of the field."[4] Their 1965 recommendations were supposedly interim, effective until more information or toxiciology and other chemical effects of oil in sea water were obtained. As of June, 1971 no further modifications to this regulation have emerged from the subcommittee. The least that can be done is to evaluate the effects on those species unique to Puget Sound and recommend to the Washington Department of Ecology (DOE) or the regional EPA office some quantitative interpretation of this ruling.

With regard to the National Contingency Plan, the significant comment was made in an article by Commander Charter, USCG. He wrote that "...non-federal plans should, if possible, be developed in such fashion as to be responsive to the Federal planning effort."[5] While this appears compatible on the surface, there are two factors that can lead to conflict. First, the federal plan is aimed at coordination of the management roles of the various departments and agencies of the federal government, and between federal, state, and local response system. The plan is weak in the areas of operational controls. Since the governmental agencies do not have the equipment or manpower to physically control an oil spill, an updated data bank of private sector capabilities must be an integral part of the national plan. Also, guidelines for recovery of costs incurred by these organizations must be set forth at the national level. The second factor is related to conflicts between local goals and national goals. While nationwide uniformity is ideal, it is the localized areas that bear the consequences of any oil spill. Thus, their efforts must receive proper sanction.

State jurisdiction over oil spills is still in its infancy. Four states enacted laws in 1970. These include Florida, Maine, Massachusetts, and Washington. Currently, the legislation enacted by Florida and Maine are not operational due to legal tieups in the federal courts. Here again, as it was demonstrated above, state and federal governments should strive to iron out jurisdictional conflicts in their control over water pollution.

B. Oil Spill Prevention, Control, and Monitoring

Based on the articles presented in this portion of the conference, oil spill prevention efforts are aimed at oil production facilities and shipboard operations, both in harbor and open seas. Methods to prevent such spills included training programs for shipboard personnel, use of oil separators on tankers, ballast water treatment, and use of skimming points at oil production facilities.

The Gulf Oil Corporation display on mammoth tankers was of special interest. The firm presented a historical sketch of the development and construction of the Universe Ireland, a 326,000 dwt tanker, incorporating many modern prevention techniques, including the load-on-top (LOT) procedure of tank cleaning and ballasting, and a collision avoidance radar computer system. Gulf has also incorporated a ballast processing plant into their operations at Bantry Bay, Ireland.

The state-of-the-art of monitoring oil-in-water is, at best, in the developmental stages. Though several articles were presented on techniques of sensing oil spills, little evidence of commercially available hardware was on display. Those displayed monitored possible tank overfills rather than monitoring the water to determine if and when significant amounts of oil were present. Devices or systems to track oil spills were not available at the exhibits, though USCG presented an article on surveillance of oil slicks.[6]

C. Treating Agents

As testimony to the containment controversy of using dispersants in treating oil spills, four of the ten papers in this portion of the proceedings addressed themselves to the effectiveness and consequences of this type of agent. The sentiment of the articles can best be summed up by the following comment by EPA:

> "..., it would appear that more than knowledge of toxicity and emulsion efficiency should guide our decisions regarding the use of chemical dispersants for oil spill cleanup."[7]

Table 9-1 is a list of treating agents displayed in the exhibit hall which have not been evaluated in Appendix 6 of this report. Since no test data was available on their performance, only the manufacturer's name is included. When performance data is obtained, these should be included in the evaluation in Appendix 6 on cleanup devices applicable to Puget Sound.

TABLE 9-1: TREATING AGENTS DISPLAYED AT THE 1971 OIL SPILL CONFERENCE (and not included in Appendix 6).

Agent	Manufacturer or Distributor
Chemo-Mechanical:	
Petro-Pak	Coastal Services 22 Industrial Park Road Hingham, Mass. 02043
Absorbent-1012	Colloid Chemical Co. 31 Perkins Street Brockton, Mass. 02402
Type 100 Type 150 Type 200 Sorbing Boom	3M Company 3M Center 220-6E St. Paul, Minn. 55101
Slikwik	Slickbar, Inc. 10 Saugatuck Avenue Westport, Mass. 06880
Chemical:	
Coastal Cleaner	Coastal Services 22 Industrial Park Road Hingham, Mass. 02043
Oil Lok	International Oil-Lok Control Ltd. 1970 Spicer Road No. Vancouver, British Columbia Canada

D. Physical Removal and Containment

Nine papers were presented in this segment of the conference. Six described various concepts of containment, while the other three developed experiements and test procedures for evaluating devices. The concepts discussed included an oleophilic belt scrubber, a free vortex system to recover oil, a disc oil recovery system, and a dynamic keel boom. However, none of these concepts has evolved into a production model as yet. The three evaluation articles outline test conditions and procedures, but do not present any test results.

Of all the concepts, the disc recovery system appears the most promising. Two firms, Atlantic Research,and Lockheed Missiles and Space Company, have separately developed a similar system.[8,9] It consists of a set of closely packed vertical discs, which are nominally half immersed in the water. Rotation of these discs through oil creates a viscous shear which attaches oil to the disc surface. The oil is then removed by a scraper, and the disc surface is again exposed to oil on its next rotation. Lockheed has tested a full scale 8 ft. diameter by 10 ft. system.

Table 9-2 presents a partial list of commercially available devices for physical removal and containment of oil spills. This list does not include displayed devices which have been evaluated in Appendices 5 and 6 of this report.

TABLE 9-2: PHYSICAL REMOVAL AND CONTAINMENT DEVICES EXHIBITED AT THE 1971 OIL SPILL CONFERENCE (and not included in Appendices 5 and 6).

Devices	Manufacturer or Distributor
Booms:	
Absorbent Boom Model 12D	Colloid Chemical Company Pollution Control & Research, Inc. 31 Perkins Street Brockton, Mass. 02402
Sea Serpent	Johns-Manville Corp. 22 E. 40th Street New York, N.Y. 10016
Sea Boom	Spiltrol 2400 W. Loop South (Suite 322) Houston, Texas 77027
Marsan Oil Barrier	Marsan Oil Recovery Service 31W060 W. Bartlett Road Box 83, Rt. 1 Elgin, Illinois 60120
Mechanical Cleanup:	
Reclam-ator	Welles Corporation P. O. Box J Roscoe, Illinois 61073
Oil Spill Reaction System	Atlantic Research 3333 Harbor Boulevard Costa Mesa, California 92626
Olea III	Pearson Bros. Co. P. O. Box 192 Galva, Illinois 61434
(Disc Recovery System)	Lockheed Missiles & Space Co. P. O. Box 504 Sunnyvale, California
Sea Boom	Submarine Engineering Associates, Inc. 278 Forest Ave. Cohasset, Mass. 02025

TABLE 9-2: PHYSICAL REMOVAL AND CONTAINMENT DEVICES EXHIBITED AT THE 1971 OIL SPILL CONFERENCE (and not included in Appendices 5 and 6). Cont'd.

Agent	Manufacturer or Distributor
Mechanical Cleanup:	
Buda II	Marine Pollution Control Corp. 9010 Roselawn Avenue Detroit, Michigan 48204
Magnecol Magnetic Broom Magnecol Oil Spill Harvester	Avco Systems Division 201 Lowell St. Wilmington, Mass. 01887
Oil Hawg	Parker Systems, Inc. Box 1652 Norwalk, Virginia 23501
Sea Skimmer	Spiltrol 2400 W. Loop South (Suite 322) Houston, Texas 77027
Mobile Beach Cleaner	Meloy Laboratories 6631 Iron Place Springfield, Virginia 22151
Floating Saucer	Acme Products, Inc. 8158 E. 44th Street Tulsa, Oklahoma 74145

E. Physical-Biological Effects

It is of significance to the study of oil spills that the two most controversial topics each warranted an entire session. Treating agents were discussed earlier, while this session dealt with the physical and biological effects of oil on water. It should also be noted that the status of the two topics did not change much as a result of the conference findings.

The first half of the session was devoted to biological effects. Studies at Milford Haven, England, and Santa Barbara, California were included. One consistent conclusion which is brought forth by the articles: there is some alteration in the biological environment as a result of the presence of oil. The articles do not identify whether this alteration has any long term detrimental effects or not. Most likely, knowledge on this topic is not fully developed at this time. A suggestion to enhance this knowledge is to continuously monitor the flora and fauna of a region such as Puget Sound to develop a biological inventory. From this, behavior and life patterns could be predicted. In addition, observed deviations from the normal pattern due to adverse conditions could be determined with a higher degree of certainty.

The second half of the session dealt with the physical behavior of oil on water. The state-of-the-art of analyzing behavior of oil on water is still in the formative stages. There are two primary reasons which have influenced this situation. First, techniques of detecting and tracking oil on water have not been thoroughly explored. Second, deliberate discharge of oil into water for study purposes is frowned upon. Thus, scientists and engineers have resorted to theoretical analysis and scale model tests. While the bulk of these are of some value, real life situations exhibit certain characteristics which cannot be duplicated in laboratory or paper models. A solution to the problem is to develop a readily deployable tracking system and obtain the necessary authorization and funds to utilize it at a major oil spill, should one occur. After several real-life tests, an evaluation of the system could determine its effectiveness. Though this approach is expensive, it is far more economical than the discharge of large quantities of oil into waters for testing purposes.

F. Oil Spill Cleanup

The final session of the 1971 oil spill conference was concerned with oil spill cleanup, including program, technique, and equipment evaluation. The session began with a British paper on cleanup programs.[10] The main conclusion of the article was that optimum cleanup operation for unique conditions of each spill requires preparedness, flexibility, and an integration of appropriate techniques.[11]

Subsequent papers discussed methods of cleaning oil contaminated beaches. Techniques included were utilization of earthmoving equipment, froth floatation cleaning, and hot water fluidization. Only the first technique has been used in a real life spill situation.

A panel discussion on oil spill cleanup experiences highlighted the session. The following trends were perceived from the dialogue:

1. Firms in the oil industry have relied on third party contractors and cooperatives to provide the oil spill cleanup capabilities required at the various facilities.

2. The biggest problem involved in responding to an oil spill is getting organized, initially.

3. On-scene commander relies on expertise of his subordinates and members of other organizations at the spill site.

4. The decision to terminate cleanup operations and the judgement of post cleanup water quality is vested solely with the on-scene commander.

Several comments come to mind with respect to the above statements. First and foremost should be a re-evaluation of contingency plans, both at the national and regional level. What portions of each are ineffective and why? Determine courses of action to improve the plans. A second suggestion is related to the operational readiness of the response. Initiation of periodic drills analogous to fire drills should improve the response time, and point up deficiencies in the operation. Lastly, increased levels of inter-agency communication is recommended to familiarize key personnel with each other's capabilities. (See Section III-B part 1, for the study group's proposal.)

The quality of preparedness for response prior to an oil spill is a key to the effectiveness of that response. Once oil is detected on water, there is no time to resolve jurisdictional or organization questions.

G. Summary and Remarks

It is interesting to note the similarities between those problems discussed in the conference and those analyzed by the Study Group in the main body of this report. What is disturbing, though, is the continued fragmentary approach to oil spill prevention. The inertias inherent to government bureaucracies and big business were evident at the conference, though tangible progress has been made since the *Torrey Canyon* incident over four years ago. But the truly remarkable revelation is that the quality of analysis and findings of the Study Group, presented in this report, can do far more to enhance the prevention effort of oil spills in Puget Sound than the generalized levels of effort presented in the Conference.

H. Endnotes for Appendix 9

1. American Petroleum Institute, Proceedings of Joint Conference on Prevention and Control of Oil Spills, Library of Congress Catalog No. 74-124324, 1971, pp
2. Captain R. I. Price, USCG, "International Activity Regarding Shipboard Oil Pollution Control, Proceedings of Joint Conference on Prevention and Control of Oil Spills, American Petroleum Institute, pp. 27-42.
3. K. E. Biglane and R. H. Wyer, Environmental Protection Agency, "Oil Pollution Control Legislation and the Water Quality Improvement Act of 1970, The Federal Viewpoint," Proceedings of Joint Conference on Prevention and Control of Oil Spills, American Petroleum Institute, pp. 11-15.
4. Ibid., p.13.
5. Commander D. B. Charter, Jr., USCG, "National Contingency Planning," Proccedings of Joint Conference on Prevention and Control of Oil Spills, American Petroleum Institute, p. 24.
6. C. E. Catoe and F. L. Orthlieb, USCG, "Remote Sensing of Oil Spills," Proceeding of Joint Conference on Prevention and Control of Oil Spills, American Petroleum Institute, pp. 71-84.
7. Dewling, Dorries, and Pence, EPA, "Dispersant Use vs. Water Quality," Proceedings of Joint Conference on Prevention and Control of Oil Spills, American Petroleum Ins. pp. 271-277.
8. Uyeda, Chuan, Connolly, and Johnson, Atlantic Research, "Concept Development of a Powered Rotating Disk Oil Recovery System," Proceeding of Joint Conference on Prevention and Control of Oil Spills, American Petroleum Ins. pp. 329-338.
9. B. Bruch and K. R. Maxwell, Lockheed Missiles and Space Corporation, "Lockheed Oil Spill Recovery Device," Proceedings of Joint Conference on Prevention and Control of Oil Spills, American Petroleum Inst., pp. 339-356.
10. W. E. Betts and H. I. Fuller, Esso Research Center, U.K., "An Integrated Program for Oil Spill Cleanup," Proceedings of Joint Conference on Prevention and Control of Oil Spills, American Petroleum Inst., pp. 497-504.
11. Ibid., p. 497.

APPENDIX 10

A THEORETICAL DISCUSSION OF OIL CONTAINMENT BOOMS

A. Introduction

The prime object of an oil containment boom is to influence the flow of surface water in such a way as to hold back the polluted upper layer. It is necessary to not only restrict the oil's natural spreading tendency but also to impede the environmental conditions (i.e., wind, current and seas) which favor spreading. The success of any cleanup operation will depend upon whether or not the oil can be confined quickly to a relatively small area in relatively thick films. Although this is important it is just an intermediate step in the overall operation. With a containment boom the spill can be encircled and removed from the surface, deflected to another area or to a beach for easier removal or be prevented from entering areas of value.

The physical characteristics and behavior of the pool of oil contained by an oil boom will first be discussed. This will also include a discussion of some oil boom failure mechanisms. Both wind and currents effect the behavior of this oil pool by increasing the load on the oil boom and by increasing the tendency for the oil boom to leak oil. This area will be discussed next along with the forces that are applied to an oil boom. A list of symbols with definitions is presented at the end of this appendix.

B. Physical Characteristics of an Oil Slick

The shape of an oil slick trapped behind an oil boom has been described by Wicks[1] to have three regions when acted upon by a slow current (i.e., a current of less than 0.6 ft/sec).

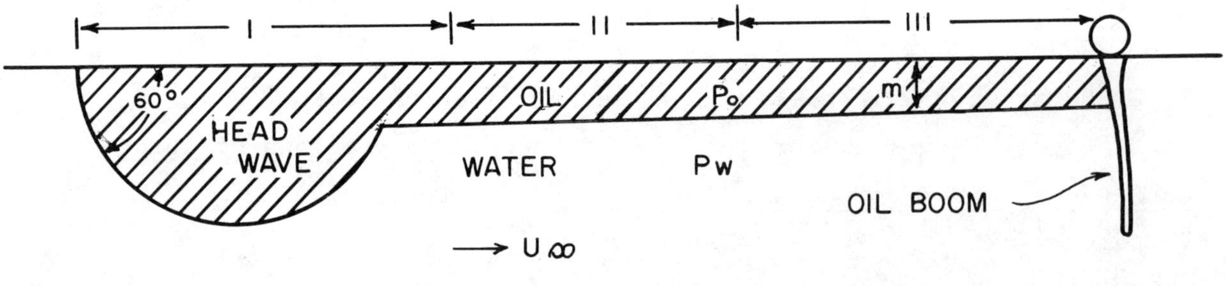

Figure 10-1

At this low velocity (less than 0.6 ft/sec) this configuration is stable. Above 0.6 ft/sec periodic traveling waves form along the oil/water interface in Region II and III. These waves travel toward the oil boom. The upper surface of the oil is quite smooth even though the lower surface is rough with these waves. Above 0.85 ft/sec oil droplets of less than 3/4" in diameter are torn off the lee side of the head wave and entrained in the flowing water. Two things can happen to these droplets. First, if they are buoyant enough, they can rejoin the oil slick before they reach Region III; second, they will be swept under the oil boom when in Region III regardless of their buoyancy. Above 1.2 ft/sec showers of droplets are torn from the lee side of the head wave. Also at these high current velocities waves begin to form at the oil/water interface of the head wave. These waves are stable in front but become unstable near the lee side of the head wave to allow droplets of oil to be torn free. There is also motion within the oil layer itself. There is a flow of oil in the direction of the current near the lower surface of the oil and a reverse flow on the upper surface.

Region I (The head wave) -
This head wave phenomenon occurs frequently in nature; it is similar to the wave shape observed when salt water is introduced into fresh water or when a submerged current of muddy water flows under clearer water.

The film thickness can be calculated from the equation:

$$m = \frac{U_\infty}{2g\ [(\rho_w/\rho_o)\ -1]}$$

at a current of about 1 ft/sec the film thickness will be 1.2 inches, and the head wave will be about twice as thick.

The rear or lee side of the head wave is the most unstable part. Waves form on the front side of the head wave and grow as they propagate to the lee side. At a critical current velocity of 0.62 ft/sec the waves become unstable and droplets begin to escape into the flowing water.

Region II -

The head wave acts as a discontinuity in the flow pattern of the current. The streamlines separate upstream of the head wave, pass under it, then reattach themselves to the surface as shown in Figure 10-2:

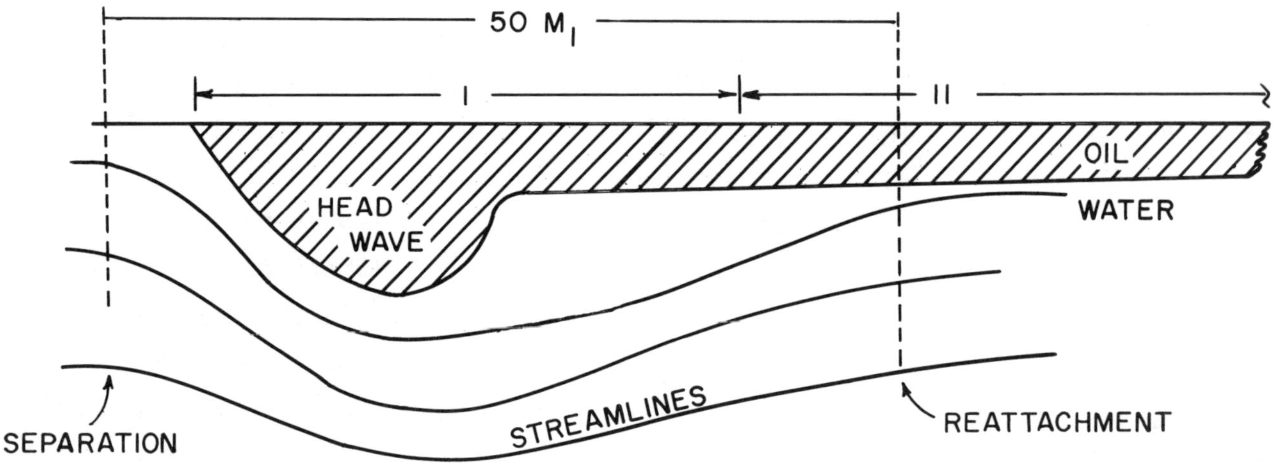

FIGURE 10-2: Boundary Layer Separation and Reattachment

Reattachment occurs about (50) (m_I) downstream. After reattachment a drag force is set up at the oil/water interface to cause circulation within the oil layer in Regions II and III.

The oil film thickness and cumulative oil slick volume are a function of the distance from the head wave. For example, in a 1 ft/sec current the slick will grow to a thickness of 8" at a distance of 20 feet from the head wave and 10" at a distance of 100 feet from the head wave.

Region III -

This region is characterized by the fact that any oil entrained in the water in this region will be swept under the oil boom regardless of the curtain's depth. As the curtain's depth increases so does Region III, although in a non-linear manner. It is usually taken to be about five times the curtain's depth beneath the oil/water interface, as shown in Figure 10-3:

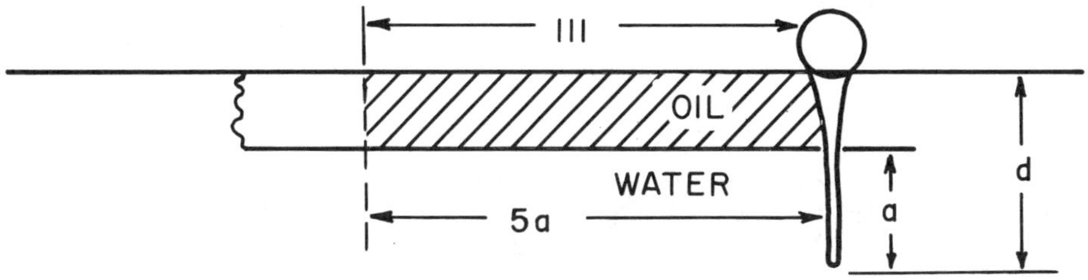

Figure 10-3

This curtain also acts as a discontinuity, so again the streamlines separate and flow under the curtain.

Now consider an oil droplet after being torn free from the lee side of the head wave. If the droplet has a large enough net upward force and sufficient time it will re-unite with the oil layer in either Region I or II. If not, it will be swept along with the water and flow under the oil boom. The terminal rise velocity, v_t can be calculated as a function of the distance downstream from the point of droplet formation. The particle will have enough time to rejoin the slick if the length of the slick in Regions I and II is longer than $2m_I U_\infty / v_t$; otherwise the droplet will be swept under the barrier.

In summary, then, there will be a velocity below which no droplets will be formed, thus, none will be entrained in the flow. At higher current velocities there may or may not be oil boom failure by droplet entrainment, depending upon the size of entrained droplets. The larger droplets with their high buoyancy force will re-unite with the oil layer, whereas the smaller droplets will be swept under the oil boom. At still higher velocities oil boom failure by droplet entrainment is certain since then even the largest particles will be swept under the oil boom.

There is still another method by which oil can escape past an oil boom and it is by a kind of draining action. Water flowing beneath the oil boom causes a pressure reduction which can pull or suck the oil under the oil boom. This is similar to what happens when a bathtub is emptied. When the bathtub is drained from the bottom a depression in the surface forms when the liquid depth reaches a certain critical height above the drain hole. This depression grows downward very rapidly, eventually forming a vortex as shown in Figure 10-4:

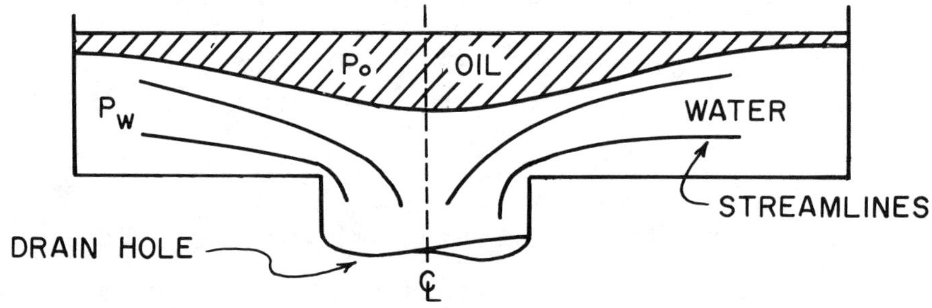

Figure 10-4

With some stretching of the imagination this radial flow into a drain hole can be re-interpreted to fit the flow of oil under an oil boom. Place the oil boom along the dotted centerline so the curtain extends into the drain hole where the streamlines will be parallel to it as shown in Figure 10-5:

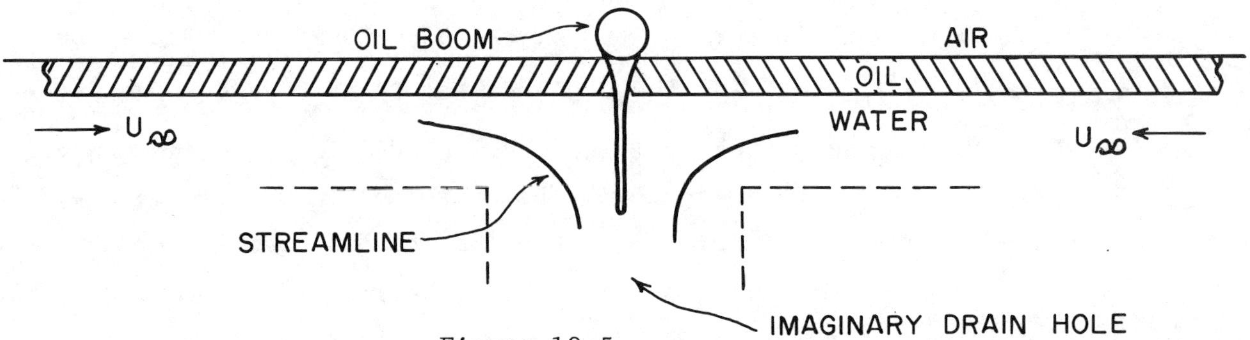

Figure 10-5

Now with an increase in the current's velocity the "critical height above the drain hole" will be reached and the oil layer will be depressed down toward the imaginary drain hole. At higher velocities a shower of droplets will be torn away from this depression and entrained into the flow, similar to the way the droplets were torn away from the lee side of the head wave in Region I. The shower of droplets will be torn away from the oil layer long before the depression shown in Figure 10-4 could have formed. Removing the imaginary drain hole and the symmetry of Figure 10-5 will lead to the method of oil boom failure by draining as shown in Figure 10-6:

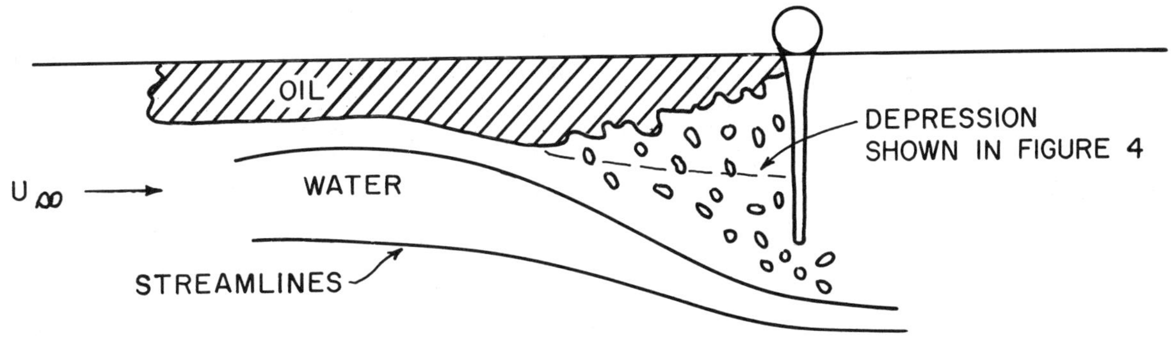

Figure 10-6

Other more obvious modes of oil boom failure include the physical destruction of an oil boom due to wind, wave or current effects; the splashing of oil over a boom by wave action and the flowing of oil under an oil boom when lifted clear of a wave trough and the flow of oil over and under an oil boom in a heavy wind.

C. The Effect of Wind

David Hoult[2] at M.I.T. looked at the problem of oil boom failure due to the action of the wind only. He looked at the effect of oil being pushed up against an oil boom's curtain of depth, d, by a wind blowing perpendicular to the oil boom as shown in Figure 10-7. With a knowledge of d the holding power of the oil boom can be determined.

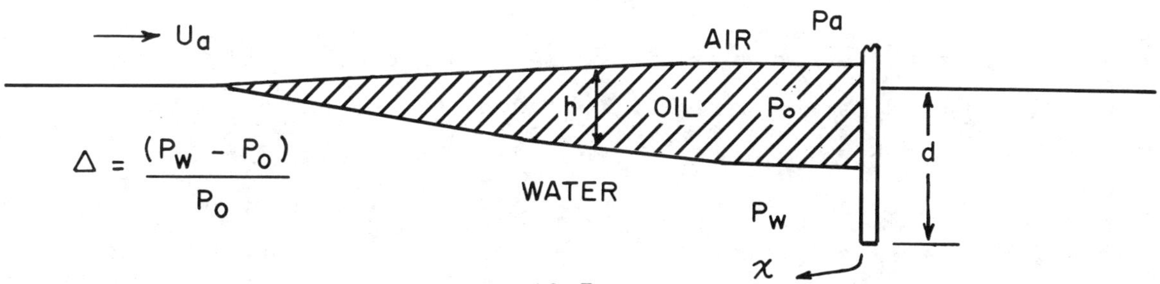

Figure 10-7

The turbulent wind stress τ acts on the oil piling it up against the oil boom until the gradient of the thickness dh/dx balances the wind stress set up. The expression relating τ to dh/dx

$$-\tau = \rho_w \Delta g h$$

The wind stress is $\tau = 1/2 \ \rho_a \ U_a^2 \ C_f$

where

$$C_f = 0.002 \text{ for oil on the open sea.}$$

The initial slope can be used to estimate the holding power of an oil boom. If Q is the volume of oil contained by an oil boom of length L, then the holding power may be expressed as:

$$\frac{Q}{L} = \frac{\rho_w \Delta \quad g\, d^3}{2\, \rho_a\, U_a^2\, C_f}$$

The major point of interest here is that the oil boom's capacity increases as the cube of the effective curtain depth. In order to double the amount of oil contained by an oil boom due to the action of the wind the effective curtain depth would have to be increased by a factor of eight. This is not a very desirable situation. The presence of waves and surface water currents will make the situation even worse.

D. The Effect of Surface Currents

Ralph Cross and David Hoult[3] of M.I.T. looked into the oil holding capacity of an oil boom in a steady current.

An oil boom will be effective in holding oil against a steady current under proper conditions. The first condition is the existence of a critical Froude number above which the oil boom holds no oil. The second condition is related to the shape and behavior of the pool of oil held by the oil boom.

A quick estimate of whether or not an oil boom will hold oil at all can be made by comparing the forces acting to draw a small column of oil under the oil boom with the buoyancy of such a column of oil. The force driving the oil downward is the difference between the stagnation pressure and the free-stream pressure near the bottom of the oil boom, i.e., $1/2\ \rho_w\ U_\infty^2$. The net buoyancy is given by $\rho_w \nabla g d$. At incipient oil boom failure these two expressions are equal. The result will be the critical Froude number.

$$F_c = \frac{U_\infty}{\sqrt{\Delta\, g\, d}} = \sqrt{2}$$

At Froude numbers above $\sqrt{2}$ the oil boom will hold no oil. Failure will occur first near the apex of the boom which is assumed to be "U" shaped.

With a Froude number of less than $\sqrt{2}$ it is next desirable to find out just how much oil an oil boom will hold under the action of a current. First an equation relating X to h is needed.
Figure 10-8 shows how h varies with X.

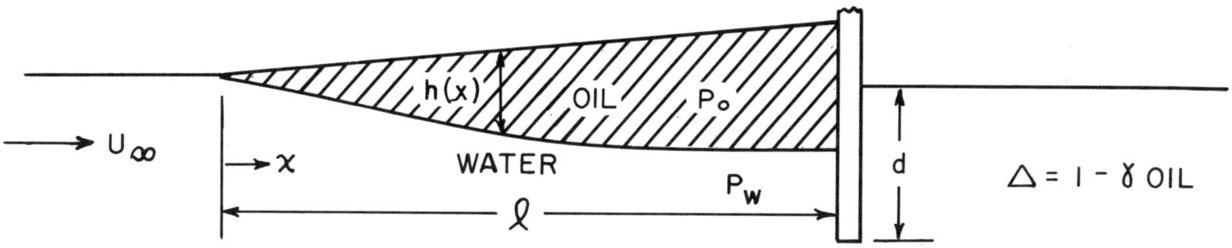

Figure 10-8

It has been shown[4] that the thickness of the oil layer, h(x), is not a strong function of the shape and construction of the oil boom. This is clearly seen by noting that the slick length may be 100 times the effective oil boom depth, d; and that the oil boom geometry can influence the motion of the oil slick up to about 5 oil boom drafts upstream. But, as a first approximation, assume only that the thickness is independent of the viscosity. This will lead to an expression relating the volume of oil held by an oil boom to the oil boom's draft, d.

The water current pushes oil up against the oil boom until the shear stress τ on the underside of the oil layer due to the current is balanced by the horizontal hydrostatic pressure gradient corresponding to an increase in thickness of the oil pool. This will result in the following equation for oil thickness as a function of slick length (h(x) vs. X):

$$h^2 = (U_\infty{}^2/\Delta g)\ C_f X$$

where

$$C_f = 0.008 \text{ and is independent of X}$$

Near the leading edge of the oil slick where the head wave occurs this equation does not hold, nor does the Froude number of $\sqrt{2}$ hold as a critical value for determining oil boom failure.

With this equation and an equation relating slick length, ℓ, to oil boom draft, d,

$$\ell = \frac{(d^2)\,\Delta g}{C_f\, U_\infty^2}$$

an expression for oil volume stores per foot of oil, boom width can be obtained. This considers the boom to be perpendicular to the current which is not always the case. The oil boom will usually take on an approximate parabolic shape as shown in Figure 10-9:

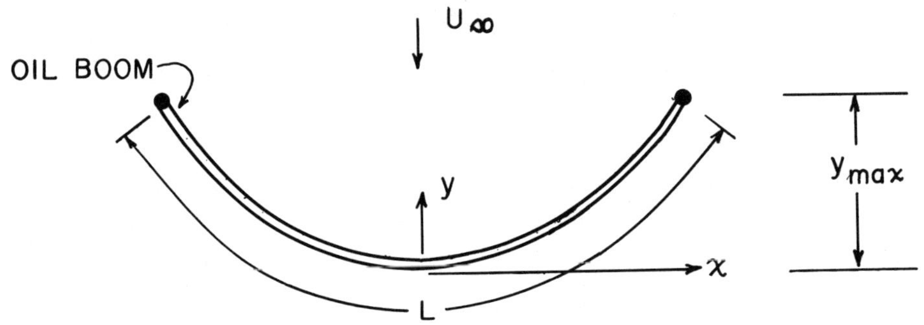

Figure 10-9

Now a more realistic calculation for the oil volume behind an oil boom can be made:

$$\text{Volume} = (1.9 \times 10^3)Ld^3 \left(\frac{\lambda d}{L}\right)^{1/2} \left(\frac{\Delta g h_o}{U_\infty^2}\right)^3$$

Where h_o is the depth of oil in the apex of the oil boom which must be less than the boom draft, d. The quantity λ is a tension factor which relates the tension in an oil boom to the normal force on a stretched-out oil boom. Values for λ are given in the following table.

$\frac{y\ max}{L}$	λ	$\frac{y\ max}{L}$	λ
.10	.028	.30	.163
.20	.078	.35	.231
.25	.115	.40	.336
		.45	.560

Again notice that if it is desired to hold twice as much oil, d will have to be increased by a factor of eight. The oil volume depends upon the cube of h_o and d. This analysis does not take into account oil boom failure by oil droplets being torn away from the oil layer. Here it is assumed that a low current drives so much oil up against an oil boom that the curtain depth d is not sufficient enough to hold it back. The combined effect of wind, waves and current on the holding capacity of a boom has not yet been investigated. Currents can be generated by both wind and waves as well as by the action of the tide.

E. Boom Forces

The mechanism of oil boom failure due to structural damage suggests that not enough information is known about the applied loads and the oil boom forces. Work done at Hydronautics Inc.[5,6] has provided basic quantitative engineering data on the performance of oil booms in a seaway. Oil boom loads (i.e., local loads and end or mooring loads) in the presence of wind current and seas and the oil boom motion (i.e., whether the oil boom will be lifted clear of a wave trough or completely immersed by a wave crest) were investigated at Hydroanutics, Inc. Their work is in the form of graphs giving forces, moments and curvatures for various oil boom configurations and mooring angles. Figure 10-10 shows the type of information available when using their data:

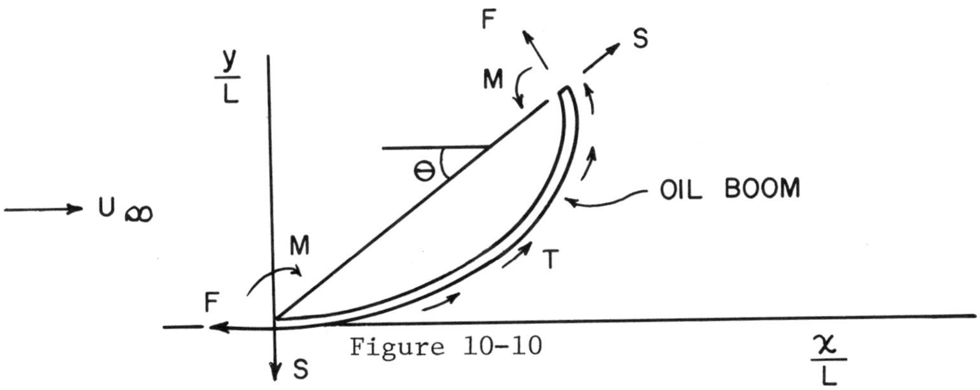

Figure 10-10

Where

θ = Mooring Angle

F = Tension or axial force

S = Shear or normal force

M = Applied moment

T = Tangential force

When an oil boom is placed perpendicular to a current it is possible to calculate the drag force on the oil boom due to that current. This will give the minimum loads on the oil boom, loads that will tend to rip the oil boom apart or pull it free of its moorings. The equation assumes that a current piles so much oil up against an oil boom that the oil layer extends down to the curtain depth, d. Any increase in current will cause oil to flow under the oil boom. If the oil boom takes on a parabolic shape due to the current, then this equation holds near the apex of the curve. The drag force, D_F is expressed by:

$$D_F = (1/2) C_D \rho_w U_\infty^2 d\ L$$

where

C_D is the drag coefficient and is equal to $C_D = 1/F$

F is the Froude number and must be less than 1.3 in order for the oil boom to hold oil.[3] With a knowledge of U_∞, the current velocity; d, the curtain depth; and L, the oil boom length it is possible to estimate the loads on an oil boom. This equation does not take into consideration oil boom failure by oil droplet formation at the head wave or elsewhere along the oil/water interface, nor does it account for the effect of wind and waves.

F. Conclusion

In many areas of Puget Sound the combined effects of wind, waves and currents would require an extremely large oil boom to hold back the flow of the polluted surface water. However, to deflect the oil rather than stop its flow would require a much smaller oil boom. The object in deflecting the flow of oil is to save valuable property (either ecologically important or valuable man-made areas) while allowing the oil to flow into other areas. This is the concept of selectively sacrificing an area (letting the oil flow into an area where it can be easily cleaned up) in an attempt to save a more valuable area.

The problem with attempting to completely confine the oil or prevent any of it from getting into an area with an oil boom is that the oil boom does not always stay perpendicular to the current or wind direction. It will bulge out in the center and collect oil in the middle of the channel where it is difficult to remove. Also when placing an oil boom in a channel, the narrowest part might look like a good place to construct it but this area has the highest current velocity. Another consideration is the shape of the beach into which the oil is to be deflected. A gently sloping beach will have a lower current velocity along its sides than a beach with cliffs for sides, plus it will be easier to clean up and support cleanup operations.

Research conducted by Milz[7] has provided some current and wind velocities necessary to cause oil to flow under an oil boom. At current velocities of over 12 ft/sec and in sea states of over 2 to 4 feet oil booms have been observed to fail. At high current velocities, above 2.0 to 3.0 ft/sec oil booms become very ineffective because most of the oil is swept under the oil boom. As mentioned before, oil booms fail by having water splash over their tops;this happens when the ratio of the wave height to the wave length is greater than 0.08. Surface winds above about 12 knots were observed to force oil to flow under an oil boom.

It might be appropriate to mention two disadvantages of oil booms. First, they prevent surface craft movement, and second, they confine volatile materials which can create a fire hazard.

G. List of Symbols

ρ_o = Oil density

ρ_a = Air density

ρ_w = Water density

Δ = $(\rho_w - \rho_o)/\rho_o$
defined as (1 - Specific gravity of oil) or as the fractional density difference between the oil and the water.

U_∞ = Water current velocity, ft/sec

U_a = Air current velocity, ft/sec

g = Local acceleration of gravity, 32.2 ft/sec^2

m = Film thickness, ft

d = Oil boom draft (curtain depth), ft

a = Curtain's depth below oil water interface, ft

h = Variable film thickness away from curtain, ft

C_f = Friction coefficient

Q = Volume of oil held by boom, ft^3

L = Total boom length, ft

F_c = Froude number

λ = A tension factor

C_D = Drag coefficient

l = Slick length

H. Endnotes for Appendix 10

1. M. Wicks, Fluid Dynamics of Floating Oil Containment by Mechanical Barriers in the Presence of Water Currents, API/FWPCA Joint Conference on Prevention and Control of Oil Spills, December 15-17, 1969.

2. D. Hoult, Containment and Collection Devices for Oil Slick, Department of Mechanical Engineering, M.I.T., 1969.

3. D. Hoult and R. Cross, Oil Booms in Tidal Currents, M.I.T., 1970.

4. D. Hoult, Containment of Oil Spills by Physical and Oil Barriers, M.I.T., 1970.

5. J. Scherer, Design Requirements for Booms, Hydronautics, Inc., Washington, D.C., 1969.

6. W. Lindenmuth, J. Scherer and P. Van Dyke, Analysis and Model Tests to Determine Forces and Motions of an Oil Retention Boom, Hydroanutics, Inc., Washington, D.C., January 1970.

7. E. Milz, An Evaluation-Oil Spill Control Equipment and Techniques, Research Development Laboratory, Shell Pipe Line Corporation, Houston, Texas, April 1970.

APPENDIX 11

EVALUATION OF POLLUTION ABATEMENT PROPOSALS

A. Introduction

There currently is no shortage of anti-pollution proposals with regard to improvements in ship design and operation.[1] Equally obvious should be the fact that the proposals are, in themselves, not an adequate solution to the oil spill prevention problem. Some will ultimately prove to be more effective than others, or perhaps a combination of two or three or more may provide the best answer under the existing constraints. These constraints presently take two forms. One is the technical ability of meeting the requirements set forth by the various proposals. The second constraint is the endowment available with which to work. Although it is desirable, the economic constraints on proposed solutions are not infinite. There is a definite amount of funds available determined primarily through a complex ordering of priority according to the perceived need. Any discussion which neglects economic constraints is not realistic and thus should not be seriously considered. For the most part, technical ability is the least imposing of the two constraints.

This leads to the following: some means must be developed to effectively evaluate proposed prevention techniques. That is, some framework needs to to be established so as to allow for an analysis of the incremental costs of each proposal or combination of proposals as opposed to the incremental benefits derived therefrom. Little work has been done in this area as specifically applied to marine casualties that result in oil pollution.

Any analysis of marine casualties is a difficult problem, primarily because of the inadequate and inconsistent collection of experience. This absence of an adequate management control and information system is what leads to the accident-related response proposals presently used for correction of marine related problems.[2] With no means to consider marine accidents from a systems point of view, a suboptimal solution is generally reached; one in which there are no supporting quantitative means available with which to evaluate the effectiveness of the change.

In this era of giant supertankers where the potential for damage to the environment from these vessels and others is significant, the present system for response to an evaluation of solutions to marine problems is grossly inadequate.

Some work has been initiated to help close the gap. Stratton and Silver present a conceptual model of an operational research and cost benefits analysis of navigation proposals toward reducing marine accidents through their implementation.[3] The Coast Guard is presently developing a comprehensive management information system utilizing the abilities of modern computers. Paralleling this information system, they are developing a model of casualty situations from which it will be possible to evaluate various proposals as solutions to reduce or minimize further casualties. At the printing date of this report, it was learned that this project was stalled as the result of the untimely death of its leading proponent two years ago. Since then the project has not been actively pursued because of the press of other requirements. However, some indication of the direction the work was assuming can be gathered from an outline prepared by the planning staff of the Office of Merchant Marine Safety, July 7, 1969.

A brief discussion of both the Stratton and Silver, and Coast Guard models follows.

B. Stratton-Silver Model

The model discussed by Stratton and Silver is aimed at determining an optimal solution to the reduction of marine casualties through analyzing how changes in navigational procedures (technical advances) will affect the incremental costs and related incremental benefits. They maintain a view of the complex interaction between navigation, vessel operation and the world at large in their model. Their ideas are equally adaptable to other proposals for maritime casualty reduction and should not be construed as being confined strictly to navigational proposals. The model forms a conceptual base which can be expanded to accommodate other casualty reduction proposals as well.

The relationship of the various costs resulting from efforts to reduce, through technology, maritime casualties is shown in Figure 11-1. The

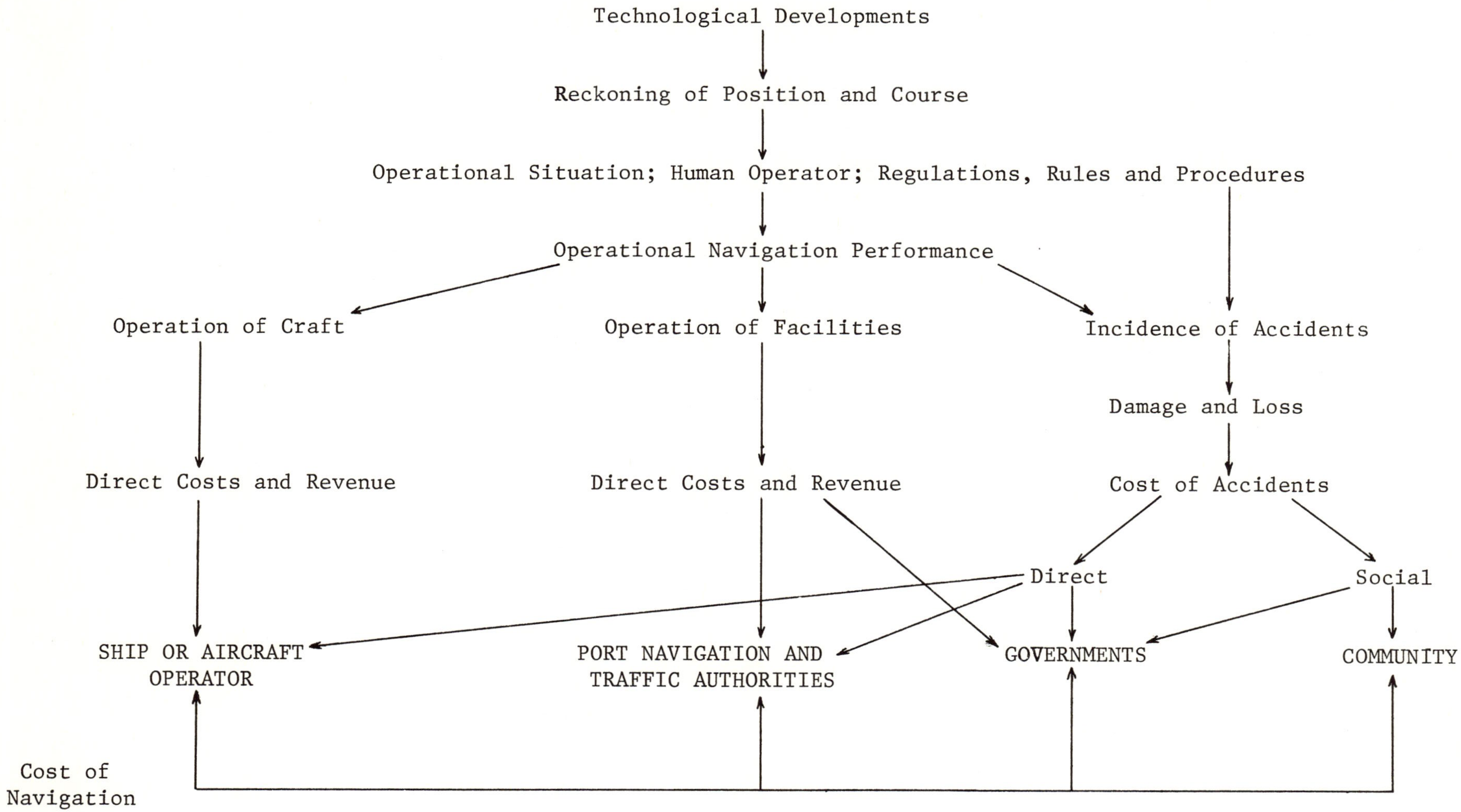

Figure 11-1: Relationship of Costs to Reduce Maritime Casualties[4]

operational performance is a function of the technical improvement and is determined in part by the interaction between the ability to effectively utilize the benefits of a design change, the operational situation and the operator's interpretation of the situation in relation to the control procedure that has been established.

The operational performance, in turn, influences the operation of the vessel, of supporting facilities, the incidents of accidents, and hence the costs and revenues associated with these happenings. Casualty costs fall on operators through loss of revenue and increased insurance rates and on society through the loss of life, property damage, loss of earnings, damage to the environment, etc. Some of this social cost falls on the government while the rest is borne directly by the community. The government also receives a load of the direct costs associated with a casualty which may somewhat offset the community cost. These costs to the government are in the form of direct payments to the injured community from insurance or damage claims. Thus, the total cost of changes in vessel operational performance as a result of technical proposals falls on the vessel operators/owners, the various operating authorities and the government. The end result cost to society is not so measurable in dollars as it is in the loss of amenity of life. Costs will be transferred among these various sectors only in relation to the extent of the external penalty imposed.

Because of complexity of these various interactions, quantification of these relationships is necessary to avoid a high degree of subjective judgement in the decision making process. This involves two related areas of study. First, it is necessary to apply operational research techniques to relate the operation of the vessel and supporting facilities and the incident of casulaties to the operational performance and situation resulting from a technological change. That is, the effectiveness of any technological change must be evaluated. An extensive data base is necessary for this type of analysis and must include such parameters as the number and size of ships lost or damages, the degree of damage, fuel consumption, time of voyage, etc. Secondly, these parameters must be translated into economic costs.

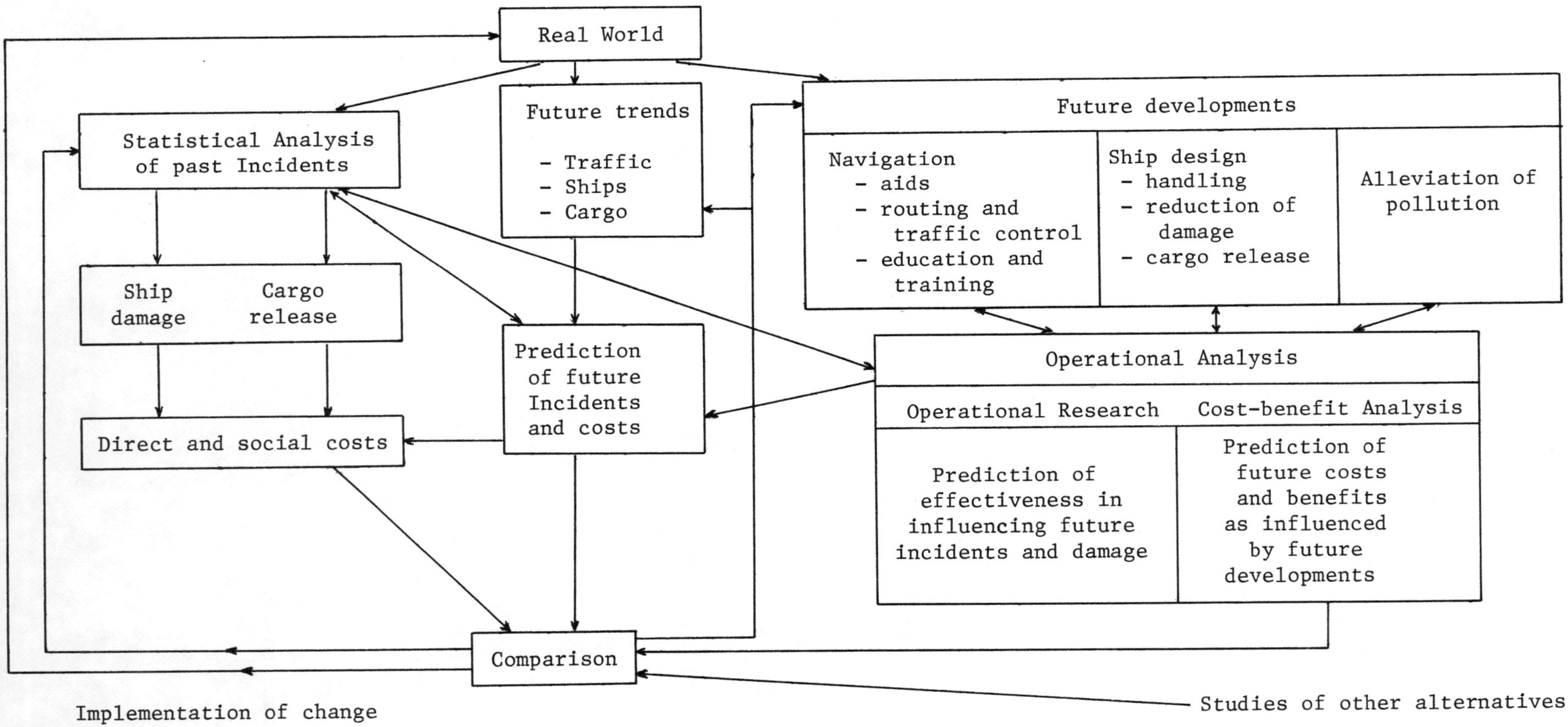

Figure 11-2: The method of conducting an operational analysis on marine accidents with particular reference to navigation[5]

The application of the overall system analysis to marine casualties is shown in Figure 11-2 in block diagram form. This diagram shows a method of conducting an analysis by abstracting the real world sufficiently to predict the effect of introducing any changes. The basis of the model is historical data and known physical laws.

C. Coast Guard Model

The Coast Guard Model is in conceptual form. The purpose of it is to compare the relative effectiveness of measures proposed to prevent or minimize potential oil pollution of the sea from ship casualties.

The basis of the model involves a consideration of the circumstances in which oil may be released, taking into account the degree of pollution which may occur and the likelihood of the incident.

Elements Involved:

I. Alternative Measures: (not a complete list)

1. Training of Ship's Officers
2. Installation of Radar
3. Tank Size Limit
4. Tank Arrangement
5. Backing Power Requirement
6. Decreased Turning Circle
7. Sea Lanes

II. Potential Pollution Incidents: (not a complete list)

a. Collision with Ship - Busy Populated Port
b. Stranding - Populated Area
c. Collision with Ship at Sea $<$ 10 miles
d. Collision with Ship 10-100 miles at Sea
e. Collision with Ship - Small Unpopulated Port
f. Stranding - Isolated Area
g. Collision with Ship $>$ 100 miles at Sea
h. Collision with Offshore Structure
i. Collision in Port with Dock

III. Severity Power

R is a factor to take cognizance of the seriousness of any one of the potential polluting incidents, as regards the quantity of oil which may be released, the difficulty of controlling the released oil, problem of cleaning up, and the extent of damage ensuing.

Thus, R_a is the severity of the incident "Collision with Ship-Busy Populated Port."

IV. Frequency: (F)

F is the relative likelihood that any potential polluting incident will occur. Because the data available to evaluate the probability of a casualty is quite limited a possible weakness may exist as to the relevance of the model.

V. Effectiveness: (E)

E is a factor to evaluate the utility of any Alternative Measures as applied to any potential polluting incident.

Thus, E_{1a} is defined as follows:

$$E_{1a} = \frac{(\text{Number of incidents of Type a in which Alternative 1 might be beneficial})}{(\text{Number of Incidents of Type a})}$$

VI. Comparative Worth: (W)

W is the comparative worth of any Alternative Measure. Thus, the worth of Alternative Measure 1 through m is:

$$W_1 = (R_aF_a \cdot E_{1a} + R_bF_b \cdot E_{1b} + \ldots + R_hF_hE_{1n})$$

$$W_2 = (R_aF_a \cdot E_{2a} + R_bF_b \cdot E_{2b} + \ldots + R_nF_nE_{2n})$$

.

.

.

$$W_m = (R_aF_a \cdot E_{ma} + R_bF_b \cdot E_{mb} + \ldots + R_nF_nE_{mn})$$

VIII. Other Factors:

It is conceivable that in some Alternative Measures there may be inherent an increase in risk of pollution. Such a factor would have to be separately identified and a means found for judging the negative contribution to Worth.

D. Summary

Two facts should become evident as a result of the implications of the Stratton-Silver Model and the C.G. work.

1. A quantitative systematic analysis of marine casualties is necessary if the potential gains from casualty reduction proposals are to be optimized.

2. A satisfactory statistical analysis of historical incidents is necessary to the relevance of the model and at present is hampered by the lack of a sufficient data base.

E. Endnotes for Appendix 11

1. See Section III-C, Part 1, for discussion of policies related to ship's design and operation standards.

2. Marine disasters where loss of life or a significant loss of property is involved are reviewed by a board to determine the cause of the disaster. Once a determination is made, effort is directed through the regulatory process to "legislate" against another accident occurring from the same cause.

3. A. Stratton and W.E. Silver, "Operational Research & Cost Benefit Analysis on Navigation with Particular Reference to Marine Accidents," Journal of the Institute of Navigation, July 1970, pp. 325-340.

4. Ibid., p. 326.

5. Ibid., p. 328.

APPENDIX 12

ANACORTES OIL SPILL

While onloading 20,000 barrels of #2 diesel fuel at the Texaco Refinery dock at Anacortes, Washington, United Transportation (UT) Barge 17 discharged approximately 5,500 barrels (232,000 gallons) of its cargo into Padilla Bay, Washington on April 26, 1971. The following paragraphs describe the events subsequent to discovery of the oil spill. Topics covered include response, surveillance, cleanup techniques, resource protection, and the damage assessment. The information utilized in this appendix has been provided by Washington Department of Ecology, U.S. Coast Guard, and United Transportation Company.[1,2,3] All times used in the discussion are designated on a 24-hour clock basis.

A. Notification

Loading operations for Barge 17 were initiated at 2030 hours on April 25. At 0130 hours Texaco notified the tankermen of Barge 17 that the amount transferred was approaching 20,000 barrels; a check of the barge revealed that space was still available. A visual inspection of the water around the vessel indicated oil in the water. At 0135 hours Texaco shut down the loading opeation. After evaluation of the source of the spill and approximate quantity discharged, the tankermen notified the home office in Seattle that approximately 1,000 gallons had been discharged. This information was relayed to the duty officer, USCG, 13th district, at 0245 hours. Under the Coast Guard contingency plan, this quantity is classified as a moderate spill.[4] USCG initiated a routine investigation, and dispatched the Captain of the Port (COTP) Pollution Officer to the scene.

At 0800 hours COTP issued a press release stating that about 5,000 gallons of oil had been spilled and closed its case on the incident at 1110 hours.[5] A radio broadcast at 1200 hours reporting a larger slick being sighted in Rosario Strait prompted COTP to reopen the case at 1556 hours. Not until 1600 hours, April 27 (over 40 hours after the spill incident) did COTP confirm that over 5,000 barrels was discharged.[6] A statement in the United Transportation report suggested that the misunderstanding between the terms "gallons" and "barrels" impaired response capability.[7]

B. Surveillance

Based on the information available, the only technique utilized in the tracking of this oil spill was visual. Aerial surveillance, supplemented by waterborne observations, and beachcombing were the primary methods. The behavior of this oil spill is shown in Figure 12-1, based on inputs from USCG and UT. By 0730 hours, April 26 (six hours after the spill), oil was sighted throughout Guemes Channel. The slick then spread into Rosario Strait, where oil was reported in the 1200 hours radio broadcast. By 1800 hours, April 26, the oil had spread as far south as Smith Island, a distance of 24 statute miles from the source. Oil had also travelled as far north as Sinclair Island, moving on both the east and west shores of Guemes Island. Aerial reconnaissance inspection on April 27 indicated oil had spread west of Smith Island, and even extended to San Juan Channel by 1900 hours.

All sightings after April 27 were of oil patches inside the 1900 hour perimeter shown on Figure 12-1. Based on the characteristics of #2 diesel fuel and the water condition of that spill area, it is surmised that the visual components of the oil had dissipated by this time.

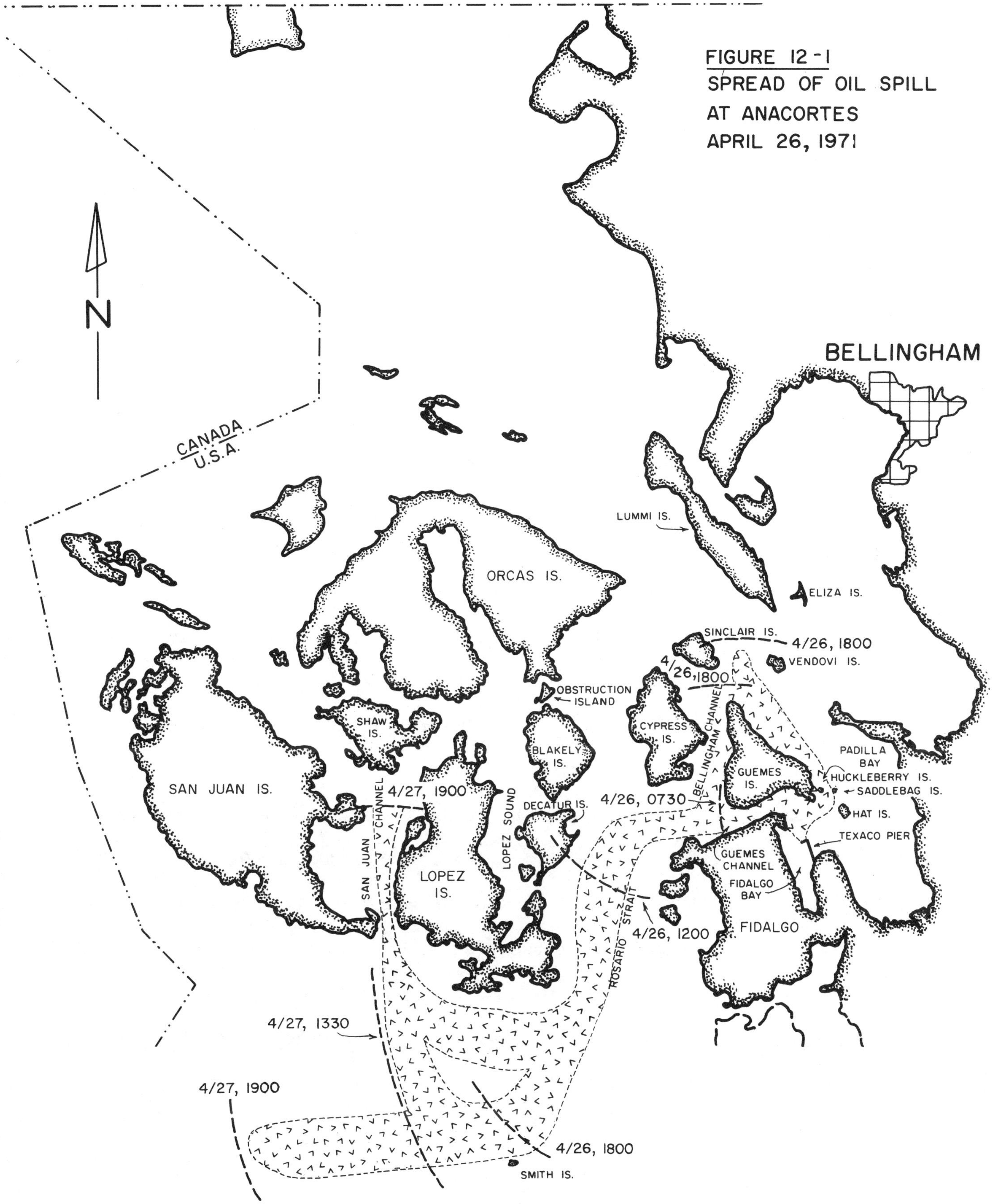

FIGURE 12-1
SPREAD OF OIL SPILL
AT ANACORTES
APRIL 26, 1971

C. Response

At the time of the spill, the tug Sea Breeze was onsite, and was activated by 0300 hours, April 26, to breakup and disperse the slick. UT management arrived on scene by 0500 hours and immediately combed Guemes Channel for evidence of oil. By 0730 surveillance by tug had been completed. Not until after the noon radio broadcast was MOPS activated. DOE personnel also reached the scene about this time.

At 1600 hours on April 26, bird care supplies were dispatched to Guemes Island. At this time, the first wave of cleanup equipment arrived at Anacortes. This included one Mop Cat, 300 feet of 6" rubberized boom, 200 sacks of Fiber-Perl absorbent, pitchforks, shovels, and mesh screens. In addition, a work force of 25-30 men was included.[8] 100 bags of chopped straw were also secured from the Texaco Refinery. Two tugs with 1000 ft. of log booms were deployed into Bellingham Channel.

At 0600 on April 27, the tug Spartan arrived on-scene with 150 sacks of Fiber-Perl. Also on-scene was the tug Put River with 300 feet of rubberized boom. A team of out-of-town Texaco officials arrived at 0930 hours to survey this spill. The USCG arrived at 1130 hours with a COTP representative. A conference with UT at 1235 hours was the first on-site coordination effort in this spill incident.[9] At 1300 the DOE representative arrived at the spill site. COTP arrived on scene at 1400 hours, and called a meeting of the subregional response team (SRT) and the regional response team (RRT) at 1600 hours. A command post, with a mobile communications center and helicopter as support, was established at the Texaco Refinery at March Point. Concurrently, ARCO and Mobil deployed equipment to fence off Lummi Island.

On the morning of April 28, Texaco scientists inspected the site and theorized that the oil slick off Lopez Island had dissipated by natural action of micro-organisms.[10] At noon a contingent of personnel from Washington, D.C. arrived, stayed one hour and then departed, except for the EPA representative. In the afternoon, the U.S. Navy supplied 50 men to aid in the cleanup activities. Also at this time, COTP and DOE staffs departed from the scene. At 1800 hours, a team of EPA biologists completed an inspection tour of the marine life and expressed their concern for the lack of effort to cleanup beach areas.[11] A request for increased cleanup efforts was initiated, though USCG implied that complete beach cleanup efforts was improbable without removal of tons of sand and gravel.[12] At 0930 April 29, COTP received a report confirming the increased cleanup action requested by EPA. Sometime during the morning of April 29, a conflict of opinion arose between EPA and COTP with respect to the cleanup efforts.[13]

At 1700 on April 29, representatives of USCG, EPA, DOE, and UT discussed cleanup status and decided to continue for one more day. At 1130 April 30, a second meeting concluded that further activities were unwarranted. Operations were demobilized at 1400, and the command headquarters at Texaco were closed.

Apparently, the COTP, as on-scene commander, was satisfied with the response and cleanup, since there is no documented evidence of making "recommendations as to the action and type of resource to be used."[14] The Coast Guard contingency plan also specifies protection of "critical water use areas."[15] It appears that no preventive efforts were made to deflect oil from areas such as Lummi Island, Bellingham Channel, or San Juan channel until <u>after</u> oil had been sighted in those areas. Given the deployment capability and the effectiveness of booms, any action initiated after sighting a slick would be almost too late to stop the oil from entering the critical area. True, the efforts probably averted further consequences, but the purpose of defining critical regions is to protect them from potential oil slicks.

The referenced reports indicate that communications networks, especially inter-agency, were far below optimum. The small number of SRT meetings may be attributed to this factor.

With respect to mobilization time, about 6 hours elapsed from the Rosario Strait report until the arrival of the first equipment. The 10.5 hours prior to the radio report are not allocated to this figure. But it must be remembered what parties were involved. The oil was spilled by UT and UT is also an integral member of MOPS, the third party contractor. What happens next time if some other party outside MOPS is responsible for a spill? A conservative estimate of adding another 2 hours for communication seems to be in line, resulting in an 8 hour mobilization time.

Lastly, all agencies involved should critically examine their standards for judging adequacy of cleanup operations. "Out of sight, out of mind" is a poor criterion.

D. Cleanup

To minimize belaboring an already long discussion, the following statement by USCG will suffice:

> "The major effort of cleanup was concentrated on the residual light slicks with the cleanup of these light slicks by MOPS being effective. Puget Sound Tug and Barge estimated the entire cleanup operation recovered only 10 barrels of oil."[16]

USCG also concluded that the cleanup action on the beaches was considered more than adequate.[17] One must ask, how much is adequate, and by whose standards? Only time will tell in this case, since there was no scientific basis used at the time of the spill.

E. Damage Assessment

The Environmental Protection Agency (EPA) awarded $45,000 to Texas Instruments (TI) to make damage assessment studies, including chemical, biological and other types of analyses on water affected by the April 26 Anacortes spill.[18] Unfortunately, the firm was not notified by EPA until May 3, 1971 - one full week after the spill incident. By that time, the immediate short-term effects of the oil on the biota and other marine life had already been established. Without a baseline definition of this marine life, TI had no basis to assess the extent of short-term damage attributed to the oil spill.

1. Sampling Techniques

A study plan prepared by TI indicated that they would take approximately 20 samples for chemical analysis.[19] The following is quoted as guideline for their techniques.[20]

> "A recent report by Dr. Max Blumer on a diesel spill on the east coast will* be examined for new analytical approaches to determine content and concentration of hydrocarbon in the samples."

It would appear from the foregoing statement that, at the time the study plan was written TI had no definite plan on how samples were to be analyzed nor how they were to be prepared. This is further indicated by their sampling techniques.

Water samples taken by the TI team were made up in polyethylene bottles.[21] An organic material such as polyethylene, contacting an oily substance, can alter the contents of the container and invalidate any analytical results. The Coast Guard contingency

*Underlining by coordinator.

plan includes the following warning:

"Glass containers always must be used because plastic containers, with the exception of teflon, have been found in some cases to absorb organic materials from water and in other cases compounds have been dissolved from plastic containers."[22]

It is known that oil, being lighter than water, will float on top. Thus, when taking bottom samples, care must be taken to isolate the container from the surface waters when immersing it in the water column. It is not clear how TI obtained their bottom samples; however it is known that standard bottom water sample bottles were used, they were lowered remotely, and the stoppers were pulled when the desired depth was reached.[23] No mention is made on how the bottle was lowered, or whether it was recapped in the water column or at the surface. Such information is vital to the integrity of an analysis.

It is interesting to note where TI set up their sampling stations. Figure 12-2 depicts the areas where TI conducted their study.[24] When compared to Figure 12-1, it is questioned why TI evidently did not include Lopez, San Juan, and Cypress Islands. (See Section II-A of this report for a pictoral view of the location of marine resources in this region.)

2. Coordinated Activities

Under the terms of the contract, TI was required to respond within 48 hours of notification by EPA.[25] The team did not arrive at the spill site until May 8, 1971, five days after receiving notice. Data was finally collected almost two weeks after the spill incident, May 9 through May 12.

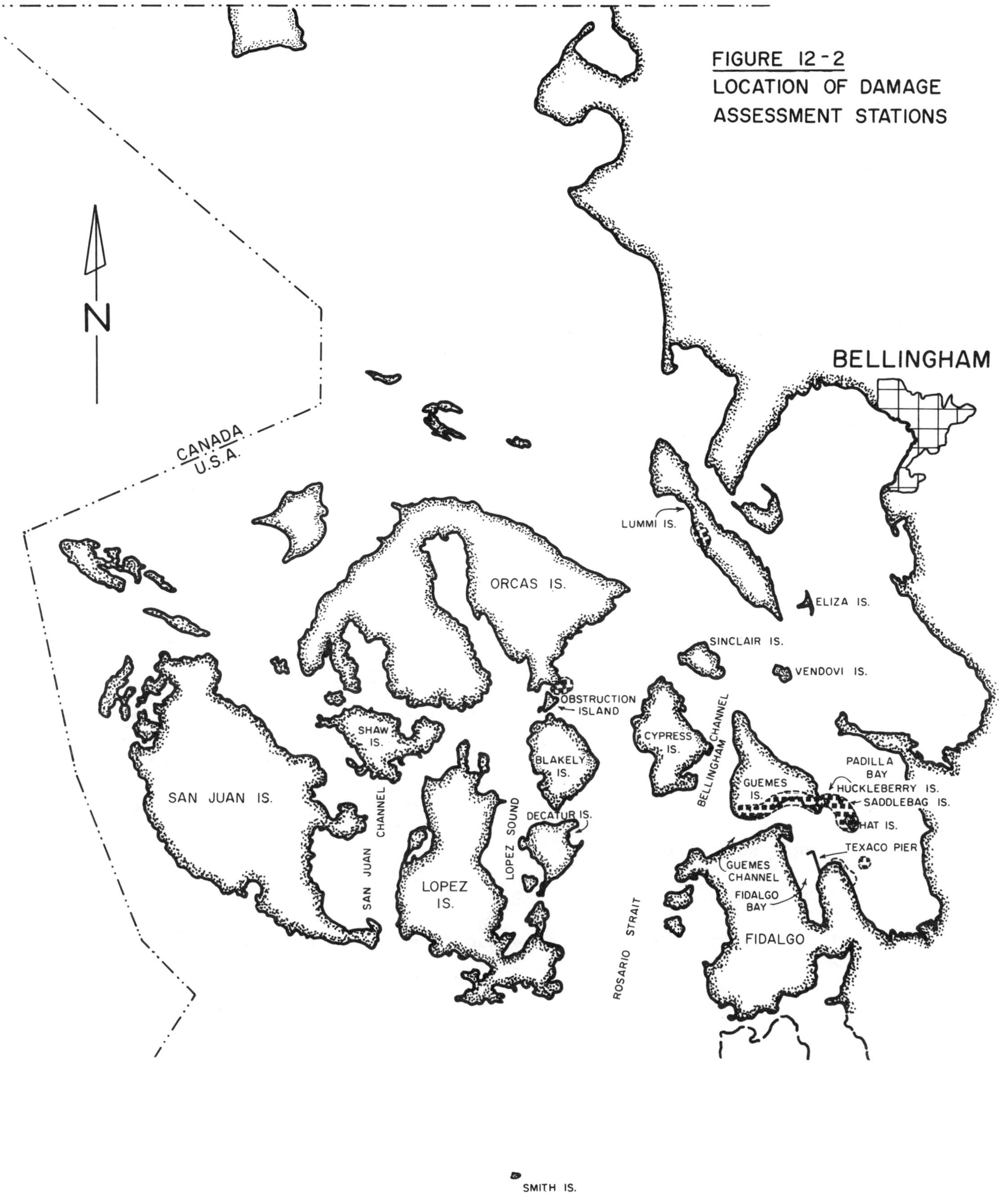

FIGURE 12-2
LOCATION OF DAMAGE ASSESSMENT STATIONS

In the area of coordination, the following are quoted from the TI study plan:

"TI will coordinate with area institutions and agencies to arrange for exchange of technical data concerning the subject spill, if available."[26]

"Additional scientific data on the spill as may be acquired for institutions and agencies (and will be included in the report)."[27]

In the preliminary report, it is stated:

"Other academic personnel from the University of Washington at Seattle carried out a subtidal inspection of the Guemes Island area by diving; they believed, however, that there was no need to return for detailed studies."[28]

Neither in the preliminary nor final reports is there further substantiation nor explanation of the results of the noted subtidal inspection. The "academic personnel" are not identified nor their credentials established.

Moreover, TI did not contact Robert C. Clark, a noted marine biologist for the National Marine Fisheries Service in Seattle, and co-author with Dr. Blumer on studies of effects of hydrocarbons on marine life.[29] The TI report did refer, however, to studies by Berg and Webber of Western Washington State College[30] and quoted some results.

F. Recommendations

Let us hope that the techniques and procedures utilized to combat and assess the Anacortes oil spill do not become standard operations for solving the oil-on-water problem in Puget Sound. But to alter these operations or to revise procedures requires evidence that changes will bring home some tangible benefit. It is not an overstatement to say that society in Puget Sound has not only incurred a loss from the spill itself, but also by the subsequent expenditures for control and assessment. Once the attitude for change is positive among responsible parties, then the mechanisms of a coordinated improvement program can be initiated. Society should insure that such improvement will be implemented if and when another oil spill occurs.

G. Endnotes for Appendix 12

1. Department of Ecology memorandum from H. B. Tracy to J. Behlke on the Anacortes spill, dated May 10, 1971.
2. Captain of the Port, Station Seattle Pollution Report #43-071, Captain of the Port, USCG, Seattle, Washington.
3. R. W. Roe, Report on Barge 17 Oil Spill, United Transportation Company.
4. U. S. Coast Guard, Seattle Coastal Region Oil and Hazardous Materials, Pollution Contingency Plan, December 1, 1970, p.5.
5. Report #43-071, Enclosure (1), p.1.
6. Ibid., p.2.
7. Roe, p.2.
8. Ibid., pp.3-4.
9. Report #43-071, Enclosure (1), p.2.
10. Roe, p. 11.
11. Ibid., p. 12.
12. Report #43-071, Enclosure (1), p. 3.
13. Tracy, p. 4.
14. Seattle Coastal Region, Annex XX, Appendix 1, Paragraph 3100.2.
15. Ibid., p. 14, paragraph 306.1-2.
16. Report #43-071, Enclosure (1), p. 6.
17. Ibid.
18. Seattle Post Intelligencer, August 15, 1971, p. 1.
19. Texas Instruments, Study Plan Anacortes Spill Survey, Texas Instruments Study Proposal, p. 5.
20. Ibid., p. 5.
21. Texas Instruments, Inc., Biological Assessment of Diesel Spill, Anacortes, Washington, May 1971 Preliminary Report, Texas Instruments, Inc., June 14, 1971, p. II-11.
22. Coast Guard Plan, p. VIII-5, paragraph 1804.2-5.
23. Preliminary Report, p. II-19.
24. Ibid., p. I-5.
25. EPA contract no. 68-01-0017.
26. Study Plan, p. 6.
27. Ibid.

28. Texas Instruments preliminary report, p. II-17.

29. Texas Instruments preliminary report lists the following document in their bibliography:

 R. C. Clark and M. Blumer, "Distribution of h-parrafins in Marine Organisms and Sediments: Limnology and Oceanography," Vol. 12, 1967, pp. 79-87.

30. H. H. Webber, _Survey of Mortality at Two Areas of the South Beach of Guemes Island between April 28 and May 15, 1971_, Huxley College of Environmental Studies.

APPENDIX 13

A PREDICTIVE MODEL FOR OIL SPILL FROM SHIP COLLISIONS ON PUGET SOUND

A. Introduction

Previous studies on the probability of a ship collision in Puget Sound have analyzed the Sound as a one body of water, yielding a single probability collision value.[1] This type of analysis certainly is of some value when attempting to predict future collision frequencies in the face of growth in waterborne commerce. However, when trying to identify portions of the Sound that are more hazardous than others, in terms of the number of potential collisions and resultant oil spills, it becomes necessary to view the region as a system comprised of a number of waterways on which vessels travel. By applying a ship collision model to each individual waterway (channel), an estimate of the collision probability for that channel and probability of a resultant oil spill can be obtained. Examination of all channels will give an indication of which areas are most hazardous.

B. Model Description

The model used in this study is based on one developed by Sperry Piedmont Company[2] for predicting the probable number of vessel collisions in a channel of a given configuration. The discussion presented in this appendix will not duplicate the model development, but will depict its final form. The interested reader is referred to the original study for a detailed derivation.

The model is based on two assumptions:

1. The decision making behavior of the watch officer onboard ship is essentially the same in all seas, and
2. The behavior of collision victims (vessels) is not radically different from that of non-collision vessels.

The model, as developed by Sperry, is comprised of four separate parts:

1. The encounter rate of vessels (E_T):

$$E_T = 1/2 \; \frac{LN^2}{KV_T}$$

where

L = average length of channel

N = total number of vessels passing through the channel in one year

K = 8,760, the number of hours in one year, a constant

V_T = velocity of encountered ship

2. The number of collisions per encounter, defined as the probability of a vessel collision P(C):

$$P(C) = \frac{\bar{C}}{E_T}$$

where

$\bar{C}$ = the number of collisions per year.

3. The probability of a collision situation in a given channel $(P(C_o))$:

$$P(C_o) = b/w$$

where

b = breadth of a ship

w = width of a channel

4. The probability of a ship collision given a collision situation $(P(C|C_o))$:

$$P(C|C_o) = \frac{P\ (C)}{P(C_o)}$$

since a collision cannot happen unless there exists a collision situation.

For the purposes of analyzing Puget Sound, only equations (1) and (2) have direct impact on the results. The goal of this analysis was to determine relative rankings, rather than attempt an exact numerical evaluation. Equations (3) and (4), when utilized with appropriate input data, will enhance the achievement of the latter objective.

The model was programmed on a CDC 6400 computer, using 1970 as a base year for input data. The relevant independent variables which were varied includes vessel speed (V_T), and the number of vessels in a given channel (N). Data was computed in each channel for five year periods between 1970 and 1990. These results are the basis from which the probable number of vessel collisions per time period, per channel, were calculated. The ultimate goal of the model is to predict future oil spills resulting from vessel collisions.

Evaluation of the range of vessel speeds led to the selection of 15 knots as the speed used in the following discussion. The number of collisions per encounter, P(C), is dependent on the channel in which the collision takes place. That is, owing to the physical dimension and the number of vessels in the various channels, there will result a unique value for number of collisions per encounter in a channel.

C. Analysis of Data Base

The initial efforts in analyzing this problem was to divide the Puget Sound Region into a number of channels. Basis for the sectioning were traffic patterns and geographic characteristics. Table 13-1 describes the five selected channels and their dimensions, and these are shown in Figure 13-1. Examination of the diagram shows two areas, one just inland of Channel 1; and the other, the Strait of Georgia, have been omitted. The behavior of vessels in these areas does not fit the Sperry model and they are analyzed separately.

Any future projections of ship collision in these waters requires an established data base on today's vessel behavior. Specifically, two parameters need to be identified, the number of collisions per year, $\bar{C}$, and the number of ship encounters per year, E_T. After extensive searching, it was concluded that accurate collision records were not readily available in the Puget Sound area.

These estimates are critical to the final results as the $\bar{C}$ values directly determine the probability of a collision in any particular channel. What is required is historical data on the number of collisions between all vessels regardless of size in each channel. At present the only known source of information that approximates these requirements is the casualty files at Coast Guard Headquarters in Washingotn, D.C. These files, however, include data pertinent only to registered vessels and only reported accidents. It would not include those vessels engaged in commerce that are not registered, the numbers of which are unknown. It is also

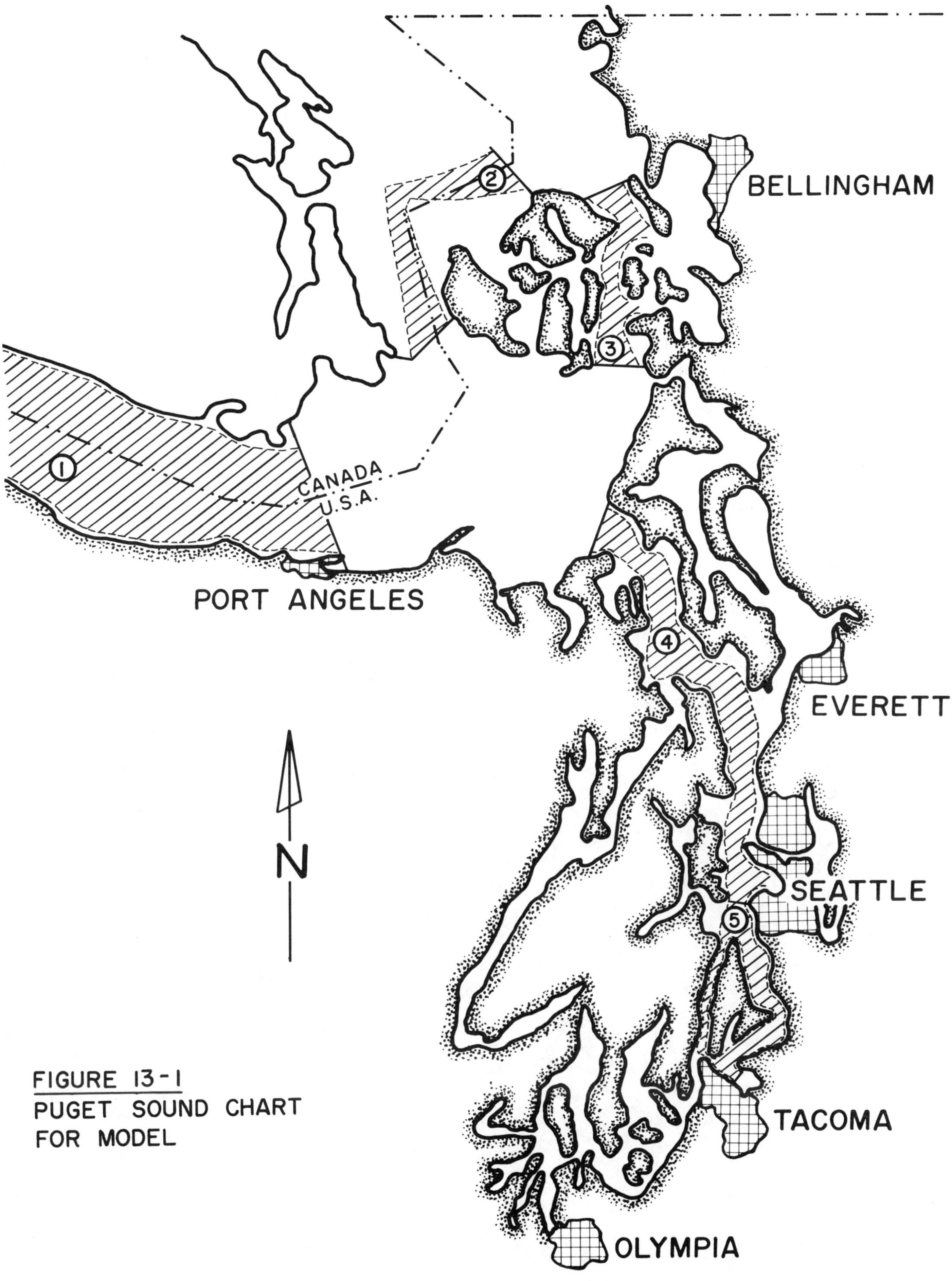

FIGURE 13-1
PUGET SOUND CHART
FOR MODEL

TABLE 13-1: CHANNEL DEFINITION FOR SHIP COLLISION MODEL

	Approximate Average Width (miles)	Approximate Average Length (miles)
Channel 1:		
Cape Flattery-Bannilla Pt. to Dungeness Spit - Trail Islands	12.2	65.0
Channel 2:		
Discovery Island - Cattle Pt. to East Pt. - Alden Pt.	4.2	28.9
Channel 3:		
Pt. Colville - West Pt. to Lawrence Pt. - Village Pt.	3.19	18.0
Channel 4:		
Pt. Wilson - Admiralty Head to Elliott Bay	3.45	39.0
Channel 5:		
Restoration Pt. - Alki Pt. to Commencement Bay (Two Passages) -		
Colvos Passage	1.01	17.3
East Passage	2.53	23.5

necessary to consider those collisions between vessels that went unreported for one reason or another. Again no specific numbers can be placed on these occurrences, only estimates.

Because of time and logistics problems the Coast Guard data could not be utilized. It has therefore been necessary to estimate the $\bar{C}$ values based on the number of major collisions in the Puget Sound region over the past five years.[3] For each channel, three figures were computed, the historical value, a high value, as well as a low value to produce a band of possible collisions rather than a specific number. These are shown in Table 13-2 below. The historical value was considered the median value and the upper bound is twice the median value, while the lower bound was one-half of the median.

TABLE 13-2: NUMBER OF COLLISIONS PER YEAR ($\bar{C}$)

Channel	Low	(Historical) median	High
1	.2	.4	.8
2	.1	.2	.4
3	.1	0	.4
4	.4	.8	1.6
5	.1	.2	.4

Where the data indicated no collision had taken place during the five year period in a particular channel a low value of .1 was assumed, a high value of .4.

In Part B, it was shown that E_T was a function of the length of the channel, number of vessel movements and the velocity, which has been assumed to be 15 knots for the purpose of presenting a set of results. Channel dimensions are shown in Table 13-1. There remains the task of defining the number of vessel movements.

The historical data for the volume of vessel traffic is derived from 1967 data accumulated by the Corps of Engineers.[4] Analysis of the vessel movement trend between 1965-1970 reveals little change in the year-to-year volume. The following assumptions were made for the analysis.

1. The number of ship movements in each channel is a fucntion of the number of ships arriving or departing each port along the channel, plus those ships that must traverse the channel to arrive or depart ports beyond the channel.

2. The total number of ship movements in each channel can be arrived at by analyzing separately the external and internal waterborne commerce which moves within the channel.

The total number of vessel movements in each channel was determined as a percentage of the total vessel movements in the Puget Sound region. The referenced data indicates 308,860 movements in 1967. This number represents all vessels entering or leaving a port in the Sound, and includes non-self-propelled as well as self-propelled vessels, cargo vessels as well as tank vessels. The following are the per channel vessel movement calculations:

Channel 1:

Of the total waterborne commerce of 49 million short tons, 57% was commerce internal to Puget Sound, while 43% was external. Since all external traffic must assume to traverse the Strait of Juan de Fuca,

.43 (308,860) = 132,000 external vessel movements are allocated to this channel. In addition, the internal traffic allocated to the ports of this channel were:

Neah Bay	5.2%
Port Angeles	1.4%

totaling 6.6%. Thus, the total vessel movement in this channel is 132,000 + (.066)(308,860) = 152,300 vessels.

Channel 2:

This channel includes Canadian ports only, since external commerce from those ports must traverse U.S. waters; in particular, this channel.

Vancouver	2.9%
New Westminister	0.5%
Other B.C. Ports	0.6%
TOTAL	4.0%

This leads to a total of 12,300 vessel movements.

Channel 3:

Anacortes	5.8%
Bellingham	6.2%
TOTAL	12.0%

This yields 37,000 ship movements.

Channel 4:

Port Townsend	3.8%
Everett	7.6%
Seattle	40.0%
Lake Washington Ship Canal	7.2%
Waterway	2.0%
Traffic from Channel 5	15.1%
TOTAL	75.7%

The total vessel movements for this channel is 233,500. It is assumed that all traffic in Channel 5 must traverse Channel 4 in arriving or leaving a port. This assumption may lead to slightly larger values for Channel 4, but based on the above values, the effect is negligible.

Channel 5:

Tacoma	13.2%
Olympia	1.9%
TOTAL	15.1%

This yields 46,600 vessel movements.

The summing of vessel activity per channel from the 1967 data, and is utilized in this model as the 1970 base:

Channel	Vessel Movements
1	152,300
2	12,300
3	37,000
4	233,500
5	46,600

PROBABILITY OF A SHIP COLLISION IN PUGET SOUND

CHANNEL NUMBER

C	1	2	3	4	5
.10	.00000002	.00000691	.00000123	.00000001	.00000059
.20	.00000004	.00001382	.00000245	.00000003	.00000118
.30	.00000006	.00002074	.00000368	.00000004	.00000178
.40	.00000008	.00002765	.00000491	.00000006	.00000237
.50	.00000010	.00003456	.00000613	.00000007	.00000296
.60	.00000012	.00004147	.00000736	.00000009	.00000355
.70	.00000014	.00004839	.00000859	.00000010	.00000415
.80	.00000016	.00005530	.00000981	.00000011	.00000474
.90	.00000018	.00006221	.00001104	.00000013	.00000533
1.00	.00000020	.00006912	.00001226	.00000014	.00000592

PROBABILITY OF A COLLISION SITUATION IN PUGET SOUND

CHANNEL NUMBER

1	2	3	4	5
.0011639	.0033649	.0044514	.0041159	.0056126

PROBABILITY OF A SHIP COLLISION ON PUGET SOUND GIVEN A COLLISION SITUATION

CHANNEL NUMBER

C	1	2	3	4	5
.10	.0000172	.0020542	.0002755	.0000035	.0001055
.20	.0000344	.0041084	.0005510	.0000069	.0002110
.30	.0000517	.0061626	.0008266	.0000104	.0003165
.40	.0000689	.0082167	.0011021	.0000138	.0004221
.50	.0000861	.0102709	.0013776	.0000173	.0005276
.60	.0001033	.0123251	.0016531	.0000207	.0006331
.70	.0001206	.0143793	.0019286	.0000242	.0007386
.80	.0001378	.0164335	.0022041	.0000276	.0008441
.90	.0001550	.0184877	.0024797	.0000311	.0009496
1.00	.0001722	.0205418	.0027552	.0000345	.0010552

VELOCITY OF VESSELS IN THE CHANNELS IS 15 KNOTS

THE ENCOUNTER RATE OF VESSELS IN CHANNEL 1 IS 4988730
THE ENCOUNTER RATE OF VESSELS IN CHANNEL 2 IS 14467
THE ENCOUNTER RATE OF VESSELS IN CHANNEL 3 IS 81537
THE ENCOUNTER RATE OF VESSELS IN CHANNEL 4 IS 7035827
THE ENCOUNTER RATE OF VESSELS IN CHANNEL 5 IS 168856

FIGURE 13-2

COMPUTER PRINTOUT, 1970, 15 KNOTS

At a velocity of 15 knots, the resultant E_T per channel is as follows for 1970:

Channel	E_T (Vessel Encounters)
1	4,990,000
2	14,800
3	81,800
4	7,036,000
5	168,900

For a given channel, this encounter rate varies inversely with the ship velocity. In a given body of water, the faster a vessel traverses that body, the lower the number of vessels he will encounter, given that all other vessels maintain their original performance.

Having computed $\bar{C}$ and E_T, the model can now generate the probability of collision, P(C). Figure 13-2 shows the computed output for 1970, with V_T = 15 knots. For each channel the probabilities corresponding to the upper and lower bound of the $\bar{C}$ values is shown in Table 13-3. The low probabilities associated with Channels 1 and 4 are the result of the enormous volume of vessel movements in those channels.

TABLE 13-3: SHIP COLLISION VALUES FOR 1970

V_T = 15 KNOTS

	Lower Bound		Upper Bound	
Channel	$\bar{C}$ collision/yr.	P(C)	$\bar{C}$ collision/yr.	P(C)
1	.2	4.0×10^{-8} (est.)	.8	1.6×10^{-7}
2	.1	6.9×10^{-6}	.4	2.76×10^{-5}
3	.1	1.2×10^{-6}	.4	4.9×10^{-6}
4	.4	5.7×10^{-8}	1.6	2.3×10^{-7}
5	.1	6.0×10^{-8}	.4	2.4×10^{-6}

The data in Figure 13-2 also gives the probability of a collision given a collision situation, $P(C|C_o)$. This figure is a constant in each individual channel regardless of the number of vessel movements or their velocity. Basically it is dependent on the actual dimensions of the channel. For the five channels chosen in this study these probabilities expressed as ratios follow:

Channel	Number of resulting collisions per 10,000 collision situations
1	12
2	34
3	44
4	41
5	56

These figures give some indication of the ship's ability to avoid a collision once it finds itself in a collision situation.

D. Exceptions

As mentioned earlier, two areas of the Sound have been excluded from the analysis using the Sperry model. In these waters, vessels exhibit more or less a random movement, as contrasted with the more predictable lanes in the five designated channels. Thus, a system analogous to two dimensional free gas model can be used to predict the probability of a collision in these two bodies of water. Unfortunately, there is insufficient information to estimate the number of collisions per year, as was done in the above analysis. Nevertheless, the probability values derived below can be used as a basis to evaluate behavior in these two waterways in the face of increasing commerce over the next twenty years.

$$P(X) = \frac{n\ a}{A}$$

where

P(X) = probability of collision between objects in the waterway

n = number of objects (vessels) in the body at any instant

a = area occupied by vessel

A = area of the body of water

It is essential here that collisions between vessels predominate over groundings, due to the physical characteristics.

For the large body of water east of the Strait of Juan de Fuca, vessels can move in five directions. Table 13-4, below, lists these paths, the length of travel, and the width of the waterway exposed to the vessel, plus the number of vessel movements. This later set of figures are based on those presented in the earlier channel analysis:

TABLE 13-4: INTER CHANNEL PATHS

Path	Length of Path (miles)	Width of Path (miles)	Number of Vessels
Channel 1-2	21	25	12,000
Channel 1-3	42	25	36,000
Channel 1-4	30	42	177,000
Channel 2-4	25	42	14,000
Channel 3-4	16	25	42,000
			281,000

The width of this body of water is the weighted average of the widths confronted by the vessels. The mean distance traveled within this area is likewise the weighted average.

TABLE 13-5: COMPUTATION OF WEIGHTED DIMENSIONS

Path	Weighting factor	Weighted Length (miles)	Weighted Width (miles)
Channel 1-2	4.3%	.9	1.1
Channel 1-3	12.8%	5.4	3.2
Channel 1-4	63.0	18.9	26.5
Channel 2-4	4.9	1.2	2.1
Channel 3-4	15.0	2.4	3.8
		28.8 miles	36.7 miles

The value 36.7 miles is the solution for W in the probability equation.

At any instant, the number of vessels in the channel, n, is dependent on the entrance and exit frequency, the speed, and the time spent in the channel. Using a velocity of 15 knots, and the weighted distance of 28.8 miles, the average time spent in this body of water by a vessel is 2.2 hours. Thus

$$n = (\#\text{ of vessels per year}) \times \frac{(\text{hrs. in channel})}{\text{hrs. per year}} = \frac{(281{,}000)(2.2)}{8760}$$

$$= 71 \text{ vessels}$$

The weighted cross sectional area of a vessel is assumed to be the sum of the weighted width and length, since these are the areas exposed to potential collisions. Using values 75 ft. and 300 ft. for the dimensions, a is equal to 22,500 sq ft, or 8×10^{-4} sq miles. The weighted area is the product of the weighted length and width of the body of water. Thus,

$$P(X) = \frac{na}{A} = \frac{71(.0008)}{(28.8)\ (36.7)} = 5.35 \times 10^{-5} = .0000535$$

The results show five chances out of 100,000 for a collision in the body of water east of the Strait of Juan de Fuca. This assumes only random movements. Given that ship navigators exercise care in piloting their vessels, the chances are even less. A detailed study of vessel movements in this area is needed to yield more accurate estimates than those presented here.

For the Strait of Georgia, only vessels from Channels 2 and 3 are considered as sources or sinks. Their destination and weighting factors are as follows:

TABLE 13-6: INTERCHANNEL PATHS IN STRAIT OF GEORGIA

Path	Length (miles)	Width (miles)	Number of Vessels
Channel 2-B.C.	16	15	12,000
Channel 2-Sandy Pt.	14	19	0
Channel 3-B.C.	23	13	11,000
Channel 4-Sandy Pt.	12	13	25,000
			48,000

The weighted Strait dimensions are thus:

TABLE 13-7: WEIGHTED DIMENSIONS - STRAIT OF GEORGIA

Path	Weighting factor	Weight Length	Weighted width
Channel 2-B.C.	25%	4.0	3.8
Channel 3-Sandy Pt.	23%	3.2	4.4
Channel 3-B.C.	52%	6.2	6.7
		13.4 miles	14.9 miles

The deadweight ton (DWT) deflators give an indication of how much the average vessel size has increased each 5 year period. This can be determined as (1 - DWT deflator). The purpose of the deflator is to reduce the total vessel movements in any given 5 year period to an amount, comparable period by period. That is, a growth in commerce will occur indicating more vessel movements, however, the size of the vessels will also increase requiring a deflation of the percentage again in total movements. Table 13-8, below, shows values for tankers (T) and freighters (F):

TABLE 13-8: DWT DEFLATORS

Year	DWT_F	DWT_T	Deflator F	Deflator T
1970	8000	45,000	1.0	1.0
1975	8500	60,000	.94	.67
1980	8800	78,000	.96	.70
1985	9000	85,000	.98	.91
1990	9300	90,000	.97	.94

An assumption is made here that the projected growth in freighter and tanker sizes is representative of the expected growth in size of all types of vessels engaged in commerce. That is, the average size vessel carrying general cargo will increase in size approximately 16 percent over the next 20 years while the typical vessel moving bulk oil will double in size over the same period of time.

Applying the DWT deflators to the percent growth in tonnage, an approximation for the percentage growth in the number of vessel movements can be estimated.

TABLE 13-9: GROWTH RATES MODIFIED BY DWT DEFLATORS, FREIGHTERS

Year	% Growth Tonnage (Other)	DWT Deflator F	% Growth Vessel Movements (F)
1970	0	1.0	0
1975	7.4%	.94	6.9%
1980	8.9%	.96	8.5%
1985	16.0%	.98	15.7%
1990	14.9%	.97	14.4%

TABLE 13-10: GROWTH RATES MODIFIED BY DWT DEFLATOR, TANKERS

Year	% Growth Tonnage (Bulk Oil)	DWT Deflator T	% Growth Vessel Movements (T)
1970	0	1.0	0
1975	23.6%	.67	15.8%
1980	16.3%	.70	11.4%
1985	14.4%	.91	13.1%
1990	14.0%	.94	13.2%

The total percentage growth in the number of vessel movements is found by taking a weighted average of the two individual growths. The weighting factors are the percentages of the total tonnage comprised of bulk oil and other commerce, with one modification.

On the average, the ratio of DWT of freighters to tankers in each of the 5 year periods is:

1970	.18
1975	.14
1980	.11
1985	.105
1990	.103

When applying the weighting factors to the percentage growth for numbers of vessels, the factors must be the weighted for growth in numbers, not tonnages. Therefore, it is necessary to modify the total tonnage weighting factors with the relative DWT ratios as follows:

TABLE 13-11: COMPUTATION OF WEIGHTING FACTORS FOR NUMBER OF VESSELS

Year	(Total Commerce) % Tanker	x	DWT Relative Ratio	=	Weight Factors Tankers	Weight Factors Freighters
1970	43		.18		.075	.925
1975	46		.14		.065	.935
1980	48		.11		.053	.947
1985	47		.105		.05	.95
1990	48		.103		.05	.95

The weighting factor for 1970 compares favorably with the ratios of numbers of tankers and freighters over 1000 net tons entering Puget Sound in 1970.

The time to travel through this body of water at 15 knots is 0.78 hours, and the resultant number of vessels in these waters at any instant, is

$$\frac{48{,}000\ (0.71)}{8760} = 4.3 \text{ vessels}$$

assuming the same vessel area, 8×10^{-4} sq. miles

$$P(X) = \frac{(4.3)\ (8 \times 15^{4})}{(13.4)\ (14.9)} = 1.71 \times 10^{-6}$$

Thus, for this body of water there is less than two chances in one million of having a collision, assuming random movements.

E. Projected Growth in Vessel Movement

To project the growth in traffic for each channel, a basic assumption was made that the base mix (allocation) will not change over time. This assumption requires approximately equal growths for all areas of Puget Sound. The total commerce projections as well as the increasing vessel size was used to arrive at a percentage rate of growth per five year period. This percentage of growth is then applied to the base number of vessel movements to obtain the projection desired.

The percentage growth per five year period in other commerce is calculated as follows:

Year	Historical Trend Projection (Short Tons)	Percent Growth
1970	28,200,000	0
1975	30,300,000	7.4%
1980	33,000,000	8.9%
1985	38,000,000	16.0%
1990	44,000,000	14.9%

The percentage growth per five year period in bulk oil commerce is calculated as follows:

Year	Historical Trend Projection (Short Tons)	Percent Growth
1970	20,800,000	0
1975	25,700,000	23.6%
1980	29,900,000	16.3%
1985	34,200,000	14.4%
1990	39,000,000	14.0%

Of the 2,485 vessels in the Sound, 235 were tankers, yielding 9.5% of the total number or .095 for vessels over 1000 net tons as compared to .075 for all vessels as determined above.[5]

The resultant weighted average growth ratio is now computed by multiplying the growth rate of each type of vessel (Table 13-9 and 13-10) by its respective weighting factor (Table 13-11).

TABLE 13-12: AVERAGE GROWTH RATES USED IN COMPUTER MODEL

Year	Total Commerce % Tanker	% Freighter	Weighted Average % Growth of Vessel Movements
1970	7.5%	92.5%	0
1975	6.5%	93.5%	7.5%
1980	5.3%	94.7%	8.7%
1985	5.0%	95%	15.6%
1990	5.0%	95%	14.4%

The average growth rate for the number of vessel movements is approximately 2.3% increase per year or 11.5% per 5 year interval.

Now that the percentage growth rate is established, the future number of vessel movements can be determined by applying the percentage growth figures to the 1970 base of vessel movements. The results are as follows:

TABLE 13-13: PROJECTED NUMBER OF VESSELS IN PUGET SOUND CHANNELS

Growth Factor	(1.00)	(1.075)	(1.087)	(1.156)	(1.144)
Channel:	1970	1975	1980	1985	1990
1	152,300	164,720	177,960	205,770	235,340
2	12,300	13,220	14,370	16,610	19,000
3	37,000	39,775	43,235	49,980	57,180
4	233,500	251,010	272,840	315,403	360,820
5	46,600	50,095	54,450	62,950	72,015

F. Projected Ship Collision Occurrences

In order to see what effect the increased commerce in each of the channels will have on the number of vessel collisions in these channels/year the number of collisions per encounter is multiplied by the new encounter rates in succeeding years and is shown as follows:

TABLE 13-14: PROJECTED NUMBER OF COLLISIONS FOR THE GIVEN YEAR

Channel		1970	1975	1980	1985	1990
1	Lo	0.20	0.23	0.27	0.36	0.48
	Hi	0.8	0.92	1.09	1.46	1.92
2	Lo	0.10	0.12	0.14	0.18	0.24
	Hi	0.40	0.46	0.55	0.73	0.95
3	Lo	0.10	0.11	0.13	0.18	0.23
	Hi	0.40	0.46	0.55	0.73	0.95
4	Lo	0.40	0.46	0.55	0.73	0.96
	Hi	1.60	1.87	2.20	2.90	3.90
5	Lo	0.10	0.12	0.14	0.19	0.24
	Hi	0.40	0.47	0.55	0.74	0.97
Total Puget Sound:						
	Lo	0.90	1.34	1.23	1.64	2.15
	Hi	3.60	4.28	4.94	6.56	8.69

Thus, the total number of collisions in each channel can be predicted in this fashion.

G. Projected Oil Spill Occurrence

One last step remains to arrive at the final objective - the probability of an oil spill. Again, by necessity, some assumption must be made concerning the number of vessels that collide. Some estimate of the conditional probability that a ship involved in a collision will spill oil is necessary. At the risk of being arbitrary, this probability is estimated by dividing the number of bulk oil vessels by the total number of vessels on the Sound. Two further assumptions are:

1. In every collision involving bulk oil carriers the vessel scantlings will be damaged such that oil is spilled from the vessel, and
2. In collisions resulting in penetration of the fuel oil tanks will not cause oil to spill.

These conditional probabilities are as follows:

Year	Probability of Collision Involving Bulk Oil Carrier
1970	.075
1975	.065
1980	.053
1985	.05
1990	.05

Finally, the number of collisions/year that will result in oil being spilled in the Puget Sound region waters over the next 20 years by channel is shown below:

TABLE 13-15: NUMBER OF COLLISIONS HAVING OIL SPILLS FOR A GIVEN YEAR

Channel		1970	1975	1980	1985	1990
1	Lo	.0150	.0156	.0143	.0180	.0340
	Hi	.0600	.0600	.0575	.0730	.0960
2	Lo	.0075	.0075	.0072	.0091	.0120
	Hi	.0300	.0300	.0290	.0365	.0475
3	Lo	.0075	.0073	.0071	.0089	.0116
	Hi	.0300	.0300	.0290	.0365	.0475
4	Lo	.0300	.0300	.0290	.0365	.0480
	Hi	.1200	.1220	.1170	.1450	.1950
5	Lo	.0075	.0076	.0073	.0093	.0120
	Hi	.0300	.0305	.0290	.0360	.0485
Total Puget Sound:						
	Lo	.0675	.0680	.0649	.0818	.0176
	Hi	.2700	.2725	.2615	.3270	.4445

The results are interesting and several observations can be made. First, the number of collisions involving the spillage of oil does not show an increase until nearly 1985, in spite of the fact that the number of total collisions is increasing in this same period. The reason for this is the result of a predicted increase in average tank vessel size which effectively reduces the ratio of tank vessels to the total number of vessels. Secondly, the total predicted number of collisions resulting in oil spills can be obtain by summing the number of collisions per year over the 20 year period. For the first 10 years, 1970-1980, the expected collisions involving oil spillage are:

Lo 0.67 collisions over 10 year period.

Hi 2.69 collisions over 10 year period.

That is, on the low side 0.67 expected collisions will occur in the next ten years which spill oil, on the high side 2.69 collisions are predicted. For the second 10 years, 1980-1990, the corresponding figures are:

Lo 0.84 collisions per 10 year period.

Hi 3.40 collisions per 10 year period.

The numbers for the 1970-1980 period compare favorably with the probable number of tank vessel collisions predicted in a study by Honeywell. This study predicted 2.4 collisions when a 3% growth rate was considered.[6]

To give some indication of the number of collisions involving oil spillage in each channel/year the values are listed in matrix form below.

TABLE 13-16: PREDICTED NUMBER OF OIL SPILL COLLISIONS PER CHANNEL

Channel		10 Years (1970-1980)	10 Years (1980-1990)
1	Lo	.152 collision	.185 collision
	Hi	.594	.749
2	Lo	.074	.093
	Hi	.300	.374
3	Lo	.074	.093
	Hi	.300	.374
4	Lo	.300	.375
	Hi	1.200	1.500
5	Lo	.074	.093
	Hi	.3	.374

This matrix indicates that the channel likely to produce the greatest number of oil spills from vessel collisions is channel 4 (Point Wilson to Elliot Bay). In the next 10 years the expected number of collisions on the low side is .300 and on the high side is 1.2. The remaining channels can be viewed in a similar manner.

H. Probability of Future Collision With the Free Gas Model

For the two bodies of water analyzed by the free gas model, the increase in vessel movements is assumed to behave similarly as for the channels. In addition, the weighted length of vessels is assumed to increase linearly from 300 ft. in 1970 to 500 ft. in 1990, while the width of vessels remains fixed. Since the geographic characteristics will not change, the only influencing factors in P(X) are n and a. Therefore, the following probabilities for collisions result:

TABLE 13-17: PROBABILITY OF COLLISION

	P(X) for Large Body Inland of Straits	P(X) for Strait of Georgia
1970	5.35×10^{-5}	1.71×10^{-6}
1975	6.77×10^{-5}	2.16×10^{-6}
1980	8.10×10^{-5}	2.58×10^{-6}
1985	1.08×10^{-4}	3.48×10^{-6}
1990	1.38×10^{-4}	4.4×10^{-6}

The probability of an oil spill resulting from a collision P(Y) is the product of P(X) and the conditional probabilities for collision involving bulk oil carries defined earlier. Thus,

TABLE 13-18: PROBABILITY OF COLLISION RESULTING IN AN OIL SPILL

	P(Y) for Body of Water Inland of Straits	P(Y) for Strait of Georgia
1970	4.0×10^{-6}	1.3×10^{-7}
1975	4.4×10^{-6}	1.4×10^{-7}
1980	4.3×10^{-6}	1.4×10^{-7}
1985	5.4×10^{-6}	1.7×10^{-7}
1990	6.9×10^{-6}	1.7×10^{-7}

These results show that the larger body of water is 3 times more susceptible to an oil spill than the Strait of Georgia. It must be emphasized that the results presented for this free gas model cannot be directly compared to the results of the Sperry model for the 5 channels.

The objectives of the free gas model presentation were to show a relationship between two bodies of water and to estimate the further trends in oil spills for ship collisions.

I. Summary Comments

It is recognized that, in the development of these predictions, many simplifying assumptions have been made as well as estimates of very pertinent data. The allocation process as to the allotment of vessel movements to each channel surely is not completely accurate, and open to question. In addition to these shortcomings one cannot even say 'let the the results speak for the model,' for all that can be compared is historical data, and what is being attempted is to predict the future. So where does this leave us? Perhaps not any better off than before the analysis! This is not exactly true. For no matter how crude the analysis may be, there is now some measure of the predicted relative collision frequencies in various areas of the Puget Sound region. These measures can be improved, perhaps significantly, through a relaxation of some of the constraining assumptions, more accurate input data as well as improved computational techniques. After all, the effectiveness of any model is dependent on its representation of the real world, and the input data. In this model, the representation of the real world (i.e., behavior of vessels) is also dependent on the input data. This input data includes the detailed information of any ship entering Puget Sound waters. Vessel dimensions, type of cargo, destinations, timetable of arrivals and departures, unusual incidents, and their causes are some of the inputs that are necessary for all vessels. In the end, it is imperative that the measures be accurately known so as to efficiently distribute the scarce resources at command in combatting future oil spills resulting from vessel collisions.

J. Endnotes for Appendix 13

1. Honeywell Marine Systems Center, A Proposed Automated Marine Traffic Advisory System for Puget Sound, Document K03/70, Appendix A, November 6, 1970.
2. Sperry Piedmont Co., Final Report Lookout Assist Device Feasibility Studies, Vol. 1, Section 3, Contract #MA-3374, August 1965.
3. Unpublished data provided by Commander G. K. Greiner, Jr., of the 13th Coast Guard District.
4. U. S. Army Corps of Engineers, Waterborne Commerce of the United States, Calendar Year 1967, Part 4.
5. Seattle Chamber of Commerce, Ship Movements and Petroleum Transportation in Puget Sound Area, State of Washington, A White Paper, 1971.
6. Honeywell.

APPENDIX 14

CHARACTERISTICS OF OIL IN THE ENVIRONMENT

This appendix is included in the report to give some insight on the complexity oil behavior technology. The initial two sections provide a glossary of terms used in describing oil and a conversion table to relate various report methods as oil production, shipment, or transfer. The last three parts document a literature suming on the effects of oil in the environment, including short and long term biological effects. No attempt was made by the Study Group to prove or disprove the published information. The sole purpose of including this data is to demonstrate the critical need for development of a coordinated program for a comprehensive study of the subject. Government agencies have the authority and funding mechanism, while industry and academic institutions have the technology. For a study to be as complete as possible, a sustained joint effort is key.

A. Glossary of Terms[1]

API Gravity: An empirical scale for measuring the density of liquid petroleum products, the unit being called the "degree API".

Ash: Inorganic residue remaining after irnition of combustible substances determined by definite prescribed methods.

Asphalts: Black, solid or semisolid bituments which occur in nature or are obtained as residues during petroleum refining.

Bilge Oil: Waste oil which accumulates, usually in small quantities, in the lower spaces in a ship, just inside the shell plating. Usually mixed with larger quantities of water.

Blowout: A sudden violent escape of gas and oil from an oil well when high pressure gas is encountered and preventive measures have failed.

Boiling Point: The temperature at which the vapor pressures of a liquid is equal to the pressure of the atmosphere.

Bunker "C" Oil: A general term used to indicate a heavy viscous fuel oil.

Bunker Fuel: A general term for heavy oils used as fuel on ships and in industry. It often refers to No. 5 and 6 fuel oils.

Bunkering: The process of fueling a ship.

Coker Feed (or Fuel): A special fuel oil used in a coker furnace, one of the operating elements of a refinery.

Crude Oil: Petroleum as it is extracted from the earth. There may be several thousands of different substances in crude oil, some of which evaporate quickly, while others persist indefinitely. The physical characteristics of crude oils may vary widely. Crude oils are often identified in trade jargon by their regions of origin. This identification may not relate to the apparent physical characteristics of the oil. Commercial gasoline, kerosene, heating oils, diesel oils, lubricating oils, waxes, and asphalts are all obtained by refining crude oil.

Demulsidility: The resistance of an oil to emulsification, or the ability of an oil to separate from any water with which it is mixed. The better the demulsidility rating, the more quickly the oil separates from water.

Density: Density is the term meaning the mass of a unit volume. Its numerical expression varies with the units selected.

Emulsion: A mechanical mixture of two liquids which do not naturally mix as oil and water. Water-in-oil emulsions have the water as the internal phase and oil as the external. Oil-in-water emulsions have water as the external phase and the internal phase is oil.

Fire Point: The lowest temperature at which an oil vaporizes rapidly enough to burn for at least 5 seconds after ignition, under standard conditions.

Flash Point: The lowest temperature at which an oil gives off sufficient vapor to form a mixture which will ignite, under standard conditions.

Fraction: Refinery term for a product of fractional distillation having a restricted boiling range.

Fuel Oil Grade: Numerical ratings ranging from 1 to 6. The lower the grade number, the thinner the oil is and the more easily it evaporates. A high number indicates a relatively thick, heavy oil. No. 1 and 2 fuel oils are usually used in domestic heaters, and the others are used by industry and ships. No. 5 and 6 oils are solids which must be liquified by heating. Kerosene, coal oil, and range oil are all No. 1 oil. No. 3 fuel oil is no longer used as a standard term.

Innage: Space occupied in a product container.

In Personam: An action *in personam* is instituted against an individual, usually through the personal service of process, and may result in the imposition of a liability directly upon the person of a dependent.

In Rem: An action *in rem* is one in which the vessel or thing itself is treated as offender and made defendent without any proceeding against the owners or even mentioning their names. The decree in an action *in rem* is enforced directly against the *res* by a condemnation and sale thereof.

Load on Top: A procedure for ballasting and cleaning unloaded tankers without discharging oil. Half of the tanks are first filled with sea water while the others are cleaned by hosing. Then oil from the cleaned tanks, along with oil which has separated out in the full tanks, is pumped into a single slop tank. The clean water in the full tanks is then discharged while the freshly-cleaned tanks are filled with sea-water. Ballast is thus constantly maintained.

Oil Films: A slick thinner than .0001 inch and may be classified as follows:

Standard Term	Gallons of Oil per Square Mile	Appearance
"barely visible"	25	- barely visible under most favorable light conditions
"silvery"	50	- visible as silvery sheen on surface water
"slightly colored"	100	- first trace of color may be observed

Standard Term	Gallons of Oil per Square Mile	Appearance
"brightly colored"	200	- bright bands of color are visible
"dull"	666	- colors begin to turn dull brown
"dark"	1332	- much darker brown

Note: Each one-inch thickness of oil equals 5.61 gallons per square yard of 17,378,709 gallons per sq. mile.

Outage: Space left in a product container to allow for expansion during temperature changes it may undergo during shipment and use. Measurement of space not occupied.

pH: Term used to express the apparent acidity or alkalinity of aqueous solutions; values below 7 indicate acid solutions and values above 7 indicate alkaline solutions.

Pour Point: The lowest temperature at which an oil will flow or can be poured under specified conditions of test.

Residual Oil: A general term used to indicate a heavy viscous fuel oil.

Scuppers: Openings around the deck of a vessel which allow water falling onto the deck to flow overboard. Should be plugged during fuel transfer.

Sludge Oil: Muddy impurities and acid which have settled from mineral oil.

Specific Gravity: The ratio of the weight of a given volume of the material at a stated temperature to the weight of an equal volume of distilled water at a stated temperature.

Spontaneous Ignition Temperature: (S.I.T.): The temperature at which an oil ignites of its own accord in the presence of air oxygen under standard conditions.

Stoke: The unit of kinetic viscosity.

Tonnage: There are various tonnages applied to merchant ships. The one commonly implied is gross tonnage although in these days tankers and other bulk carriers are often referred to in terms of deadweight.

Gross Tonnage - 100 cubic feet of permanently enclosed space is equal to one gross ton - nothing whatever to do with weight. This is usually the registered tonnage although it may vary somewhat according to the classifying authority or nationality.

Net Tonnage - The earning capacity of a ship. The gross tonnage after deduction of certain spaces, such as engine and boiler rooms, crew accommodation, stores, equipment, etc. Port and harbor dues are based on this tonnage.

Displacement Tonnage - The actual weight in tons, varying according to whether a vessel is light or in loaded condition. Warships are always spoken of by this form of measurement.

Deadweight Tonnage - The actual weight in tons of cargo, stores, etc. required to bring a vessel down to her load line, from the light condition. Cargo deadweight is, as its name implies, the actual weight in tons of the cargo when loaded, as distinct from stores, ballast, etc.

Ullage: The amount which a tank or vessel lacks of being full. (See also Outage.)

Viscosity: The property of liquids which causes them to resist instantaneous change of shape, or instantaneous rearrangement of their parts, due to internal friction. The resistance which the particles of a liquid offer to a force tending to move them in relation to each other. Viscosity of oils is usually expressed as the number of seconds at a definite temperature required for a standard quantity of oil to flow through a standard apparatus.

Viscous: Thick, resistant to flow, having a high viscosity.

Volatile: Evaporates easily.

B. Conversion Table for Oil

Approximate Conversions:

Material	Barrels per Ton (long)
Crude oils	6.7 - 8.1
Aviation gasolines	8.3 - 9.2
Motor gasolines	8.2 - 9.1
Kerosenes	7.7 - 8.3
Gas oils	7.2 - 7.9
Diesel oils	7.0 - 7.9
Lubricating oils	6.8 - 7.6
Fuel oils	6.6 - 7.0
Asphaltic bitumens	5.9 - 6.5

1 barrel = .154 short tons (rule of thumb)
= 42 gallons (U.S.)
= 34.9726 gallons (Imp)
= 5.6146 cu ft.
= 158.984 litres
= .137 long tons (rule of thumb)
= 308 lbs. (rule of thumb)
=
=

C. Behavior of Oil in the Environment

The change of the oil properties once it is exposed to the environment is a complex chemical reaction. The time dependent behavior of oil in water is speculated by Dean:

Evaporation:

"There is no doubt that the one big change which occurs rapidly when oil is spread on the sea is the evaporation of the light fractions. The rate at which this occurs is very uncertain and undoubtedly very variable, depending on the nature of the crude oil, the rate of thinning of the film, violence of wave action, strength of the wind, temperature, etc., etc. However, it seems probable that, after a few days on the sea, the amount of the oil left will be of the order of that of the residue greater than 370°C. (700°F.)."

Solution:

"It is known that the lower hydrocarbons, e.g. methane, butane, heptane and benzene, have finite solubilities in water. Undoubtedly, with increase in molecular weight there is a decrease in solubility but, on the other hand, we are considering very large volumes of sea water. At a solubility of the order of 1 ppm. the whole cargo of the TORREY CANYON could have been dissolved in a patch of sea 20 miles square and 500 feet deep."

Oxidation:

"Hydrocarbons are generally considered resistant to oxidation but, in actual fact, when oxygen is readily available and oxidation inhibitors are absent, they oxidize quite rapidly. This is particularly true when irradiated by light of short wavelengths of less than 4,000 Angstroms. Moreover, many of the sulphur compounds in crude oils are inhibitors of this oxidation although, in the process of inhibition, they are themselves converted to water soluble materials. Some of the metallocompounds present in the crude oils are, on the other hand, promoters of oxidation, thus it is an understatement to call the problem complex."

Another factor to consider is the biological oxidation of oil. A good review of this phenomena is by Holcomb.[2]

"Bacteria found in open seas tend to degrade only straight-chain hydrocarbons of moderate molecular weight so branched-chain hydrocarbons of high molecular weight in the form of tarry chunks may persist for a long time. In still waters, a series of complex events results in almost complete degradation. In 1950, Soviet microbiologists showed that, after the lighter fractions of oil spilled on the Moskva River evaporated, the remaining oil was absorbed by particles and sank. Bottom-dwelling microorganisms produced a new mixture of organic substances that were carried to the surface with bubbles of methane and other light gases. The new compounds again were absorbed, sank, and the cycle was repeated. A number of cycles, repeated over several months, were necessary to degrade most of the oil.

Studies on the thoroughness of degradation have produced conflicting results. Research at Terrebonne Bay, Louisiana, in 1966 showed that essentially complete degradation occurs within a period of several months. Oil has been a consistent pollutant in the Bay since the 1930's, but analysis of bottom mud showed that significant concentrations of petrochemicals could be found only in areas that had received oil relatively recently. However, studies in the French Mediterranean, which will be discussed later, indicate that important chemicals are persistent in bottom sediments.

There is now a considerable body of literature on the interactions of microorganisms and petroleum products, most of it based on laboratory studies and much of it dating back to Soviet work in the 1930's and American work in the 1940's. However, these studies are scarcely past the descriptive stage, and, even when combined with field studies, they are not adequate for predicting the course of degradation of an oil spill."

Even less well understood is the tendency of some oils to form 'water-in-oil' emulsions. Dean notes that:

> "The consideration given to the significance of these emulsions has not been nearly sufficient. Even the reason for and the mode of formation of these emulsions is obscure. Are they stabilized by the products of bacterial action or by collecting plankton from the surface of the sea? Why do they not readily revert to the oil-in-water emulsion which is so desirable for dispersal?"

These emulsions tend to retain their identity in the marine environment. Their formation is therefore significant with respect to the long term effects of a spill.

Persistent 'lumps' of oil have been observed by Horn[3] and co-workers. These may be connected with the formation of stable emulsions. Horn reports:

> "Lumps of crude oil residue floating on the sea surface have been observed widely. Samples were taken with surface-skimming nets in the Mediterranean Sea and eastern North Atlantic Ocean; their displacement volumes were as large as 0.5 milliliter per square meter. An isopod, Idotea metallica, appears to be associated with the lumps, and a barnacle, Lepas pectinata, grows upon them. Lumps were found in stomachs of Scomberesox saurus, a surface-feeding fish important in ocean food webs. Films on the lumps, presumably consisting mostly of bacteria, consumed oxygen at the rate of 4 cubic millimeters per hour per square centimeter of lump surface. Chemical analysis suggested that certain lumps had been at large for only a few weeks; data from barnacle size and growth rate suggested that other lumps were at least two months old."

D. Short Term Biological Effects of Oil and Oil Spills

Any discussion of the effects of an oil spill should distinguish between the damage done by the oil and the damage done by subsequent spill treating agents. Once this distinction is made, further division of the short and long term consequences of the spill is meaningful.

In order to assess the effects of an actual spill, there must be knowledge of local marine life existing prior to the spill. Although this knowledge is available in part, it is fragmented among many different interest groups. Coordination and compilation of this information has not been attempted successfully. Therefore, it is strongly suggested that this information be gathered for Puget Sound. Section II-A of this report presents a preliminary inventory on marine life.

1. Oil Spill Treating Agents

a. Toxic Effects

Studies following the Torrey Canyon spill emphasize the danger of dispersing agents to varied forms of marine life. Since that time, a large amount of information on the toxicity of spill treating agents has been gathered by various institutions. Some of those groups active in this data collection are the FWQCA's Edison Water Quality Laboratory and National Marine Water Quality Laboratory, the Licensing Division of the California Fish and Game Department, the Richland Labs of Battelle Northwest, and various universities.

A review of the toxicity of various spill treating agents was sponsored by the American Petroleum Institute and completed by Battelle in May of 1970.[4] This review included over 70 commercially available treating agents and accompanying toxicity tests when available.

Battelle also made recommendations where appropriate on test procedures for determining toxicity. Those commercial agents listed, whose toxicity is unknown or questionable, should undergo tests such as those outlined by Battelle. Before this is done, this list should be updated to 1971 (something which Battelle is currently doing) and new data from any of the above institutions should be incorporated. When this is done, a more informed statement can be made on what treating agents should be used during various conditons.

b. Sub-Lethal Effects

Although much attention has been given to the acute toxic effects of treating agents, relatively little attention has been given to the long term toxic and sub-lethal effects. This is mainly due to the fact that these effects must be determined under field conditions rather than in the laboratory. The following discussion is taken from the Battelle report:[5]

"Recently, long-term studies have been conducted where low levels of toxicant have been maintained over critical periods in the life cycle of selected organisms. The results have demonstrated that the application factors to be applied to 96 hr TL_m values to indicate safe concentrations during chronic or continuous exposure are not adequate to protect aquatic life. Application factors in current use, as recommended by The National Technical Advisory Subcommittee on Water Quality Requirements for Fish, Other Aquatic Life and Wildlife, range from 1/10 to 1/100 of the 96 hr TL_m value.

While ultimately research should be conducted to determine absolute toxicity of many such materials on the diversity of both fresh-water and marine organisms, it is recommended that the following problem areas be given priority:

1. Short-term bioassay studies to identify the most sensitive species and the most sensitive life stage to a selected toxicant.

2. Long-term bioassay studies with those species and life stages which are the most sensitive to a particular toxicant to determine the following:

a. The consequential effects of non-lethal damage at any stage on successive stages in the life cycle.

b. The effects on all life stages of sub-lethal exposures during the most sensitive stages.

c. Comparative studies of the effects of exposure during the most sensitive stages and of exposures during the entire life cycle.

3. Related to investigation of the long-term effects of selected toxicants, the following studies should be undertaken:

a. Determination of the mode of toxicity, of the selected toxicant, their metabolities and degradation products.

> b. Determination of the interaction of two or more toxicants; both synergistic and antagonistic effects, and those materials with similar modes of action."

The report goes on to recommend a currently possible approach to the measurement of sub-lethal effects:

> "For the determination of sub-lethal effect or chronic exposure, it is perhaps more important to examine the communities in the environment, either aquatic or terrestrial, than it is to examine the individual organisms within the community.
>
> The community is modified as a result of pollution, in structure and in size, numbers of species tend to shift their relative frequency; or they can be eliminated entirely. It is the community of organisms that reflects the total effect of contamination over the protracted period of time."

Currently, however, due to the lack of universally accepted measurement techniques, studies concerned with the problem of these effects have not been conclusive. They include those conducted by the Warren Spring Laboratory after the Torrey Canyon spill, and those conducted by the U.S.Geological Survey and the University of Southern California following the Santa Barbara spill. The main conclusion from all of these studies was that the effects of the treating agents were difficult, if not impossible, to separate from the effects of the oil and all other phenomena. Yet, for now, the acute effects of those treating agents that have been studied in the laboratory should be the important variable in determining use.

Despite the lack of complete data, it is felt that the use of dispersants and sinking agents is to be avoided. The State DOE prohibits use of chemical dispersants in combatting oil spills.[6] It is recommended that use of absorbing and biological agents have been shown to be safe to Pacific Northwest marine species. If toxicity data on these agents does not include tests on Puget Sound species (and in most cases id does not cover many), use of absorbing and biological agents otherwise shown to be safe are recommended. However, these agents should only be used when the spill is heading toward a high priority area. (See Section III-B, part 2.) If the decision is made to direct a spill to an available sandy beach relatively sterile in marine life, avoiding the use of treating agents is stronly recommended.

2. Oil

The least studied and least understood consequence of an oil spill is the effect of the oil itself on marine life. Research on this aspect of the problem is really just beginning to develop. For the purposes of this report, initial effort has been directed to current literature and to current unpublished studies. These data raise more questions than they answer as is the case with most research in a new area.

As described earlier, the interaction of oil with the marine environment is highly complex and variable. For this reason, a quantitative prediction of what would occur if a certain oil were mixed with waters of a given area is not possible. Those reactions which are known to occur under controlled laboratory conditions may not occur in the rapidly changing environment of the sea. Even if one could say that certain classes of reactions would occur, this fact alone would not be enough to describe the toxic and sub-lethal effects of oil on marine organisms. Our limited knowledge of the mechanisms of toxicity to various marine organisms and of the range of sub-lethal effects that actually occur restricts our ability to predict the consequences of known chemical reaction.

For these reasons, the most useful data are the measured effects of actual spills. Unfortunately, these measurements have not been made in the past due to limited interest, facilities, and techniques. Only in the past few years has there been any reliable investigation of the sub-lethal effects of oil. We are just now beginning to interpret the results of these studies with the result that the direction of future research is as yet unclear. The recent grant to Battelle from Standard Oil to study the effects of oil, oil with treating agents, and treating agents alone, is another step in the right direction, but is by no means the ideal scientific investigation. Battelle's initial constraint will be time, which is one constraint that should be eliminated in any study of long term effects.

Although we can't accurately predict all of the effects, we can infer some of the effects characteristic of the general chemical classes. This has been done by Clark, et. al... and is quoted below:

> "Low-boiling, saturated hydrocarbons are related to the fat-soluble anaesthetics and have a narcotic effect on a wide variety of animals. In lower animals at low concentrations they often cause cell damage and death, especially in larval and juvenile stages of marine organisms. The solubility of these low-molecular weight hydrocarbons in water is of the order of 0.1 to 1.0 grams per liter, comparable to that of many drugs (Goldacre, 1968).
>
> Higher-boiling, saturated hydrocarbons occur naturally in many organisms and, while probably not toxic at low levels, they may interfere with chemical communications used by some organisms for locating food, as sex attractants, and as indicators of migration routes (Blumer, 1969).
>
> Olefinic hydrocarbons are not generally found in crude oils but are plentiful in gasolines and other refined products. The fate of olefins in the marine environment is poorly understood but this class of compounds is quite reactive and will combine readily with hydrogen, oxygen, chlorine, sulfur and other elements to produce toxic substances. Also, once incorporated into organisms, olefins are stable.
>
> Aromatic hydrocarbons comprise the most dangerous of petroleum fractions. The low-boiling aromatics are acutely poisonous to man and marine organisms alike. Benzene, toluene and phenols, such as found in crude oil, produce in man reactions similar to those of alcohol. The initial reaction is restlessness, then excitement, inebriation, drowsiness, depression and sleep. Death may follow from respiratory failure as the concentration rises (Goldacre, 1968). Chronic exposure to low concentrations of some aromatics, especially benzene, may cause bone marrow aphasia, chromosome aberration and leukemia (Finkel, 1960).
>
> The higher-boiling aromatics are suspected of having carcenogenic properties. Benzpyrene, 1, 2-benzanthracene and aklylbenzanthracenes have been isolated from crude oils and their carcenogenic effects on animals and man demonstrated. Some years ago, a high incidence of skin cancer in some refinery personnel was observed. The cause was traced to prolonged skin contact by these persons with petroleum and with refinery products. Better plant design and education, aimed at preventing the skin contact, have since reduced or eliminated this hazard (Eckart, 1967). However, these incidents have demonstrated that oil and oil products can cause cancer in man."

Non-Hydrocarbons: "In their behavior and toxicity, the non-hydrocarbon components of crude oil (nitrogen, oxygen, sulfur, and metal compounds) closely resemble the corresponding aromatic compounds (Blumer, 1969)."
(From "Environmental Statement on the Proposed Leasing of Puget Sound Shorelands and Beds of Navigable Waters for Oil and Gas Exploration." Dolan, Savage and Clark. Oct. 1970.)

a. Direct Effects

A good summary of research in this area up until November, 1967, is found in Battelle's literature search study.[7] Due to the length of this search, a summary of the important effects is presented here.

Due to a lack of detailed information on the relative effects of various crude oils, all crude oils have been grouped together. Possible exceptions to this general category include those crudes which have been mixed with an unusual amount of water soluble and high boiling components.

Similar reasoning leads to a second grouping of more highly refined products. Possible variation of effects within this group could be attributed to the variation in chemical structures. Kerosene and gasoline are mainly composed of saturated straight and branched chains, whereas the jet fuel and other highly refined products have a large portion of ring structured (aromatic) components.

Some oils do not readily fall into either category. The heavy residual oils approach the crude oils in physical properties, but are not necessarily similar chemically. This is due to the mixing of high and low boiling components to achieve the desired properties of handling and combustion, as is done with Bunker C. The distillate oils contain more refined components which are more water soluble and low boiling than the heavier components. Yet, their physical properties do not compare as closely with the more refined products, and they tend to be more persistent physically in the marine environment.

At the time of the survey done by Battelle, the most complete before and after study of an oil spill was reported by

W. J. North of California Institute of Technology.[8] He studied the effects of the oil spilled by the Tampico, near Baja, California. He reports:

> "The cargo of about 60,000 barrels of diesel oil began leaking from the stricken ship and spread along the shore, carried by wind and waves. The hull of the Tampico formed a breakwater across the cove, reducing the oxygenation of these waters from breaking waves. The luxurious submarine gardens in this cove and along the surrounding beaches perished. Soon the shore was littered with piles of seaweed and dead and dying animals.
>
> Only a few animal species survived. One was a tiny snail, the periwinkle Littorina planaxis. This creature inhabits rocks well above the high water mark and depends on spray from waves to satisfy its water requirements. Unfavorable conditions apparently did not extend up to the periwinkle's domain. Several large green anemones, Anthopleura xanthogrammica, were found alive in the tidepools. These animals, which sometimes grow in salt water lines of oil refineries that use seawater as a coolant, are believed to be very tolerant of oil pollution. Among the dead species were lobsters, abalone, sea urchins, starfish, mussels, clams, and hosts of smaller forms.
>
> We have visited the area each year since catastrophe struck this little cove, recording plants and animals at a dozen different locations. Gradually the various forms of life returned. Sixty-nine species of animals have recently been observed, compared to only two species immediately after the shipwreck. Among plants the tally is fifty-seven now, versus four a month after the accident. In spite of the reestablishment of lost species, the cove is still not the same as we believe it was before the wreck. Many species have just appeared within the last year or two and often only one or two specimens have been seen. Prior to the wreck these species must have been abundant, judging from the numbers cast up on shore after the accident. Conversely, other species are now plentiful, of which only a few individuals were present before the wreck."

Another good summary of the current thinking at that time is found in a collection of papers edited by Arthur and Carthy.[9] This collection begins to indicate that the sub-lethal effects of spilled oil should be investigated. The following excerpts from Goldacre's paper point out this fact:

> "The lower members of the paraffin hydrocarbons have nevertheless a strong effect on living material. ...Changes produced in the cell membrane by hydrocarbons may plan an important part in causing malignant change. The cell membrane of cancer cells is different from that of the corresponding normal cell (Ambrose, 1967) (and) the work of Sonneborn (1964) has shown that it is possible for membrane changes to be inherited without any change in the nucleus. Hence hydrocarbon carcinogens could cause permanent changes in the cell membrane leading to a breakdown in cell-cell communication and to cancer."

Clark summarizes another paper in this collection by Arthur:

> "In the marine environment, oil pollution at low levels, where direct toxicity cannot be measured, is believed capable of interfering with biological function in two ways: (1) by blocking chemoreceptor organs and (2) by producing fake stimuli leading to unproductive behavior. Oil pollution at higher levels can cause death of many organisms in the food chains of higher animals such as birds and fish, causing an upset in the ecological balance. Oil brings about the death of tiny organisms when the oil coats their surfaces either by preventing molecular exchange across cell membranes or by causing incorporation of toxic components of the oil into the tissue (Arthur, 1968)."

At that time (July 1968) very little work on these sub-lethal effects had been done, although other researchers starting with Mallet (1964) had known about the carcinogenity of the heavier oil fractions and their presence in sediments. Then in late 1969 a spill occurred about six miles from Wood's Hole Oceanographic Institute. This led to the continuing work of Dr. Max Blumer and his co-workers on the sub-lethal effects of a spill. They were fortunate to have a documented biological baseline of Buzzard's Bay and the facilities to conduct sensitive analytical measurements. The first report appeared in January 1970 and the latest report updates that study by eight months (The West Falmouth Oil Spill-persistence of the Pollution Eight Months After the Accident - September 1970.) As the title implies, this latest study emphasizes that the oil has persisted in the ecosystem causing a wide variety of sub-lethal effects.

The abstract from the first report[10] is quoted below:

> "A spill of 650,000 to 700,000 l of No. 2 fuel oil has contaminated the coastal areas of Buzzards Bay, Massachusetts (USA). Gas chromatography demonstrates the presence of this oil in the sediments of the affected area. Two months after the accident, essentially unchanged oil is still being released from the sediments. The presence of the same pollutant is demonstrated in whole oysters Crassostrea virginica and in the adductor muscle of the scallop Aeguipecten irradians. A presumably biochemical modification leads to a gradual depletion of the straight chain and, to a lesser extent, of branched chain hydrocarbons. This does not result in detoxification, as the more toxic aromatic hydrocarbons are retained in the organisms several months after the accident. Scallops from an uncontaminated area contain hydrocarbons in lesser amounts and of very different molecular weight and type distribution; they are accountable entirely from biological sources."

The report itself details the precision and care of the measurement system - a precision not used before in other spill areas.

Portions of Dr. Blumer's congressional testimony[10] reflect the findings eight months later:

> Persistence and Spread of the Pollution:
>
> "Oil from the accident has been incorporated into the sediments of the tidal rivers and marshes and into the offshore sediments, down to 42 feet, the greatest water depth in the sea.
>
> The #2 fuel oil involved in the spill is a relatively volatile oil product. In spite of this, volatilization has been minimal and the oil remains in the marshes, exposed to air, and in the offshore areas, 10 months after the accident.
>
> The pollution has been spreading on the sea bottom and now covers at least 5000 acres offshore and 500 acres of marshes and tidal rivers. This is a much larger area than that affected immediately after the accident.
>
> Bacterial degradation of the oil is slow; degradation is still negligible in the most heavily polluted areas and the more rapid degradation in outlying, less affected, areas has been reversed by the influx of less degraded oil from the more polluted regions.

The kill of bottom plants and animals has reduced the stability of marshland and sea bottom; increased erosion results and may be responsible for the spread of the pollution along the sea bottom.

Bacterial degradation first attacks the least toxic hydrocarbons. The hydrocarbons remaining in the sediments are now more toxic, on an equal weight basis, than immediately after the spill.

Oil has penetrated the marshes to a depth of at least 1-2 feet; bacterial degradation within the marsh sediment is still negligible eight months after the accident."

Biological Effects of the Pollution:

"Where oil can be detected in the sediments there has been a kill of animals; in the most polluted areas the kill has been almost total. Control stations outside the area contain normal, healthy bottom faunas.

The kill associated with the presence of oil is detected down to the maximum water depth in the area.

A massive, immediate kill occurred offshore during the first few days after the accident. Affected were a wide range of fish, shellfish, worms, crabs and other crustaceans and invertebrates. Bottom living fishes and lobsters were killed and washed up on the beaches. Trawls in 10 feet of water showed 95% of the animals dead and many still dying. The bottom sediments contained many dead clams, crustaceans and snails.

Fish, crabs, shellfish and many other animals were killed in the tidal Wild Harbor River; and in the most heavily polluted locations of the river almost no animals have survived.

The affected areas have not been repopulated, nine months after the accident.

Mussels that survived last year's spill as juveniles have developed almost no eggs and sperm.

(From: Hampson and Sanders, 1969; Sanders and Hampson, 1970; Sanders and Hampson, personal communication.)

Effect on Commercial Shellfish Values:

"Oil from the spill was incorporated into oysters, scallops, soft-shell clams and quahaugs. As a result, the area had to be closed to the taking of shellfish.

> The 1970 crop of shellfish is as heavily contaminated as was last year's crop. Closure will have to be maintained at least through this second year and will have to be extended to areas more distant from the spill than last year.
>
> Oysters that were removed from the polluted area and that were maintained in clean water for as long as six months retained the oil without change in composition or quantity. Thus, once contaminated, shellfish cannot cleanse themselves of oil pollution.
>
> The tidal Wild Harbor River, a productive shellfish area of about 22 acres, contains an estimated four tons of the fuel oil. This amount has destroyed the shellfish harvest for two years. The severe biological damage to the area and the slow rate of biodegradation of the oil suggest that the productivity will be ruined for a longer time.
>
> The presence or absence of an "oily smell" is no clue for the presence of oil pollution in fish or shellfish. Only a small fraction of petroleum has a pronounced odor; this may be lost while the more harmful long-term poisons are retained. Boiling or frying may remove the odor but will not eliminate the toxicity."

Many of these effects have not been reported elsewhere because of the difference in precision of various sampling techniques. Perhaps if Dr. Blumer and his co-workers would have been monitoring the Santa Barbara spill, the following would not have appeared in U.S. News and World Report:[11]

> "This study, by Dr. Dale Straughan of the Allan Hancock Foundation of the University of Southern California, was released on Jan. 19, 1971. It confirmed many of the findings of the Geological Survey report ... In conclusion, said Dr. Straughan: "Three main points emerge from assessment of the data: (1) It is often difficult to isolate the effects of oil pollution from other phenomena. (2) Damage to flora and fauna in the Santa Barbara Channel was much less than predicted. (3) The area is recovering well."

Holcomb[12] reports the ensuing conflict:

> "Dr. Robert Holmes, the first director of a major study after the Santa Barbara blowout, has stated that plankton populations were unaffected, and, although his remark was

challenged at the Massachusetts Institute of Technology symposium where it was made, it is generally agreed that visible damage to organisms other than birds has been relatively light."

b. Indirect Effects

i) One important indirect effect of spilled oil is the increased concentration of pesticides dissolved in stable oil sediments.

In May, 1970, Hartung and Klinger[13] report:

> "In the Detroit River, areas containing near 1% sedimented oils reach p,p'-DDT concentrations near 1 p.p.m. The combined levels of chlorinated hydrocarbon insecticides approach levels similar to those reached in insecticidal applications. These areas are also devoid or very deficient in benthic arthropods (Vaughan and Harlow, 1965).
>
> Since sedimented polluting oils are a significant constituent in aquatic environments in areas with a history of oil pollution, the partitioning behavior of these oils must be taken into account when one seeks to evaluate the impact of pollutants on benthic organisms, or the food-web that is supported by this benthos."

In March, 1970[14] similar findings are reported:

> "Pesticide concentrations 10,000 times greater than those found in surrounding waters have been discovered in oily surface slicks in Florida's Biscayne Bay. According to Drs. Eugene F. Corcoran and Douglas B. Seba of the University of Miami's Rosenstiel School of Marine and Atmospheric Sciences, the slicks are natural oceanographic phenomena that appear as calm streaks or patches on the rippled surface of the open oceans, lakes, or coastal waters. ...said Dr. Corcoran. "In Biscayne Bay, for example, we have observed gulls and pelicans diving into surface waters to feed on the dense schools of small fish that feed on plankton concentrated in the slicks. Since these sea birds eat many times their weight in fish, they eventually have more pesticide concentrated in their tissues than the fish did."

The growth of the bottom area contaminated by oil sediments (as previously reported by Blumer) and the presence of stable oil sublayers magnifies the importance of these findings. Evidence of the presence of a stable oil sublayer below the surface where these pesticides could concentrate is described in an unpublished report by the National Marine Fisheries Service. This report describes oil coated specimens fished up from 10 to 50 fathoms by the research vessel Aruba.

ii) Another important indirect effect is the disruption of the marine food chain. Clark summarizes this interaction:

> "A marine shoreline has a series of zones relating to exposure to different combinations of environmental conditions (waves, currents, surge, desiccation, light, substrate, etc.). Within each zone, ecosystems are composed of populations of primary producers (attached algae), browsers and grazers (herbivores) and predators (carnivores and omnivores), delicately balanced with respect to the environmental conditions prevailing within the zone (Bellamy and Whittick, 1968). If one step in the food chain is eliminated by oil pollution, the balance of the entire ecosystem is destroyed."

Holcomb[15] elaborates on this delicate web, emphasizing the concentration of carcinogens and pesticides in higher life forms through this process.

> "Pollutants tend to enter the food chain more easily and to pass through it with fewer changes in aquatic environments than they do in terrestrial environments. They can be introduced in solution through bottom sediments and even in dispersed droplets that are ingested by the numerous filter feeders that constitute an important part of aquatic food.
>
> The presence of DDT in Lake Michigan's coho salmon drew public attention to the fact that some hydrocarbons pass through the aquatic food chain relatively unchanged. Work at Woods Hole has demonstrated that the ratios of olefinic hydrocarbons in zooplankton to those in livers of basking sharks and herring that feed on the plankton are so constant that they can be used to determine the feeding grounds of these species.
>
> Obviously, studies should be conducted to see what concentrations of 3,4-benzopyrene and other potentially dangerous hydrocarbons must be in seawater or sediments before they are introduced into the food chain and whether the chemicals persist as they pass through the chain."

iii) Yet, another indirect effect arises from the chemically stimulated behavior of certain marine species. Holcomb summarizes the findings of Blumer:[16]

> "Blumer points out that very small amounts of certain chemicals are used by many species of sea animals as behavior signals in the vital activities of food finding, escaping from predators, homing, and reproduction. He has shown, for example, that starfish are attracted to their oyster prey by chemicals in concentrations of a few parts per billion. The responsible chemicals have not been identified, but Blumer believes that in many cases they may resemble the high-boiling, saturated hydrocarbons found in petroleum products. Because of the extreme sensitivity of the response and the similarity of the animal and petroleum chemicals, he thinks it is possible that pollution interferes with chemically stimulated behavior "by blocking the taste receptors and by mimicking natural stimuli."

iv) Other indirect effects to be considered include:

- Exposure of the marine ecosystem to possible cancer causing chemicals.
- Consumption of dissolved oxygen by degrading oil.
- Evaporation of volatile components can degrade short term air quality.

c. Long Term Effects

Long range effects of oil pollution are often difficult to detect, even more so to determine precise effects. Effects such as decreased growth, decreased viability, tumor initiation may have obscure complex causes which are difficult to determine. The additions of petroleum products to water may not be directly harmful but they may initiate other changes which are detrimental.

Life in the marine environment is very complex. One species during its life cycle can live in many environments and thus have varying vulnerability to oil pollution. The adult English sole, for instance, is a bottom dwelling species living at a depth of 20-50 fathoms. Its eggs float free in the water column at varying depths. The young spend their first year in very shallow water, often in the intertidal zone. The adults may be considered to be relatively safe from surface oil but the eggs and young are very susceptible.[17]

The richness of the Puget Sound biota is well known. The biological basis of this richness lies in the plankton that covers the waters. It is this segment from zero to 5 cm. deep that appears to be most sensitive to damage by an oil spill. In addition to being in an area where contact with oil is most likely, plankton has the disadvantage of not being able to move a sufficient distance to avoid the oil. The sensitivity of many of these organisms to oil is very great. Certain zooplankton, for example, die in 24 hours at a concentration of 0.1 ppm crude oil. A concentration of 0.01 ppm oil is sufficent to kill in 72-96 hours. In addition to direct mortality, oil may have effects by altering reproduction and growth rates.[18] Diatoms, for example, experience a reduced reproductive capacity when exposed to sub-lethal amounts of oil.[19] It is not known if this occurs in other groups of organisms that comprise the plankton of Puget Sound.

In addition to algae and zooplankton, the larvae of many economically important organisms such as crabs, clams, and some fish, are found at or near the surface of the water. In addition to being in a more vulnerable position to oil, the larvae appear to be more sensitive than the adults to oil pollution. This is probably due to the smaller size of the larvae. Very roughly, the larval stages appear to be 100 times more sensitive to oil than the adults. Plaice larvae, exposed to oil at 0.1 to 0.01 ppm experienced a 40-100% deformity upon development.[20]

Larger, stationary plants such as kelp play a crucial role in the ecology of Puget Sound. They provide food and shelter to many organisms. Since many of these plants reach to the surface of the water, they may easily become coated with oil which may have detrimental consequences though not immediately evident. A 0.02 mm. film of diesel oil on kep beds has little effect for 24 hours. However, after 72 hours photosynthetic activity is completely halted. Emulsions of oil are much more harmful to plants than surface oil. An emulsion of 1.0% oil reduces photosynthesis in kelp 73% in 24 hours and 100% in 77 hours. An emulsion of 0.01% diesel oil has some inhibitory effects in 72 hours. The oil appears to act on fatty constituents of the cytoplasmic membrane. Irreversible damage occurs after 6-12 hours of exposure to emulsified oils. In addition, the phenols and cresols are very toxic to plants, a concentration of 0.1 ppm being fatal.[21]

In addition to the more immediate effects of oil on plants, there are long term effects that are not understood. In one experiment, 65% of plant mortality, due to oil, occurred two months to one year after exposure to oil.[22] This effect could have very important ramifications both biological and economical.

Oil, when it remains in water for any length of time is oxidized by bacteria, or undergoes autoxidation, leading to end products of carbon dioxide in water. However, this breakdown of oil requires a tremendous amount of oxygen. One gm. of oil requires approximately 3.3 gm. of oxygen. One liter of oil would require all of the oxygen from 400 cubic meters of saturated sea water at 15° C.[23] At the surface, this huge oxygen demand would present no problem. However, sunken or emulsified oil could deplete the surrounding waters of their oxygen. Indeed, anerobic conditions are fairly common in areas such as loading docks that have chronic oil pollution. Such a situation could perhaps develop near the proposed Cherry Point refinery, an area which is a biological preserve for juvenile flatfish.

The rate of breakdown of oil depends to a great extent on the temperature. Below 10°C., the rate is very slow. In Puget Sound, the water temperature may be below 10°C. for months at a time which could have detrimental effects and increase the amount of time the oil remains.

When bacteria oxidize petroleum products, they assimilate a great deal of the material. Approximately 30-40% of the oxidized hydrocarbons are stored in the bacteria. This provides the hydrocarbons an entry into the food chain. It is very possible that these compounds are concentrated by the organisms as they pass through the food chain, very similar to the chlorinated hydrocarbon pesticides. The effect of such a concentration is not known. Even if it is not lethal or harmful to the fish, it may give their flesh an oil texture and thus decrease their economic value which could have far reaching effects on the Puget Sound economy.

It has been suggested that petroleum products may initiate cancerous growths in fishes. Tumors may be initiated by 4 and 5 ring aromatic hydrocarbons similar to those found in oils.[24] While such action by oil is yet to be proven, even the possibility may be enough to depress the values of fish taken from oil polluted waters.

Besides the direct delitarious effects of oil on fishes, there may be other effects. Large scale oil spills could disrupt migrations of fish, especially salmon. Spawning or feeding areas could be lost. Chemical senses play an important role in the lives of aquatic organisms and the addition of hydrocarbons could disrupt these sensory mechanisms. Many fish attract mates by chemical clues. These mechanisms are very sensitive and the fish are able to sense chemicals at less than 1 ppb.[25] The effects of large amounts of oil dissolved on the water are not known but they almost certainly would be detrimental to these chemical sensory mechanisms.

The greatest potential hazard from oil pollution would be a change in the equilibrium of species now present. By selectively destroying or decreasing the viability of certain species on a large scale, changes would occur in the abundance of almost all organisms. Organisms which are seemingly unaffected may find a vital food source gone and they themselves disappear; or a predator may decrease, and the prey increase in abundance accordingly. Such effects would be almost impossible to predict. This would probably occur only in localized areas with heavy, chronic pollution and poor water circulation. It would be accompanied by decreased species diversity and hence decreased stability of the system.

The possible long range effects of oil pollution are many and it is very difficult, if not impossible, to determine the full extent of the damage done by an oil spill. Table 14-1 presents a summary of known effects of oil on marine life. In the long run, the costs of an oil spill may be much greater than any benefits the importation of oil may have.

TABLE 14-1: KNOWN EFFECTS OF OIL POLLUTION IN THE MARINE ENVIRONMENT[26]

	Crude Oil	Refined Products	Dispersants + Oil
Plankton/ Bacteria	Toxic to some and utilized by others.	Toxic to some and utilized by others.	Inhibits some larval development, bacterial degradation.
Fish	Suffocation, surface fouling, hemorrhaging or capallaries, non-fatal tainting, disruption of food chain.	Carcenogenic tumors (refinery wastes), increased apparent toxicity of oil with a decrease in oxygen, non-fatal tainting loss of equilibrium.	Loss of equilibrium and orientation, destruction of gills and olfactory epithelium, breakdown of protein, inhibits reproduction.
Shellfish	Surface and ingested tainting, suffocation, poisoning, incorporation of oil into egg yolks, breaks down protective mucous coating of gill epithelium.	Tainting, suffocation, incorporation of carcenogenic compounds, changes in amino acid content.	Breaks down protective mucous and water coatings, inhibits growth and normal development.
Marine Plants	Suffocation, inhibition of flowering, leaf damage during photoinductive period.	Breakdown of fatty constituents of cytoplasmic membrane.	Increases the toxic effects of the oil and the rate of uptake of the oil.
Sediments	Incorporation of oil into sediments causes long-term tainting of shellfish and bottom-dwelling organisms, high concentration of chlorinated pesticides in oil-rich muds.	Same as for crude oil.	Incorporation of oil into sediments is enhanced by most dispersants and sinking agents.
Waterfowl	Physical coating, poisoning, starvation, destruction of protective habitat.	Poisoning by ingestion during preening, possible changes in reproduction (thin egg shell), concentration of certain fractions from contaminated food chain organisms (deposited in fatty tissues).	Removes natural protective oily body coatings, increases susceptibility to pneumonia.

E. Endnotes for Appendix 14

1. U. S. Coast Guard, Seattle Coastal Region Oil and Hazardous Materials Pollution Contingency Plan, December, 1970, Annex XV.
2. Science, Vol. 166, p. 204.
3. Science, Vol. 168, April, 1970.
4. Battelle Memorial Institute, Oil Spill Treating Agents - Test Procedures: Status and Recommendation, Mayl, 1970.
5. Ibid.
6. Department of Ecology, Laws and Oil Spill Emergency Procedures, p. 10.
7. Battelle Memorial Institute, Oil Spillage Study Literature Search and Critical Evaluation for Selection of Promising Techniques to Control and Prevent Damage, November 20, 1967, Chapter 6.
8. Sea Frontiers, Vol. 13, July-August, 1967.
9. Carthy and Arthur, editors, Biological Effects of Oil Pollution on Littoral Communities, June, 1968.
10. Marine Biology, Vol. 5, No. 3, March, 1970, pp. 195-202.
11. U.S. News and World Report, February 8, 1971.
12. Science, Vol. 166, p. 205.
13. Environmental Science & Technology, Vol. 4, No. 5, May, 1970.
14. Sea Secrets, Vol. 4, No. 2, March, 1970.
15. Science, Vol. 166, p. 205.
16. Ibid.
17. W. Clemens and G. Wilby, "Fishes of the Pacific Coast and Canada," Bulletin of Fish Reserve Board, #168, Canada, second ed.
18. O. Mironov, "Hydrocarbon Pollution of the Sea and its Influence on Marine Organisms," Hologolander Wiss Meersunters, 17, p. 335.
19. C. Zobell, "The Occurrence, Effects, and Fate of Oil Polluting the Sea," Advanced Water Pollution Research, Vol. 5, p. 83.
20. Mironov.
21. Zobell.
22. J. Mackin, "Effects of Crude Oil and Bleedwater on Oysters and Aquatic Plants," Progress Report, Texas A & H Research Foundation, 1950.

23. Zobell.

24. M. Blumer, "Oil Pollution of the Ocean," in Oil on the Sea by Hoult, 1969.

25. Ibid.

26. Dolan, Savage, and Clark, Environmental Statement on the Proposed Leasing of Puget Sound Shorelands and Beds of Navigable Waters for Oil and Gas Exploration, 1970, Table 5.